Cheeney
2000 February-

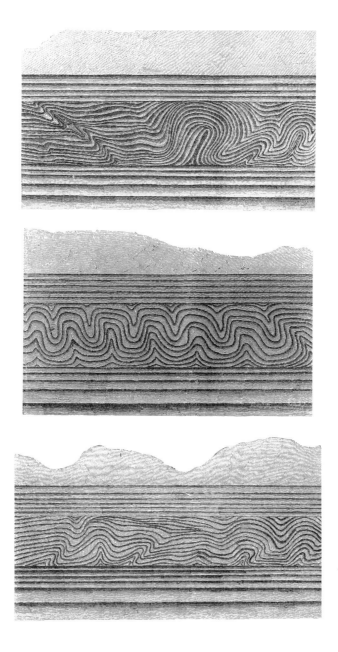

Intra-stratal deformation in Quaternary deposits along the Black River, near Ironton, New Jersey, as portrayed by Vanuxem (1842). The controversy surrounding these and similar structures in nearby Ordovician limestones is outline in section 1.3.2.

The Geological Deformation of Sediments

The Geological Deformation of Sediments

Edited by Alex Maltman
Institute of Earth Studies, University of Wales, Aberystwyth

CHAPMAN & HALL

London · Glasgow · Weinheim · New York · Tokyo · Melbourne · Madras

Published by Chapman & Hall, 2–6 Boundary Row, London SE1 8HN, UK

Chapman & Hall, 2–6 Boundary Row, London SE1 8HN, UK

Blackie Academic & Professional, Wester Cleddens Road, Bishopbriggs, Glasgow G64 2NZ, UK

Chapman & Hall GmbH, Pappelallee 3, 69469 Weinheim, Germany

Chapman & Hall Inc., One Penn Plaza, 41st Floor, New York NY 10119, USA

Chapman & Hall Japan, Thomson Publishing Japan, Hirakawacho Nemoto Building, 6F, 1-7-11 Hirakawa-cho, Chiyoda-ku, Tokyo 102, Japan

Chapman & Hall Australia, Thomas Nelson Australia, 102 Dodds Street, South Melbourne, Victoria 3205, Australia

Chapman & Hall India, R. Seshadri, 32 Second Main Road, CIT East, Madras 600 035, India

First edition 1994

© 1994 Chapman & Hall

Typeset in Times, 10 pt on 11 pts manual by Interprint Limited, Malta
Printed in Great Britain at the University Press, Cambridge

ISBN 0 412 40590 3

Apart from any far dealing for the purposes of research or private study, or criticism or review, as permitted under the UK Copyright Designs and Patents Act, 1988, this publication may not be reproduced, stored, or transmitted, in any form or by any means, without the prior permission in writing of the publishers, or in the case of reprographic reproduction only in accordance with the terms of the licences issued by the Copyright Licensing Agency in the UK, or in accordance with the terms of licences issued by the appropriate Reproduction Rights Organization outside the UK. Enquiries concerning reproduction outside the terms stated here should be sent to the publishers at the London address printed on this page.

The publisher makes no representation, express or implied, with regard to the accuracy of the information contained in this book and cannot accept any legal responsibility or liability for any errors or omissions that may be made.

A catalogue record for this book is available from the British Library

Library of Congress Catalog Card Number: 93-74887

∞ Printed on acid-free text paper, manufactured in accordance with ANSI/NISO Z39.48-1992 (Permanence of Paper).

Contents

List of contributors	ix
Editor's preface	xi
Acknowledgements	xv

Chapter 1 Introduction and overview
Alex Maltman

1.1	General considerations	1
	1.1.1 Terminology: sediment, lithification and deformation	1
1.2	Mechanical aspects	6
	1.2.1 General	6
	1.2.2 Volume changes due to burial	6
	1.2.3 Sediment strength	8
	1.2.4 Sediment deformation	9
	1.2.5 Mechanical role of pore fluids	12
	1.2.6 Experimentation	16
1.3	Causes of deformation	16
	1.3.1 Ice	16
	1.3.2 Disturbance in place	17
	1.3.3 Gravitational mass movement	23
	1.3.4 Fluid–sediment movements	27
	1.3.5 Tectonism	29
	1.3.6 Igneous activity	32
1.4	Mélanges as a case history	34

Chapter 2 Mechanical principles of sediment deformation
Mervyn Jones

2.1	Introduction	37
2.2	Mechanics of particulate media in theory and experiment	39
	2.2.1 Force, stress and strain	39
	2.2.2 States of stress and strain in a body	42
	2.2.3 Pore pressure, effective stress and porosity	44
	2.2.4 Primary consolidation and creep	48
	2.2.5 Shear stress and deformation	51
	2.2.6 Other stress paths and associated strain states	60
	2.2.7 Over- and underconsolidation	64
2.3	Natural stress, strain and pore pressure	66
	2.3.1 General	66
	2.3.2 Earth pressure and one-dimensional compaction	67
	2.3.3 Natural shear stress and deformation	68
	2.3.4 Origin of overpressures in sediments	70
2.4	Conclusions	71

Chapter 3 Glacial deformation
Tavi Murray

3.1	Introduction	73
3.2	Subglacial conditions	74
	3.2.1 General	74
	3.2.2 Bed type	74
	3.2.3 Thermal regime	75
	3.2.4 The ice–bed interface	76
	3.2.5 Bed thickness	77
	3.2.6 Realistic basal conditions	77
3.3	Stresses arising from overlying ice	78
	3.3.1 General	78
	3.3.2 Compressive stresses	78
	3.3.3 Shear stresses	79
3.4	Sediment properties	79
3.5	The sediment transport system: production, alteration and loss	81

3.6	Models of sediment properties and deformation	82
	3.6.1 General	82
	3.6.2 Deformation of a homogeneous sediment body	82
	3.6.3 Deformation of a structured sediment body – anisotropy and inhomogeneity	84
3.7	Basal processes as a control on deformation	85
	3.7.1 Hydraulic processes	85
	3.7.2 The consolidation–dilation competition	86
3.8	Effects of deformation	87
	3.8.1 General	87
	3.8.2 Features arising from consolidation	87
	3.8.3 Features arising from shear deformation	88
	3.8.4 Development of sediment form	89
	3.8.5 Sediments beneath surge-type glaciers	91
3.9	Preservation of features	91
3.10	Other types of glacial deformation	92
	3.10.1 Proglacial deformation	92
	3.10.2 Deformation of frozen-substrates–basal-ice	92
3.11	Conclusion	93

Chapter 4 Sedimentary deformational structures
John Collinson

4.1	Introduction	95
4.2	Principles of physical disturbance	95
4.3	Physical deformation structures	99
	4.3.1 Partial loss of strength and density inversion	99
	4.3.2 Structures due to progressive loading of cohesive sediment	103
	4.3.3 Partial loss of strength and applied shear	105
	4.3.4 Structures related to upwards escape of pore water and sediment–water mixtures	108
	4.3.5 Synsedimentary faults	113
	4.3.6 Structures due to sediment shrinkage	114
	4.3.7 Structures due to sediment wetting	117
	4.3.8 Deformation related to compaction	117
	4.3.9 Deformation related to early chemical precipitation	118
4.4	Conclusion	124

Chapter 5 Mass movements
Ole Martinsen

5.1	Introduction	127
	5.1.1 General	127
	5.1.2 Classification schemes	127
	5.1.3 Basic theory	128
5.2	Falls	130
	5.2.1 Introduction	130
	5.2.2 Rock falls	130
	5.2.3 Debris falls	131
5.3	Fluidal flows	133
	5.3.1 Introduction	133
	5.3.2 Turbidity currents	133
	5.3.3 Flows related to volcanic eruptions	135
	5.3.4 Snow- and ice-generated (avalanching) flows	138
	5.3.5 Fluidized flows	140
5.4	Flows with plastic behaviour	140
	5.4.1 Introduction	140
	5.4.2 Debris flows	140
	5.4.3 Liquefied flows	142
	5.4.4 Grain flows	143
5.5	Slumps	144
	5.5.1 Introduction	144
	5.5.2 Process	145
	5.5.3 Products	147
5.6	Slides	152
	5.6.1 Introduction	152
	5.6.2 Process	153
	5.6.3 Products	157
5.7	Creep	162
	5.7.1 Introduction	162
	5.7.2 Process	162
	5.7.3 Products	164

Chapter 6 Tectonic deformation: stress paths and strain histories
Dan Karig and Julie Morgan

6.1	Introduction	167
6.2	Stress paths during burial and uplift of sediments in basins	168
	6.2.1 General	168
	6.2.2 Laboratory studies of consolidation	172
	6.2.3 Effects of geological processes not duplicated in laboratory experiments	175
	6.2.4 Theoretical stress paths during sediment deposition and unloading	177
	6.2.5 *In situ* measurements of stress in sediments	181
	6.2.6 Geological implications from theoretical and measured stresses	185
6.3	Stress paths associated with deformation in accretionary prisms	187
	6.3.1 General	187
	6.3.2 Lagrangian description of sediment accretion	188
	6.3.3 Large-scale variations among prisms and mechanical implications	189
	6.3.4 Diffuse strains in prism toes	190
	6.3.5 Experimental studies	195
	6.3.6 Implications for the deformational histories of accreted sediments from observation and experiment	200
6.4	Conclusions	203

Chapter 7 Fluids in deforming sediments
Kevin Brown

7.1	Introduction	205
7.2	Some basic hydrogeological concepts	207
7.3	Fluid sources and the nature of the tectonic processes driving fluid flow	209
	7.3.1 General	209
	7.3.2 Consolidation and swelling	210
	7.3.3 Transient fluid sources and sinks generated by faulting	212
	7.3.4 Response of poorly-lithified sediments	214
	7.3.5 Response of well-lithified sediments	215
	7.3.6 Chemical sources of fluid: cementation, hydrocarbon generation and mineral dehydration	218
7.4	Control of lithology and burial-related consolidation on the permeability of sedimentary units	220
	7.4.1 General	220
	7.4.2 Permeability anisotropy resulting from consolidation	222
	7.4.3 Equivalent permeabilities and gross permeability anisotropy	224
7.5	Permeability variations due to deformation in active tectonic systems: fractures, faults and gouge	224
	7.5.1 General	224
	7.5.2 Permeability changes in muddy fault zone materials	225
	7.5.3 Permeability changes in fault zones in sands and sandstones	226
7.6	Permeability changes at low effective stresses	230
	7.6.1 General	230
	7.6.2 Extensional failure where the least principal effective compressive stress is tensional	230
	7.6.3 Open-fracture development in the absence of regional tensional stresses: load-parallel extensional fractures	233
	7.6.4 Stress amplification mechanisms	234
7.7	Effect of deformation on the tortuosity of flow paths at different scales	235
7.8	Discussion: transience and the intimate coupling of hydrogeological and tectonic processes	236

Chapter 8 Sediment deformation, dewatering and diagenesis: illustrations from selected mélange zones
Tim Byrne

8.1	Introduction	239
8.2	Progressive deformation and dewatering in the Nankai accretionary prism	240
8.3	Progressive deformation of coherent sediments in the Kodiak accretionary prism	245
8.4	Progressive deformation of mélange terranes in the Kodiak accretionary prism	249
8.5	Deformation and fluid evolution in an accretionary sequence in western Washington	259
8.6	Conclusions	259

Chapter 9 Deformation structures preserved in rocks
Alex Maltman

9.1	Introduction		261
9.2	Techniques of examination		261
9.3	Microfabrics		265
9.4	Micro- to macroscopic structures		268
	9.4.1	Shear zones	268
	9.4.2	Slickensides	273
	9.4.3	Scaly clay and related features	275
9.5	Macro- to mesoscopic structures		277
	9.5.1	Faults, folds and related structures	277
	9.5.2	Liquefaction structures	285
	9.5.3	Dewatering structures	287
9.6	Recognition of sediment deformation structures		292
	9.6.1	General	292
	9.6.2	Importance of recognizing pre-lithification deformation	300
	9.6.3	Outline of possible criteria	302

Appendix: List of symbols   309

References   311

Index   355

Contributors

Kevin M. Brown
Scripps Institute of Oceanography, Geological Research Division, University of California, San Diego, 9500 Gilman Drive, La Jolla, California 92093, USA.

Timothy B. Byrne
Department of Geology and Geophysics, University of Connecticut, 354 Mansfield Road, Storrs, Connecticut 06269-2045, USA.

John D. Collinson
Collinson Jones Consulting, 56 Shropshire Street, Market Drayton, Shropshire TF9 3DD, UK.

Mervyn E. Jones
Department of Geological Sciences, University College, Gower Street, London WC1E 6BT, UK.

Dan E. Karig
Department of Geological Sciences, Snee Hall, Cornell University, Ithaca, New York 14853-4780, USA.

Alex J. Maltman
Institute of Earth Studies, University of Wales, Aberystwyth, Wales SY23 3DB, UK.

Ole J. Martinsen
Geologisk Institutt, Avd. A, Universitetet i Bergen, Allegt. 41, 5007 Bergen, Norway.

Julie K. Morgan
Department of Geological Sciences, Snee Hall, Cornell University, Ithaca, New York 14853-4780, USA.

Tavi Murray*
Department of Geophysics and Astronomy, University of British Columbia, 2219 Main Mall, Vancouver V6T 1W5, Canada.

*Present address: School of Geography, University of Leeds, Leeds LS2 9JT, UK.

Editor's preface

Sediments are now known to undergo deformation in a wide variety of geological circumstances. The deforming processes can happen on a vast scale and at all stages before the material becomes fully lithified. In fact, as exploration of the earth continues, the widespread extent and importance of sediment deformation is still being revealed, for example, below the oceans and beneath ice sheets. At the same time, it is still being realized just how varied are the resulting structures, and how strikingly similar they can be to those produced by the deformation of deeply buried rocks.

However, there are few precedents to guide the geologist in interpreting structures that formed in unlithified sediments, or in understanding the mechanisms through which they arose. This is largely because structural geology has traditionally been predisposed towards the deep-seated deformation processes that operate on rocks long after they have been lithified. The concern has been with the mechanics of rocks, particularly under elevated temperatures and pressures, almost to the exclusion of considering how unlithified sediments deform. Structures formed in sediments have tended in the past to have been viewed merely as localized, surficial oddities that were not an integral part of the structural evolution of an area. This is now changing, at least in the geological settings where the scale and significance of pre-lithification deformation have been realized. There is a new appreciation of the importance of these early structures and a concern with discovering how they are produced. There is a realization that they reflect a long and subtle gradation between near-surface conditions and those at greater depth. Hence the need is growing for a more rigorous understanding of the mechanics of the shallow processes and for the pursuit of more quantitative relationships. With these goals in mind, workers are increasingly drawing on the principles and methods of the well-established engineering discipline of soil mechanics.

All this is beginning to attract wider geological interest. Yet to the newcomer, because progress has been rapid in recent years, the literature is already formidable. The information is scattered, so even an expert on sediment deformation in a certain setting may be unaware of analogous problems and successes in other environments. At the same time, although the same basic principles apply in the various geological regimes, a subtly different terminology is evolving, which can make the subject boundaries hard to cross. The divergent approach and nomenclature used in soil mechanics add further to the confusion.

This book is a first attempt to approach these problems. It is written for structural geologists and others who wish to know more about sediment deformation – i.e. the principles, processes, terminology and the interpretational possibilities associated with preserved examples of the structures. Few of the works in each of the environments of sediment deformation refer to parallel enquiries in other contexts: I have attempted in this book to bring together some of these disparate studies. The worker concerned with sediment deformation in a particular geological setting can glimpse comparable studies in other regimes. The geologist used to considering rock deformation may see that structures preserved in sedimentary and metasedimentary rocks cannot simply be assumed to have originated through tectonic stresses operating after lithification, at depth in the earth's crust. At the very least, the book should provide an entry into the voluminous relevant engineering and geological literature.

The literature references cited are merely a representative selection, but their number in some chapters provides an illustration of the growing interest in this field of geology. I have given priority to the most recent works, and tried to minimize duplication between chapters. Even so, references on some of the topics are so numerous that the selection in places is quite arbitrary. Also rather subjective is the range of sediment behaviour and structures included here. Some of the processes that affect sediment at the same time as or immediately after deposition could easily be viewed as deformation, but they are largely excluded from the present treatment. To give just two examples, processes such as debris flows and structures such as convolute laminations tend to fall into the realm of sedimentology rather than structural geology, and so are given less attention here than the deformations that arise substantially after sedimentary processes have ceased. Some triggering agents, such as magmatic and volcaniclastic activity, have not as yet been sufficiently analysed to warrant detailed review. A particular problem has been presented by the role of diagenesis in deformation. The mineralogical evolution of a sediment is no doubt critical to its deformational behaviour, and the latter in turn is probably highly influential on the further progress of chemical change. Yet what information there is in the book on this crucial interplay is meagre and scattered among various chapters. This comes about because knowledge on this complex topic is so far glaringly sparse. Our understanding is simply still too poor to justify a more substantial treatment.

I have written the first chapter as an overview of the material presented in the book as a whole, and have tried to give a simple introduction to the basic terms and concepts before they are amplified and applied in later chapters. In these ways, the introduction is designed to provide integration between the following chapters, which naturally reflect the differing approaches and priorities of each of the authors. Also, the introductory chapter, despite the newness of much of the information it reviews, is pervaded by a consciousness that many of the ideas have been around for a very long time. Geology is as guilty as any science of a certain arrogance about what is being done now. To give a couple of illustrations of this in the present context, consider first the existence of structures called sedimentary dykes. These structures were among the very first deformation features of any kind to be recorded, and their origin as earthquake liquefaction effects was essentially understood over a century ago. Yet articles are still published today merely reporting the existence of these structures at a particular place, and perhaps suggesting that they probably formed as a response to seismic activity. Most other structures seem not to warrant such special treatment: it seems that some authors are simply oblivious of the pedigree of studies into sedimentary intrusion. As a second, terminological, example, a feature recognized back in 1914 was referred to then as ball-and-pillow structure, and the term remains in use today. Yet along the way different authors have neglected the precedent and invented new names. They have seen fit to call what appear to be essentially similar structures such things as: flow rolls; slump rolls; birds-eye structure; slump casts; balled-up structure; roll-up pebbles; flow casts; flow structure; kneaded structure; pseudonodules; hassock structure; slump balls; and storm rollers. Therefore, in an effort to avoid this repeated 'reinvention of the wheel', I make a point in Chapter 1 of outlining the historical framework, and of mentioning some older review references.

Each of the chapters following the introduction has been planned as a state-of-the-art review of a particular aspect of sediment deformation. Chapter 2 focuses on theoretical principles and experimental studies. The following chapters progress roughly from settings at the earth's surface and shallow levels of burial to the generally deeper environments where lithification becomes complete. So by chapters 6 and 7, for example, much of the discussion deals with materials where the distinction between a sediment and a rock is blurred. Some of the ideas presented there are equally applicable to both materials. Chapter 8 discusses case histories of the close interaction between some of the processes identified in earlier sections.

Some overlap and repetition between chapters are unavoidable as each one is meant to be fairly complete in itself. In places, different nuances of emphasis or meaning serve to illustrate contrasts in the approach of different workers. Within the confines of each chapter, each contributor has dealt with their subject matter largely as they have felt appropriate. Taken together, the chapters should provide a blend of the theoretical, experimental and descriptive approaches to the subject. I have tried to avoid any conflict of information, even though the thinking of some of the contributors was evolving as the material was being written! After all, despite the great progress during the last 20 years or so in recognizing and understanding sediment deformation, the subject is still in its infancy.

Acknowledgements

Drs Jon Arch, Bobb Carson, Richard Cave and Antony Wyatt made helpful comments on early drafts of various parts of chapters 1 and 9. Ruth Cripwell, Earth Sciences Editor at Chapman & Hall, helped guide the book through its sometimes tortuous route to completion. Jeff Schmok and Gordon Hamilton are thanked for comments on Chapter 3.

The following publishers kindly gave permission for illustrations to be reproduced, from the sources indicated in the figure captions: Académie des Sciences, Montrouge, France; American Association of Petroleum Geologists, Tulsa, Oklahoma; Association of Engineering Geologists, Palo Alto, California; A.A. Balkema, P.O. Box 1675, Rotterdam; Blackwell Scientific Publishers, Osney Mead, Oxford; Elsevier Science Publishers BV, Academic Publishing Division, Amsterdam; Geological Society of America, Boulder, Colorado; Geological Society Publishing House, Bath, Avon; Societé International des Geologistes Ingineurs, Paris; Ocean Drilling Program, College Station, Texas; Pergamon Press Ltd, Headington Hill Hall, Oxford; Society for Sedimentary Geology, Tulsa, Oklahoma; Springer-Verlag, Berlin; University of Chicago Press, Chicago, Illinois.

The editor compiled much of the material while on sabbatical leave at the University of Minnesota. The help of the Geology Department staff and students there, and Peter Hudleston in particular, is gratefully acknowledged. The Hartshorn family of Minnetonka made the stay of the Maltman family both practicable and pleasurable. Finally, Jo, Alastair and Emily are thanked for their support throughout – and for time on the home computer.

CHAPTER 1

Introduction and overview

ALEX MALTMAN

1.1 GENERAL CONSIDERATIONS

1.1.1 Terminology: sediment, lithification and deformation

Sediments are accumulations of particles at the earth's surface, the result of physical deposition or, in some cases, biological or chemical activity. **Deformation** is a change in the bulk shape of the aggregate. The transformation of these loose, particulate aggregates into rock is termed **lithification**. (One traditional use of the word rock to include all earth materials whether loose or not is unhelpful in the present context.) The change from sediment into rock comes about through some combination of mechanical and chemical processes. The mechanical aspects, essentially the moving of the grains closer together so that there is increasing interaction, go hand-in-hand with the deformation processes that form the basis of this book and consequently are dealt with in detail in the following pages.

The chemistry of lithification, however, although studied intensively in many ways, has had little systematic analysis in the context of deformation, and forms only a minor part of the present book. The various chemical processes that help turn a sediment into a rock are here referred to collectively as **diagenesis**. In this usage, diagenesis is not synonymous with lithification but a fundamental component of it. Diagenetic reactions gradually increase the chemical bonding between grains, whether or not the particles are being brought physically closer (Lade and Overton 1991). The processes are dominated by **cementation** – precipitation into a pore space; **recrystallization** – solid-state reorganization of the grains; and **diffusion mass transfer** – solid-state movement of grain material to sites of lower stress. The last process, given the water-rich environment of most sediments (Schutjens 1991), is commonly fluid-assisted and called **pressure solution**. This process becomes more significant in the later stages of lithification (e.g. Bjørlykke, Ramm and Saigal 1989; Ruilsback 1993), where it can also help generate cementing material (Tada and Siever 1989). Cementation operates in parallel with the mechanical effects of burial by helping to reduce the sediment porosity (Lasemi, Sandberg and Boardman 1990), a central aspect of much lithification. In geotechnical (soil mechanics) usage, all types of diagenesis are commonly referred to as 'cementation' and the result is a 'structured' material. This terminology is usually unhelpful in geology.

As indicated above, how such chemical changes interact with deformation processes has been explored only in a sketchy manner (Bayly 1993). At present the topic is insufficiently understood to justify a separate treatment. Consequently, it is emphasized at the outset that this book does not deal explicitly with diagenetic aspects of deformation. Some illustrations are scattered throughout the book (section 4.4, for example, discusses some effects associated with growing concretions and section 6.3.5 documents some mechanical effects of grain bonding), but the progressive change of a sediment into a rock and the deformation that arises before complete lithification is treated here very largely in terms of mechanical processes.

For many purposes of deformation studies, therefore, the diagnostic property of a sediment is that it changes its bulk shape principally in a particulate or granular way, by frictional grain-boundary sliding. The particles are generally the strongest part of the aggregate and simply slide past each other, undergoing negligible deformation themselves. Any delicate clay domains probably undergo some modification, and there are circumstances in which even quartz grains are affected (section 9.5.3), but most sediment deformation is dominated by what has been called **independent particulate flow** (Borradaile 1981). In contrast, rocks deform by utilizing a range of intragranular deformation mechanisms (Figure 1.1). Hence, in this view, lithification consists essentially of the processes that act to curb progressively the role of grain-boundary sliding.

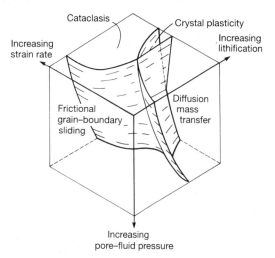

Figure 1.1 The relation of frictional grain-boundary sliding, characteristic of the deformation of unlithified sediment, to the three main intragranular deformation mechanisms that dominate post-lithification deformation, as conceptualized by Knipe (1986b), from which this diagram is simplified. (Used by permission of R. Knipe.)

The highly gradational nature of these processes means that materials can also be in transitional states, that is, **partially lithified** or **semi-lithified**. In these situations, grain-boundary sliding remains dominant, but operates in conjunction with a greater proportion of intragrain deformation. Engineering geologists use analytical methods and laboratory tests that are different for sediments (e.g. Mohr-Coulomb parameters and shear tests, section 1.2.3) and rocks (e.g. unconfined compressive strength and joint analysis), and often refer to the intermediate state as **weak rock**. Mechanical repercussions of this spectrum of behaviour are amplified in section 2.2.5, and some specific illustrations from nature are given in Chapter 8. The converse situation, where rocks disintegrate through weathering and return to the particulate state, has received little attention from structural geologists but is of great practical importance for engineering geologists. Salt, although regarded for most purposes as a sediment, moves via mechanisms that largely fall outside the definition of sediment deformation offered above (Jackson and Talbot, 1993), and is not dealt with here.

Such a mechanistic view of lithification has shortcomings in addition to that of ignoring the chemical processes. For example, during the course of a geological deformation event, physical conditions, such as strain rate and fluid pressure, may fluctuate such that the relative importance of intergrain and intragrain processes varies. A sediment may find itself temporarily under stress conditions that drive grain breakage. The material could conceivably be deforming at one time as if it were lithified, and at some later stage, when conditions have altered, it could go on to behave as a sediment. Moreover, a particular difficulty for geologists analysing structures preserved in rocks is that the grain behaviour at the time of deformation may have been obscured by later events. Therefore, the definition used here does little to aid the difficult and frustrating business of distinguishing between structures developed in rocks and those formed in sediments that were subsequently lithified (section 9.6). However, a more practical definition remains elusive. The mechanically based approach used here is powerful in that it allows geologists to capitalize on the principles well established in the science of soil mechanics. In time it may become possible to incorporate chemical changes also.

A cameo of the mechanical effects of lithification, wholly qualitative but of practical use with modern sediments, is provided simply by the ease with which materials can be indented and cut! Unlithified sediment can be carved with a cheese-knife. Firm sediments require greater finger pressure to indent them than do soft materials, and very firm sediments require thumb-nail pressure. Bladed knives, progressively sturdier with increasing lithification, are needed to cut partially lithified materials. Rocks have to be sawn.

Deformation is bulk shape change, usually quantified, just as in rocks, as a strain (e.g. Williams, Goodwin and Ralser 1994). However, with sediments the accompanying volume changes are usually significant and have to be considered along with the deformation. Measurements of volume change are referred to as **volumetric strain**. These concepts are dealt with systematically in Chapter 2.

Deformation and bulk volume changes normally come about as a response to an applied force. **Gravity** is the overridingly important body force in sediment deformation. Changes within a sediment, in particular a fluid-pressure increase, can reduce the strength of the mass such that it is no longer stable in the earth's gravity field, and deformation may ensue. This is the basis of the various mass movements discussed in section 1.3.3 and Chapter 5. Also, if sediments are deposited such that their density does not increase uniformly downwards in line with gravity, instabilities can result which also induce deformation. A similar effect arises where the sediment density gradient becomes modified after deposition, for example by fluid changes, freezing, or the addition of igneous material. These in-place processes are mentioned further in section 1.3.2 and amplified in Chapter 4. The results can be striking in weak, water-rich sediments, but the potential energies resulting from these gravity driven instabilities are normally too small to be of significance for deeply buried sediments.

Surface forces causing deformation can arise locally, due, say, to a nearby igneous intrusion or from passing ice. They can also operate regionally. In this case, the forces are usually of deep-seated origin – the result of lithospheric physico-chemical changes – and they are called **tectonic** forces. The resulting effect is tectonic deformation. This is not confined to rocks. As is made clear in section 1.3.5 and in later chapters, tectonic forces, albeit initiated deeply, can be transmitted through loose, water-rich sediments and cause their deformation even at surficial levels. Some geologists have drawn a contrast between tectonic and so-called 'soft-sediment' deformation. However, the former refers to a force and the latter to a material condition, and so the two cannot be compared (Maltman 1984). Also, the term soft-sediment lacks precision in that it may or may not include partial lithification and because softness is not a quantified parameter. The term hydroplastic is used by some workers, but also lacks rigour (section 4.2). Terminological aspects of these localized, early deformational processes are discussed by van Loon (1992). One intriguing example of 'soft-sediment' being used purely to describe the state of material is its application to early structures formed in igneous rocks, by P.E. Brown, Chambers and Becker (1987).

In general, gravitational deformation is dominant in the upper parts of a sedimentary pile and tectonic forces are more significant deeper down, but there is much overlap and gradation. For example, buried sediments that have remained weak, perhaps because their porosity was unable to decrease, can be subjected to gravitational deformation at depths of several kilometres (e.g. M.E. Jones and Addis 1985). Conversely, structures forming through tectonic forces can form in near-surface sediments and can outcrop on the ocean-floor (Maltman *et al.* 1993a).

A semantic awkwardness surrounds the term tectonic, because it can also refer to the form of a geological mass. Thus geologists talk about 'tectonic style' or the 'tectonics' of a district. The meaning is usually clear from the context. However, the situation can be confusing in the glaciological context, where structures that originated through flow of the ice have been described as tectonic (Sharp, Lawson and Anderson 1988). The term glaciotectonic is used for both glacial structures and their production from ice, with the 'glacio' sometimes being dropped. Lea (1990), for example, discusses 'glacial tectonism', and Ruszczynska-Szenajch (1988) mentions processes called 'tectonic detachment' and 'tectonic redeposition', all of which result solely from ice action. This can lead to situations, for example in the Himalayas (Owen 1989), where some sediment structures are due to the tectonic processes associated with mountain uplift and others are the result of glacial processes, yet both are referred to as tectonic! Similarly, introduction of the term synsedimentary tectonics (Thomas and Baars 1988) seems unhelpful in the interests of clarifying the forces responsible for deformation.

The notion that tectonic deformation can take place in surficial, unlithified sediments may be a revelation to some, in that it is out of line with much structural geological thinking of the last 150 years. Yet once, much deformation was thought to occur in association with water. In order to place traditional structural geology in its full historical context, Historical vignette 1 considers briefly why the preoccupation with the deformation of buried rocks came about.

Historical vignette 1: the influence of early thinking

When geological deformation structures were first observed, during the eighteen century, their origin tended to be ascribed to one of two groups of processes. They were thought either to be due in some way to the effects of heat or they were the result of processes associated with water. It is interesting to speculate that it is because, in general, the former group won out – ideas in which heat was invoked to engender the deforming forces and therefore to operate at depth, and on rocks rather than subaqueous sediment – that a predisposition arose towards the deformation of buried, lithified rocks.

The importance ascribed by some early workers to water-based processes came about not only because of their significance in the lithological and chronological classification systems of the time, and a literal belief in the biblical Flood, but also because they allowed a ready visualization of how deformation could happen. Disturbances of strata from their original horizontal position were generally envisaged as collapse phenomena rather than forced uplift, and erosion by water currents could easily be pictured as generating channels into which the sediments could fall and become disrupted. Who knows, if this approach had been allowed to evolve, gravitational sliding and its significance, for example, may have been acknowledged much sooner by geologists, and may even have become a conventional explanation for deformation structures seen in sedimentary rocks? However, as it happened, the importance of heat gradually began to dominate geological thinking. James Hutton, for example, successfully demonstrated the igneous origin of intrusive rocks such as basalt and granite, and he convinced many of the role of the earth's heat in other geological processes. He argued that variable thermal expansions and contractions could best explain the deformation structures seen in rocks.

More and more writers of the time were reporting deviations of beds from the horizontal and the various associated contortions. They usually invoked some sort of lateral pushing as an explanation, and it became increasingly common to call on thermal changes as the source. In this way deformation became increasingly envisaged as occurring at depth, where the sediments were sufficiently buried to enter the realm of thermal effects. Hutton's colleague James Hall was a pioneer of studying such distorted beds and he was probably the first to attempt laboratory simulations of the processes. In 1812 he reported his experiments to replicate folded structures in the laboratory, and the title is telling: 'On the vertical position of convolutions of certain strata and their relation with Granite'. Right from the start, his thinking involved thermal effects at depth, and his experiments simulated the load that was supposed to overlie the deforming materials (Figure 1.2).

The die was cast. As time progressed, processes such as subaqueous gliding received occasional mention, but they gradually fell into abeyance. When, towards the end of the nineteenth century, experiments demonstrated that rock subjected to confining pressure and elevated temperature was capable of flowing, there seemed even less need to question the belief that deformation structures necessarily occurred after burial. Deformation commonly became assumed to occur at depth.

This thinking also fitted with the stratigraphical ideas of the time: sediments accumulated in huge troughs, and only after a sufficient thickness had been reached was orogeny prompted – long after most of the deposits had been buried and lithified. In Wales, for example, vast thicknesses of sediments were envisaged as accumulating throughout Cambrian, Ordovician and Silurian times, and only then, at the end of the Early Palaeozoic, did the deforming stresses arise, to affect materials that could have seen as much as 200 million years elaspse since their deposition.

The state of lithification of the deforming materials was equally hazy, largely because the

General considerations 5

Figure 1.2 The pioneering experimental set-up of Sir James Hall involving, from the outset, the simulation of a presumed burial load and deformation at substantial depth. (Reproduced from Hall 1812.)

relevant processes were very poorly understood. The word 'rock' was used very generally – loose sands and clays, and even ice, were counted as rocks; Lyell (1838) remarked on the 'insensible passage from a soft and incoherent state to that of a stone' all being encompassed by the single term rock. It is still used this way by some geologists, but for the present purposes the distinction between loose sediment and lithified rock is essential, as the mechanics of deformation of the two materials are different. Interestingly enough, the term 'consolidation' was used by Hutton and other early workers in a way approximating lithification, as some geologists would use it today (section 1.2.2), but ideas on how it happened were vague. Nebulous processes, with names like congelation, petrification and concretion, were variously mentioned, but, again following Hutton's lead, heat was normally invoked as the chief lithifying agent. It therefore followed that the deformation went hand-in-hand with or followed lithification. It is instructive to quote directly from Hutton, in view of the far-reaching influence of his ideas: 'The strata formed at the bottom of the ocean are necessarily horizontal ... and continuous in their horizontal direction or extent. There cannot be any sudden change, fracture or displacement. But if these strata are cemented by the heat of fusion, and erected with an expansive power acting below, we may expect to find every species of fracture, dislocation and contortion' (Hutton 1788, p. 229).

Modern thinking, of course, corroborates the fundamental role of thermal changes in the earth, but heat is not normally invoked today to explain directly either lithification or deformation. However, the legacy of the earlier thinking was profound. Structural geology increasingly focused on the deformation of rocks at depth, almost to the exclusion of other kinds of deformation. Ironically, engineering geology originated through the need to consider the deformation of surficial materials under the influence of gravity! The fact that the two disciplines evolved in opposite directions helps explain why cross-fertilization has been delayed for so long.

1.2 MECHANICAL ASPECTS

1.2.1 General

The physical principles of how sediments deform are well established in the engineering discipline of soil mechanics, through efforts to understand the behaviour of foundations and slopes. (Geotechnical engineers and engineering geologists refer to particulate materials as soils rather than sediment.) Because sediments are particulate or granular materials – the solid particles are more or less dispersed within gaseous or liquid phases – the concepts of soil mechanics are of greater relevance to understanding sediment deformation than, say, fluid or continuum mechanics.

Many of the concepts and terms outlined below and developed in Chapter 2 derive from soil mechanics principles, and they are applied and illustrated in later chapters. Recent introductions to soil mechanics and its basic terminology include Atkinson (1987), J.A. Barker (1981), F.G. Bell (1987), enkins (1984) and Whittow (1990). Further sources are given in Chapter 2. Extensions of the ideas to the submarine situation, which have come about with the growth of various aspects of offshore engineering, were reviewed by Bouma *et al.* (1982), Denness (1984), Demars and Chaney (1990), Sangrey (1978, 1982) and Poulos (1988). The behaviour of particulate aggregates from the point of view of applied mechanics and material science, respectively, was reviewed by Satake and Jenkins (1988) and Shahinpoor (1983).

1.2.2 Volume changes due to burial

In geology, words such as compaction, consolidation and lithification are often used more or less interchangeably, but if we are to understand how sediments deform it is necessary to adopt a more rigorous usage. Lithification has already been defined (section 1.1.1). **Compaction** is useful as a general term for all permanent reductions in the bulk volume of an aggregate, including adjustments both to the grains themselves and the pore spaces between them. A compacted sediment has simply lost volume, irrespective of how it was achieved. Compaction could result from thermal changes adjacent to an igneous intrusion, for example, or through diagenetically induced shrinkage of the sediment. **Consolidation** on the other hand, in line with soil mechanics usage, is restricted to time-dependent mechanical reductions in volume. Normally it is pore volume that is lost, in response to a burial load. The time-dependency comes about because the rate at which consolidation can proceed is governed by the time it takes to dissipate the appropriate amount of pore water to maintain equilibrium (see section 1.2.5). Because of this, Schmoker and Gautier (1989) were prompted to treat consolidation in terms of time, but most analyses of consolidation involve the magnitudes of gravitational load (see below and section 2.2.4). Squeezing out of the pore fluid dominates the early stages, called **primary consolidation**, but the later, fine-scale reductions – **secondary consolidation** – involve some adjustments to the grain framework. As noted earlier, numerical manipulation of any or all of these volume changes is referred to as volumetric strain.

Therefore, in the terminology encouraged here, consolidation is one kind of compaction. Both processes help convert sediment into rock. They are therefore important contributors to lithification but they are not synonymous with that term, which, as emphasized in section 1.1.1, also includes diagenetic (chemical) effects. Consolidation can come about in response to tectonic forces (e.g. Minshull and White 1989), but it is normally the result of sediment burial. It is in the latter context that the mechanics of consolidation are usually considered, but analogous processes operate in sediments being loaded by ice (Boulton and Dobbie 1993).

As sediment becomes progressively overlain by further additions of material, in the earth's gravity field it becomes subject to an increasing load, of a magnitude that depends on the thickness and density of the overlying sediment. The load is variously referred to in geology as **burial, overburden** or **lithostatic pressure**. In experimental work it is simulated by the **cell** or **confining pressure**. In nature, it tends to be dominated by the vertical, downward component, and, being directional, is better called a stress, although it can be of equal magnitude in all directions. The latter situation, in analogy with the stress configuration that exists in water at rest, is referred to as **hydrostatic**. It is useful to use pressure where no direction is implied, and stress where a directional component is meant. Deviatoric, or differential stress, where shear stresses arise because the stress magnitudes in the three, orthogonal, principal directions are unequal, becomes significant when considering deformation (section 2.2.2).

The magnitude of **horizontal**, or **lateral**, **burial stress** depends on the ratio referred to as the **K value**, where:

$$K = \frac{\text{magnitude of horizontal stress}}{\text{magnitude of vertical stress}}. \quad (1.1)$$

If K = 1, the sediments are subject to hydrostatic stress and undergo isotropic consolidation. A more realistic situation is for the horizontal stresses to be just sufficient to prevent lateral strain in the sediment, a condition termed K_0, the coefficient of earth pressure at rest (sections 2.2.6 and 6.2.3; Mesri and Hayat 1993). K_0 consolidation tends to induce anisotropy in the sediment, with corresponding effects on strength and stress–strain behaviour (Atkinson, Richardson and Robinson 1987). Other implications of the K_0 condition have been explored by Daramola (1980), Garga and Khan (1991) and Kitamura and Shinchi (1988).

Increasing burial pressure tends to progressively pack the grains closer together. Assuming that pore fluids are able to dissipate adequately, a critical matter that is developed later, this decreases the **porosity** (η) or **void ratio** (e) of the aggregate, where:

$$\eta = \frac{\text{volume of voids}}{\text{volume of aggregate}}; \quad e = \frac{\text{volume of voids}}{\text{volume of grains}}$$

$$\eta = \frac{e}{1+e}; \quad e = \frac{\eta}{1-\eta}. \quad (1.2)$$

In much of geology, η is multiplied by 100 to express porosity as a percentage. Assuming water is occupying the pores, there takes place a concomitant decrease in the **water content** (w), normally expressed in per cent by weight:

$$w = \frac{\text{weight of pore water}}{\text{weight of dry grains}} \times 100 \quad (1.3)$$

although occasionally **wet water content** is used:

$$w_{\text{wet}}\% = \frac{\text{weight of pore water}}{\text{weight of wet aggregate}}. \quad (1.4)$$

The latter has the advantage of allowing porosities to be readily calculated from it if the relevant densities are known (Boyce 1976). Modern methods of determining water content are outlined by Gilbert (1991).

Water content governs the **consistency** of sediments. Most fine-grained materials, because of the presence of clays and organic matter, are naturally plastic and easily achieve unrecoverable deformation without fracturing, that is, they exist within their **plastic range**. A drier sediment, however, may fall below its **plastic limit**, and a wetter one can exceed its plastic range by having a water content above its **liquid limit**, and flowing under its own weight like a fluid. The difference in water contents between the two limits is called the **plasticity index**. Such limiting values to plasticity, sometimes called the **Atterberg limits**, can be of geological relevance, for example, in assessing the stability of sediments on slopes (Einsele 1989) and the form of mass flows (section 1.3.3). Coarse sediments, especially those without clay, tend to lack plastic properties. Note that this usage of the term plastic differs slightly from that describing a rheological behaviour (section 1.2.4), in that no yield stress or strain-rate requirement is implied. Plasticity tends to increase with greater carbon content (Bennett *et al.* 1985): Booth and Dahl (1986) documented in a clayey silt from offshore southern California, a 20% and 14% increase in the liquid limit and the plastic limit, respectively, with a mere 1% increase in organic carbon.

Porosity reduction that maintains mechanical equilibrium with the burial load is termed **normal consolidation**. Numerous theoretical and empirical attempts have been made to quantify the relationships (Baldwin and Butler 1985; Dzevanshir, Buryakovskiy and Chilingarian 1986; Bayer and Wetzel 1989; section 6.2.1). Observations on porosity loss with burial on the modern ocean floor are reported by Faas (1982) and Shepard and Bryant (1983). It is clear that in nature porosity loss does not always keep pace with progressive burial, which leads to **underconsolidation**. The most common explanation involves the sediments being unable to expel the pore water at a sufficient rate. **Overconsolidated** sediments, with porosity less than that expected from their burial level, are also well known in nature (Mayne 1988; Rafalovich and Chaney 1991). A common explanation of this effect invokes diagenesis, as cementation can independently reduce the porosity, though without an overall compaction (Ehrenberg 1989). Subaerial sediments commonly become overconsolidated through geological uplift and erosion of the overlying material. In detail, consolidation is a complex phenomenon, interacting with compression of the sediment framework and proceeding at variable rates (Sridharan and Rao 1982; Znidaric and Schiffman 1982; Yong and Townsend 1986; Nagaraj, Vatsala and Sriniva Murthy 1990; Griffiths and Joshi 1991). The influence on consolidation progress of different lithic grains in a sand framework was investigated by Pittman and Larese (1991). The various states of sediment consolidation are addressed in greater detail in sections 2.2.4, 2.2.7, 2.3 and 6.2.

1.2.3 Sediment strength

As sediment particles are being deposited they increasingly interfere with each other (Martinez et al. 1987; Tan et al. 1990). The gradation between the stage of hindered settling (Pane and Schiffman 1985; Toorman and Berlamont 1991) and final deposition has been the subject of sedimentological study (Mehta 1991) and much engineering research, in contexts ranging from sludge disposal to dredging (Yong and Townsend 1984). As the particles are brought closer together, electrostatic bonds are initiated between them, imparting a physical **cohesion** to the sediment (Wetzel 1990). Also, movement between adjacent grains begins to involve **intergrain friction**. For either or both reasons the sediment acquires some resistance to an applied shear stress. (The concept of shear stress, and its difference from a principal stress, is explained in section 2.2.2). The maximum shear stress the sediment can sustain before failure is called its **shear strength**. The time-honoured view of the relation between these parameters utilizes the Mohr-Coulomb equation:

$$\tau = c + \sigma_n \tan \phi, \qquad (1.5)$$

where τ is the shear stress sustained at the point of maximum resistance, σ_n the stress normal to the plane of failure, c is termed cohesion and ϕ the angle of internal friction. The continuing usefulness of this equation is reflected by its use in several of the following chapters. (Some other criteria for defining the failure of a sediment are mentioned in section 3.6.1). Strictly, the last two terms of the equation are solely mathematical quantities and not physical attributes of the sediment, but there are numerous observations that point to a relation between the friction coefficient and the nature of grain surfaces (e.g. Koerner 1970; Frossard 1979). For this reason, angular sediments, such as carbonates, give high friction angles and are relatively strong (Semple 1988). In unsaturated sediments, increasing water content decreases the magnitude of cohesion and friction, and hence shear strength (Yoshida, Kuwano and Kuwano 1991). Note that discussion of the role of fluid pressure in these relationships is deferred until section 1.2.5.

The shear strength of sediments is sometimes expressed in words such as hard, stiff, firm and soft. Sediments that have undergone intense disturbance are termed **remoulded**. The **sensitivity** of a sediment is the ratio between shear strengths in the undisturbed and remoulded states, a difference that can be marked in clays. The ratios can exceed 4 in sensitive clays, and values in excess of 16 are a characteristic of **quick clays**. To give some quantitative view of strength: very weak, near-surface sediments typically have shear strengths of a few kilopascals (kPa), increasing to hundreds of kPa with tens of metres of normal burial, and perhaps to a few MPa on lithification. A strong sandstone might sustain a shear stress of several hundred MPa. To complete the picture, a shear strength of around 100 kPa is achieved at the base of a 10-m pile of dry sugar! The strength varies by about 20 kPa depending on whether the sugar is loosely or densely packed (Feda 1982).

Progressive packing of the sediment particles during burial provides further opportunities for intergrain bonding and increases the normal stress acting across the grain boundaries. In general, and still ignoring fluid pressure effects, it is the effect of the normal stress on the frictional term that is responsible for increasing sediment strength with burial. An observable repercussion of this is that clayey sediments, with their electrostatically charged large surface areas, acquire greater cohesion than clastic materials and can be stronger at shallow burial levels, whereas the latter are normally the stronger material at depth, as the frictional contribution begins to dominate. For example, some surficial volcanic sediments weather to give clay minerals that appear to impart a particularly strong bonding to the sediment, hence generating strong sediment at shallow levels of burial (Belloni and Morris 1991). The proportion of clayey material in the matrix of glacial sediments is a paramount influence on their deformational response to ice (section 3.4). However, in general, increasing burial reduces the ability of the grains to overcome the frictional resistance to their sliding and the sediment grows in strength – the sediment undergoes lithification.

1.2.4 Sediment deformation

The deformation of a sediment probably begins with a small amount of elastic strain, an aspect considered in some detail in Chapter 6. However, for most sediments, the elastic range is distinctively smaller than that of their lithified equivalents (section 2.2.1) and its effects become obscured by the grain-boundary sliding that rapidly ensues (Hardin and Blandford 1989). Many sediments, and particularly dense sands and overconsolidated clays, show a high point in the stress–strain curve, known as the **peak strength**, normally equivalent to the shear strength. This is followed by an overall strain softening as they drop to some value of **residual strength** (Lupini, Skinner and Vaughan 1981), where additional strains are achieved at virtually constant stress. Loose sediments show no peak, and gradually attain a steady state; others show some degree of strain hardening. In the laterally confined axial-shortening arrangement known as **triaxial testing**, the sediment behaviour depends crucially upon whether the pore fluid is trapped in the specimen – an **undrained test** – or allowed to escape – a **drained test** (see Figure 2.26). A synopsis of typical behaviours is given in Figure 1.3.

In the residual state, any cohesion possessed by the aggregate will have become negligible and the behaviour governed by its frictional characteristics (Burland 1990). The drop to a residual strength is well shown by clays, where it comes about through the reorientation of the platelets into zones of alignment. Skempton (1964) showed that sediments with 20% clay content behave essentially like sands or silts, proportions between 20

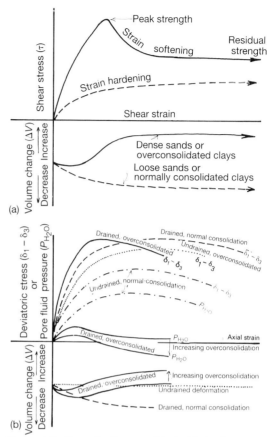

Figure 1.3 Synoptic diagram of typical behaviour patterns of deforming sediments. (a) Densely packed and loose sands, and normally and overconsolidated clays subject to shear. (b) Normally and overconsolidated clays subjected to drained and undrained triaxial deformation.

and 50% give a transitional behaviour, the exact nature of which depends on both the precise nature and amount of the clays. A 50% clay content gives a marked drop to residual, which does not increase in conspicuousness with yet higher clay proportions. The role of friction is illustrated by montmorillonite clays ($\phi = 5°$) showing the lowest residual strength, illites ($\phi = 10°$) having intermediate values, and kaolinites ($\phi = 15°$) showing the highest residual strength values of these three minerals (Skempton 1985). R. Moore (1991) pointed out the significant influence of pore-water salinity on these values, and on the cohesion and plasticity of clays. Moore's measurements, on subaerial clays, showed an increase in residual strength of up to 21% as the pore fluid was changed from fresh to salt-water.

Attempts to document the quantitative stress–strain behaviour of sediments during deformation, largely by laboratory testing, have led to the concept of **stress paths** (Shibuya and Hight 1987; Vaid, Chung and Keurbis 1990) and their portrayal on **stress path diagrams**. These are presented in two or three dimensions, and various parameters are plotted on each of the axes (Jamison 1992). The familiar Mohr-circle construction uses a two-dimensional example in τ–σ space, as pursued in section 2.2.5. More common in soil mechanics treatments are p–q diagrams, as explained and applied in section 2.2.5. Further applications of p–q stress path diagrams arise in sections 6.3.5 and 8.2.

One of the chief shortcomings of the Mohr-Coulomb approach to sediment deformation is that volumetric strain is not directly taken into account, although significant volume change is characteristic of many deforming sediments. For example, sands and normally consolidated clays tend to pack more tightly in response to stress (e.g. Sladen and Oswell 1989), thus showing **contractive** behaviour, whereas materials such as dense sands and overconsolidated clays show marked **dilative** behaviour as they approach failure (Hardin 1989; Shimizu 1982). Some geological repercussions of dilation were considered by Brodzikowski (1981).

As a consequence of the importance of volume change, p–q stress paths are commonly combined with the volumetric responses of a deforming sediment into a three-dimensional diagram (section 2.2.5). This approach has led to a series of synoptic models, mostly based on that originally developed at the University of Cambridge for clays, and referred to as Cam-clay theory (Srinivasa Murthy, Vatsala and Nagaraj 1991). The concepts are now extended to sands (Been, Jefferies and Hachey 1991) and carbonate sediments (Coop 1990). One aspect of such models is their prediction that sediments reach a certain volume–stress combination that remains constant during additional deformation, a condition known as the **critical state**. Hence these stress–volume analyses are commonly called **critical state models**. The principles are explained in section 2.2.5 and, for example, by Wood (1990). The applicability of critical state models to various submarine sediments is illustrated in sections 6.3.5 and 8.2. Many structures of geological interest, however, appear to have involved much greater strains than are likely to be achieved in nature at the critical state, because of irregularities and fabric changes inducing volume fluctuations, so it may be that they developed in the residual condition. Here also the stresses remain approximately constant with increasing strain, but the fabric collapse that typically allows the high strains to be achieved in turn engenders volume loss.

Modifications to fabric are an integral part of sediment deformation but so far the contribution that soil mechanics results can make to geology is restricted, because the work is still largely qualitative. There are numerous descriptions of the interplay of consolidation and deformation with microscopic fabric (Clark and Gillott 1985; Rothenburg and Bathurst 1989; Wesley 1990) and macroscopic features such as joints (Kinkaldie 1992), fissures (Costa-Filho 1984), and bedding surfaces (Eigenbrod and Burak 1991), but attempts to quantify fabric changes and integrate them with models of stress and volume changes remain elusive. A further complication is the effect of time, for it is well known in soil mechanics that fabrics can change even when the sediment is at rest – a condition known as **ageing** (Schmertmann 1991; Tsuchida, Kobayashi and Mizukami 1991).

A further condition becomes relevant where long periods are considered, namely the phenomenon of **creep**. Here is a good example of a term being used in slightly different ways, according to context. Sedimentologists and geomorphologists tend to use it to indicate any very slow – perhaps centimetres per year – gravitational downslope movement (e.g. McKean et al. 1993), without any particular implication of the stress conditions, strain rate evolution or volume change. The treatment of mass movements in Chapter 5, for example, uses the term in this way (see, especially, section 5.7). In addition, although the movement tends to differ from slope failure by being more distributed, an analogous localized creep is mentioned in section 9.5.1 as an inferred explanation for certain detachment structures. In some soil mechanics usage, on the other hand, creep has a stricter meaning of continuing long-term strain, not necessarily downslope, and typically showing a three-stage response to low-level stresses. Silva and Booth (1985) suggested that sediments moving slowly downslope may actually be following this latter pattern, in that their laboratory testing of slope sediments from offshore northeastern USA reproduced the expected three-stage creep behaviour. Some implications of this work for analysing slope stability effects are mentioned in section 1.3.3. Yet a further variation in the meaning of creep is illustrated in sections 2.2.4 and 6.2.3, where the term is used synonymously with secondary consolidation, for the slow volume loss that may continue after the pore fluid appropriate for the burial load has been expelled.

Much of the analysis of sediment deformation outlined above is based essentially on a Mohr-Coulomb view of the behaviour. The value of this stress-based approach is thoroughly established in soil mechanics and it will be used repeatedly in the following chapters. It has proven an appropriate treatment for many geological situations. An alternative approach, however, is to view the deforming sediment as some kind of fluid and to consider its rheological properties. This method also has engineering proponents (Keedwell 1984, 1988; Vyalov 1986), but has so far been less widely used in geology. Examples of the rheological approach arise elsewhere in this book; a brief introduction follows.

Settling sediments are commonly regarded as approximately **Newtonian fluids** (no shear strength, and a shear–strain response proportional to strain rate):

$$\frac{d\gamma}{dt} = \frac{\tau}{\mu}, \qquad (1.6)$$

where γ is shear strain and t is time. The proportionality term, μ, is termed the **viscosity**. In integrated form (1.6) becomes:

$$\tau = \mu \dot{\gamma} \qquad (1.7)$$

or

$$\dot{\gamma} = \frac{1}{\mu}\tau \qquad (1.8)$$

where $\dot{\gamma}$ is the change in shear strain with time. In reality, as much work on materials such as paint, drilling mud and wet cement has shown, the behaviour of particle suspensions is much more complex than this. As the settling approaches completion the particles increasingly interfere with each other and the aggregate acquires significant strength (section 1.2.3). If this strengthened sediment is still able to show a linear response to shear stress it is termed a **Bingham fluid** (see Figure 3.4):

$$\tau = \tau_y + \mu_p \dot{\gamma} \qquad (1.9)$$

where μ_p is the plastic viscosity, which incorporates the excess shear stress sustained over that in Newtonian behaviour, and τ_y is the resistance to shear.

Most sediments do not show such a linear response, partly because the viscosity itself varies with increasing strain rate:

$$\tau = \mu(\dot{\gamma}) \cdot \dot{\gamma} \qquad (1.10)$$

and so more complex relationships are required (see Barnes, Hutton and Walters 1991). According to whether the viscosity increases or decreases with strain rate, the behaviour is referred to as **shear thickening** or **thinning**, respectively. The change may be gradual (Figure 3.4) or form a series of plateaux. Wet clayey sediments, for example, commonly show progressive shear thinning (Philip Harrison, personal communication).

Many mechanical analyses of geological sediments use an elastic–plastic rheology. Here, an initial **elastic** (recoverable) strain gives way at a yield point to bulk **plastic** behaviour (unlimited, rate-independent strain beyond the yield stress). The elastic strains, by definition, are not recorded in the sediments (section 6.1), and the latter mode dominates in any prolonged deformation. The range of elastic behaviour is usually considered to be minimal in poorly lithified sediments, but its importance grows with progressive increase in interparticle bonding. The strain response of a wholly elastic solid to a given stress will, following **Hooke's Law**, depend on the proportionality constant of the material, known as **Young's modulus**. The elastic strain at right-angles to the direction of maximum strain, for example the lateral extension accompanying greatest shortening in the vertical, is given by the **Poisson's ratio** of the sediment. These parameters are pursued further in section 6.2.3; other parts of Chapter 6 and section 7.6.2 analyse the contribution of elastic behaviour to sediment deformation.

Much sediment deformation beyond the yield point is somewhat rate dependent – most stress–strain curves are curved, with the slope at any particular point dependent on the rate of stress increase – and so some rheological models use a **viscous** rather than a plastic behaviour. Here the strain rate is related to the stress level. Such non-linear viscous fluids, provided the viscosities are not extreme, are commonly represented by a **power law** of the form:

$$\mu = x\dot{\gamma}^{n-1}, \qquad (1.11)$$

where x is called the consistency of the sediment and n the power law index.

Realistic modelling of geological sediments may well have to combine several models of rheological behaviour, and the power laws can be complex. Kutter and Sathialingam (1992), for example, attempted to incorporate rate-dependency into their analysis of clay deformation, leading to an elastic–viscoplastic model. Other, more or less complex formulations are illustrated in the following chapters. Section 1.3.3, for example, mentions viscous-based modelling of landslides, section 3.6.1 discusses several flow laws that have been used for the base of ice sheets, and section 5.1.3 describes some variations on the basic models in the context of mass movements. Thus, for many geological deformations, the crucial step in a rheological analysis is the choice of the most appropriate model or combination of models. The value of the analysis depends on its realism and on the simplicity or otherwise of the resulting flow law.

1.2.5 Mechanical role of pore fluids

Much of the analysis of fluid movement in geological material has been in the realm of groundwater studies (e.g. Nielsen and Johnson 1990), petroleum geology (e.g. Robinson in press), and, more recently, waste disposal (e.g. Devgun 1989; Valent and Lee 1976). Earlier literature on water resources from unlithified sediment was reviewed by Dudgeon and Yong (1969). However, the crucial importance of pore fluids for soil engineering was established long ago, and it is clearly the same in geological sediments. The foregoing outline of sediment behaviour has ignored this fundamental effect only for the sake of introducing topics systematically. General principles of fluids in geology are discussed by Bear (1972) and Fyfe, Price and Thompson (1978).

In most geological sediments the fluid is water of some sort, and this supports some portion of any applied load, giving rise to a **pore** or **fluid pressure**. Both terms, together with pore fluid pressure, are widely used in geology and throughout this book. In normally consolidating sediment (i.e. the pore water is continually expelled to keep pace with progressive burial), it is called **normal fluid pressure** – and is that portion of the burial load arising from the overlying column of pore fluid (plus sea water in the case of submarine sediments). The remaining component is that sustained by the sediment particles and is called the **effective stress,** σ':

$$\sigma' = \sigma - P_{H_2O} \qquad (1.12)$$

where σ is the total stress and P_{H_2O} the fluid pressure. This symbolism for fluid pressure is used throughout this book, in line with much geological usage. In soil mechanics literature the symbol u is commonly used for fluid pressure. It is the effective rather than the total stress that controls deformation (see later), and it is therefore best to modify the Mohr-Coulomb relation (1.5) in terms of parameters with respect to effective stress:

$$\tau = c' + \sigma'_n \tan \phi'. \qquad (1.13)$$

A prime symbol added to a term indicates that allowance is made for pore fluid pressure. Sediment that shows contractive behaviour during deformation (section 1.2.4) will undergo a simultaneous, more or less transient, fluid pressure increase (effective pressure decrease) while dilative materials will experience an accompanying decrease in pore pressure. Such effects on the pore fluid pressure and their embodiment in the concept of effective stress is a fundamental aspect of the mechanical behaviour of sediments and is amplified below. The importance in soil mechanics is emphasized by Wood (1991), and some further ramifications are discussed by Mayne and Stewart (1988), Sridharan and Rao (1979) and in section 6.2.3.

With burial, the pore fluid has to be expelled in appropriate amounts if the pore volume is to reduce in equilibrium with the additional load. Possible variations in porosity and void ratio with vertical effective stress are given in Figures 6.1–6.3. The ability of a sediment to dissipate pore fluid is a function of its **permeability**, as explained in section 7.2. This in turn depends on the configuration of the pores and particles in the sediment (Beard and Weyl 1973; Ahuja *et al.* 1989; Bloch 1991; Bryant, Code and Meller, 1993). Although most quantitative models of consolidation make the simplifying assumption of constant permeability, Schiffman (1982) has emphasized that this critical parameter is likely in nature to vary sensitively with different amounts of burial. The expulsion of pore water during laboratory tests is known as **drainage** in soil mechanics and in geological sediments as **dewatering** (Burst 1976). The process is central to consolidation under the effect of gravity (Al-Tabba and Wood 1991), but in geology, tectonic dewatering is being increasingly recognized in addition (Carson and Berglund 1986; Shi and Wang 1985).

If, after a given increment of loading, a sediment lacks sufficient permeability for the pore fluid to escape sufficiently, the fluid, being virtually incompressible, has to sustain a disproportionate part of the load (Gretener 1981). This pressure excess over the normal fluid pressure is termed **excess fluid** (or **pore**) **pressure** or, in geology, **overpressure, geopressure, hydropressure, abnormal** or **supernormal fluid pressure**. **Subnormal** or **negative fluid pressure**, sometimes referred to as **underpressure** (section 6.2.4), is also known in nature (Belitz and Bredehoeft 1988). The amount of excess pressure is often expressed as the fluid pressure ratio (λ), where:

$$\lambda = \frac{\text{pore fluid pressure}}{\text{total burial pressure}}. \quad (1.14)$$

Excess pore pressure is sometimes defined with respect to steady-state flow conditions (Gibson, Schiffman and Whitman 1989); Audet and McConnell (1992) presented a mathematical analysis of the process. Normal fluid pressure is also commonly called the **hydrostatic pressure**, in both geology and engineering. The types of pressure gradients likely in a sedimentary sequence are represented schematically in Figure 1.4. Real *in situ* measurements are difficult to obtain and are still in their infancy (e.g. Hurley and Schultheiss 1990; Carson *et al.* in press).

The pressure head in overpressured fluid is no longer balanced out by the elevation head (section 7.2), as in the equilibrium condition, and so the fluid attempts to move to lower values of the potential gradient (Chapman 1981). There are numerous geological repercussions of this, especially if the fluid is able to flow with sufficient velocity to entrain the host sediment. Note that the fluid potential gradient, although upwards overall, can be downwards locally. This is why sedimentary dykes, for example, do not necessarily intrude upwards (section 1.3.4). Section 7.4.3 shows that interbedded sands and muds should develop a higher proportion of lateral drainage than more homogeneous sequences (Magara 1976). Bredehoeft, Djevanshir and Belitx (1988) have documented lateral dewatering on the regional scale.

We come now to the profound implication that overpressuring has for sediment mechanics. Overpressuring weakens sediments. Because the fluid is sustaining an extra part of the stresses acting across the aggregate framework, the effective stress decreases, the intergrain friction is reduced ($\sigma'_n \tan \phi'$ in equation 1.13), and hence so is the sediment strength. This is why, as stated above, it is the effective rather than the total stress that governs deformation. Overpressured horizons are commonly sites of shear failure (e.g. section 1.3.3 and Chapter 5). Overpressured sediments may undertake intrusion (section 1.3.4) and diapirism (sections 4.3.2 and 2.3.3). The reduction of effective stress and strength by overpressuring is a crucial factor in all settings of sediment deformation. Two examples follow.

Lash (1985), in his analysis of Ordovician rocks in the central Appalachian tectonic orogen of eastern Pennsylvania, used fluctuating effective stresses to explain the differing structural evolution of neighbouring sand and mud horizons. Initial deformation increased pore pressures in sand beds contained within mud-rich horizons, and the reduced strength led to flow and stratal disruption. The interlayered mud, because of its greater cohesion, became the stronger sediment and deformed only along discrete zones. These eventually interweaved to produce a scaly foliation, which facilitated dewatering of the mud, thus further increasing its strength. Meanwhile sand beds that lacked mud envelopes had drained adequately and maintained strength and coherence throughout the early deformation. As deformation continued, the contrasts in drainage, and hence effective stress, strength and deformation behaviour, progressively decreased, and the various lithologies converged to a more uniform behaviour. Abbott, Embley and Hobart (1985) found that those sediments in the South Pass district of the Mississippi Delta that gave permeability measurements of only 4×10^{-14} m^2 (expressed by those workers as hydraulic conductivity, see section 7.2), had shear strengths around 3 kPa and were highly deformed. In contrast, sediments with permeabilities of 1.8×10^{-13} m^2

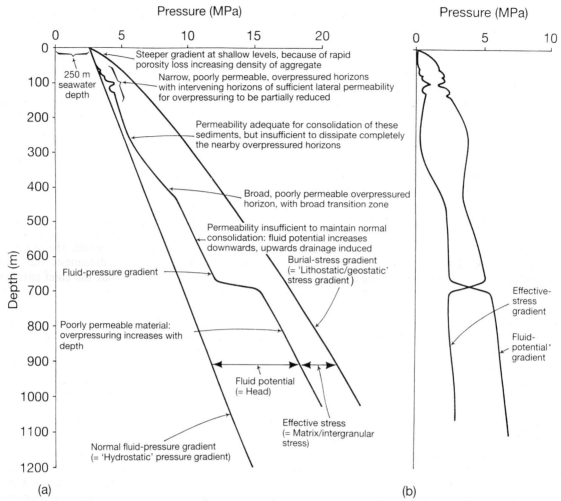

Figure 1.4 Patterns of pressure increase with depth. (a) Rates of increase of pressure and stress due to burial. The gradient for normal fluid pressure is based on an overlying seawater depth of 250 m and a constant fluid density in the entire column of 1025 kg m^{-3}. The burial-stress gradient is based on a grain density of 2650 kg m^{-3} and a porosity decrease with depth following the average values for sand–silt–clay of Einsele (1989). The hypothetical fluid-pressure gradient shown here illustrates the kinds of effects that can arise in a sequence of varied sediments. Compare with Figure 7.5. (b) Effective-stress gradient and fluid-potential gradient derived from the curves in (a). Note that the fluid potential and its gradient should properly be expressed in units of metres (the height that could be supported by a column of specified fluid, normally water or mercury) rather than pascals. The fluid potential, or total head (H), is here a pressure head (Φ), arising from the excess fluid pressure ($P_{H_2O_e}$). See Figures 7.1 and 7.2, and section 7.2 for further explanation. (From Maltman (1994) Reproduced with permission of Pergamon Press.)

showed strengths around 8 kPa and were virtually undeformed. In the former materials, the temporarily trapped pore water was overpressured and the reduced effective stress and sediment strength were allowing greater strains. Overpressuring as a cause of hydraulic fracture has been given great attention in rocks, but in sediments this aspect has so far been little addressed (Murdoch 1992).

Although it is a long time since Karl Terzaghi clarified how these kinds of factors prompted slope failure rather than any effect of lubrication, it is surprising how often this latter notion is still invoked in geological explanations of moving sediment masses. Most silicate minerals are not lubricated by water, and most unlithified sediments contain too much water for the thin-film phenomenon of lubrication to be of relevance. In fact, Savina (1983) remarked that invoking lubrication in mass movement betrays those geologists who lack an understanding of the mechanical principles involved. It is fluid pressure that holds the key, and the geological repercussions of this relationship between permeability, overpressure, effective stress and sediment strength – and hence intensity of deformation – are enormous, and further examples arise frequently throughout this book.

Overpressuring can originate in numerous ways in geological situations, such as through mineral dehydration (Colten-Bradley 1987), biopressuring resulting from organic decay (Nelson and Lindsley-Griffin 1987), and gas liberation from clathrate decomposition (Carpenter 1981). Further mechanisms are indicated in section 2.3.4. Where the dissipation of pore fluid is greatly curbed, or increments of loading arise in addition to burial, say from the transmission of seismic waves, the pore fluid may have to temporarily sustain the entire stress acting on the sediment. In this situation the internal friction is effectively reduced to zero, and the aggregate, now lacking resistance to shear, behaves as a liquid. The resulting phenomenon of **liquefaction** causes a number of geological structures in sediments (section 1.3.2) and has been much studied by engineers (R.D. Davis and Berrill 1983; Seed 1985; Richards, Elms and Budhu 1990). Similar results can arise in sediments with the particularly delicate fabric known as quick clays (Torrance 1983). Marine geotechnology has been much concerned with overpressuring and liquefaction arising through the height of storm waves temporarily increasing the fluid head acting on the pore fluid (Kraft *et al.* 1985; Okusa 1985). In addition to these direct consequences of overpressuring, at sites of high fluid-pressure gradients, the drag force exerted on a grain framework by a flowing fluid, termed the **seepage force**, may be sufficient to exceed the sediment strength and to prompt failure (Orange and Breen 1992).

Nevertheless, despite all these repercussions, the behaviour of fluids in sediments is far from being fully understood. Although water is the common fluid in nature, it can have variable chemistry, the effects of which are little explored. Moreover, biogenic gas will often be present in near-surface materials (Wheeler 1989; Sills *et al* 1991). Further complications arise from the fluid interacting with the sediment grains, an aspect at best simplified in current quantitative models (Stevenson and Scott 1991), and from the fluid comprising more than one physical phase (Parker 1989). Permeability remains poorly investigated in many geological situations. Vast numbers of on-land measurements have been collected in engineering hydrology, but the effects on permeability of, for example, inhomogeneities in the sediment (Atkinson and Richardson 1987; Bethke 1989), anisotropy (Arch and Maltman 1990) and dilatancy (Harp, Wells and Sarmiento 1990) remain to be understood. *In situ* submarine measurements are in their infancy (Fang, Langseth and Schultheiss 1993).

Finally, fluids are undoubtedly central to diagenetic reactions and their effect on sediment mechanics. As mentioned earlier, the interplay between deformation and diagenesis has been little explored. It may be that the chief effects of deformation on diagenetic reactions are not direct but through the production of features that affect fluid distribution. For example, rates of ionic diffusion through pore water have little direct sensitivity to pressure, whereas the permeability reductions associated with fabric changes during consolidation will reduce diffusion magnitudes by a half to a twentieth of the value in free solution (Manheim 1971).

In contrast, there is more documentation of the progress of diagenesis directly affecting sediment deformation. Boudreau, Brueckner and Snyder (1984), for instance, noted diagenetic influences on the pre-lithification deformation of Upper Palaeozoic rocks in Nevada. The silica in clay-rich siliceous sediments tends to remain as diagenetically immature opal-A and opal-CT longer than in clay-poor sediments, which convert rapidly to quartz. Therefore, the former material tended to deform in a more ductile manner than the latter, quartz-enriched and hence more brittle sediment. At the same time, the diagenesis was enhancing the ductility contrast between the adjacent materials, thus promoting boudinage and a flexural-slip mechanism of folding. Moreover, because loading during burial caused preferential dissolution of silica from the clay-rich layers, the remnant, now clay-enhanced, laminae were able to act as the slip surfaces during the folding. An analogous example from Miocene rocks is given in section 9.5.1. Maliva and Siever (1988) suggested that some mineral textures produced during lithification are the result of stresses generated by diagenetic reactions – a force of crystallization induced by new mineral growth. Tribble (1990) documented the progress of clay diagenesis in the Barbados accretionary prism and, through its effect on sediment physical properties, its possible influence on the location of thrust faults and the basal décollement of the prism. Sample (1990) argued that carbonate cementation in the Upper Cretaceous sandstones of Kodiak, southern Alaska, played a major role in the dynamics of the plate convergence. Sample envisaged a 'cementation front' moving through the evolving prism, which influenced sediment strength and the styles of deformation, and which may even have led to the large amount of underplating postulated for this plate junction.

1.2.6 Experimentation

Soil mechanics principles have been derived largely from laboratory testing. However, the vast majority of this kind of work is devoted to routine evaluations for engineering construction purposes, using the routine tests described in handbooks such as Bowles (1978) and Vickers (1983). The specifications for executing standard tests are stated in publications such as ASTM (1991) and BSI (1991). The testing methods are, however, under continuous review (Donaghe, Chaney and Silver 1988; Toolan 1988; Jewell 1989; Burland and Hight 1990; Atkinson, Lau and Powell 1991). Advances in instrumentation are expanding the role of field testing, which offers less control but has the marked advantage of minimizing disturbance to the measured materials (Demars and Nacci 1978; Graham, Crooks and Bell 1983; Chaney and Demars 1985). Techniques for permeability measurement are discussed by Aban and Znidaric (1989), Dunn and Mitchell (1984) and Fernuik and Haug (1990).

The great value of laboratory testing in the present context is to provide quantitative limits for the interpretation and understanding of natural processes. To give one example, Lucas and Moore (1986) observed cataclastically deformed quartz grains in near-surface marine sediments off southern Mexico. Any possibility that the grain breakage had arisen purely through burial was eliminated on the basis that laboratory tests had shown the available burial stresses to be insufficient. More importantly, steepened bedding dips at the sample site raised the possibility that the deformation was due to submarine sliding. However, comparison of the shear stresses in the likely slope conditions with the appropriate mechanical parameters derived from laboratory testing also eliminated this possibility. The deformation had to involve tectonic stresses. More elaborate illustrations of applications of laboratory testing to understanding sediment deformation are given in sections 2.2.5, 6.2.2 and 6.3.5.

1.3 CAUSES OF DEFORMATION

1.3.1 Ice

Historically, ice was one of the first direct agents of sediment deformation to be documented, as a result of noticing structures in deposits recognized as glacial in origin (e.g. Sorby 1859). Exactly how the ice produced deformation structures was debated for a time: Salisbury and Atwood (1897), for example, invoked the grounding of icebergs and ice rafts to explain examples of intrastratal folds, whereas Johnston (1915, p. 43) reported sand laminae that were 'minutely folded and overthrust by the overriding of the ice sheet'. The evolution of early views on glacial deformation is summarized by Aber, Croot and Fenton (1989). However, despite the impressive range of ideas, the notion of ice bulldozing into sediments and forcing various 'ice-thrust' structures became, for a long time, the dominant concept. The influential textbook of Flint (1971, p. 121), for example, remarks that most deformation of glacial sediment 'seems to have been made at or near glacial termini'. In this view, the deformation of sediment by ice is confined to some of the peripheral areas of the glacier, and consequently the process received limited attention.

A major discovery of recent years is that some glaciers and parts of major ice sheets rest not on bedrock, but on a layer of sediment (section 3.1). This seems particularly true of the fast-flowing zones of ice sheets known as ice streams, and it seems that shearing of the basal sediment layer is largely responsible for the high flow rates that have been observed (e.g. Alley 1991). This 'deformable bed' hypothesis forms the basis of Chapter 3. There, soil mechanics principles are applied to an understanding of the subglacial dynamics. The chapter also outlines the rheological approach to quantifying the behaviour of these weak and highly strained subglacial sediments. The principles are remarkably parallel to those involved in understanding other environments of sediment deformation. For example, glaciologists are currently investigating sediment consolidation (Boulton and Dobbie 1993), shear strength and viscosity (Humphrey et al. 1993), and hydrogeology (Murray and Dowdeswell 1992): all matters of topical interest at convergent plate margins and elsewhere (section 6.3). This new view of ice-sheet motion represents a major change in glaciological thinking (Boulton 1986) and the way in which structures in glacial deposits are interpreted. Although caution has been advised against applying the deforming substrate hypothesis too widely (Robin 1986; Clayton, Mickelsen and Attig 1989), it appears to be gaining support, prompting a new interest in how sediments deform in the glacial environment.

In addition to the shearing effects expected at the base of a moving ice sheet, sediment deformation comes about in response to ice loading (Drozdowski 1982) and a host of glacial and periglacial processes, such as iceberg drop and grounding (G.S.P. Thomas and Connell 1985), repeated freeze–thaw cycles (Coutard and Mucher 1985) and deglaciation (Vesajoki 1982). The processes are complex and closely interrelated.

Not surprisingly, therefore, the structures resulting from glacial deformation are extremely varied (e.g. Fischbein 1987). Their similarity with structures formed in completely different environments has long been noted. Salter (1866, p. 566), for example, compared structures in the glacial drift of southern England with 'those produced in our mountain ranges'. More recent examples include comparisons of glacially induced structures with thin-skinned thrust belts (Croot 1987) and with accretionary prisms (Aber, Croot and Fenton 1989). Aber (1988) compared the origin of some ice-related hills in Saskatchewan to the mudlumps of the Mississippi Delta, which are themselves analogous to the diapirs associated with accretionary prisms. At the microscopic scale, authors such as Hart and Boulton (1991) have compared highly sheared till fabrics to mylonites, although in view of the crystal–plastic processes that dominate in rocks, the grain-scale mechanisms must differ. Fabric studies have long been an important aspect of glacial deposits. This approach has helped interpret the origin of structures (Menzies and Maltman 1992), distinguish till genesis (van der Meer 1987), and even to verify the glacial origin of deposits (Mahaney 1990). Soil mechanics methods have been applied to understanding some of the structures, such as joints (Feeser 1988) and shear zones (Tsui, Cruden and Thomson 1989). Sauer and Christiansen (1988) measured the consolidation states of glaciolacustrine clays from central Saskatchewan in order to find likely burial loads and hence ice thicknesses at the time of glaciation, and in the Netherlands the likely glacier dynamics were deduced from geotechnical measurements by Schokking (1990).

It follows from the interaction of the kinds of glaciological processes indicated above that many structures are not the result of a single agent. For example, Owen (1989) reported structures in which both glacial and neotectonic deformations have been involved, and in the active Kleszczów Graben of central Poland, Brodzikowski, Krzyszkowski and van Loon (1987) have documented structures due to tectonic, glacial and sedimentological interaction. Pederson (1987) described structures resulting from both glacial ice and gravitational mass-flow, and the glacially generated gouge zones described by Stauffer, Gendzwill and Sauer (1990) governed both further glacial deformation and post-glacial landslides. Brodzikowski and van Loon (1991) reviewed effects of the interplay between sedimentological and glacial processes.

A number of glacial landforms and related features are now thought to involve various kinds of sediment deformation. Although geomorphological phenomena are outside the scope of this book, examples are mentioned in section 3.8.4 because their genesis is related so closely to the structures preserved within them. Recent collections of works on landforms, their internal deformation structures and related features include Croot (1988), van der Meer (1987) and Goldthwait and Matsch (1989). Some further examples of deformation structures from the glacial environment are mentioned in Chapter 9.

1.3.2 Disturbance in place

It is still being learned just how gravitationally unstable many near-surface sediments are – on the sea floor, in lakes and on land. Material lying on a slope has long been known to be vulnerable to downslope motion, and the deformational implications of these mass movements are dealt with in the next section and in Chapter 5. However, sediment instability without substantial lateral transport is also now widely recognized, particularly in the near-shore setting, where it has immense repercussions for marine geotechnology. Such in-place disturbance gives rise to a variety of deformation structures, and these are described in section 4.3.1. The separation between these and the structures dealt with in Chapter 5 is arbitrary: many natural structures involve a small amount of lateral motion, and are part-way between in-place and true mass movements. In the geological record, their precise origin might well be unclear.

The kinds of geological structures that would now be ascribed to these processes have been known for some time. William Logan (1863), for example, reported a 'singularly wrinkled structure' that occurred within otherwise uniform Devonian limestones at Gaspé, Quebec (Figure 1.5). The vexing nature of such 'intrastratal' disturbances is illustrated by the fact that the precise origin of some examples would still be debated today, although, hopefully, understanding of the possible formative processes is now more rigorous. In earlier times, some of the statements on inferred processes were merely speculative and intuitive, criteria were vague, and even the observations lacked the necessary precision. Some of the discussions, however, were more informed and perceptive. In the interests of learning from such past experience, Historical vignette 2 summarizes the exchanges of views on a particular example of these structures: those illustrated in the frontispiece to this book.

Figure 1.5 Sir William Logan's illustration of localized deformed beds at Gaspé, Quebec. (From Logan 1863.)

Historical vignette 2: Intrastratal disturbances at Trenton, New York

Lardner Vanuxem, at various times a farmer in Pennsylvania, a university professor in South Carolina, and a gold-miner in Mexico – and during his scientific training in Paris an associate of such luminaries as Brogniart and Haüy – was for 4 years part of the official geological survey of New York. His 1842 report which concluded this last enterprise mentioned some 'extraordinary contortions' in the Ordovician limestones near Trenton Falls, in Oneida County, central New York State. He supplied a rather stylized representation, reproduced here as Figure 1.6a. He also reported a series of similar structures in Quaternary deposits along the nearby Black River and for these latter structures he provided more accurate illustrations – beautifully engraved wood-cuts, two of which form the frontispiece herein. 'The layers show a series of contortions of different kinds', Vanuxem declared, 'for which no cause can reasonably be assigned but different degrees of lateral pressure' (1842, p. 213). He was unsure of the origin of the pressure, and vaguely invoked either some differential crystallization between the layers or some accommodation response to 'lateral pressure ... forcing whatever soft or yielding material was subjected to its power' (p. 216). Vanuxem emphasized the contrast

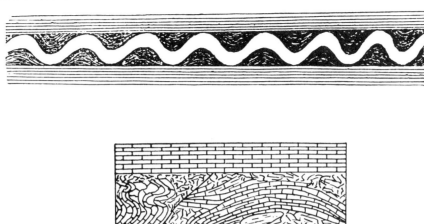

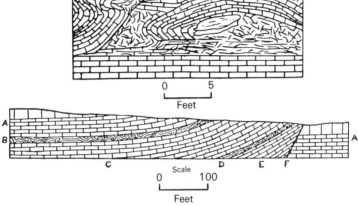

Figure 1.6 (a) Lardner Vanuxem's stylized representation of the intrastratal contortions at Trenton, New York. (From Vanuxem 1842.) (b) Reproduction of William Miller's depiction of the intrastratal disturbances at Trenton, New York. (As shown in Miller, 1908, 1909 and 1922.)

between the disturbed and the normal horizons, but did not specifically mention the critical observation of the truncated upper and lower contacts of the disturbed horizon, although they are recorded in the illustrations.

The same structures caught the attention of Theodore White during his palaeontological work in the area for his doctoral thesis at Columbia University. He thought Vanuxem's explanation to be inadequate, as well as a suggestion he had heard from W.O. Crosby, a professor at the Massachusetts Institute of Technology (MIT), that the cause may have been the yielding of this layer under the tremendous weight of strata above' (White 1895, p. 90). He cited observations such as the undeformed nature of the fossils and the insufficient degree of metamorphism as incompatible with the hypotheses offered, though he seemed unable to specify any alternative explanation. He merely commented that 'he was inclined to think that the contortion took place, from some cause, very shortly after the period of deposition'.

About 50 years after Vanuxem's project, W.J. Miller was engaged in a survey of the same area, and was also struck by the 'highly twisted or contorted' structures at Trenton and the similar phenomena along the Black River (Miller 1908, p. 429). However, the standard of precision in illustrating the structures was noticeably poorer (Figure 1.6b), and Miller came to a different interpretation. He found no difference in the crystalline state of the different limestones, as implied by Vanuxem, and proposed that 'it is thought that the folded structure ... was in reality caused by a differential movement within the mass of the Trenton limestone', as a result of accommodation to nearby thrusting. Some layers were able to 'move more easily and consequently faster' (p. 431), and the 'folded zones thus merely indicate horizons of weakness along which the differential movement took place'. The idea, of course, is perfectly feasible in general, but Miller made no attempt to assess the truncations of the Trenton structures, or, especially in view of the uniform 'crystallinity' he had emphasized, why the various limestone layers should have been so mechanically contrasting. Miller also noted that the 'locally folded clay layers between non-folded layers along the Black River are to be explained in a somewhat similar manner', although here 'the movement of the upper over the lower masses may have been caused by ice action or by having been pulled down the hillsides by gravity' (p. 432).

Later, Miller (1909) repeated his descriptions, and reproduced again the drawing in Figure 1.6b. In considering the origin of the structures, he recalled Logan's (1863) remarks on the similar features of Gaspé (Figure 1.5): 'it would appear as if the layers after their deposit had been contorted by lateral pressure, the underlying stratum remaining undisturbed, and had been worn smooth before the deposition of the next bed. Where the inverted arches of the flexures occur, some of the lower layers are occasionally wanting, as if the corrugated bed had been worn on the underside as well as the upper side' (Logan 1863, p. 392). Miller commented that 'it is difficult to see how such a lateral pressure could cause certain layers to become highly folded and broken while the layers immediately below them are undisturbed' (Miller 1909, p. 32). Logan had put his finger on the crucial matter of the truncated boundaries, but although Miller had reproduced the remarks, the significance eluded him. 'The rubbed or worn character' was simply one repercussion of the entire sequence being differentially moved by faulting.

In contrast with these intuitive speculations, the next contribution was based on the results of observed processes. Albert Heim in 1908 had produced very detailed descriptions, in German, of two disastrous landslides that had entered Lake Zurich, and from which falling lake levels had given the opportunity for the resulting structures to be directly observed. Fritz Hahn had read the accounts while working in Stuttgart, but had subsequently become temporary Curator at New York's Columbia University, from where he happened to visit the Trenton structures. He immediately perceived their likely subaqueous pre-lithification origin, and wrote a detailed argument in support of this interpretation. Amadeus Grabau, a giant of the time in eastern North American geology, took up Hahn's theme in his monumental 1200-page textbook (1913).

Grabau, fluent in German, was also conversant with Heim's writings. Oddly enough, he had been a student of Crosby at MIT, but he perceived the significance of Heim's observations and espoused the importance of subaqueous sliding. 'As the result of such gliding the strata must suffer much deformation' (1913, p. 660), he wrote, and went on, 'such deformations have all the characters of orogenic disturbances due to lateral pressure, and indeed it has been suggested that some extensive mountain folds and overthrusts may have originated in this manner'. Grabau reviewed the Trenton structures, mentioning one explanation that was 'tectonic – lateral pressure having resulted in the folding of certain strata while others took up the thrust without deformation', and another 'that they were due to squeezing out of certain layers under the weight of the overlying rock masses'. Grabau concluded that in his view 'both are unsupported by the detailed characteristics of the folds and their relationships to the enclosing layers' (p. 784) and argued in support of the subaqueous sliding hypothesis.

Miller (1915) could not accept this explanation. Moreover, he was aggrieved that Grabau 'neither states nor presents arguments against the present writer's hypothesis' (p. 135). Apparently he felt it was not represented by Grabau's statement on tectonic causes. Miller's list of evidence in support of his hypothesis is rather subjective, and provides two general illustrations of the way in which geologists work. First, previous experience tends to colour perception. Miller's geological experience and perspective contrast with those of his opponents. Having been born near the Lassen Peak volcano, in California, and having worked mainly on igneous and metamorphic phenomena, he approached the structures from a 'hard rock' viewpoint and may have been predisposed towards these kinds of processes. At this stage at least, Miller seems to have had little feel for the sediment sliding process. Ironically, he was later to work on some California landslides! Second, intuition alone can mislead. To Miller, the deformed horizons appeared too thin to be due to slumping, given their areal extent, and he felt that 'the slope of the sea bottom ... was altogether too slight, a fact 'well-nigh fatal to the hypothesis of submarine slumping' (p. 143). Regarding his feel for the rates of lithification of limestone he felt that 'the corrugations could not have been produced before deposition of the overlying masses because the limestone layers at least were comparatively hard and brittle when contorted, as shown by the numerous sharp breaks'. He concluded that 'it would seem that Hahn has fallen into a common error of making a single hypothesis or explanation altogether too inclusive in its scope'. Hahn met a premature death, in 1914 at the age of 29, and never had the opportunity to reply.

The differential weighting idea continued to receive some favour in interpreting structures like those at Trenton, and Kindle (1917) provided some experimental justification. He pointed out that although 'various explanations have been offered, the subject appears to be still open to experimental investigation'. From the results of irregularly piling different-density wet sediments in a tank, he argued that 'disturbances of the bedding in strata of sand may originate in the same general way through the juxtaposition of beds of quicksand and ordinary sand' (p. 382). Miller, however, remained unconvinced, and in 1922 reaffirmed that 'in the opinion of the writer differential movement, not always as an accompaniment to thrust faulting, is involved in many, if not most, cases of intraformational disturbance'. He also published for a third time the drawing reproduced here as Figure 1.6b, and repeated his complaint that Grabau had discussed the Trenton structures 'without even mentioning the tectonic differential slipping hypothesis' (p. 590). Miller then reviewed intraformational structures more generally, together with a range of possible origins, including igneous processes. He went on to a long and illustrious career, becoming, for example, the first head of the new geology department at the University of California at Los Angeles, but there is no evidence that he ever became converted to the importance of subaqueous movements.

Near-surface sediments are at their most rapid stage of consolidation, and if the pore fluid is unable to escape adequately they are particularly vulnerable to two effects. First, layers of differing permeability will undergo different rates of porosity decrease, so that at any one time denser sediments may be overlying less dense materials. Primary deposition can sometimes produce the same arrangement. Because gravitational potential results from height above a datum times mass, the overlying layer has the higher potential energy and will attempt to mobilize to produce the more stable, greater density with depth, configuration. Should the shear strength of any of the layers be approached, through further pore-pressure increments, which lower the effective stress, or the addition of some extra load, the potential mobility will be activated. The kinds of deformation structures described in section 4.3.1 will be generated. The best analysis of these kinds of processes remains the work of Anketell, Cegla and Dzulinski (1970) and Ramberg (1981).

Second, lowering of the effective stress is facilitated in these loose, near-surface materials, because they are likely to deform initially by contracting, thus temporarily increasing the fluid pressure and promoting further weakening. In addition, pulses of greater pore pressure may arise from mineral dehydration, aquathermal effects, etc. at depth, as discussed in sections 2.3.4 and 7.3.6. Such processes that are internal to the sediment give rise to **static loading**. Addition of an extra external stress is called **dynamic loading**; individual loading events are described as **monotonic**. They can arise in nature for a host of reasons, including: an increment of rapid sedimentation (G. Owen 1987); emplacement above of a moving sand dune (B.G. Jones 1972), slump sheet (Eva 1992) or igneous mass (Einsele 1982); even the footstep of a mammal (Lewis and Titheridge 1978) or shock (supersonic) loading due to meteorite impact (Read 1988).

Cyclic loading comes about most commonly through the oscillatory transmission of seismic waves (Seed and Idriss 1982) and, in subaqueous sediments buried (in typical water conditions) no more than about 200 m, from the passing of travelling pressure waves at the sea surface (Seed and Rahman 1978). Each stress rise in these repetitive processes may well be much less than in a monotonic event, but if each fluid pressure response is incompletely dissipated before the next pulse, the incremental accumulation of overpressure can be large. The idea is shown schematically in Figure 1.7. A variant on this behaviour, called **cyclic mobility**, can induce strength loss even in relatively dense sediments. It arises where the periodic loads oscillate in nature between compression and tension and create transient pulses of overpressure (Castro 1975, 1987). These loading phenomena are discussed further, in the geological context, by Decker (1990).

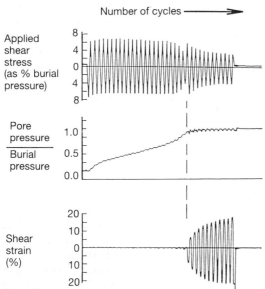

Figure 1.7 Schematic diagram of cyclic loading leading to liquefaction. The horizontal axis represents the number of loading cycles with time. The fluid pressure accumulates incrementally until it equals the burial load, whereupon the sediment liquefies and the shear strain response is instantaneous. (Based on Decker (1990). Used by permission of P.L. Decker.)

The rapidity of these loading processes can easily outstrip the ability of the sediment to dissipate the nearly instantaneously pressured pore fluids, such that their pressure may rise to equal the burial pressure, hence reducing the effective stress to zero. In the absence of significant cohesion, the sediment now lacks shear strength and temporarily exists as a fluid. This is the phenomenon of liquefaction. The sediment is available to generate the kinds of injection structures discussed in sections 4.3.4 and 9.5.2. It will remain in this condition until the pore pressure is reduced below the burial load. Cakmak and Heron (1989) illustrate the engineering approach to dealing with this crucial phenomenon. For some geological purposes (e.g. sections 4.2 and 5.1.2), it is necessary to distinguish between liquefaction, in which the particles are able to settle, and **fluidization**, in which the upward flow of the fluid is sufficient to maintain the particles in suspension. An effect analogous to liquefaction comes about in the delicate but cohesive sediments that show **thixotropy**. Here, constant-rate loading destabilizes the sediment framework and disperses the particles in the pore fluid so that strength is lost. The effect is marked where the loading is abrupt and rapid. This kind of behaviour is commonly included with liquefaction although, strictly, overpressuring is not involved, and irrespective of the loading conditions thixotropic sediment retrieves its strength when it returns to rest and the particles reflocculate. Liquefied sediments that find themselves in a reversed density gradient will be highly unstable. Any perturbations at the interface between the overlying, denser material and the underlying less dense sediment will, seeing as they are behaving as fluids, act as Raleigh–Taylor instabilities, causing the irregularities to amplify until gravity driven overturn can be achieved (Ronnlund 1989). There are similarities with movements in salt (Jackson and Talbot 1994).

Hird and Hassona (1990) showed experimentally that, in principle, any sediment can liquefy, but that rounded grains promote the effect and that the presence of platy minerals, because of their cohesion and compressibility, retard it. Bornhold and Prior (1989) discussed the role of gas hydrates in promoting liquefaction and associated subsidence. Vaid et al. (1990) argued that because particle grading affected the ability of a sediment to show contractive deformation it therefore influenced the tendency to liquefy. For example, in their experiments, high-porosity, poorly graded sand resisted cyclic loading better than a well-graded equivalent, whereas lower porosity material showed the opposite effect.

In-place disturbance can come about for a wide variety of reasons, but most involve pore-fluid effects in one way or another. The fundamental mechanical role of pore water is well illustrated by the commonness of deformation structures in wet sand dunes (Horowitz 1982) as opposed to their scarcity in desert sands (McKee and Bigarella 1972). The other major role of pore fluid is, of course, in promoting diagenesis, and this is discussed in sections 1.2.5 and 4.3.9.

Another type of in-place deformation arises through differential consolidation, such as in producing the faults that are increasingly being reported from the modern sea floor (e.g. Buckley and Grant 1985). Unusual variations on this theme come about through the removal of material in some way, and the consequent foundering of the beds above. Upright folds in the Eocene sediments of the South Carolina coastal plain were claimed by Johnson and Heron (1965) to be due to the 'uneven let down' of the beds into the material below, from which carbonate, or possibly opaline silica, was being preferentially dissolved by groundwater. Wardlaw (1972) described deformation in an evaporite sequence that was prompted by the preferential dissolution of interlayered halite, anhydrite and carnalite beds; pure halite remained unaffected. Mohl and Bakken (1968) illustrated what they call slump structures that are due to collapse into underlying, burned coal-beds, thought to have been ignited by lightning strikes!

1.3.3 Gravitational mass movement

The importance to society of slope movements on hillsides is obvious (e.g. Brabb and Harrod 1989), and the effects have long been witnessed. Even geological reports of the phenomena go back at least as far as Silliman's (1829) descriptions from the White Mountains of New England. Today, similar gravity driven processes are being increasingly discovered on the ocean floors, and they are even observed on other planets (Shaller *et al.* 1989). The attendant terminology has had an intricate evolution, via 'landslip' – still used for some English examples – through to 'landslide'. At its simplest, the latter is merely 'the movement of rock, earth, or debris down a slope' (Cruden 1991). In this usage there is no restriction on the mode of movement or on the setting – landslides can flow and they can be submarine. As Cruden (1991, p. 28) put it: 'like cowboy, landslide is another North American word formed by two words which together mean something entirely different'. Terminology for the components of a landslide, from an engineering viewpoint, is given by the IAEG (1990). Geologists prefer the term mass movement, perhaps because it is more obviously all-encompassing. The great variety of processes by which earth masses move downslope in subaerial and subaqueous environments is reviewed in Chapter 5. Most attention is paid there to the processes that induce the deformation of existing sediments rather than to those that give rise to fresh deposition, and are hence essentially sedimentological in nature.

To many geologists, it is the submarine phenomena that are of greatest interest, because of their widespread occurrence, vast scale and relatively recent discovery. In fact, progress in recognizing submarine mass movement closely mirrors the exploration of the ocean floor. Benest (1899, p. 394) noted the frequent breakages in the submarine cables that were beginning to be laid on the floor of the ocean and inferred 'unsuspected forces constantly in action and altering the features of the sea bottom'. Continuing exploration of sea-floor bathymetry revealed all kinds of new features, and three decades later Shephard (1932, p. 226) was describing submarine canyons and explaining them as the result of 'landslides caused largely by earthquakes accompanying faulting-action along the continental slopes'. Modern oceanographic work has detected the gravitational movement of sediments on an enormous scale (e.g. Bugge, Belderson and Kenyon 1988), and in all kinds of materials (e.g. J.G. Moore *et al.* 1989). The Agulhas slump off southeast Africa, for example, occupies an area of $80\,000\,km^2$ – considerably more than, say, the area of England and Wales or the state of Oklahoma (Dingle 1977). The Neogene slides in the Gulf Coast basin described by Morton (1993) reach 600 m in thickness, but a striking aspect of many modern submarine slides is their thinness relative to the areal extent. The slide northwest of the Canary Islands, for example, covers around $40\,000\,km^2$ of the Madeira abyssal plain, yet is mostly less than 20 m in thickness (Masson *et al.* 1992). Section 5.5.2 draws attention to the apparent discrepancy between the magnitude of these modern features with those so far recognized in the geological record.

Following the demise long ago of inferring aqueous processes as agents of deformation (section 1.1.1), the modern realization that some major deformation features in rocks may after all be the result of subaqueous instability was prompted largely by reports on landslides into Swiss lakes (section 1.3.2). Shortly after, T.C. Brown (1913, p. 224) interpreted disoriented and folded Lower Palaeozoic conglomerates in Pennsylvania as having been 'slumped or slid along the (sea) bottom due to gravity'. Two decades later, the importance of these long-discarded ideas was being restored. Archanguelsky (1930) emphasized the importance of mass movements in the Black Sea, and illustrated sequences of repeated stratigraphy and folds in rocks that he interpreted as the result of subaqueous processes. Hadding (1931) remarked on the commonness of the results of subaqueous sliding, and presented both a workable classification scheme and a perceptive analysis of various possible causes of instability. De Terra (1931) described folds, shearing planes and faults in rocks from Wyoming and the Himalayas that he thought were due to 'sub-aquatic gliding'. He also made the premonitious remark that 'the fact that gliding can set free a tangential pressure which causes structural forms similar to those which are usually considered to result from contraction of the crust seems to give us a useful hint for any analysis of related features' (de Terra 1931, p. 213). Geology, however, has been slow to return to recognizing the importance of large-scale subaqueous deformation. The conceptual difficulties and the problem of distinguishing between the results of large-scale mass movement and post-lithification tectonic deformation are encapsulated in a famous controversy between two leading British geologists, which is recalled in section 9.6.1.

The mechanics of downslope movement have received much analysis, particularly through the importance of hazard prediction (Zaruba 1987), and such engineering studies have provided a quantitative basis for understanding the geological processes. Historical reviews of the engineering approach to the stability of on-land slopes are given by Graham (1984) and Fleming and Varnes (1991). Interestingly enough, Graham's discussion includes a further illustration of the 're-inventing the wheel' point made in the present Preface, as certain fundamental contributions concerning curved failure surfaces on clay slopes were virtually forgotten for almost 80 years, before their resuscitation to become a keystone of some modern slope analysis techniques.

Most analyses of slope stability are based on Mohr-Coulomb principles; the most widely used are outlined by Kenney (1984). They assume that failure comes about when the shear stress acting along the slope exceeds sediment strength, and that the resulting movement takes place on a discrete plane of failure. A common specification of slope stability uses a safety factor F – the ratio of available shear strength to the likely shear stresses (e.g. Graham 1984). Slopes are unstable where F falls below unity (see, for example, Figure 1.8). A variety of analytical models are used for subaerial slopes, for example, according to whether the geometry of the failure is analogous to a sliding block or some circular

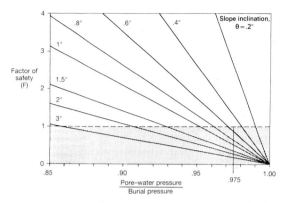

Figure 1.8 Relations between factor of safety (F), slope inclination (θ) and overpressuring, as deduced by Prior *et al.* (1981) for the Mississippi Delta. The fluid pressure was calculated on the basis of

$$P_{H_2O} = \frac{c' - F(\rho z \sin\theta \cos\theta)}{\tan\phi'} + \rho z \cos^2\theta,$$

where c' and ϕ' are the likely effective Mohr-Coulomb parameters (see equation 1.13), and ρ and z are the density and thickness, respectively, of the sediment. A factor of safety <1 (stippled area) implies conditions of slope failure. A pore fluid pressure/burial pressure ratio of 1 implies liquefaction. (Reproduced with permission from the American Association of Petroleum Geologists.)

arc (Espinoza, Repetto and Muhunthan 1992). Analyses of submarine slopes are most commonly approached by an 'infinite slope analysis', in which the failure plane is treated as planar and parallel to the free surface, and a small volume is taken to adequately represent the total sliding mass. Such a treatment is in some ways simplistic, but it provides a reasonable approximation, at least with our present knowledge of sea-floor properties. An example of this approach is incorporated in Figure 9.15.

In the simplest view of the stresses acting on a slope, if the sediment has an area A and a weight W (area × thickness × bulk density × gravitational acceleration), and slopes from the horizontal at an angle β, then the shear stress operating along the slope will be:

$$\tau = \frac{W}{A}\sin\theta, \quad (1.15)$$

and the stress normal to the slope will be:

$$\sigma_n = \frac{W}{A}\cos\theta. \quad (1.16)$$

Therefore, assuming that a potential detachment parallels the planar slope, the overlying load acts both to strengthen the sediment by increasing the stress normal to the failure plane and to promote failure along the plane by increasing the shear stress. Although for slopes of low and moderate inclination the strength (increasing through the cosine of the slope angle) will grow more rapidly with increasing burial thickness than the shear stress (increasing through the sine of the slope angle), with steeper slopes (in which sine values increase more rapidly than cosines) the opposite will be the case. Hence, sediments on slopes above a certain inclination will become increasingly unstable with progressive burial.

Ross (1971) measured shear strengths of sediments and slope angles around the Middle America Trench. In line with the above relationships, the data suggested that whereas the shallow slopes of the adjacent shelf bore sediments stable during burial – because the extrapolated shear-strength increase with burial continued to exceed the calculated shear-stress increase – on the steeper slopes of the trench flanks, shear stresses exceeded the strength at a depth of a few tens of metres. That failure is in fact occurring at such depths is supported by the seismic profiles across the trench flanks, whereas there are no indications of failure on the sections across the shelf.

For many near-surface sediments, the contribution of cohesion to their strength is negligible. In this case, incorporating (1.15) and (1.16) into the basic Mohr-Coulomb relation (1.5) gives:

$$\frac{W}{A}\sin\theta = \frac{W}{A}\cos\theta \tan\phi. \qquad (1.17)$$

At the failure condition of shear stress equalling strength, the load terms cancel out, so that:

$$\theta = \phi, \qquad (1.18)$$

that is, the slope cannot exist at an inclination (sometimes called the **angle of repose**) greater than the friction angle of the sediment. As discussed below, wet sediment may be unstable at considerably lower inclinations.

In these ways the potential for failure depends on the interplay between slope angles and the sediment strength. In a real geological situation, both the slope angles and the mechanical properties are likely to be continuously evolving (Baraza *et al.* 1990). Slope inclinations can change in response to diapirism (Martin and Bouma 1982), fault reactivation (Holler 1986), salt movement (Almagor 1984), and because of erosion at the slope foot (Schwab *et al.* 1988). Strength parameters will be evolving through progressive consolidation and diagenesis (Charles 1982). Various effects owing to sea-level fluctuations have also been invoked to explain submarine slope instabilities (Morton 1993).

However, it is likely that most slope failure comes about through overpressuring. Essentially, this reduces the sediment strength, as outlined in section 1.2.5, and hence the resistance to failure. Overpressuring in subaerial slopes commonly comes about through artesian conditions, or after heavy rainfall (Reid and Baum 1992), which temporarily increases the elevation head of the pore fluid (Hutchinson 1989; van Genuchten and de Rijke 1989; Reid, Nielsen and Dreiss 1988). Vegetation may sometimes play a role (Terwilliger 1990). In a subaerial mudslide in Dorset, southern England, Allison and Brunsden (1990) documented the rainfall–fluid-pressure relation and its influence on permeability. This in turn determined whether the style of sliding was dominated by what they called stick-slip, graded-slip, surge or 'random' modes of motion.

In the subaqueous environment, biogenic gas (Esrig and Kirby 1978), gas-hydrate melting (Kayen and Lee 1990), and storm waves (Lu *et al.* 1991) may be important, but many mass movements are surmised to have been caused by earthquakes (Poulos 1988). Propagation of seismic waves adds to the bulk shearing stress acting on the sediment but a more crucial effect is the instantaneous transmission of the stress to the pore fluid. This creates spontaneous overpressuring and perhaps liquefaction, as explained in the previous section. The oscillating seismic waves may produce an incremental accumulation of fluid pressure or cyclic loading (setion 1.3.2) and further promote slope instability (Jibson and Keefer 1993). A further cause commonly invoked to explain overpressuring is a rate of sedimentation that outstrips the ability of the sediments to dissipate the pore water adequately, as outlined in section 1.2.5. As an illustration, Aksu and Hiscott (1989) pointed out that large-scale sliding is more significant at higher latitudes than near the Equator, because of the higher rates of sediment accumulation adjacent to continental ice sheets.

Prior et al. (1981) illustrated the importance of overpressures in the Mississippi Delta. Calculations on typical properties showed that an average slope would reach failure if fluid pressure ratios exceeded 0.85, and that the delta-front slope of 0.5° would fail at a ratio of 0.97 (Figure 1.8). Because these fluid-pressure values agree directly with field piezometric measurements, the authors argued that much of the delta must be in an inherent state of failure, due to the overpressuring. They went on to argue that the strain softening associated with initial failure would lead to in-place collapse, followed by upslope retrogression of the subsiding area. Their conception is shown diagrammatically in Figure 1.9. They further argued that conditions for subsidence would readily be met in the Mississippi Delta, thus explaining the common observations of collapse depressions in the region.

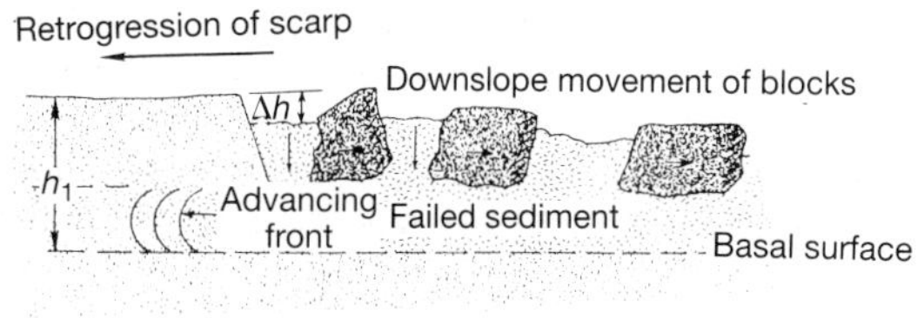

Figure 1.9 Schematic representation of the upslope retrogression of failure conditions and the block subsidence that arises where the minimum subsidence ratio,

$$\frac{\Delta h}{h_1} = 1 - \left(1 - \frac{4\bar{C}_u}{UWh_i}\right)^{\frac{1}{2}}$$

is exceeded. $\bar{C}_u$ is the average undrained strength of the sediment and UW its bulk unit weight. In the Mississippi Delta, common values of h are 0.3–3.0 m and of h_1 are 7–10 m, so that h/h_1 is in the range 0.05–0.03 m. Using a bulk density of 1400–1500 kg m^{-3} and $\bar{C}_u$ values of 4.8–9.6 kPa for burial depths of 10 m, the critical subsidence ratios are 0.07–0.15, suggesting that slope failure in the Mississippi may enlarge by a retrogression mechanism. (From Prior et al. (1981). Reproduced with permission from the American Association of Petroleum Geologists.)

Another cause of slope failure is seepage at sites of groundwater outflow, which causes erosion (Robb 1990) and chemical dissolution (Paull & Neumann 1987), leading to undercutting and eventual slope failure. The process is commonly referred to as **sapping** (Dunne 1990). It may be responsible, for example, for the headless submarine canyons that appear not to have experienced current erosion (Orange and Breen 1992).

Although slope failure in general comes about when the sediment exceeds its peak strength (section 1.2.4), Skempton (1985) has argued that many unstable slopes will have already undergone some failure and mass movement, and therefore the sediment will be exercising its residual strength. This is very likely the case for the intermittent, far-travelled movements common in the oceans. However, verification in the latter situation is hampered by the difficulty of acquiring reliable strength measurements.

An alternative approach to analysing slope stability is to treat the sediment not in terms of Mohr-Coulomb failure but as a viscous material (section 1.2.4). Such treatments view the mass movement either as fluid-like flow or as taking place on a fluid-like substrate (Ohlmacher and Baskerville 1991). In analogy with other contexts, this approach does involve the problem of choosing the most appropriate flow model to use. Sousa and Voight (1992), for example, modelled a landslide using two different viscosities, to represent the mass deforming below and beyond its yield point. Some treatments do not explicitly take into account the crucial effects of fluid pressure. Iverson (1985) took a non-linear viscous flow model and included a Mohr-Coulomb term in addition, together with some allowance for effective rather than total stress. He suggested that the required data on parameters such as strain rate, fluid pressures and viscosity could be estimated adequately from field observations. However, other workers have argued that even if such a flow model is reasonably realistic, knowledge of these properties in nature is still too poor to justify a viscosity based approach (e.g. Hutchinson 1986).

The same shortcoming prompted Einsele (1989) to use Atterberg limits (section 1.2.2), measured from retrieved cores, to assess stability on submarine slopes. His data indicated that even normally consolidated marine sediments commonly have sensitivities between 4 and 20, or more, implying a high liquefaction potential even at hundreds of metres below the sea floor. Hutchinson (1986) combined this approach with Mohr-Coulomb principles, suggesting that the masses termed slides are likely to form in low sensitivity clays or friction-dominated clastic sediments. The failed material would still be around its peak strength for localized, rotational slides but at the residual state for more far-travelled blocks. On the other hand, material forming slumps would exhibit lower strengths, and commonly comprise higher sensitivity, perhaps normally consolidated material. (Terms for the morphology of mass movement are explained in Chapter 5.) Flows showing plastic behaviour must have operated at a water content within the plastic range of the sediment, whereas those with fluid behaviour are probably of high-sensitivity sediment which must have exceeded its liquid limit. Material with a relatively high plasticity index will tend to give thicker flows with a lesser run-out distance.

In addition to the relatively sudden slope failure implied in all the above models, it may be that much deformation associated with mass movement is due to downslope creep (section 1.2.4). The process is little recognized so far in submarine deposits because the appearance of its results are undocumented in sediments or rocks, and they are probably too subtle to appear on most seismic records. The experiments of Silva and Booth (1985) suggested that creep is a significant process off the northeastern USA coast. Annual rates of movement are several centimetres a year or more, with slopes of over $10°$ undergoing metres of movement, almost certainly leading to eventual failure. Moreover, the work revealed a degradation of the sediment strength brought about by the creep, which probably further engenders the likelihood of ultimate failure. This notion received support from field measurements carried out on Ruanda hillsides by Moeyersons (1989). In addition, he pointed out differences in orientation of friction-influenced surfaces of failure and the slip lines that reflected shearing while the mass was behaving plastically. Local adjustments to these differences were taken by Moeyersons to account for certain small-scale structures seen in the landslide such as en échelon microfissures and 'microdistortions'. If such features are preserved in ancient sediments and rocks, they appear not to have been recognized yet.

Some varieties of mass movement, such as falls and flows, appear to produce little in the way of recognizable internal structure, although there may be some alignment of clasts and features in any cohesive clay that is present (section 9.4.3). Those masses of greater strength, however, and especially slumps and slides, can develop a wide range of striking structures, some remarkably analogous with other environments of deformation. Examples of these are presented in Chapter 5. Further instances of the kinds of structures produced by mass movement as they are preserved in rocks are illustrated in Figures 9.17 and 9.37.

1.3.4 Fluid–sediment movements

The escape of pore fluids in response to the loading of sediments can itself produce deformation structures, especially if the sediment is in a liquefied state. Some principles of fluid behaviour in sediments were sketched out in sections 1.2.5 and 1.3.2. The theoretical background to fluid movement and examples of how this interacts with deformation are presented in Chapter 7. Many of the principles outlined there are of relevance to any deforming wet sediment, but, as the discussions in Chapter 7 illustrate, the drainage–deformation interplay is particularly important in those materials that are sufficiently bonded to allow some degree of brittle fracture. Section 4.3.4 describes some of the structures that result from near-surface deformation.

Historically, a theoretical analysis of how fluids move through sediments had to wait for Henry Darcy's classical work on permeability in 1856, after which the early efforts were dominated by groundwater (King 1899) and, later, petroleum geology (e.g. Fraser 1935). However, observations concerning natural structures greatly preceded all this – in fact reports of sedimentary dykes form some of the earliest records of structures in sediments. In 1821, W.T.H.F. Strangways published descriptions of structures along the River Pulcova, near St Petersburg, and drew attention to two dykes, labelled x and y on the map inset reproduced here as Figure 1.10. These were 'infilled with diluvian gravel' (Strangways 1821, p. 386). Charles Darwin noted sedimentary dykes in Patagonia while on his voyage on the *Beagle* in 1833. He described four examples in some detail and remarked of one of them: 'the structure of this dyke shows obviously that it is of mechanical and sedimentary origin', and, regarding its genesis: 'if we reflect on the suction which would result from a deep-seated fissure being formed, we may admit that if the fissure were in any part slightly open to the surface, mud and water might be drawn into it' (Darwin 1846, p. 150).

Early workers had difficulty coming to terms with the igneous-like form of these structures and

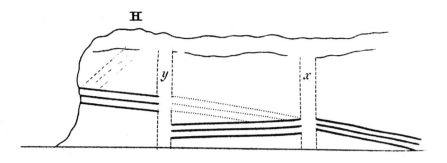

Figure 1.10 Inset to a map by Strangways (1821) of the Pulcova River area, near St Petersburg, depicting the two sedimentary dykes described in the accompanying report.

the sedimentary content. Strickland (1840), for example, remarked of dykes in Jurassic rocks near Brora in northeastern Scotland (first reported, though not interpreted, by Murchison 1829) that 'however much this rock might resemble an aqueous product ... it forms genuine intrusive dykes' (p. 599). He went on, 'the sedimentary structure of this rock forbids us to refer it to igneous injection from below, and notwithstanding the complete resemblance of these intrusive masses to ordinary plutonic dykes, we have no recourse left but to refer them to aqueous deposition, filling up fissures which had been previously formed' (p. 600). James Dana discovered a series of sandstone dykes in 1841 during his explorations around the mouth of the Columbia River, in what was then British Oregon Territory. He, too, interpreted the dykes as fissure-fills, 'probably formed after or during the deposition of the sandstone while the region was yet underwater', and added that 'their number and irregularity evince that these regions have been often shaken by subterranean forces' (Dana 1849, p. 654).

Diller (1890), reporting Tertiary sandstone dykes west of the Sacramento River in California, deduced that the sand was injected into the fissures, upwards as well as downwards, and probably as a result of seismic activity. This last inference was facilitated by the contemporary direct observations of fluid and sediment expulsion at the earth's surface during the Sonora earthquake of 1887 (Goodfellow 1887). Although many of the contributions appearing around this time were concerned with sandstone intrusions, Gresley (1898) described what he called clay veins, associated with the Upper Palaeozoic coals of Pennsylvania, and Shaw (1914) described the extrusion of mud diapirs at the surface of the Mississippi Delta – the so-called mudlumps – and interpreted them as responses to differential burial loads. Reports of sedimentary dykes continued to proliferate, and they continue to this day, in some cases without adding much to an understanding of their origin (Dionne and Shilts 1974; see M.Y. Williams (1927) for a historical review). The structures are sometimes also referred to as clastic dykes. Although it has been advocated that the term 'dyke' be restricted to those discordant sheets known to have been forcefully injected, because it is not always possible in the geological record to distinguish these from passive fissure-fills it seems practical to allow the term to be used in a broad sense. Some workers distinguish passive fissure-fills as **neptunian dykes**. This is more useful than using the term, as has been done, merely as a further synonym for all sedimentary dykes.

Modern discussions on sedimentary intrusions include the significance for mining operations (Bauer and Hill 1987); the importance of distinguishing between dykes and periglacial wedge-like structures for environmental reconstructions (Elson 1975); horizontal injection (Aspler and Donaldson 1986) and the downward injection of sands (Chown and Gobeil 1990) and diamictites (von Brunn and Talbot 1986). Sedimentary diapirism is known on a substantial scale, from both ancient rocks and the modern sea floor (e.g. Brown and Orange 1993; section 2.3.3). Sedimentary dykes have been used by Snavely and Pearl (1979) to deduce palaeostress patterns, and Borradaile (1977) used folded examples as estimators of post-dyke compaction. Clastic sills were discussed by Hiscott (1979), and the difficulties of distinguishing between sills and normal beds were considered by Archer (1984). All the postulated mechanisms by which the sediment was able to flow in order to generate these various intrusions involve some combination of the liquefaction and thixotropic processes outlined in section 1.3.2. Where such mobilized sediments reach the surface they are said, continuing the analogy with igneous features, to extrude (e.g. Schwan *et al.* 1980), from where the sediment may undergo recycling (Talbot and von Brunn 1987). The extruded sediment may form volcanoes, in both sands (Okada and Whittaker 1979) and muds (Dionne 1973). Neumann (1976) reported sand volcanoes actually forming during modern construction operations – apparently the load that induced the overpressuring was a caterpillar tractor!

Escaping pore fluid may travel pervasively, particularly in shallow-buried sediments (Sills and Been 1984) or via more localized routes. The increasing recognition on the modern sea floor of localized biological communities has been interpreted to indicate the widespread existence of channelized fluid escape (e.g. Fujioka and Taira 1989). Many of these seem to be sited at the outcrop of faults, which are likely to provide conduits if they are undergoing dilative shear (e.g. Buckley 1989), but not all the seismic traces indicate the presence of faults. The sea-floor features known as pockmarks – circular depressions commonly a few metres across – are in some cases thought to be sites of water seepage (e.g. Whiticar and Werner 1981; Harrington 1985), although the overall genesis is probably more varied (Hovland and Judd 1988). The influence of pre-existing fractures on sediment permeability has been explored in the soil mechanics context by Bosscher, Bruxvoort and Kelley (1987); Walsh (1981) gave a mathematical analysis of the effects of fracture width and roughness. Chapter 7 explores further the role of localized strain zones for sediment drainage.

Preserved structures show that the velocity of the fluid expulsion is sometimes sufficient to entrain some of the host sediment, although the physics of the process has received little attention in the present context. It has been much studied in fluid mechanics and sedimentology (Allen 1984), but the slow, non-turbulent conditions relevant to buried sediment are probably more analogous to the entrainment of crystals in a moving magma (e.g. S. Blake 1987). In on-land sediments, where such localized drainage conduits are referred to as pipes (e.g. J.A.A. Jones 1990), entrainment and transport of sediment has been shown to progressively modify the permeability (Harp, Wells and Sarmiento 1990). It may be that the various enigmatic cylindrical structures and unusually long 'trace fossils' that have for long perplexed geologists (e.g. Wnuk and Maberry 1990) may in some instances have a dewatering origin. Further examples of structures associated with the transfer of fluid and sediment are given in sections 4.3.4 and 9.5.3.

1.3.5 Tectonism

Recognition of the tectonic deformation of sediments has a relatively short history, perhaps because of the kinds of reasons surmised in section 1.1.1. However, recently there have been major changes in appreciating the extent to which tectonic stresses affect sediments as well as deeply buried rocks. The change in outlook has come about for three chief reasons:

1. a realization of the close interplay in many geological settings between sedimentation and tectonics;
2. a new interest in neotectonics;
3. the advent of ocean-floor exploration and plate tectonics concepts.

Each of these points is commented on below.

In a sedimentary sequence undergoing burial the transition from gravitational to tectonic stress (section 1.1.1) as the dominant cause of deformation is highly gradational. In many settings the highest levels of a sedimentary pile are very largely under the effect of gravity. This is certainly the case with the high-level structures considered in Chapters 4 and 5. However, the role of tectonism in governing the patterns and processes through which sediments are first laid down is now widely accepted (Ingersoll 1988; Pickering and Taira 1994). Gravity and regional tectonics work hand-in-hand to influence sedimentation.

Almost immediately after deposition a sediment becomes subject to a smaller scale but equally close interplay between gravitational and tectonic stress. Gravity is mainly responsible for driving the consolidation of the sediment, at least in extensional sedimentary basins, but lateral tectonic stresses can play a significant part in contractional settings. Horizontal tectonic stresses influence the extent to which the sediment follows a K_0 or some other consolidation stress path during burial (section 2.2.6). The importance of these deep-seated stresses grows as sediments undergo increasing lithification and extend the range of their elastic behaviour. Such interactions between burial consolidation, lateral tectonic stresses, and the elastic–viscous response of lithifying sediments are considered in detail in section 6.2. Section 7.3 discusses the tectonic processes that are responsible for driving fluids out of the sediments as they are buried, deformed and progressively lithified.

Appreciation of the role of active tectonism in influencing sedimentation leads to the notion that the tectonic stresses generating the regional structures can also have some deforming effect on the sediments that are being deposited (Higgs 1988). As an example, Rascoe (1975) related localized deformation breccias and features that he termed gravity-slide faults and sand-blows, preserved in Upper Palaeozoic rocks in Oklahoma, to the growth at the time of sedimentation of regional anticlines and domes in the underlying basin floor. He suggested that analysis of the pre-lithification deformation structures should aid in evaluating the hydrocarbon potential of the area, as they helped detect the positive structures that were influencing the nature of the sedimentation. Rumsey (1971) provided a general review, in the petroleum context, of how fractures in an unlithified sedimentary cover can reflect regional features in the underlying basement rocks.

Understanding of pre-lithification tectonic structures preserved in rocks is probably most advanced in convergent-margin settings. As examples: Agar (1988) analysed faults, sheath folds and broken formations in the Cenozoic Shimanto Belt in order to decipher the kinematics of convergence between the Phillipine Sea and Eurasian plates; G.F. Moore and Karig (1980) described shear structures that formed during plate convergence in the Sunda Trench; Kimura, Koga and Fujioka (1989) reported veins, foliations and kink bands developed during bedding-parallel shear and consolidation of sediments between the Mariana and Yapp trenches; and Knipe and Needham (1986) discussed structures in the Southern Uplands of Scotland interpreted as the result of plate convergence in Lower Palaeozoic times. The precocity and complexity of the structures that can develop in a compressive regime were emphasized by Labaume, Bousquet and Lanzafame (1990) as a result of their investigations south of Mt Etna, Sicily.

Structures preserved from extensional tectonic regimes have also been documented. An example is the work of Montenat, Barrier and Ott d'Estevou (1991) around the Strait of Messina, Italy. Part of their reconstruction of the regional extensional movements was based on distinguishing between neptunian dykes in the sense of infilled fissures (section 1.3.4) and those dykes resulting from sediment liquefaction. Regarding a strike-slip setting, the importance of subhorizontal fault motion in governing Triassic sedimentary basins in Morocco was identified by Laville and Petit (1984) on the basis of pre-lithification fault features such as slickenline orientations. Genna (1988) measured folds, sandstone dykes, cleavage and microshears to derive the stress regimes that were operative during early, prelithification tectonism in the Pyrenees. These too were indicative of strike-slip motions, perhaps accompanying regional flexure.

The subject of neotectonics includes tectonic processes and structures that have arisen in recent geological time up to the present. Slemmons (1991) and Pavlides (1989) discuss the definition of the term, and Becker (1993) assesses the length of period covered. The kinds of structures dealt with in this field are related to stresses of lithospheric origin but are chiefly developed in surficial, poorly lithified sediments (Stewart and Hancock 1994). Many of the relevant observations so far have been on subaerial processes but it is now proving possible to measure directly the patterns and magnitudes of tectonically induced strain accumulating in modern submarine sediments (Larsen, Agnew and Hager 1993).

Neotectonic faults and associated features, such as joints, form particularly intensively in areas of earthquake activity, where they can impinge on various aspects of society and provide indications of seismic risk. Russ (1979), for example, analysed cross-cutting fractures and associated folds and sand dykes in the New Madrid earthquake zone of the central USA in an effort to assess the seismic periodicity of the region. Bullard and Lettis (1993) emphasized the importance of ductile features, such as folds, in their analysis of earthquake risk in the Los Angeles basin. There, folds on the scale of several kilometres are forming on surficial sediments, in association with blind contractional thrusting at depth. The interaction of deep-seated stresses with surficial sediments has even been argued to be the fundamental influence on the geomorphology of the earth's surface (Scheidegger and Ai 1986). Attempts to use joints and faults as indicators of the stress system of sedimentary basins are reviewed in section 6.2.6. Of course the most direct way of determining current stress configurations is to make *in situ* measurements. However, there are severe problems in applying the methods to particulate materials. Section 6.2.5 analyses the efforts that have been made.

Regarding the repercussions of oceanographic exploration and plate tectonics thinking, there has been some direct sea-floor documentation of extensional tectonic features. S.J. Williams (1987), for example, has used seismic and shallow coring methods to investigate extensional faulting in the east Atlantic, and Masson (1991) has discussed extensional fracture patterns in sediments on the outer walls of subduction-related trenches, as seen on GLORIA (side-scan sonar) surveys. Masson concluded that the fracture orientation almost always reflects either the strike of the subducting slab or an inherited oceanic-spreading fabric.

Nevertheless, it is at the sites of plate convergence where tectonically induced sediment deformation has emerged as a topic of importance. In his historical review, Carson (1983) remarked that a principal result of geological research at convergent margins has been the realization that deformation occurs remarkably early, such that older, highly deformed sequences from this setting almost certainly have an early history, which so far has been largely unappreciated. When seismic sections across what have come to be known as accretionary prisms were first interpreted as a series of landward-dipping thrusts, it followed that these had to be affecting high-level sediments. Sampling of the materials by the Deep Sea Drilling Project (DSDP) verified that the stresses arising from plate convergence were being transmitted through the wet, unlithified sediments and producing a variety of deformation structures.

At first, diagnosis of the structures in retrieved cores as being of tectonic origin rather than being gravity- or drilling-induced was frustrating (von Huene and Kulm 1973). Consequently, early work was directed at assessing the effect of tectonism on consolidation rather than deformation (e.g. Lee, Olson and von Huene 1973). However, the importance of tectonic stresses for both dewatering and deformation rapidly became clear (Carson, von Huene and Arthur 1982). J.C. Moore and Karig (1976), reporting folds and foliations from the active Nankai accretionary prism, presented one of the first systematic documentations of the tectonic origin of such structures. An outline of the overall structural style of accretionary prisms was provided by J.C. Moore and Lundberg (1986) and a summary is given here in section 6.3.3. Lundberg and Moore (1986) reviewed the deformation structures recovered to that time by DSDP drilling at convergent margins. Examples of tectonic structures in sediments from the exceptionally well-documented Nankai prism, off southwest Japan, are described in sections 6.3.4 and 8.2, and illustrated in Figure 9.5.

The tectonic deformation of sediments is a field that has proved especially fruitful for the application of soil mechanics principles. Both the DSDP and Ocean Drilling Program (ODP) have given some priority, in all tectonic settings, to investigating the geotechnical behaviour of the sediments encountered. The physical properties of sediments are analysed routinely during the shipboard operations, and large repositories of data are to be found in DSDP and ODP cruise reports.

Nowhere have geotechnical relationships been pursued more vigorously than in the study of accretionary prisms. For example, Karig (1986) synthesized the data available at that time in an analysis of the mechanical state of sediments accreted into the Nankai prism. Examples of newer data include the results of ODP Leg 110 to Barbados (Mascle *et al.* 1988), Leg 131 to the Nankai prism (Taira *et al.* 1991), and Leg 146 to the Cascadia prism (Carson *et al.*, in press). In some cases knowledge is sufficient to enable stress path diagrams (sections 1.2.4 and 2.2.5) to be drawn, though with varying degrees of quantitative rigour. Such diagrams are used in sections 6.3.5 and 8.2 to summarize two slightly different synoptic models of progressive sediment deformation in the Nankai prism. The latter example is set in the context of incipient mélange generation. Section 6.3 presents a more extended treatment of stress paths for tectonically contracting sediments and introduces new experimental data. In so doing it illustrates the new solicitude for the quantitative understanding of sediment deformation, in addition to summarizing the current state of knowledge on the mechanical behaviour of sediments in the compressional tectonic environment.

1.3.6 Igneous activity

The intrusion of magma into sediments has various effects, and a major control is the degree of lithification – and hence mechanical resistance – of the sediments. This influences the intensity of any diapiric effect the magma may induce, but it also controls the extent of magma–sediment interaction. For example, in Pliocene sediments of Inyo County, California, Bacon and Duffield (1978) noted that a basaltic intrusion passed cleanly through several horizons of differing lithology, and only interacted with the mud–silt layers. The intrusion formed a flat-lying wedge, with a progression towards it of incipient structures in the more distant mud–silt, through asymmetric isoclinal folds, to a chaotic zone at the igneous contact. From a judgement of the difference in structural dimensions and geometry the authors suggested that at the time of intrusion the silt and mud layers had a viscosity ratio reaching as much as 20, and were therefore in a delicate state. Leat and Thompson (1988), reporting the results of Miocene volcanism in northwestern Colorado, noted that basalt had no interaction at all with the Cretaceous rocks that were lithified at the time of intrusion, but that it mixed violently with the overlying, unlithified Miocene deposits.

Explosive reactions between sediment and magma have received greater attention from volcanologists than other kinds of interaction, because of their importance for generating volcanic features at the land surface. Factors that control the degree of explosivity include the sediment to magma ratio (Kokelaar 1986) and the geometry of the magma–sediment interface. Theory and experiment suggest that a water–magma mass fraction of about 0.35 generates maximum explosive energy, with greater ratios leading to less dynamic features such as pillow lavas. Smaller ratios, which would be expected if the magma is entering buried sediment, have been investigated less systematically, and much of the information is merely descriptive.

The interface between the two materials is influential in both explosive and quiescent behaviour because of the influence of geometric boundary instabilities that can be generated (Wohletz 1986). The presence at the interface of highly active perturbations can induce rapid instability and violent mixing, leading to explosivity. In more ordered intrusion, the perturbations are less unstable but serve to nucleate sites of attempted mixing. The situation is analogous to the *in situ* disturbances of differing density sediments mentioned in section 1.3.2. Needham (1978) described circular and rectilinear depressions, up to 250 m and 50 m across, respectively, in middle Proterozoic sandstones of the Northern Territories, Australia, which he interpreted as being exhumed from below basalt flows. The instabilities that arose from the reversed density gradient across the base of the basalt flow had induced 'giant load moulds', the crest areas of which show 'giant convolute laminations'. In other situations, the emplacement of igneous sheets can evidently cause the sedimentary substrate to fail, for Dixon (1990) recorded both magma–sediment brecciation, below an Ordovician andesite intrusion in east-central Wales, and syn-emplacement, probably listric, gravity slides. Density instabilities coupled with downslope slumping were invoked by Lorenz (1982) to explain complex 'three-limb'-style folds in the Dunnage mélange of Newfoundland.

Kokelaar (1982) has emphasized the importance of sediment fluidization at the margins of some igneous intrusions. He argued that although fluidization would be curbed by burial pressures greater than about 30 MPa, equivalent to a depth of 3 km of sea water or 1.6 km of wet sediment, at levels higher than this, the fluidization process will induce complex deformation structures in the contact region while allowing the bulk of the host sediment to undergo minimal disturbance. A case history of fluidization and other interactions of silicic volcanism with shallow-buried sediments is provided by the features preserved in the Lower Ordovician rocks of Ramsey Island, southwest Wales (Kokelaar, Bevins and Roach 1985).

One of the few attempts to analyse the quantitative controls on magma entering buried sediment was prompted by the successful drilling by the DSDP, in the Guaymas Basin, central Gulf of California, of a series of active sills within wet, unlithified sediments (Einsele 1982). One of the sills encountered was still hot! The two major effects of the sills on the adjacent sediment were thermally driven diagenetic changes and load-induced consolidation. The two are interrelated. Not only does the increased temperature of the pore water promote diagenetic reactions, especially the dissolution of opaline silica and the partial dissolution of carbonate, but above a critical threshold of burial pressure and temperature the pore water transforms into steam. The expansion associated with this reaction increases the fluid-pressure gradient away from the igneous contact, thus accelerating water expulsion. The steam film provides some insulation, so that the rate of heat loss is reduced sufficiently for mineral and organic reactions to reach completion, producing baking within the contact zone. Also, the effective stresses may be sufficiently reduced to prompt hydraulic fracture in the contact zone, leading to some localized drainage and enhanced diagenesis.

Einsele (1982) attempted some approximate quantitative assessment of the consolidation induced by these sills, using the relation:

$$h_p = h_1 - h_2 = h_2 \left(\frac{100 - \eta_2}{100 - \eta_1} \right)^{-1} = f \cdot h_2, \quad (1.19)$$

where, expressing the fluid pressures as heights of water column supported: h_p = height of expelled pore water; h_1 = original height of pore water; h_2 = present, reduced height of pore water; η_1 = original porosity; and η_2 = present, reduced porosity. Some representative values are given in Figure 1.11. The results suggest that the water expelled from the sediment is very roughly the same as the sill volume, so that the mechanical emplacement of the sill (as opposed to the thermal effects) does not have to significantly deform the host material, nor produce topographic effects at the sea floor. Some intrusions, of course, are diapiric; the strain patterns around a forceful, laterally invading sheet are probably very similar to the effects around salt modelled by Yu, Lerche and Lowrie (1991). As an illustration of the huge amounts of fluid expulsion that igneous emplacement can entail, according to Einsele (1982), a sill

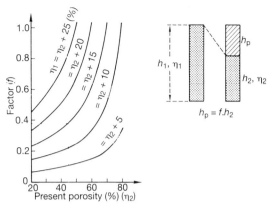

Figure 1.11 Diagram based on equation (1.19) to relate the amount of water loss (expressed as a column height, h_p) from a sediment undergoing igneous intrusion (and where the known initial and present porosities are η_1 and η_2, respectively) to the thickness of an invading sill. For example, if the porosity were reduced from 80 to 70%, then from the $\eta_2 + 10$ curve and the present 70% porosity value, $f = 0.5$, and $h_p = 0.5 h_2$. If h_2 (the present height of the relevant sediment section) has been reduced from 15 to 10 m by the introduction of a 5-m sill, the column of pore water expelled will equal $0.5 h_2$, giving a similar value of 5 m. Thus, in conditions such as this, and providing the intrusion is sufficiently slow, pore water can be expelled to make room for the igneous material. (From Einsele (1982). Used with permission of G. Einsele.)

30–40 m in thickness and covering a 1 km^2 area will expel 3×10^7 m^3 of pore water. The drainage probably takes place along induced fractures rather than completely pervasively, leading to hot springs at the sea floor. Einsele calculated, using the physical properties measured at the Guaymas location and other likely values, that at burial depths greater than about 200 m the pressures would only allow the magma to intrude as dykes, which would use the tensional fissures induced by ocean spreading. Above this level the pressures would allow horizontal intrusion and, following the processes outlined above, a sequence of sills to be built up, separated by sediment hardened by the heat of the underlying sill. Einsele's conception of this successive emplacement is encapsulated in Figure 1.12. In summary, although some intrusions are manifestly diapiric, in poorly consolidated sediments it is the thermal effects in the vicinity of the magma that commonly induce the greatest physical effects on the host material; consolidation can allow magma to be emplaced with little mechanical disturbance.

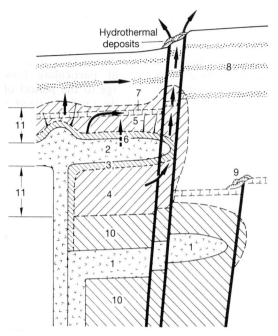

Figure 1.12 Einsele's conception of igneous emplacement and water loss from shallow-buried sediment, based on DSDP drilling of the Guaymas Basin, Gulf of California. 1 = lower, older sill; 2 = upper, younger sill; 3 = possible steam zone, at contact with wet sediments; 4 = zone of porosity reduction due to upper sill; 5 = shrinkage cracks acting as vents for escaping steam; 6 = possible transfer of dissolved chemical species from magma to sediment; 7 = permeable layer with lateral water flow; 8 = sandy base of overlying turbidites; 9 = buried older hydrothermal deposits; 10 = zone of earlier porosity reduction in association with emplacement of older, lower sill; 11 = zones of water loss. (Adapted from Einsele (1982). Used with permission of G. Einsele.)

1.4 MÉLANGES AS A CASE HISTORY

None of the agents outlined in the preceding sections operates in isolation. Mass movements, for instance, can be prompted by and interact with glacial movements and tectonic stresses; fluid motions are involved in all the deformations. These interactions between different agents and processes are perhaps nowhere better illustrated than in the materials known as mélanges. Indeed, it is the very closeness of the interplay, and the difficulty with ancient examples of isolating the contributing factors, that have engendered the long-standing debate on the origin of mélanges. For example, as well as tectonic shearing and downslope mass movement, in recent years diapirism has also been invoked to explain the origin of mélanges. Geological settings such as convergent plate margins commonly involve all three groups of processes, acting on materials at various stages of lithification. The interaction of a range of mechanical processes is intimate.

Mélange studies have a long history. Geologists have long been intrigued by the occurrence of unusually coarse breccias and conglomerates, commonly with a mixture of clast types, more or less isolated, and perhaps of exotic provenance. Many examples showed some degree of deformation, and this was usually taken to be symptomatic of the generating processes. In line with the thinking on deformation alluded to in section 1.1.1, some form of deep-seated crushing was normally invoked by way of origin. Murchison (1829), for example, interpreted the Jurassic 'boulder beds' at Helmsdale, northeast Scotland, as resulting from upheavals associated with the nearby Helmsdale granite intrusion: 'in breaking through those submarine deposits which might not perhaps have been originally in contact, it has fractured and dislocated their beds as to have prepared them for reconsolidation in the state of a brecciated rock' (p. 354). Early explanations of exotic blocks in Alpine deposits also involved deeply buried shearing. Subsequent thinking invoked ice as the transporting agent, but the belief gradually returned that subaqueous instabilities most readily accounted for many of the examples. The historical development is summarized by Bailey, Collet and Field (1928) in the light of their studies on what they interpreted to be submarine landslips, preserved in Palaeozoic rocks near Quebec City, Quebec.

As appreciation of the extent of submarine instability in the modern oceans grew, more formations of these mixed rocks were interpreted in this light, and in 1956 the term olistostrome was coined. However, a purely gravitational origin was not always clear in practice, and in some quarters the term began to incorporate the possibility of tectonic processes also having been involved. Back in 1919, Edward Greenly had coined the term mélange for a unit of mixed rocks in Anglesey, North Wales, which he believed were the result of intense, tectonically induced shearing. The term has enjoyed a renaissance in recent decades, but with a variety of implications, together with debate over the extent to which Greenly meant it to be confined to situations known to be of tectonic origin. Some authors feel its value is as a purely descriptive term, so that there can be sedimentary mélanges. Indeed the 'type' melange in Anglesey is now interpreted as a sedimentary rather than a tectonic product. Part of the interpretational difficulty is that disruption and mixing of sediments by gravitational processes and diapirism are promoted in active tectonic environments, where the products are likely to be overprinted by tectonism, which can act on the deposit at any stage of its progress to lithification. The problem is particularly well shown at convergent plate margins, in the materials known as shale-matrix mélanges. The matter is discussed in Chapter 8.

The application of soil mechanics principles to understanding mélange formation has helped explain some of the features observed in these materials and has also emphasized the paramount role of pore-fluid pressure. Brandon (1989), for example, working in the Pacific Rim mélange complex of Vancouver Island, was able to distinguish between a unit of layer-parallel fragmentation lacking in primary sedimentary structures and units comprising disordered, dismembered asymmetric folds with well-preserved sedimentary structures. He interpreted the latter horizons as the product of retrogressive shear failure (section 1.3.3) and relatively coherent downslope mass movement, and the former as resulting from *in situ* liquefaction and intense lateral flowage under conditions of cyclic loading (section 1.3.2).

As a further example, K.M. Brown and Orange (1993) recognized three stages in the evolution of an on-land diapiric mélange complex associated with the Cascadia accretionary prism. Each period resulted from a major change in conditions of pore-fluid pressure, causing the material to deviate from critical state failure. The first stage took place under low effective stress, which allowed primary disruption of the sediments and turbulent intrusion. The subsequent development of broad shear zones involving cataclasis evidenced a period of higher effective stress and strain hardening. A later rise in fluid pressure allowed dilation and strain softening, especially at the diapir margins, and the final emplacement of the mélange mass. The intricate geometry shown by mineralized veins in the mélange (section 8.5) suggests that the fluid pressure fluctuations may be even more spasmodic than this three-stage sequence.

K.M. Brown and Orange (1993) further argued from experimental and natural evidence that the transition from independent particulate flow (section 1.1.1) of the first stage to the period of cataclasis took place while the sands retained a porosity of between 31 and 41% and that therefore the effective stresses could not have exceeded 5 MPa. Stratigraphical evidence requires that the processes occurred at burial depths greater than 2 km. Therefore, in order to produce such low effective pressures at that depth, the fluid pressures must have approached lithostatic conditions. Such values might be expected in conditions of diapirism, and serve once again to emphasize the crucial importance of pore-fluid pressures in governing sediment deformation.

It is not relevant here to explore the terminological minefield of mélanges, diapirs and olistostromes. The importance here revolves around two points.

1. The terminological difficulties arise because of the very intimacy, in the kinds of environments in which mélanges are likely to form, of gravitational and tectonic processes. That is: mélanges illustrate the range, interaction and complexity of the kinds of processes by which natural sediments deform.
2. The interpretational problem is exacerbated when dealing with ancient examples, because criteria for distinguishing between gravitational and tectonic mélanges and identifying the state of lithification are at best subtle – mélanges illustrate the kinds of difficulties faced by field geologists attempting to interpret deformed sedimentary rocks.

These two matters are taken up in Chapter 8. There, examples of mélanges are presented to act as case histories of the difficulties of dealing with sediment deformation in nature. The second concern, of recognizing pre-lithification deformation long after the materials have been turned into rocks is pursued further in section 9.6. How does the geologist go about diagnosing structures as the result of sediment deformation? After all, application to the geological record of the principles and ideas outlined here and amplified in the following chapters assumes that the geologist knows he is dealing with structures formed in unlithified sediments.

CHAPTER 2

Mechanical principles of sediment deformation

MERVYN JONES

2.1 INTRODUCTION

The majority of sediments originate as weak materials that deform readily, but which become stronger during burial and residence at depth. This process of lithification eventually converts the sediments into rocks (section 1.1.1), which exhibit pronounced strength. Lithification is a progressive change: the transition from unlithified through partially lithified sediment to rock is highly gradational.

Sediments are of great concern to geotechnical and foundation engineers, who normally refer to the materials as soils (Lambe and Whitman 1979). The mechanical behaviour of such materials is described in the literature of soil mechanics (Schofield and Wroth 1968; Atkinson and Bransby 1978), a subdiscipline of civil engineering. The principles of soil mechanics form the basis of this chapter. However, although the term soil is very widely used in the engineering literature, sediment is preferred here and throughout the present book because this term is so firmly established in geology. In this context, sediments can be regarded as aggregates of mineral particles in frictional contact (section 1.1.1), although some examples, such as clays, may have weak interparticle bonds.

In this chapter, only the mechanical behaviour of sediments will be considered, although this closely interrelates with diagenetic changes. For example, lithified strength often develops because compression of grain boundaries enhances solution and diffusion of the mineral material, allowing grain interpenetration, serration of grain boundaries and the redistribution of mineral material as grain overgrowths (Rutter 1976, 1983; Gratier 1987). In such cases, lithification not only enhances rock strength, it is also a consequence of rock deformation. Sediments may, however, compact (lose bulk volume) during burial without the development of strength due to intergranular bonding.

The term consolidation is used here in its specific geotechnical sense, as defined in section 1.2.2. For a particular stress state the sediment consolidates until an equilibrium pore volume is developed (Figure 2.1a). The term normal consolidation is commonly used to describe the reduction in equilibrium pore volume that accompanies progressively increasing stress (Figure 2.1b). This terminology is used throughout this book, although some authors also refer to this process as 'compression'. The reader is referred to Atkinson and Bransby (1978) for a comprehensive discussion of the use of these terms in geotechnics. Consolidation results in a progressive densification of the sediment in the presence of increasing stress, and typically accompanies sediment burial (Burland 1990). Consolidation is only inhibited during sediment burial by low permeabilities (which prevent or substantially retard the dissipation of excess pore fluid pressures) and integranular cementation (which resists frictional sliding of grains). Although consolidation has been described extensively in soil mechanics (Atkinson and Bransby 1978; Lambe and Whitman 1979), it has received little attention in the literature of sedimentology and structural geology (but see Rieke and Chilangarian 1974; M.E.

The Geological Deformation of Sediments Edited by Alex Maltman Published in 1994 by Chapman & Hall ISBN 0 412 40590 3

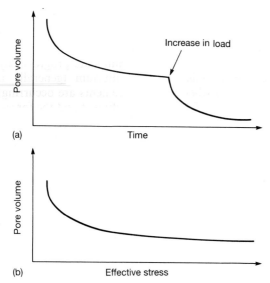

Figure 2.1 Illustrations of consolidation and normal consolidation. (a) The response of the pore volume of a saturated permeable clastic sediment to the application of two increments of load. This is the change in pore volume that accompanies the time-dependent dissipation of an excess pore fluid pressure, and is termed consolidation. (b) The relationship between increasing effective stress and equilibrium (fully consolidated) pore-volume states in a similar sediment. This is the compression curve that is the locus of the end-point pore-volume states for each consolidation increment. See also Figure 2.20.

Jones and Addis 1986; M.E. Jones et al. 1991). Consolidation constitutes a major facet of mechanical lithification and in many sediments it is the major mechanism responsible for pore-volume changes, the development of stress and pore-pressure regimes, and the development of early fabrics (most notably in mud rocks). It brings the mineral grains closer together, increasing the frictional strength (section 1.2.3) and prompting chemical bonding.

Sediments, from the moment of deposition, are subject to body forces, especially the action of gravity, but with increasing burial tectonic activity can also become important. The response of any sediment to these forces depends upon its physical properties, the magnitude of the forces and its physical location. For example, a clastic sediment deposited slowly on a flat sea floor will compact under gravitational loading through expulsion of pore fluid and grain rotation (normal consolidation). If the sediment contains clays or other platy minerals then a fabric will develop which reflects this passive state of gravitational compaction. If the same material were to be deposited on a steeply inclined slope, then the action of gravity would promote downslope movement rather than compaction. In geomechanical terms the material would tend to fail and the downslope sliding would be a shear deformation. Shear deformation in particulate materials is not normally accompanied by compaction, so that early slumping will cause a different fabric to develop. Thus, two sediments that were identical at the moment of deposition can, by virtue of the physical environment in which they were deposited, evolve in quite different ways and develop very different fabrics (Figure 2.2). A good example of this can be found in the Danian Chalks in the Central Graben of the North Sea. Here, deposition on slopes has led to reworking of some parts of the sequence through slumping, debris flows and turbidity currents (Kennedy 1986). Fabrics developed at this early stage have influenced the subsequent compaction history, even though the rock is now buried to a depth of approximately 3 km. Today, the reworked chalks still preserve larger primary pore sizes which, where not infilled with sparry calcite, provide the best hydrocarbon reservoir quality, such as in the large Ekofisk oil field.

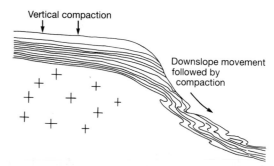

Figure 2.2 The influence of location on the response of a sediment to gravitational loading. These early fabrics will be preserved during the geological history of the resulting sedimentary rock and will influence subsequent deformation.

It follows from this introductory discussion that irrespective of depositional environment and sediment type there are two ubiquitous responses to the action of applied forces. From the moment of deposition, all sediments will develop coexisting and related states of stress and strain. These stress–strain states are diverse and complex, and evolve during burial, diagenesis and tectonism. They generally exert a strong influence on the nature of any fabrics that develop in the sediment, and are themselves strongly influenced by the evolving nature of those fabrics. To understand the deformation characteristics of sediments and weak sedimentary rocks it is necessary to understand this complex interrelationship between the applied forces, stress state, strain and fabric.

2.2 MECHANICS OF PARTICULATE MEDIA IN THEORY AND EXPERIMENT

2.2.1 Force, stress and strain

When a sediment or sedimentary rock is subject to forces, states of stress and strain develop simultaneously within the material (M.E. Jones et al. 1991). As the forces change in magnitude and/or orientation (as may occur during sediment burial), the stress–strain state progressively changes, maintaining equilibrium (generally a condition where no displacements are occurring) between the fabric of the sediment and the forces. The action of a force thus causes the sediment to undergo changes in shape, which constitute the state of deformation or **strain** (Figure 2.3a) However, the sediment is able to resist the development of strain to an extent that depends upon both its strength and its geometrical factors (Figure 2.3b). This resistance constitutes the state of **stress** (Janbu 1985). Thus a weak, unlithified sediment may achieve a particular strain state in response to a weak force and the resulting

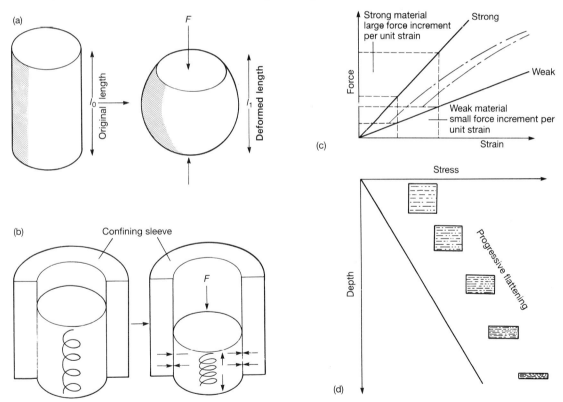

Figure 2.3 Illustrations of force, stress and strain. (a) The development of strain in an unconfined, right-cylindrical specimen due to the application of a force (F). l_0 is the original length and l_1 the deformed length. (b) The effects of the application of a force (F) to a confined specimen. The confining sleeve prevents lateral strain and a horizontal compressive stress develops to maintain equilibrium. The vertical stress in the rock sample is also related to the strain state and is illustrated schematically in this figure as the progressively compressing spring. (c) Idealized force (and hence stress) and strain relationships for a strong and a weak sediment. This shows the variation in compressibility and can be likened to changing the stiffness of the spring in (b). (d) Idealized development of vertical stress and vertical shortening in a sediment (stippled) deposited on a horizontal surface and subject only to gravitational loading. This is equivalent to the strain states that develop in the specimen illustrated in (b) as the axial force is increased.

stresses will be small, whereas a lithified sedimentary rock will require a much larger force to achieve the same state of strain, and the associated stresses will then be large (Figure 2.3c).

In formal mechanics, stress is defined as the force acting per unit area:

$$\sigma = \frac{F}{A}, \quad (2.1)$$

where σ is the stress, F is the force and A is the area over which the force is acting. Strain (ε) is most simply expressed as elongation and is defined (to keep contractional strains positive) by equation (2.2):

$$\varepsilon = \frac{l_0 - l_i}{l_i}, \quad (2.2)$$

where l_0 is the original length and l_i is the deformed length of the object subject to strain (Figure 2.3a). In geomechanics it is the convention that stresses that act in a compressive manner and strains that are compressive are regarded as positive. The states of strain and stress that develop when a sediment is subject to the action of a force are related by a physical property termed the **compressibility**. This is an inverse measure of strength and is defined as the strain per unit stress:

$$C = \Delta\varepsilon / \Delta\sigma, \quad (2.3)$$

where C is the compressibility; $\Delta\varepsilon$ is the incremental increase in strain and $\Delta\sigma$ is the incremental increase in stress.

Thus when gravity acts on a sequence of unlithified sediments, a vertical compactional strain and a vertical stress develop, both of which increase with depth. This burial stress is a product of the density of the sediment and the depth of burial (Figure 2.3d) and is given by equation (2.4):

$$\sigma_v = \rho \cdot g \cdot z, \quad (2.4)$$

where σ_v is the vertical stress in the sediment column, ρ is the density of the sediment, g is gravity and z is the depth. If for simplicity we assume that the deformation resulting from the action of gravity is one-dimensional and in the vertical direction, then the vertical strain will also increase with depth as the vertical stress increases, and its magnitude will depend upon the compressibility of the sediment.

It is possible to reproduce this relationship between stress and strain in laboratory experiments using an **oedometer** (a one-dimensional compression apparatus, Figure 2.4a). Experiments on the initial compaction of sediments have been conducted (Been and Sills 1981; Edge and Sills 1989; Leddra, Petley and Jones 1992) and reveal relationships between stress and strain similar to that shown in Figure 2.4b. This curve shows that as the sediment achieves equilibrium strain states for the increasing load, the increment of stress that develops per increment of strain increases. This means that as the sediment deforms (through compression of its pore space) its compressibility progressively decreases. As the sediment equilibrates to the increasing load it is becoming stronger and more able to resist subsequent deformation. This is why the stress increase is greater with increasing strain.

If the force acting upon the sediment in the oedometer is removed, then the stress decays with the force but some strain state will generally be preserved, giving an unloading curve of the type illustrated in Figure 2.4b. The deformation resulting from the application of the force is largely irrecoverable. This is a feature of the deformation of sediments and other particulate materials (Atkinson and Bransby 1978; Burland 1990).

If a similar compression test were to be conducted on a lithified sedimentary rock, and providing that the applied force does not promote failure of the bonding responsible for lithification, then a relationship between stress and strain of the type shown in Figure 2.4c will be observed. This is a linear relationship (generally only quasi-linear for real materials) and shows that the compressibility is nominally constant. When the force is removed the strain is seen to recover as the stress decays. This style of materal behaviour is normal for strong materials and is described as **elasticity**. Elastic materials obey Hooke's Law (Jaeger and Cook 1979). All rocks exhibit elasticity at small strains, but in highly compressible materials such as unlithified sediments the range of this kind of behaviour is extremely limited (Burland 1989). This leads to the important observation that unlithified sediments and weak sedimentary rocks normally exhibit strongly non-linear stress–strain relationships (Uriel and Serrano 1973; Vaughan 1985; M.E. Jones and Leddra 1989; M.E. Jones et al. 1991) over the range of stresses encountered during deposition and initial burial.

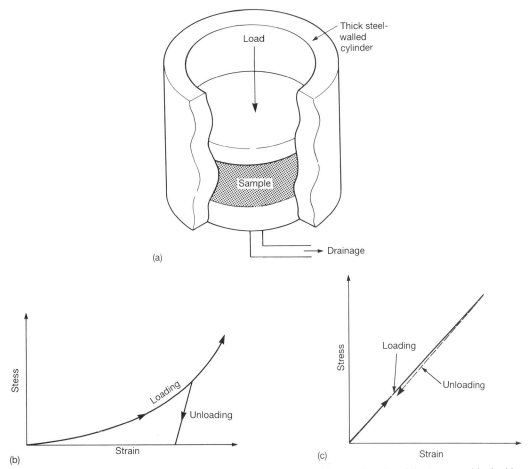

Figure 2.4 Consolidation, stress and strain. (a) Schematic illustration of an oedometer of the type used by Leddra, Petley and Jones (1992) and others to study the gravitational compaction of sediments. Load is applied to the top piston and pore fluids are expelled through a porous plate at the base of the sample. Apparatus of this type is routinely used in geotechnics to study the consolidation of sediments. (b) A typical stress–strain curve for a weak saturated clay deforming under oedometric loading. (c) Stress–strain curve for a sedimentary rock deforming within its elastic range.

The difference between the linear elastic deformation of lithified sedimentary rocks and the non-linear deformation behaviour of particulate sediments is due to the location and nature of the strain in the material. Elastic deformations occur at an atomistic scale. The bonds between individual atoms or molecules in parts, or all, of the rock are stretched or shortened like springs (Figure 2.5a). The strain is located in these distorted bonds and the rock must store a considerable amount of work energy to maintain the strain. This is facilitated by the stress that is mobilized with the strain. When the force promoting the strain is removed, the stress decays and the strain energy is released as the distorted bonds revert to their original positions. In an unlithified sediment, the frictional contacts between the grains are far weaker than the atomic bond strengths within the individual mineral crystals. The action of a force causes grains to slide past one another (Figure 2.5b), mobilizing only small stresses so that elastic distortion of atomic bonds is minimal. This strain is accomplished with little work energy being stored in the system, so there is little or no potential for the strain to recover when the force (and associated stress) is removed.

This mechanical distinction between unlithified sediments and sedimentary rocks is fundamental to the behaviour of the materials and also accounts for the almost independent development of the otherwise related geotechnical disciplines of soil mechanics and rock mechanics. However,

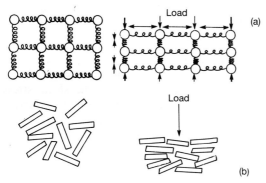

Figure 2.5 Recoverable (elastic) and irrecoverable (inelastic) strain. (a) Schematic representation of the location of strain in the interatomic bonds of elastic rock materials. The bonds are represented by springs, which lengthen or shorten according to their orientation in response to the applied load. (b) Schematic representation of the location of strain in a particulate sediment subject to increased load. The grains slide to new positions and, when the load is removed, remain there because there is no potential to return them to their original sites. Elastic behaviour in deforming sediments is discussed more fully in Chapter 6.

the behaviour of a completely unlithified sediment and of the equivalent fully lithified sedimentary rock represent the ends of a broad spectrum. Most naturally occurring sediments and sedimentary rocks fall within this spectrum and exhibit material behaviour which involves a combination of elastic and particulate deformations depending upon the stress magnitude (Uriel and Serrano 1973; Vaughan 1985; M.E. Jones and Leddra 1989; M.E. Jones et al. 1991). Elastic deformation is discussed further in section 6.2.

2.2.2 States of stress and strain in a body

Sediments are generally deposited in situations where they constitute part of a laterally extensive continuum. It is therefore incorrect to treat stress and strain as one-dimensional quantities because in real systems the reaction of a sediment to the application of a force is three-dimensional and will involve the generation of stresses and/or strains in all directions.

The formal description of the stresses acting upon a cube that is at rest can be found in any number of texts (e.g. Ramsay 1967; Jaeger 1969; Jaeger and Cook 1976; Bayly 1992). The treatment leads to the definition of the stress tensor in terms of principle stresses and shear stresses. There is no need to repeat the analysis here, but it is necessary, before considering in detail how sediments deform, to review what is meant by the terms **principal stress** and **shear stress** and how they are related. The significance of principal stresses, which act normal to planes that do not allow the mobilization of shear stresses, is illustrated in the following idealized example.

Consider the situation illustrated in Figure 2.6a. A rigid crystalline rock is resting on its flat base on a smooth horizontal surface of a similar material (this could, for example, be the idealized situation of a glacial erratic deposited on a smooth glaciated surface). The rock is at rest, and neither it, nor the surface below, have strained appreciably as a result of its presence. None the less, the erratic is exerting a load on the rock which is mobilized as a stress that is proportional to its mass acting over the area of contact. Both the boulder and the rock below will be subject to small elastic strains. This stress is acting normal to the contact surface and has no component that can be resolved to act parallel with it because in this situation no forces are acting to move the boulder across the surface. It therefore remains in position pressing down on the rock below. There is no shear stress acting on the surface and the gravitational stress acting normal to it is, therefore, a principal stress. Now consider the situation in Figure 2.6b. Here the surface is inclined to the direction of the gravitational force. The vertical stress acting through the boulder is still a principal stress but intersects the surface

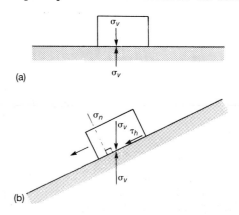

Figure 2.6 Principle stresses and shear stress. (a) The development of a normal stress that is also a principal stress between a gravitationally loaded block and the horizontal surface upon which it is resting. (b) The development of normal and shear stresses between a gravitationally loaded block and the inclined surface on which it is resting. In this case, the greatest principal stress is still vertical.

obliquely. Now two components can be resolved, one acting in the plane of the surface (the shear stress) and the other normal to it (the normal stress). The shear stress develops because a component of the gravitational force is trying to cause the boulder to slide downhill, whilst the normal stress is due to the component of the gravitational force acting across the surface, trying to compress it. The magnitudes of the shear and normal stresses vary with the inclination of the surface with respect to the principal stress. In all stress systems there are three principal stresses which are mutually perpendicular. These may have the same magnitude, or can be very different, and one or more may be zero. In the example in Figure 2.6a, as one principal stress (the greatest principal stress) is acting normal to the contact surface the other two lie on that surface and will be small.

The magnitude of normal and shear stresses depends not only on orientation but also on the difference between the principal stresses. If all three principal stresses are equal, the shear stress is zero and the normal stress is equal to the principal stress. The greater the difference between the maximum and minimum principal stresses, the greater the shear stress. The magnitudes of the shear and normal stresses acting on a surface as illustrated in Figure 2.6 are given in equations (2.5) and (2.6):

$$\tau_\theta = \frac{\sigma_1 - \sigma_3}{2} \sin 2\theta, \quad (2.5)$$

$$\sigma_\theta = \frac{\sigma_1 + \sigma_3}{2} + \frac{\sigma_1 - \sigma_3}{2} \cos 2\theta, \quad (2.6)$$

where τ_θ is the shear stress resolved on a surface whose normal is inclined at angle θ to the greatest principal stress, σ_θ is the stress acting normal to the surface and σ_1 and σ_3 are, respectively, the greatest and least principal stresses. Inspection of equations (2.5) and (2.6) reveals that only the greatest and least principle stresses are considered. This two-dimensional analysis is a useful approximation that will be used throughout this chapter. It must, however, be realized that where natural stress systems have been measured, it is generally the case that all three principal stresses have different values (Teufel *et al.* 1992). In these real systems, deformations occur in three dimensions and the magnitude of the intermediate principal stress is then important. By making the assumption that only the greatest and least principal stresses are important, it is possible to represent the state of stress by a Mohr's circle construction. An example of such a plot is given in Figure 2.7, which shows clearly how the magnitude of the shear stress depends on the difference between the principal stresses and the orientation of the surface on which the stress is resolved.

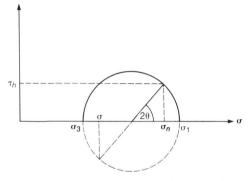

Figure 2.7 The standard Mohr's circle construction defining the states of magnitudes and normal stress acting on surfaces inclined at an angle θ to the greatest principal stress within a body subject to principal stresses $\sigma_1 > \sigma_3$. (See Atkinson and Bransby (1978) or almost any structural geology or geomechanics textbook for a discussion and description of this construction.)

The states of strain associated with the mobilization of normal and shear stresses are different. Normal stresses accompany changes in dimension in the direction parallel to the stress. Generally they are associated with a reduction in volume when the deformation is compressive (Figure 2.8a). Shear stresses are associated with changes in shape and need not be accompanied by volume change (Figure 2.8b). Thus a spectrum of strain states exists with entirely volumetric strain and entirely shear (distortional) deformations as its end members. During deformation a sediment will experience some combination of volumetric and distortional deformation, which will depend upon the character of the stress system. This diversity of natural stress systems means that in nature any strain state may exist. Thus sediments deposited on unstable slopes or affected by syndepositional faulting will develop fabrics associated with large shear deformations but may experience little or no compaction (volumetric strain) at the time (Figure 2.2), whereas those deposited and buried in flat-lying quiescent

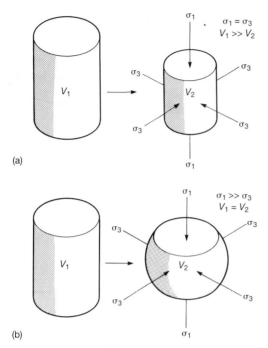

Figure 2.8 (a) The action of isotropic stress, leading to a reduction in the volume of a sediment but without the distortional components of strain associated with shear stress. (b) The deformation developed in an identical sample to that in (a) due to the development of the same principal stress σ_1. In this case, however, the lateral stress has been kept small and the sample has developed large resolved shear stresses. The result is that the sample has shortened, but has distorted (through lateral expansion), so that although the axial and shear strains are large, the volumetric strain is negligible.

basins will compact under gravitational loading but experience little bulk shear deformation (Figure 2.2).

At this point it is sensible to summarize the main points of this discussion.

1. Principal stresses act normal to planes on which the shear stress is zero.
2. There are always three principal stresses, which are mutually perpendicular.
3. The magnitude of a shear stress depends on the difference between the greatest and least principal stress and on orientation with respect to the principal stresses.
4. Principal (and other normal) stresses cause compressional deformations (or elongations) whereas shear stresses promote sliding or distortion (shear strains).
5. At or near to the earth's surface, the stress induced by the gravitational force is always a principal stress. It should, however, be noted that except where the earth's surface is horizontal, gravity will induce shear stresses parallel to the surface. These shear stresses are responsible for downslope movements in large sediment bodies, such as deltas, and for many of the fabrics associated with landslides, slumps, debris flows and sediments deposited by turbidity currents (Chapter 5).

2.2.3 Pore pressure, effective stress and porosity

Porous sediments and sedimentary rocks are two-phase systems, consisting of a skeleton of mineral grains and pore spaces filled with water, air, brine, hydrocarbons or some combination of these fluids. Generally, except at very shallow depth, the pore space is filled with water, brine or a liquid hydrocarbon with a compressibility that is comparable with or less than that of the mineral skeleton. When these materials are subject to load, the resulting total stress (σ) is partitioned between the effective stress that acts through the mineral grains (σ') and the pore fluid $P_{(H_2O)}$. Thus, in liquid-saturated sediments there is generally an effective stress that acts through the mineral grains and a pore fluid pressure acting in the pores (Figure 2.9). In extreme cases, when the pore pressure is equal to the total stress, the effective stress can be zero. For a sediment which is permeable and porous, and subject to gravitational loading, the pore pressure increases with depth in a similar manner to the vertical stress, except that the rate of increase is somewhat less because the density of water, brine or oil is less than the density of the mineral compo-

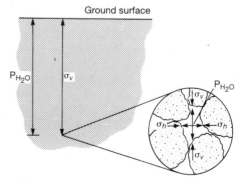

Figure 2.9 Representation of the state of effective stress in a saturated porous sediment in the subsurface.

nents of the rock (Figure 2.10; see also Figure 1.4). All compressional deformations of this sediment are accompanied by changes in the effective stress (σ'), and are therefore associated with either an increase in total stress (σ) (which may be due to progressive burial or tectonic loading) or a change in pore fluid pressure $P_{(H_2O)}$ (Terzaghi 1936, 1943). The effective stress is most simply expressed as the difference between the total stress and the pore fluid pressure according to the Terzaghi equation:

$$\sigma' = \sigma - P_{H_2O}. \tag{2.7}$$

This equates with equation (1.12); the concepts were introduced in section 1.2.5.

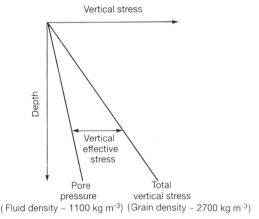

Figure 2.10 Schematic illustration of the development of pore pressure, total vertical stress and vertical effective stress with depth in a brine-saturated, porous sediment subject to gravitational loading beneath a laterally extensive horizontal surface. (See also Figure 1.4.)

An increase in total stress will therefore cause a change in effective stress only if the stress change is experienced by the mineral skeleton of the sediment. This will occur readily in sediments that are of high permeability and able to drain. Under these conditions, an increase in the load acting on the sediment will lead to a reduction in pore volume as the total and effective stresses increase and pore fluid is expelled. If, however, drainage of pore fluid is prevented, the pore volume of the sediment can decrease only if the pore fluid compresses in the pore spaces. For aqueous fluids under small loads, the compressibility is insignificant, and an increase in pore fluid pressure will occur. In this case strain will accumulate only if the excess component of the pore pressure caused by the increased load dissipates, allowing the effective stress to increase (Figure 2.11). This process is dependent upon the permeability of the sediment and is termed primary consolidation. It is described more fully below.

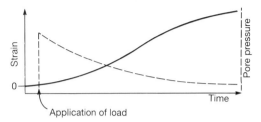

Figure 2.11 The relationship between pore-pressure dissipation and the development of strain with time in a low-permeability sediment. For thick layers of shale, the time axis of this type of plot may extend to several million years.

Pore fluid pressure therefore determines the magnitude of effective stresses acting within a sediment. It has two components:

$$P_{H_2O} = P_{H_2O_h} + P_{H_2O_e}, \tag{2.8}$$

where $P_{H_2O_h}$ is the hydrostatic component of the pore pressure caused by the overlying column of fluids extending to the earth's surface; and $P_{H_2O_e}$ is the component of pore fluid pressure that is in excess of the hydrostatic gradient. At equilibrium:

$$\sigma' = \sigma - P_{H_2O_h}, \tag{2.9}$$

but during sediment burial, equilibrium may not be attained and the effective stress changes continuously with time:

$$\frac{d\sigma'}{dt} = \frac{d\sigma}{dt} - \frac{dP_{H_2O_h}}{dt} + \frac{dP_{H_2O_e}}{dt}. \tag{2.10}$$

Sediments undergoing burial may sustain large pore fluid pressures, or may become overpressured through other mechanisms (see sections 2.2.3 and 2.3.4), with the result that even in the presence of large total stresses (due to substantial burial or the action of large tectonic forces), the effective stress may still be very small.

When sediments are subject to increased effective stress, pore volume is lost and although the actual detail of this process varies with the type of sediment and with the stress path (see below), the basic relationship is that shown in Figure 2.12. This is a plot of porosity (the ratio between pore volume and total rock volume) against the mean effective stress (defined as the mean

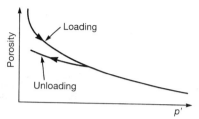

Figure 2.12 The relationship between porosity and mean effective stress (p') for a porous clastic sediment deforming by compression of the pore space. See also Figure 6.1.

of the three principal effective stresses $p' = (\sigma_1 + \sigma_2 + \sigma_3)/3$). Similar compaction curves have been determined for many sediments during geotechnical studies. These are commonly represented as straight-line plots of void ratio (the ratio between pore volume and the volume of mineral grains) against the natural logarithm of the mean stress (Figure 2.13). These plots show that the major mechanism of compaction is loss of pore volume, and that for a given sediment under particular effective stress conditions, there is a maximum pore volume that can exist. As the effective stress is increased the pore volume is decreased. Thus, increase in effective stress is accompanied by an increase in sediment strength due to progressive densification. Because sediments of the type considered here are virtually inelastic, unloading does not recover the pore volume lost during loading. Figures 2.12 and 2.13 both illustrate a typical unloading path, sometimes referred to as an **elastic swell line** (e.g. Figure 7.4), which reveals the limited elasticity of particulate materials. The only exception to this general behaviour is where the sediment contains a high concentration of swelling clays. However, as bonding between the particles grows and the sediment increases in degree of lithification, the range of elastic behaviour increases (Chapter 6).

For sedimentary rocks that have become lithified and are elastic at low stresses, the change in pore volume with increasing effective stress is complicated by the bonding stiffness. The behaviour of these materials is best explored by regarding the effective stress as a resistance to deformation that is mobilized in the mineral skeleton as a consequence of its rigidity (Janbu 1985). This resistance may be viewed as containing three components:

$$\sigma' = r_c + r_i + r_m, \qquad (2.11)$$

where r_c is the resistance due to strong intergranular bonding, r_i is the resistance due to intergranular friction, and r_m is the resistance due to weak intergranular chemical effects, such as van der Waals bonds, water absorption and osmosis. In particulate sediments the effective stress is dominated by the frictional resistance (sands) or by the chemical resistance (clays) (Atkinson and Bransby 1978; Lambe and Whitman 1979). In porous lithified rocks, strong intergranular bonding is present and resistance to deformation is greater while these bonds are intact (M.E. Jones et al. 1990).

The elasticity produced by strong intergranular bonding at low effective stresses is destroyed once the strength of the mineral skeleton (either the cement bonds or the component grains) is exceeded. When this occurs the sedimentary rock yields (Figure 2.14), and its compressibility increases. At yield, strong bonding resistance is progressively destroyed by fracturing of grains or intergranular bonds, and the source of the effective stress must progressively change to become located in intergranular friction and/or the cohesion due to weak chemical bonding. During yield

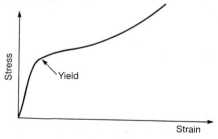

Figure 2.14 Stress–strain curve for an oedometric experiment conducted on a bonded but porous sedimentary rock. This curve shows the linearly elastic response (initial steep section) at low stress and the post-yield compaction behaviour with the strength increasing as the sediment densifies.

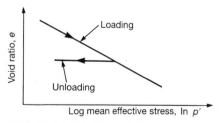

Figure 2.13 The data in Figure 2.12 replotted in the standard void ratio (e)–natural log p' diagram widely used in soil mechanics. See also Figures 6.2 and 7.4.

the rock is effectively changing from a strained material of the type illustrated schematically in Figure 2.5a to a strained material of the type illustrated in Figure 2.5b. This is a change of state. The weakened sediment is unable to support the effective stress and must either harden (through compaction) or reduce the effective stress. Generally, the strain energy released by the change in state is transferred to the pore fluid as an excess pore pressure, which serves to support the porous fabric of the sediment by reducing the effective stress it must sustain. This transient pore pressure then dissipates and the sediment consolidates until it is sufficiently densified to again support the effective stress present at yield. The process is shown schematically in Figure 2.15 and in the stress–strain, compressibility–strain and pore-pressure–strain curves in Figure 2.16. After yield, porosity changes with the effective stress in a manner similar to the equivalent unlithified sediment (Figure 2.17).

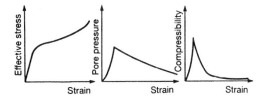

Figure 2.16 Stress–strain, pore-pressure–strain and compressibility–strain curves for a cemented porous sediment that yields. These curves show how the compressibility increases to a maximum at yield. This causes the increase in pore pressure because the effective stress cannot be instantaneously maintained. Dissipation of the pore pressure is accompanied by compaction of the pore volume and densification, leading to a recovery of low compressibility. The effective stress increases as the sediment densifies.

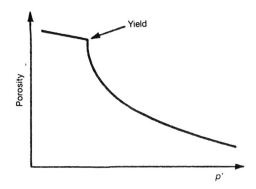

Figure 2.17 The relationship between porosity and mean effective stress for a porous sedimentary rock that yields. The substantial loss of porosity following yield is due to the densification necessary to sustain the effective stress state (Figures 2.15 and 2.16).

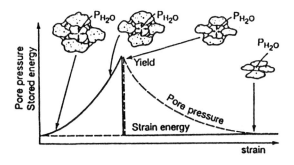

Figure 2.15 Mechanistic interpretation of the change in state that accompanies yield in a cemented porous sediment. Pre-yield strain energy increases in the intergranular bonds, which behave like springs (this energy is stored in the interatomic bonds (Figure 2.5a)). At yield the bonds fail, but the energy in the system must remain unchanged, so a transient increase in pore pressure occurs as the instantaneous response. This pore pressure then dissipates (if the rock is sufficiently permeable) and the grains compact (Figure 2.5b).

To see the effect of equation (2.11), it is necessary to rewrite equation (2.7) in a more precise manner in order to consider the compressibility of both the mineral particles (C_m) and the mineral skeleton (C_s) of the sediment (Bishop 1976):

$$\sigma' = \sigma - \left(1 - \frac{C_m}{C_s}\right) P_{H_2O} \qquad (2.12)$$

The term C_m/C_s is very small in unlithified sediments, where the effective stress is controlled by r_i and/or r_m, and equation (2.7) is valid. However, in fully lithified sediments this term may approach unity. Destruction of cement bonding in a sediment at yield causes $[1-(C_m/C_s)]P_{H_2O}$ to approach P_{H_2O} causing an excess pore fluid pressure with a maximum value $P_{H_2O} - \{[1-(C_m/C_s)] P_{H_2O}\}_{\text{pre-yield}}$. The magnitudes of the greatest effective stresses sustained by a lithified sediment are therefore dependent upon the nature of the deformation resistance mobilized by the mineral skeleton, the ability of the sediment to sustain an excess pore fluid pressure (which depends on permeability of the sediment and adjacent rocks), and the volumetric strain.

A thorough familiarity with the effective stress principle is essential to an understanding of the deformation of fluid-saturated sediments. In unlithified sediments, unique relationships exist between the maximum sustainable pore volume and the maximum sustainable effective stress. Put simply, compacted pore-volume states can be sustained at low effective stresses, but larger than equilibrium pore volumes are never sustained at any maximum effective stress. If the sediment is on its normal consolidation line (Figure 2.1b), then any increase in effective stress is always accompanied by a decrease in pore volume. Similarly, the pore volume can decrease only if the effective stress is increased.

If the sediment has become lithified, the situation is more complex. Strong intergranular bonds and pore-filling cements allow a range of pore volumes to be sustained by the sediment prior to yield. Often quite porous, lithified, sedimentary rocks will sustain large pore volumes until the effective stress increases to a sufficient degree to promote yield. At yield, such rocks undergo a significant reduction in strength. Either the effective stresses or the pore volume – or, most typically, both – must evolve rapidly to maintain mechanical equilibrium. After yield these rocks behave in a manner similar to the equivalent particulate material. If the porosity of such sediments has been reduced by pore-filling cements, then the yield stress is progressively increased (Figure 2.18), although if the grain fabric is not appreciably altered by the cementation the post-yield behaviour remains similar (Figure 2.19).

2.2.4 Primary consolidation and creep

The progressive decrease in pore volume that accompanies an increase in effective stress nor-

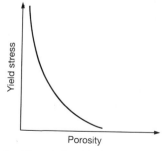

Figure 2.18 The relationship between yield stress and porosity for a sediment showing progressive infilling of its pore space with a mineral cement.

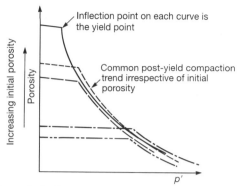

Figure 2.19 The effect of pore-infilling cement (without other fabric change) on the post-yield behaviour of a sediment.

mally represents an equilibrium condition (Figure 2.1b). This generally has been regarded as independent of the rate of change of effective stress and of the duration over which the particular effective stress state is maintained (Atkinson and Bransby 1978). Change in effective stress is not, however, rate insensitive. Increase in effective stress occurs either because of increase in the total stress acting upon the sediment, or because of a decrease in the pore fluid pressure (equation 2.7). Generally, in geological situations, it is progressive growth of the total stress due to increasing burial depth or tectonic processes that causes effective stresses to increase. The total stress acts on both the mineral skeleton and the pore fluids, but the instantaneous response of a saturated sediment to an increase in total stress is an increase in the pore fluid pressure. This increased pressure causes fluids to flow from the sediment and its pore volume to progressively decrease, whilst the effective stress increases until the equilibrium state of consolidation is re-established. The development of a new equilibrium effective-stress condition is thus dependent upon the permeability of the sediment, the permeability of the drainage path to a low-potential drainage boundary (such as the earth's surface) and the length of the drainage path. Changes in effective stress and the associated volumetric strains are therefore sensitive to, and dependent upon, the rate of pore-pressure dissipation. This process is primary consolidation and is illustrated schematically in Figure 2.20. The original formal description of this phenomenon was by Terzaghi (1936), and it has subsequently been widely investigated in soil mechanics (Atkinson and Bransby 1978; Lambe and Whitman 1979).

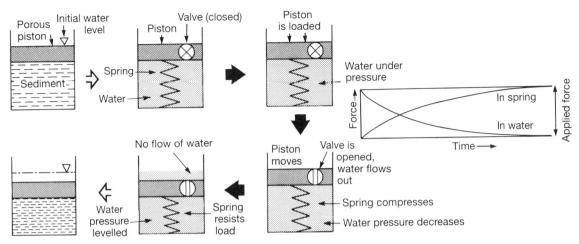

Figure 2.20 Schematic representation of consolidation. In this diagram the sediment is represented by a sealed container filled with water (the pore fluid) and a spring (the mineral skeleton). A force applied to the movable lid of the container causes a reaction in the pore fluid, and the force is resisted by the generation of pore pressure. When a valve in the lid of the container is opened (representing the permeability of the rock), fluid flows from the container and the pressure decreases while the spring compresses. The stress in the spring increases through time (the effective stress) while the pore pressure decreases. Eventually the applied force is balanced entirely by the stress in the spring and the system is in equilibrium. The 'sediment' is now fully consolidated for the applied conditions.

Terzaghi identified that the magnitude of the excess component of the pore pressure ($P_{H_2O_e}$) caused by an instantaneous increase in total stress from time, t_0, to any time, t (t greater than t_0) and depth (z) in a sediment layer draining to its upper surface, is described by the equation:

$$C_v \frac{d^2 P_{H_2O_e}}{dz^2} = \frac{d P_{H_2O_e}}{dt}. \qquad (2.13)$$

C_v is termed the **coefficient of consolidation** and is given by

$$C_v = \frac{K}{C} U W_w \qquad (2.14)$$

where K is the hydraulic conductivity of the sediment, C is its compressibility and UW_w is the unit weight of the pore fluid. Solutions to equation (2.13) of the form $P_{H_2O_e} = P_{H_2O_e}(z, t)$ provide the variation of excess pore pressure with time in the consolidating layer, from which the variation in effective stress can be determined. Equation (2.13) requires that the total stress remains constant with time. During sediment burial, and/or tectonism, this is unlikely to be the case. Under these circumstances the consolidation equation takes the form given below:

$$C_v \frac{d^2 P_{H_2O_e}}{dz^2} = \frac{d P_{H_2O_e}}{dt} - \frac{d\sigma_v}{dt}. \qquad (2.15)$$

For a single layer draining in one direction, normal to its surface, the pore-pressure distribution with time follows the form illustrated in Figure 2.21, and compaction proceeds at a rate determined by the associated progressive changes in effective stress. The sediment is described as fully consolidated (for the *in situ* stress) when all excess pore pressures are dissipated. The approach to this condition is seen as a progressive reduction in layer thickness as the volumetric strain increases (Figure 2.22).

Consolidation accompanies the deposition and burial of all clastic sediments owing to the

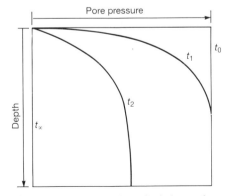

Figure 2.21 Pore pressures in a single layer of consolidating clay at different depths (z) for times ($t=0$, $t=t1$, $t=t2$, and $t=$infinity) after the onset of consolidation.

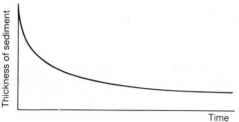

Figure 2.22 Change in thickness of the consolidating layer (Figure 2.21) with time.

progressive increase in vertical (gravitational) stress as burial depth increases (Figure 2.10). In sands and other highly permeable sediments, the time required for full consolidation is extremely short and burial is not accompanied by the development of excess pore fluid pressures, unless the sand becomes sealed by overlying sediments of low permeability. In clays and other sediments of low permeability, the rate of pore-pressure dissipation may be slower than the rate of increase in vertical stress due to burial. These sediments (and also more permeable sediments sealed by low-permeability layers) will become progressively overpressured as they are buried, and the resulting excess pore pressures may be sustained over appreciable periods, both during and following burial. This mechanism of overpressure development is widely recognized by geologists (Yassir 1989a) but is far from being the only cause of excess pore fluid pressures in sediments (section 2.3.4).

Consolidation of a sediment under gravitational loading has been described as self-weight consolidation and has been investigated experimentally by Been and Sills (1981). They have concentrated on the very low stress (large pore volume) conditions that accompany initial deposition and have observed the changes in effective stress as slurries deposit sediment particles under gravitational loading. The consolidation of clays and silts at somewhat larger stresses has been widely studied in soil mechanics (Gibson 1958; Gibson, England and Husey 1967; McClelland 1967; Skempton 1970; Burland 1990). However, consolidation under the larger stresses more typically encountered in geology, such as may exist in sedimentary basins, has received less attention (Yassir 1989; Charlez and Heugas 1991; Leddra, Petley and Jones 1992). In investigations of natural geological consolidation, it should be noted that experimental observations may not be completely representative of consolidation in nature (section 6.2.3).

Moreover, Burland (1990) reports that naturally consolidated materials are able to sustain larger pore volumes than materials consolidated in the laboratory. This may be a rate effect, or it may be a consequence of natural hardening of the mineral skeleton due to chemical interactions between grains and cementation of the material. As emphasized in section 1.1.1, the effect of diagenesis on these mechanical changes has been little explored.

It has been the common experience of experimentalists that sediments and sedimentary rocks can accumulate additional volume strain even after the sediment is fully consolidated to the applied stresses (Atkinson and Bransby 1978; de Waal 1986; Johnson, Rhett and Seimers 1989; Rhett 1990; Andersen, Foged and Pedersen 1992). This additional strain is variously referred to as secondary consolidation or **creep**. Bishop and Lovebury (1969) undertook experiments on remoulded London clay which revealed that creep was still continuing, although at an extremely slow rate, 3 years after primary consolidation was completed. Compaction experiments on chalk (Johnson, Rhett and Seimers 1989; Rhett 1990; Andersen, Foged and Pedersen 1992) have revealed that creep continues for periods of the order of 1 week to 2 months after the sample is equilibrated to the applied stresses. The mechanisms of secondary consolidation or creep are uncertain and probably diverse. Possible candidates include: grain surface diffusion; time-dependent crack generation associated with a redistribution of stored strain energy; and diffusion in microfractures, with stress corrosion weakening the fracture tips. Note that the term creep in this usage differs from that in other geological contexts (e.g. section 5.7; see section 1.2.4).

Thus, although consolidation is a widely recognized and well-understood process in geotechnical engineering, its application to natural sediment burial needs careful consideration. Although natural sediments clearly experience consolidation as they are buried, the natural process is complicated by:

1. Chemical interactions between grains, which may strengthen the sediment and reduce the equilibrium volumetric strain;
2. long-term creep effects, which may contribute appreciable additional volumetric strains over long periods;
3. other sources of pore pressure (see section 2.3.4).

The results of these complications are manifest when real sediments and sedimentary rocks are examined. In many cases, consolidation fabrics are overprinted by the effects of pressure solution, whereas in other examples extremely high porosities are preserved at great burial depths, either as a result of strong grain bonding, or through the presence of high pore fluid pressures. Although laboratory data from geotechnical studies indicate that even very thick (2–3 km) sedimentary sequences should be fully consolidated within 5 million years, in nature, high-porosity, highly overpressured formations are encountered in sediments that have been buried to these depths for 50 million years or more (van den Bark and Thomas 1980; Chapman 1983). Thus the evolution of effective stress conditions in basins and other large accumulations of sediments is far more complicated than indicated by simple gravitational consolidation models for burial and compaction (M.E. Jones and Addis 1984, 1986).

The topic of consolidation history is considered further in sections 2.3 and 6.2. However, it is worth reiterating the statement made in the previous section that sediments which become overpressured will be subject to small effective stresses irrespective of the total stress state. The behaviour of such sediments will depend upon the stress history but, in the case where small effective stresses have been maintained throughout the history of the sediment, the strength (in the absence of diagenetic cements) will remain very low. Consolidation state, in the absence of intergranular bonding, has a crucial influence on strength.

2.2.5 Shear stress and deformation

Principles

Isotropic consolidation leads to reduction in bulk volume of a sediment but not to deformation in the sense of a change of shape (see section 1.2.2). Although volumetric strains are very common in nature, they are by no means ubiquitous, and they do not necessarily accompany changes in shape. Many of the common expressions of sediment deformation (such as faults, landslips, diapirs and slumps) occur with little or no volumetric strain. These changes in the shape of the sediment body are due to the mobilization of shear stress and strain. The present section turns to a consideration of these fundamental aspects of deformation. Almost all natural deformations involve some component of shear.

The magnitude of the shear stress sustained by a sediment or sedimentary rock is proportional to the difference (q) between the greatest (σ'_1) and least principal effective stresses (σ'_3) (section 2.2.2). This difference is a stress invariant referred to variously as the **differential** or **deviatoric stress** (equation 2.16). It is proportional to the diameter of the Mohr stress circle and is therefore, at the specified stress conditions, a direct measure of the shear stress that will develop in favourable orientations within the material (Atkinson and Bransby 1978). Only when the material is subject to purely isotropic stress (a state that rarely exists in nature) will the deviatoric stress be zero.

$$q = \sigma'_1 - \sigma'_3 \qquad (2.16)$$

As the deviatoric stress is the difference between two effective stresses, and as it is directly proportional to the shear stress, it follows that the magnitude of the shear stress is independent of the magnitude of the pore fluid pressure. However, shear-stress magnitudes are limited by the shear strength of the sediment or sedimentary rock. Materials can never mobilize or sustain shear stresses that exceed their shear strengths because shear failure will serve to maintain the shear stress at or below the failure condition. This has been demonstrated dramatically in a number of laboratory studies on weak, saturated sediments and leads to the important concept of the critical state.

Laboratory studies on weak sediments, such as remoulded London Clay (Schofield and Wroth 1968), remoulded Gault Clay (Parry 1960) and mud volcano clays (Yassir 1989a), reveal that at any given mean effective stress (p') (equation 2.17) for the case where $\sigma_2 = \sigma_3$, shear failure of the material occurs at constant deviatoric stress (q), irrespective of the strain. At this stress condition, the sediment is theoretically able to deform at constant pore volume to infinite strain, or, in the case of real systems, until geometrical changes force a change in the stress system or stress magnitude. This condition is termed the **critical state**. The stress–strain curves of mud-volcano clay samples deforming in this manner are shown in Figure 2.23. These curves illustrate the accumulation of strain at constant stress. The

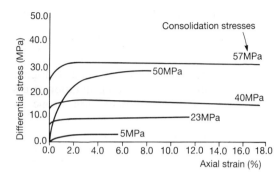

Figure 2.23 Stress–strain curves for a mud-volcano clay subject to undrained shear deformation following consolidation to the isotropic consolidation stresses indicated against each curve. (After Yassir 1989a.)

experiments were conducted such that the state of volumetric strain was held constant during shear deformation. This conforms to ideal critical state behaviour (Atkinson and Bransby 1978). The sediment is behaving as a perfectly plastic material in the sense that it is undergoing a permanent change in shape without fracturing and without hardening or softening. While at this critical state condition, any attempt to increase the deviatoric stress supported by the material simply results in an increase in strain rate.

$$p' = \frac{\sigma'_1 + 2\sigma'_3}{3}. \qquad (2.17)$$

If identical samples of a saturated sediment that is capable of exhibiting critical state behaviour are consolidated to different mean effective stresses (p'), the deviatoric stress (q) required to induce shear failure at the critical state is observed to increase with the consolidation stress. In other words, consolidation to greater effective stresses serves to increase the shear strength of the sediment. The results of such experiments can be represented by their stress–strain curves (Figure 2.23) or as a family of Mohr's circles describing the stress states at failure (Figure 2.24), but the most convenient representation, and the common practice in soil mechanics, is to use a diagram in which the deviatoric stress is plotted against the mean effective stress (Figure 2.25a). Plots of this type are often referred to as **stress path diagrams**, because they trace the stress history of the sediment in terms of both the effective and shear stresses it sustains. A stress path in p'–q space is directly equivalent to a family of Mohr's stress circles except that each circle is represented by a single point with the coordinates p' and q. The quantity p' defines the position of the circle, and q its diameter (Farmer 1983). (See Atkinson

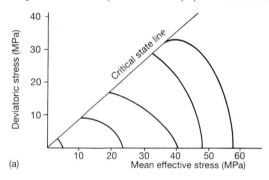

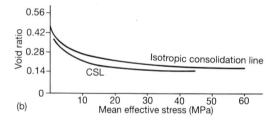

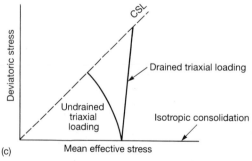

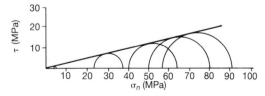

Figure 2.24. Mohr's circles at failure for the experiments represented in Figure 2.23, showing the failure envelope.

Figure 2.25 (a) The stress paths in deviatoric-stress–mean-effective-stress space for the experiments represented in Figure 2.23, and the critical-state line (CSL) defined by those experiments. (b) Pore-volume states at failure as a function of the mean effective stress. (c) Representation of the stress paths followed during (i) isotropic consolidation followed by drained triaxial loading and (ii) isotropic consolidation followed by undrained triaxial loading.

and Bransby (1978) for a comprehensive discussion of this representation of effective stress history.) Stress path diagrams form the basis of the approach adopted in Chapter 6 for analysing the tectonic deformation of sediments.

The shear strength of a sediment can be determined by a series of laboratory experiments, each conducted at a different consolidation stress (the effective confining pressure), in which the deviatoric stress is increased until failure occurs. The results of these experiments can be represented by the family of Mohr's circles that represent the stress states at shear failure, and a line can be constructed that is tangential to and enveloping these circles and which defines the failure envelope (Figure 2.24). The sediment cannot sustain stress states that lie outside this envelope. Alternatively, the same experimental results can be represented by a line in the stress-path diagram. This is the locus of the ends of all stress paths on which the sediment attained the critical state (Figure 2.25a). This (normally) straight line which connects the ends of the individual stress paths at the critical state is referred to as the **critical state line**. The critical state line is a representation of the shear-failure criterion for the sediment.

Strictly, the critical state concept applies only to homogeneous sediments that behave in a perfectly plastic manner. This is rarely true of natural materials. Most natural sediments exhibit an initial shear strength that lies above the critical state (at least at low consolidation stresses). This is generally referred to as the peak strength. Once the peak strength has been exceeded, the deviatoric stress sustained by the sediment is seen to decay until the critical state is attained. For many real sediments the critical state corresponds to a residual strength condition (Farmer 1983; Leddra, Petley and Jones 1992).

Sediments can only normally sustain stress states that lie below the critical state line. Any attempt to increase the effective stresses acting upon the sediment to shear-stress states outside of the critical state simply results in increased strain. The critical state line for any sediment is thus part of a **state boundary surface**, dividing stress–space into attainable and unattainable regions. It illustrates how the maximum deviatoric stress that can be sustained by a sediment is limited by its physical characteristics and consolidation state. Sediments subject to stress states at the critical state will deform until either the causative forces dissipate or are relieved by the deformation, or until changes in geometry or deformation mechanism promote a change in either the stress state or the sediment's strength.

Inspection of the critical state line in Figure 2.25a illustrates how, at low consolidation stress (low effective stress), sediments are prone to large-scale shear deformation. Obvious examples of this type of behaviour in nature include the failure surfaces developed at shallow depths in landslips and syndepositional faults, the mass movement of unconsolidated and lightly consolidated materials in flows, slides and slumps. Under larger effective stresses, greater deviatoric stresses are necessary to promote shear failure at the critical state, but deformation under these conditions may take place in some fault gouges and account for the mobilization of shales to form diapiric structures.

Similarly, anomalously high pore pressures will allow low effective stresses to be maintained, so that deformation at the critical state under low consolidation stress may occur even in environments where the total stresses are large. Thus, although the magnitude of the deviatoric stress is independent of the pore fluid pressure, the magnitude of q at which a sediment is able to deform plastically to large strains is strongly dependent upon the mean effective stress. Low shear strengths will characterize sediments that have sustained large excess pore pressures throughout burial, irrespective of the burial depth. Conversely, sediments that develop excess pore pressures owing to some secondary mechanism of pore-pressure generation may fail at the critical state if the value of the deviatoric stress remains unchanged while the mean effective stress is reduced by the increase in pore pressure.

The stress path diagram (Figure 2.25a) allows the relationship between shear stress and mean effective stress to be illustrated along with the conditions for shear failure of the sediment (the critical state line). It is also the case that increase in mean effective stress leads to a reduction in pore volume (see sections 2.2.3 and 2.2.4) and hence the critical state line can also be represented on a plot of pore volume against mean effective stress (Figure 2.25b). This is equivalent to Figure 2.12, except that in this case the recorded pore-volume states are those sustained during shear failure of the sediment. This curve therefore represents the locus of end-point pore-volume states for samples consolidated to different mean effective stresses and subject to deviatoric stresses corresponding to the critical state. Figure 2.25b represents an important relationship, as it illustrates the mechanism (pore-volume reduction) by which consolidation to greater effective stress increases the shear strength of a sediment, thereby reducing the likelihood of shear failure.

In Figure 2.25c, two paths are shown linking the end point of consolidation with the critical state line. One of these shows an increase in mean effective stress with increasing deviatoric stress and is a straight line; the other is a curve showing decreasing mean effective stress. The former corresponds to the case where loading to failure is accompanied by full dissipation of any excess pore pressures, so that the deformation can be described as drained. Typically this path is followed in a laboratory triaxial experiment (Figure 2.26a) where the sample is consolidated under an

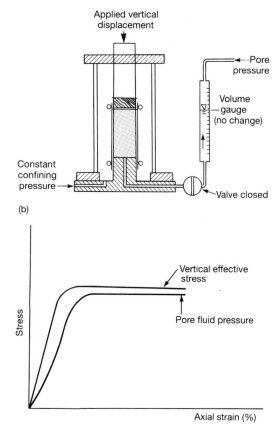

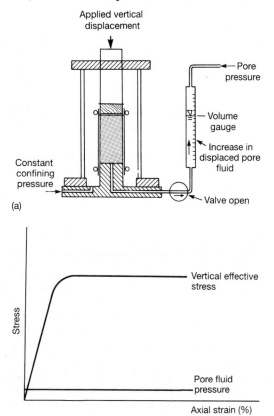

Figure 2.26 (a) Schematic representation of a drained triaxial experiment. (b) Schematic representation of an undrained triaxial experiment.

Figure 2.26 Contd.

isotropic effective stress ($\sigma'_1 = \sigma'_3$) at low pore pressure. The consolidation stress (cell or confining pressure, σ'_3) is then held constant while the axial stress (σ'_1) is increased. The rate of deformation is adjusted until pore-pressure dissipation (which depends on the coefficient of consolidation (C_v) (equations 2.13 and 2.14)) equals or exceeds the rate of pore-pressure generation. Any pore pressures associated with the consolidation that accompanies axial deformation are thus fully dissipated. The second curve corresponds to the situation where drainage of pore fluids is completely prevented during the application of the deviatoric stress – undrained deformation. Here, axial deformation of the sample is not accompanied by continued consolidation (Figure 2.26b). In this experiment, pore pressure increases with the deviatoric stress. The result is that the mean effective stress is reduced, the effective stress being dependent upon the pore pressure, while the deviatoric stress increases as the axial stress increases until shear failure occurs at the critical state.

Figure 25 (a and b) represents projections of a three-dimensional surface on to a pair of the three axes: mean effective stress, deviatoric stress and pore volume (void ratio). The surface is shown in three-dimensions in Figure 2.27. It is referred to as a **yield surface** (Atkinson and Bransby 1978; M.E. Jones, Leddra and Potts 1990; M.E. Jones *et al.* 1992), and is seen to consist of two complex curving surfaces that intersect at the critical state line. These surfaces are known in soil mechanics as the **Roscoe** and **Hvorslev surfaces** (Atkinson and Bransby 1978). They are geometrically and, to a degree, physically equivalent to a Mohr-Coulomb failure surface for porous rocks (M.E. Jones, Leddra and Addis 1987; M.E. Jones and Leddra 1989; Abdulraheen, Roegiers and Zaman 1992; Loe, Leddra and Jones 1992).

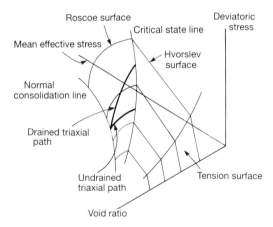

Figure 2.27 The state boundary surface for a particulate sediment (after Atkinson and Bransby 1978) showing the Roscoe surface (where normal consolidation occurs), the critical-state line and the Hvorslev surface. The paths followed in a drained and an undrained triaxial experiment starting from the same stress–pore-volume condition are illustrated. Paths that track into the interior of this figure, or those that reach the Hvorslev surface, are associated with overconsolidation.

The yield-surface diagram (Figure 2.27) provides the means to represent the potentially complex deformation path that a sediment may follow during its geological history, in terms of progressive effective stress, deviatoric stress and pore-volume states. Deformation at the critical state has already been described and represents shear deformation that proceeds at constant deviatoric stress. All normally consolidated sediments deform along paths that lie on the Roscoe surface and deformation on this surface may involve components of both volumetric and shear strain (Loe, Leddra and Jones 1992). Only when the sediment has been unloaded (either by increase in pore pressure or through reduction of the total stress) can paths within the internal volume of this figure, or on to the Hvorslev surface, be followed. These states are known as overconsolidated (section 2.2.7) and contrast with the normally consolidated states that define the Roscoe surface. The figure defines the only stress–pore-volume states in which the sediment can exist and its surface is therefore a state boundary surface for the sediment.

The critical state concept and the yield surface figure provide a framework within which all aspects of the deformation of porous sediments can be described and interrelated effectively. The Roscoe surface links the normal-consolidation line for the sediment (which plots on the zero deviatoric stress plane), where strain is entirely volumetric, with the critical state line, where deformations are entirely through shear. Unloading is not normally accompanied by significant recovery of the volumetric strain (but see section 6.2.4), so unloading paths leave the Roscoe surface to proceed to lower mean effective and deviatoric stresses without expansion of the pore volume. These 'overconsolidated' states also influence reloading. During reloading of an overconsolidated sediment, little change occurs in the pore volume until the previous maximum stress state is attained. At this condition the stress path will have arrived at a point on the Roscoe surface, or will have attained the critical state. These relatively simple concepts of consolidation and shear, and the relationships between them, are quantitatively valid for clays and fine carbonates, and for sands where grain size and angularity do not cause dilation during shear deformation. For coarse, angular grained, clastic sediments, the critical state concept must be modified to allow for the hardening that accompanies dilation during shear. When such materials maintain high porosities, increases in pore volume are minimal, but as the structure becomes more densely packed, the effects of dilation become more pronounced. None the less, the yield surface figure still preserves the geometry shown in Figure 2.27, although the deformation paths for dilatant materials will tend to be more complex. This approach to describing sediment deformation thus provides a unified model for material behaviour from the moment of sediment deposition, through repeated cycles of burial and uplift, until diagenetic cementation renders the sediment a lithified rock.

Experimental deformation of a real sediment

Shear deformation of many real sediments is complicated by the elastic stiffness caused by intergranular bonding at low stresses. This aspect of material behaviour has received attention only in recent years, but studies have shown a consistent style of deformation behaviour irrespective of the sediment type or strength. Such investigations include those of: Vaughan (1985), Vaughan, Maccarini and Mokhtar (1988), Addis (1989), M.E. Jones and Leddra (1989), Leddra (1990), Leddra, Goldsmith and Jones (1990), Leddra and Jones (1990), M.E. Jones, Leddra and Potts (1990), Charlez and Heugas (1991), Steiger and Leung (1991), Leddra, Petley and Jones (1992) and Loe, Leddra and Jones (1992). A family of stress paths recorded from deformation experiments on a weak but bonded sedimentary rock (chalk) is illustrated in Figure 2.28. The curves show a marked change in behaviour with increasing consolidation stress, which is more complex than the ideal critical state behaviour described above. The result is typical of bonded mudrocks, carbonates and sands, and can be explained in terms of the effects of stiffness due to intergranular cementation, and the change of state from

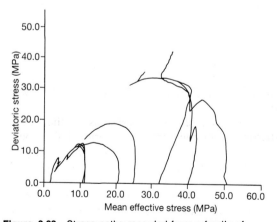

Figure 2.28 Stress paths recorded from a family of undrained triaxial experiments on chalk samples with a similar pre-test porosity, which were consolidated to different mean effective stresses prior to undrained loading. The origin of each curve indicates the consolidation stress.

cemented to particulate material as the rock yields. The following interpretation is offered as an explanation of this variable behaviour, and is believed to be a more realistic model for the deformation behaviour of most natural sediments (Leddra, Petley and Jones 1992) than the idealized critical state model described above. Nevertheless, it is the critical state model that provides the framework for this interpretation.

The stress–strain curves for the chalk experiments that produced the stress paths in Figure 2.28 are shown in Figure 2.29. The curves have been divided into three sets according to the shear-failure behaviour exhibited by the sample. For samples consolidated under low mean effective stresses, the stress paths pass beyond the critical state line, and then return to it before reaching the condition where deformation can proceed to large strains at nominally constant stress. After deformation, these samples always show evidence of brittleness, with fractures containing disrupted coccolith material separating areas of undeformed rock (Figure 2.30a). The paths followed by samples consolidated to intermediate mean effective stresses reach the critical state line, but then show strain softening, following the critical state line towards the origin before the critical state condition of deformation at constant stress is attained. These samples are rather more pervasively deformed than those illustrated in Figure 2.30a, but still contain discrete failure surfaces along which strains became localized (Figure 2.30b). The third group of samples shows strain hardening at the critical state. Deformation is pervasive with few, or no, obvious failure surfaces and severe disruption of the coccolith fabric occurs throughout the sample (Figure 2.30c), which becomes barrel shaped.

The three different responses during loading to shear failure exhibited by these chalk samples are a consequence of the stress conditions pertaining when the bonding responsible for the elasticity exhibited by this rock underwent yield. Under low consolidation stresses, the volumetric strains are small, a consequence of the elastic stiffness imparted by the intergranular bonding. Distortional strain associated with the increasing shear stress as the samples approached failure was also largely resisted by the bonding stiffness of the fabric. The sample thus reaches the critical state in an intact, bonded condition. The critical state defines a failure characteristic of disaggregated

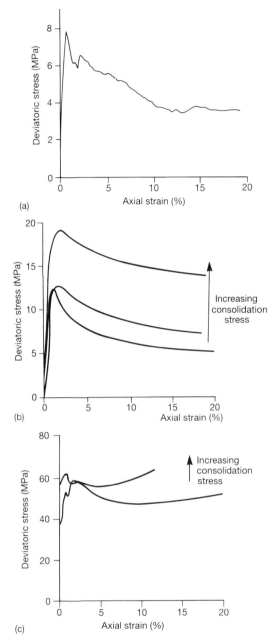

Figure 2.29 Stress–strain curves for the samples of Buster Hill chalk that were deformed in undrained triaxial compression at (a) low consolidation stress, (b) intermediate consolidation stress and (c) high consolidation stress. (After Leddra 1990.)

chalk and is not a property of the intact material. The stress paths are thus able to pass beyond the critical state line, and shear failure occurs on the Mohr-Coulomb envelope for the intact chalk.

Because intact chalk is stronger than disaggregated chalk, the Mohr-Coulomb envelope lies above the critical state line, in the low consolidation stress regime. Loading to the Mohr-Coulomb envelope necessitates the rock storing appreciable strain energy in elastic distortion of the intergranular bonds (refer to Figure 2.5a). At brittle failure, this energy is dissipated on the fracture surface, promoting disaggregation of adjacent coccoliths and sliding. A chalk gouge rapidly develops on the fracture surfaces. This consists of disaggregated chalk, which is unable to support stresses beyond the critical state condition. As this gouge forms it dominates the stress–strain characteristics of the entire sample, and the stress state evolves towards the critical state. Shear strain becomes strongly localized about the shear failure surfaces, and the intact sections of the sample are effectively unloaded. They remain in the bonded state and respond in a passive manner during subsequent deformation. The large strain behaviour that is seen in the stress–strain curves (Figure 2.29a) is entirely due to frictional sliding in the disaggregated material within the failed zones.

The behaviour exhibited by this chalk is a typical response of a rock deforming by sliding on a weak fracture surface. The initial failure at the Mohr-Coulomb envelope can be equated to the peak strength, whereas the steady-state condition that occurs at large strains (the critical state) is equivalent to the residual strength (see section 1.2.4). Thus shear failure under low consolidation stress corresponds to a change in state, with part of the rock material evolving from an intact, bonded material to a particulate frictional material. This change of state is seen in the stress path diagram and the stress–strain curve as the transition from peak- to residual-strength conditions.

The peak strength corresponds to the initial disruption of the cemented fabric of the rock, which has developed as a consequence of diagenesis and compaction during its geological history. On the time-scale of the laboratory experiment, or a natural deformation such as a rock fall or slump, the change is irreversible. Once the rock has attained a residual strength (critical state) condition, this will dominate its shear strength until other (diagenetic?) processes impose further changes in state. For example, diagenetic cementation or mineralization of the fractures would

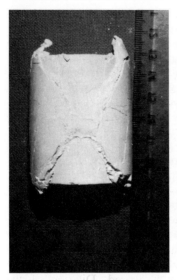 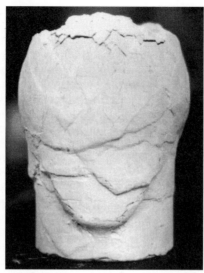 

Figure 2.30 (a) Macroscopic appearance of Buster Hill chalk specimen that was deformed in undrained triaxial deformation experiments at low consolidation pressures (confining pressures). (b) Macroscopic appearance of Buster Hill chalk specimen that was deformed in undrained triaxial deformation experiments at intermediate consolidation pressures (confining pressures). (c) Macroscopic appearance of Buster Hill chalk specimen that was deformed in undrained triaxial deformation experiments at high consolidation pressures (confining pressures). (After Leddra 1990.)

re-establish the intact (peak) strength of the material.

Laboratory experiments conducted on sandstones and cemented mudrocks (Leddra, Petley and Jones 1992; Petley et al. 1992) show that these materials behave in a manner similar to the chalk reported above. In each case, peak strength pertains to the failure of the intact, bonded rock fabric, and is followed by reduction in deviatoric stress as strain becomes localized into conjugate shears deforming at the critical state. Strain localization of this type is important in most natural deformations. At the scale of a landslip, the strength of the entire system becomes dominated by the shear-failure surface at its base. Activation from the intact rock occurs at peak strength but with increasing shear strain, displacements within the slipped mass become controlled by the residual strength of the failure surface. At a larger scale, in deltas and extensional basins, strain becomes localized on syndepositional faults. Once established, the residual shear strength of these fault surfaces will control the magnitude of the deviatoric stresses throughout the basin.

The chalk samples consolidated to intermediate pressures (Figures 2.28, 2.29b and 2.30b) were still in an intact, bonded, state at the onset of shear loading. However, the consolidation stress will have caused a considerable elastic compression of the mineral framework so that shear loading proceeds in the presence of appreciable, stored, strain energy. As the deviatoric stress is increased, the bonded coccolith framework continues to resist distortional strain but the additional strain energy associated with this resistance promotes yield of the framework before the critical state is attained. These chalk samples reach the critical state line in a partially disaggregated condition and are therefore unable to support deviatoric stresses that exceed the critical state. At the critical state, strain is localized initially in the failed regions of the rock and is accompanied by pronounced strain softening. Unlike the samples deformed after consolidation at low stresses, this softening is not caused by a change in the failure condition but because the proportion of disaggregated chalk increases with the shear strain. As more of the material passes from intact to disaggregated state, the deviatoric stress needed to sustain shear failure at the critical state is reduced, and the sample strain softens along the critical state line. These samples show a progressive transition from a peak to residual condition, but in this case the transfer occurs entirely on the critical state line and is far less dramatic than observed in samples consolidated under low stresses.

In many respects the style of shear failure exhibited by these samples, which underwent progressive disaggregation at the critical state, is the most dramatic expression of shear deformation. In the North Sea, chalks developing this type of failure have been reported to flow into oil wells (Leddra and Jones 1990), almost to the total exclusion of hydrocarbons; high-porosity chalks involved in rock falls mobilize a similar style of deformation and become debris flows, which behave as if the chalk had undergone liquefaction. In laboratory experiments, weakly cemented, highly porous mudrocks behave in an equivalent manner, and these are also known to invade oil wells in a manner akin to the North Sea chalk. Geologically, this type of shear failure is probably a less frequent occurrence than the strain-localized failures described above. However, the development of flows and slides, the pervasive restructuring of North Sea chalk during resedimentation (Kennedy 1986), and the progressive mobilization of mudrocks into diapiric structures, almost certainly involve this particular type of strain-softening behaviour.

In sandstones, the equivalent failure state is less easy to observe because in laboratory experiments pervasive disaggregation renders the test material an incohesive powder. If such failures occur in sandstones they may be associated with phenomena such as sand invasion of oil wells and the mobilization of weak sand failures in landslips. The former is known to be a frequent occurrence; the latter has been reported by Brunsden (personal communication) for the westward extension of the Black Ven landslip, Lyme Regis, Dorset.

The third group of chalk samples (Figures 2.28, 2.29c and 2.30c) were consolidated to large stresses. The consolidation stresses exceeded their yield strengths (section 2.2.3), and as a consequence the samples experienced a permanent volumetric strain prior to shear loading. These samples were therefore in a compacted, particulate state (although not necessarily completely disaggregated) prior to the application of the deviatoric stress. Because of the high consolidation stresses, the increase in deviatoric stress necessary to promote failure at the critical state is also large, so that loading to the critical state will be accompanied by further deformations and, at least in the case of chalk, some brittle deformation of the individual mineral grains. At the critical state, shear deformation is accommodated by grain sliding and rotation, and is completely pervasive. The samples are seen to harden as they deform, because the deformation improves grain packing in the disaggregated material, reducing the ability of the grains to slide. This hardening promotes further grain fracturing in chalks (Addis, personal communication). The change to a stiffer arrangement of grains appears to be accompanied by slight dilation of the test sample. The hardening is observed during loading under undrained conditions and necessitates an increase in the mean effective stress. This requires that the increasing load on the sample is preferentially supported by the mineral grains and not by the pore fluid. A proportion of the excess pore fluid pressure generated by the load must therefore be dissipated. Dilation of the sample due to distortional strain provides this mechanism. The increase in effective consolidation stress that the hardening represents is an attempt, on the part of the material, to maintain its consolidation state, while deforming at the critical state.

The response of these chalk samples to increased deviatoric stress, following appreciable compaction due to the high consolidation stress, illustrates the importance of consolidation in promoting sediment strength. Sediments that have come to equilibrium with large effective stresses have low porosities and are therefore stiff materials. Increased consolidation reduces the probability that shear failures will be dramatic events. Conversely, the results presented above can be interpreted as indicating that sediments and sedimentary rocks that support large pore volumes and large pore pressures are those most vulnerable to large-scale shear deformation at low effective stress. Cement bonding can facilitate the preservation of high porosities during consolidation, rendering weakly cemented porous sediments highly sensitive to deformation.

In summary, shear deformation of sediments is normally accompanied by large strains, and in many cases steady-state plasticity (the critical state) dominates behaviour. Existing fabrics cause deviation from the ideal behaviour (compare, for example, the results of Yassir (1989b) with those of Leddra, Petley and Jones (1992) in Figure 2.31), whereas intergranular cementation

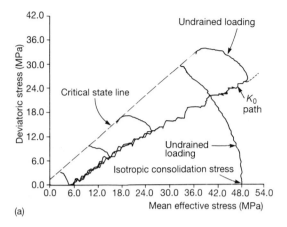

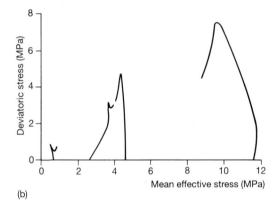

Figure 2.31 The stress paths associated with undrained shear deformation of (a) a remoulded clay (after Yassir 1989a) and (b) a naturally, compacted, weakly cemented clay. The origin of each curve indicate the consolidation stress (after Leddra, Petley and Jones 1992).

has a more pronounced effect, and may cause catastrophic failures under certain conditions (Leddra and Jones 1990; Figure 2.28). Thus lithification progressively complicates the shear behaviour of sediments and sedimentary rocks and will exert a major influence on the specific expressions of shear strain, as for example along a fault passing through rocks with different fabrics and bonding strengths. Generally, shear deformations are facilitated by large pore-fluid pressures and large pore volumes, although grain size, shape and strength also exert a profound influence. Shear deformations tend to produce pervasive fabrics in materials that are both weak and uncemented but exhibit strain localization and brittleness in intact, bonded materials.

2.2.6 Other stress paths and associated strain states

The strain states associated with consolidation under isotropic stress and shear deformations at the critical state represent two extremes. The former are entirely volumetric and occur in the absence of shear stress whereas the latter are purely distortional. The stress paths associated with the development of these different strain states reflect the differences. Isotropic consolidation is accompanied by increasing isotropic effective stress whereas the critical state represents deformation in the presence of the maximum shear stresses the material can support for its consolidation state. Between these extremes lies a continuous spectrum of possible stress–deformation paths, any of which can be followed by sediments deforming in nature. It is inappropriate here to examine the characteristics of all the deformations that accompany loading along these different paths. The reader is referred to Atkinson and Bransby (1978) for a rigorous treatment with regard to sediments, and to Loe, Leddra and Jones (1992) for a detailed investigation of the relationship between stress path and strain states for a weak, porous rock (chalk). At the time of writing, equivalent data for other weak sedimentary rocks are limited.

In the remainder of this section discussion will be restricted to the one geologically important stress path not considered above. This is the uniaxial strain or K_0 **condition**. It is the path characterized by one-dimensional consolidation with zero lateral deformation of the material. The path lies on the Roscoe surface between the isotropic consolidation and the critical state lines and represents the condition where consolidation proceeds in conjunction with sufficient distortional strain to continuously recover any lateral compaction (Figure 2.32). Thus a state of zero lateral strain is maintained.

The K_0 stress path is of particular importance because it is widely regarded as the path followed by most sediments undergoing burial in the presence of uniform gravitational loading. Under such circumstances, the sediment is constrained to compact in a one-dimensional manner because all of the adjacent sediment is compacting in an equivalent manner at an equivalent rate. Lateral strains are thus unable to develop and the stress state is forced to follow the K_0 condition. Near to

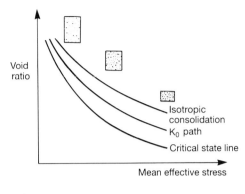

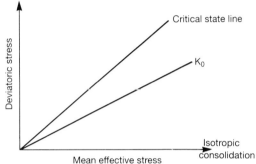

Figure 2.32 (a) The uniaxial strain or K_0 stress path in relation to the isotropic consolidation and critical-state lines. The inset sketches illustrate how bulk strain would accumulate in a laboratory specimen deforming on the K_0 stress path. (b) The relationship between the K_0 path and the critical-state line in the coordinates of the stress-path diagram.

the earth's surface, in all sedimentary basins where there is not appreciable surface relief, the sediments are likely to be under stress conditions close to K_0. With increasing depth the presence of non-gravitational forces associated with basin extension or contraction will progressively move the sediment away from this stress path.

The quantity K_0 is often referred to as the **coefficient of earth pressure at rest** (Brooker and Ireland 1965). It has important practical applications in soil mechanics in the design of excavations and retaining walls (Figure 2.33). K_0 is defined as the ratio between the horizontal (lateral) and vertical effective stresses (where such stresses are principal stresses, as is normally the case close to the earth's surface) which maintain the condition of no lateral deformation (equation 2.18). The ratio is normally determined in the laboratory in a one-dimensional compression experiment on the material of interest. It is found to be a constant for most uncemented, normally consolidated sediments.

$$K_0 = \frac{\sigma'_3}{\sigma'_1}. \qquad (2.18)$$

The magnitude of the K_0 ratio depends upon the type of sediment. Generally, the finer grained and more clay rich the material the larger the K_0 value. In sands, the ratio commonly has values around 0.3–0.4, whereas in clays it may be 0.7 or greater. Silts lie between these extremes. This variation with grain size, and to some extent with grain shape, leads to an obvious association between K_0 and the friction angle for the sediment. Typically, for uncemented, normally consolidated sediments the K_0 ratio is found to vary according to:

$$K_0 = 1 - \sin \phi', \qquad (2.19)$$

where ϕ' is the effective friction angle.

Normally, K_0 is greater in overconsolidated sediments (see below), and in clays K_0 ratios may exceed 1.0, depending upon the degree of overconsolidation (Figure 2.34). This means that lateral stresses may exceed vertical stresses in sediments that have been subject to unloading during uplift, for example as a consequence of erosion. This occurs because although the vertical stress decays as the overburden is removed, the lateral stress is to a degree 'locked' into the grain framework of the sediment, so that although it does decrease as the vertical stress decreases, it does so less efficiently. This effect is most pronounced in clays and shales, where unloading and associated overconsolidation may cause the K_0 ratio to exceed 4.0, giving rise to large incremental changes in lateral stress for small changes in vertical stress. Such effects are even more pronounced when heavily overconsolidated clays are subject to reloading.

The fact that the K_0 ratio is dependent upon the lithology of the sediment, its grain fabric, and previous stress history, has important consequences for *in situ* stress states in uncemented sediments. Under gravitational loading, an alternating sequence of sands, silts and clays will develop a vertical stress that is a function of their density and the burial depth. Vertical stress therefore increases in a fairly regular manner with increasing depth, usually shown as a straight line (Figure 2.35). By way of contrast, the horizontal stress varies from bed to bed as the K_0

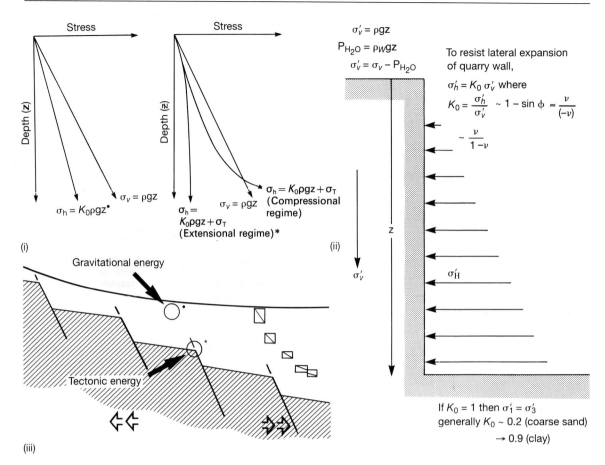

Figure 2.33 Schematic representation of the quantity K_0 (i) with respect to a deep excavation (ii) and the burial of an element of sediment in a subsiding and filling sedimentary basin (iii). (ϕ = friction angle; ν = Poisson's ratio)

ratio changes, so that the horizontal stress is strongly dependent upon the lithology. In addition, if effective stresses are considered, the presence of overpressured horizons will give rise to further irregularities (Figure 1.4). Thus geological boundaries may also be important stress discontinuities (Figure 2.35). These differences in lateral stress are of little consequence while the sedimentary succession is subject to one-dimensional gravitational loading. With the development of a uniform horizontal component of stress due to the onset of tectonism, however, these same variations in lateral stress will play an important role in the sequential development of shear failure, and in the initiation of deformation fabrics, folding and faulting.

The deformation of sedimentary rocks under K_0 conditions has been studied less than the deformation of completely unlithified sediments. The reasons for this are twofold. Firstly, the technology required to conduct reliable, routine, K_0 experiments on stronger sedimentary rocks has been available only in recent years. Secondly, the number of occasions where the compaction behaviour of porous lithified rocks has been the subject of engineering investigation has been few. The exception to this has been in the field of hydrocarbon reservoir engineering. Reservoir engineers have for a long time concerned themselves with the isotropic compression of porous sedimentary rocks, and in recent years this work has been widened to consider compression of these materials under conditions of uniaxial strain.

A further difficulty with investigating porous rocks is that the K_0 ratio changes with the state of strain. Under low stresses, when the majority of the intergranular bonds in the material is

Mechanics of particulate media

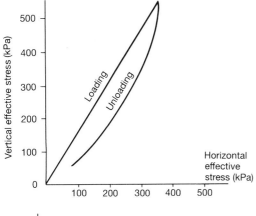

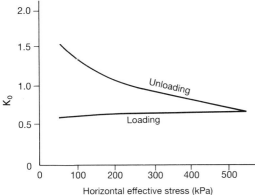

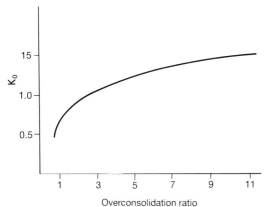

Figure 2.34 The relationship between K_0 and overconsolidation ratio (ratio between present effective stress and past maximum effective stress). (Based on Atkinson and Bransby 1978.)

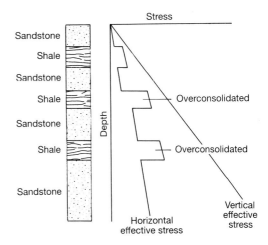

Figure 2.35 *In situ* stress states in a sequence of sediments of varying lithology subject to gravitational loading under K_0 conditions and assuming normal pore fluid pressures.

intact, the rock is stiff and able to resist distortional deformation. Under these conditions K_0 is low and the vertical stress can increase appreciably for small changes in lateral stress. However, as the elastic strain increases, the proportion of intact intergranular bonds decreases, and the sedimentary rock becomes appreciably less stiff. Lateral deformation can now be prevented only if the lateral stress increases in proportion to the vertical effective stress until the yielding process is complete. After this, the population of free grains is sufficient to allow the rock to consolidate. Further deformation occurs on the Roscoe surface and is accompanied by large compactional strains. During this process, the K_0 ratio is typically about 0.3 or less in the elastic regime, 1.0 at yield, attaining a value similar to the equivalent uncemented sediment in the post-yield regime (Figure 2.36a–c).

The behaviour represented by this path is explained in Figure 2.37 (a and b). Here it is seen that the initial behaviour corresponds to an elastic loading path that reaches the Mohr-Coulomb failure surface. The imposed condition of zero lateral strain prevents shear failure and appreciable destruction of the intergranular cement bonds occurs. This weakens the sediment but not to the extent that pore-volume compaction can begin (Figure 2.37a). This condition of progressive failure with small volumetric strains is maintained by the imposed condition of zero lateral deformation, and the path moves through the space contained by the yield surface. Destruction of intergranular bonding continues until sufficient grains are freed to allow a more catastrophic loss of pore volume. This increase in volumetric strain corresponds to the onset of deformation on the Roscoe surface. The rock

64 *Mechanical principles of sediment deformation*

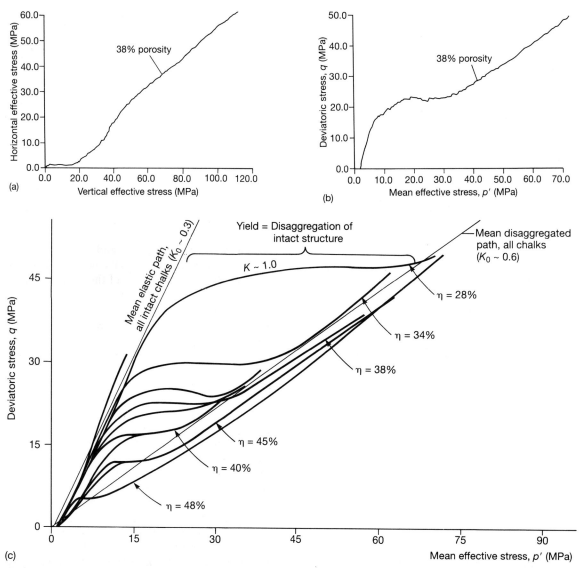

Figure 2.36 (a) A plot of greatest against least principal effective stress for a K_0 experiment conducted on a high-porosity chalk sample. (b) The same data set recorded as a stress path in the coordinates p' and q. Note that the plots in both (a) and (b) show the change in slope of the stress path that accompanies the change in K_0 as the sample passes through yield and its deformation resistance changes. (c) Stress-path plots for a family of chalk samples with different initial porosities (η) due to variations in the amount of pore infilling cement. Note that the pre-yield and post-yield stress paths are common to all samples, but that the stress path during yield is dependent upon the porosity. Samples with the lowest porosities yield at the highest stresses and require the greatest range of yield stress to transfer to the post-yield regime.

has now become sufficiently disaggregated to allow it to once again behave as if it were the original particulate sediment. For a more detailed discussion of this behaviour the reader is referred to M.E. Jones and Leddra (1989), Goldsmith (1989), M.E. Jones, Leddra and Potts (1990), M.E. Jones *et al.* (1991) and Loe, Leddra and Jones (1992). Natural K_0 stress paths are discussed in further detail in section 6.2.

2.2.7 Over- and underconsolidation

The term overconsolidation, as mentioned briefly in the previous section, refers to the state that a

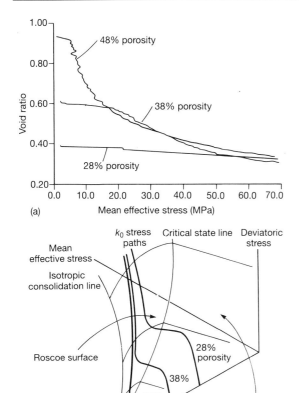

Figure 2.37 (a) The change in void ratio that accompanies K_0 deformation of a porous sediment (chalk). (b) The data in Figures 2.36c and 2.37a replotted in the coordinates of p', q and void ratio (after Leddra 1990). This diagram shows how the elastic path intercepts the Mohr-Coulomb surface (equivalent to the Hvorslev surface of fully particulate sediments) of the yield figure and then moves into the volume defined by the surface, passing beneath the critical-state line before emerging on the Roscoe surface. It is important to note that it is not until the path reaches the Roscoe surface that appreciable volumetric strain occurs.

sediment attains when it is unloaded from some previously greater effective stress. All sediments that have been buried and then returned to the surface are therefore overconsolidated. Underconsolidation is a term used in geology that has no equivalence in engineering. It is, however, a most useful geological concept and refers to the specific situation where a sediment has preserved high pore pressures during burial and has therefore not attained the equilibrium effective stress for its burial depth. Underconsolidated sediments thus behave as if they are buried to a shallower depth than that prevailing. This is a direct consequence of the preservation of pore fluid pressure and pore volume during burial. Such sediments are overpressured, undercompacted, and support large pore volumes. They are therefore amongst the weakest of geological materials (see Yassir 1989b), and easily attain shear failure at their critical state because they have experienced only limited consolidation (Figure 2.25a).

In contrast to underconsolidated sediments, overconsolidated materials tend to be brittle. They have pore volumes and mineral frameworks that have been compacted to a greater effective stress than they now sustain and this imparts a stiffness. The brittle behaviour of overconsolidated clays deforming in shear has been reported by Skempton (1966) and Skempton and Petley (1967). These results have been used subsequently in an interpretation of the natural deformation fabric of the Joes River Formation scaly clay from Barbados (Enriquez-Reyes and Jones 1991). Further laboratory studies on the relationship between previous-consolidation fabric and shear behaviour of clays have been reported by Leddra, Petley and Jones (1992).

Overconsolidation results from a reduction in the effective stress experienced by a sediment, and this, in addition to imparting a brittleness, causes the increase in the K_0 ratio discussed in the previous section. It is important to note that whereas reduction in effective stress may be due to uplift and erosion, and therefore removal of the total stress, it can also be due to the introduction of pore fluid at elevated pressures. This means that where overpressures in sediments result from the secondary generation of excess pore fluid pressure (section 2.3.4), the sediment will become overconsolidated. Overpressured sediments can therefore be either underconsolidated, in which case they will behave in an extremely weak and ductile manner, or overconsolidated, in which case they will be brittle. Secondary overpressures are probably far more common in sedimentary basins than has been generally recognized, and the widespread occurrence of deformation textures associated with overconsolidation in mudrocks may be an indication of how frequently shear deformation of overconsolidated, overpressured sediments occurs at depth.

In many respects the mechanical characteristics of overconsolidated sediments are indistinguishable from those of weakly cemented sedimentary rocks. However, overconsolidation is sedimentary rocks. However, overconsolidation is simply a mechanical process – a consequence of stress history – whereas cementation (or other causes of intergranular bonding) can result from any one of a spectrum of diagenetic processes. Overconsolidation is always a consequence of reduction in effective stress, although normally or underconsolidated sediments can become cemented and behave, at least until yield, as if overconsolidated. Preservation of pore space due to bonding in underconsolidated materials can produce spectacular deformation, with dramatic failure if the yield strength of the material is exceeded. Such materials are especially sensitive to energetic small strain events, such as the passage of seismic waves. Failure of the porous, weakly cemented structure produces high transient pore pressures and reduces the effective stress to a low value. The disaggregated rock develops large shear displacements in an extremely short period, behaving almost as if it had undergone liquefaction. Examples of such deformations are: the slope failures associated with highly sensitive, quick clays (Lambe and Whitman 1979); some slope failures triggered by earthquakes (Lambe and Whitman 1979); the Aberfan coal tip flow in Wales (reviewed in Bromhead 1992); and the mobilization of very high-porosity chalk to flow, both in the toe regions of rock falls and during transient loading during fluid production in some North Sea oil wells (Leddra and Jones 1990). In the last type of occurrence, the chalk can flow into, and fill, a large part of the well volume before the shear deformation event stabilizes (Leddra and Jones 1990). These are all expressions of extremely dynamic shear failure provoked by the extreme sensitivity of the material to small strain deformation. It is probably not correct to regard this behaviour as liquefaction (section 1.3.2), although phenomenologically it is extremely similar.

No reliable criteria are available for distinguishing overconsolidated from bonded materials, indeed it may be the case that no undisturbed natural materials are ever devoid of intergranular bonding (see Burland (1990) and the discussion in Leddra, Petley and Jones (1992)). In the absence of microscopic identification of bonding it is possible that differences in the K_0 behaviour may provide evidence (Figure 2.38). Bonded materials tend to have a low K_0 before yield whereas overconsolidated materials exhibit a high K_0 before they reach their pre-

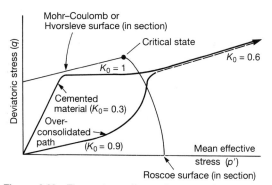

Figure 2.38 The various effects of cementation and overconsolidation on the K_0 stress path for a compacting sediment (see text for full explanation). This diagram shows the two paths projected on to a constant void ratio section of Figure 2.37b.

consolidation pressure (Addis 1987). The pre-consolidation pressure is the stress at which an overconsolidated material upon reloading returns to a normally consolidated condition. It is equivalent to the previous maximum effective stress that the material had experienced. The change from overconsolidated to normally consolidated material causes a change of slope in the stress–strain curve at the pre-consolidation pressure. This is geometrically equivalent to the yield stress of a bonded material.

The interplay between consolidation fabric, bonding, stress history and the timing and extent of overpressure development in sedimentary materials is an area in which much important and fruitful research waits to be done. The deformation fabrics we observe in sedimentary rocks depend fundamentally on the interaction of fabric and effective stress history. At present, fabric interpretation is hampered by a lack of quantitative laboratory measurements. Such experiments, and appropriate numerical modelling, are now possible, and the integration of field and laboratory data should provide for important advances in our understanding of natural sediment and sedimentary rock deformation.

2.3 NATURAL STRESS, STRAIN AND PORE PRESSURE

2.3.1 General

The styles of deformation being introduced in this chapter are those associated with volume change (compaction) and shape change (shear). Compaction typically occurs as a porous sediment is buried, though it may be resisted during natural loading by cementation and pore-pressure generation (Skempton 1970; Been and Sills 1981; Vaughan 1985; Addis 1987; Yassir 1989a). Shear deformations can be a consequence of tectonism, where they give rise to faulting, folding and fabric development in the material. However, gravitational energy is responsible for most of the shear deformation experienced by near-surface, unlithified sediments.

Gravitationally induced shear deformations include the following.

1. Mass movements in materials unable to sustain the modest shear stresses generated by surface relief (Brunsden and Prior 1984; Bromhead 1986; Hutchinson 1988; Chapter 5).
2. Where compactional displacements in the subsurface become localized on to pre-existing planes of weakness as they are transmitted through the sediment mass. This has occurred during compaction of oil fields resulting from fluid extraction (Hamilton and Mechan 1971; Castle, Yerkes and Young 1973; M.E. Jones, Leddra and Addis 1987) and, on occasions, has given rise to seismic events.
3. When changes in surface relief due to sediment deposition, uplift or erosion alter the equilibrium between gravitational stress and strain state within mountains, accretionary prisms and deltas (M.E. Jones 1991).
4. During one-dimensional compaction of sediments, because the K_0 ratio is rarely 1.0 and the equilibrium state of stress must therefore involve a deviatoric component (Brooker and Ireland 1965; M.E. Jones and Addis 1986).

At greater depth, gravity becomes relegated to providing the passive component of the stress system, inducing compaction but with major shear deformations being generally associated with the mobilization of tectonic stress (M.E. Jones and Addis 1986; Chapter 6).

An understanding of the stress–strain characteristics of sediments compacting under passive gravitational loading is aided by examination of what happens when humans extract fluid from the subsurface and cause depletion of the natural pore fluid pressure. Fluid extraction from hydrocarbon and water reservoirs, which causes a reduction in pore fluid pressure, will give rise to changes in the effective stresses acting upon the reservoir material and may mobilize compactional deformations. The most spectacular, recent example of human-induced compaction has occurred in the Ekofisk Reservoir in the North Sea, where the associated sea-floor subsidence has proved to be a major engineering problem (Aam 1988; Potts, Jones and Berget 1988). The pattern of displacements that have been mobilized as a consequence of changes in pore fluid pressure in the Ekofisk Reservoir is illustrated in Figure 2.39. This deformation is entirely a consequence of the passive loading of the reservoir by its overburden, due to dissipation of the excess pore fluid pressure in the chalk reservoir. It is one-dimensional, and occurring at a strain rate of about 10^{-12} to 10^{-13} s^{-1}. Transfer of the compactional displacements from the reservoir to the surface mobilizes shear and stretching deformation of the overlying materials (M.E. Jones, Leddra and Potts 1990).

2.3.2 Earth pressure and one-dimensional compaction

Compaction under gravitational loading generally occurs under conditions of one-dimensional consolidation – the K_0 condition described above. Under this condition the vertical stress increases regularly with depth, as does the pore fluid pressure, although if permeability is low, excess pore pressures may also develop. During burial, the strain condition will be dominantly one-dimensional, with flattening in the vertical direction and lateral expansion restricted or prevented by the earth pressure ($= K_0 \cdot \sigma_1'$). The major fabrics that result depend strongly on the rock type. In sands, silts and carbonates with fairly equant grains, compaction is simply accompanied by pore-volume reduction, whereas in clays and other sediments composed of platy minerals a pronounced planar, bedding-parallel fabric may develop (Petley *et al.* 1992; section 9.3). The mechanism by which this strain is accommodated is expulsion of the compressed pore fluid. This is accompanied in the pre-yield regime by elastic distortion of any intergranular bonds, at yield by breaking of the bonds (or bonded grains if they are weaker), and in the post-yield regime by grain-boundary sliding, including grain rotation.

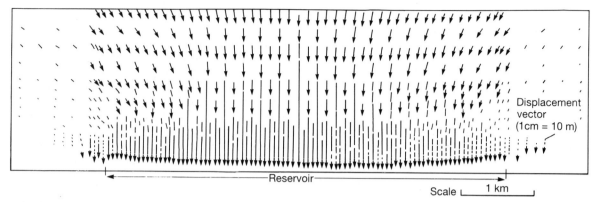

Figure 2.39 Finite-element prediction of the pattern of displacements occurring over the compacting Ekofisk oil field in the North Sea.

Compaction of this type must affect all clastic sediments subject to burial and should account for a large proportion of the pore volume lost during diagenesis. Laboratory experiments have demonstrated that sands, fine carbonates and clays will continue to compact due to pore-space compression even when the vertical stress exceeds 70 MPa. Such deformations are potentially rapid, providing excess pore pressures are able to dissipate. Once excess pore pressures are substantially dissipated, or if dissipation is prevented by low permeabilities, slower deformation mechanisms, such as diffusive mass transfer (Rutter 1983; Gratier 1987), will tend to drive continuing compaction. Given sufficient time, these mechanisms may overprint the fabrics of primary consolidation and establish or re-establish intergranular bonding. Further consolidation will then be prevented until burial stresses have increased sufficiently to promote yield of the bonded material. These concepts are pursued further in section 6.2.

2.3.3 Natural shear stress and deformation

Viewed simplistically, shear deformation can be regarded as a consequence of either reduction or increase in the lateral earth pressure with respect to the equilibrium K_0 condition. Reduction in earth pressure will cause extension in basins and mass movements on slopes whereas increase in earth pressure will result in lateral compression of a basin. In active extensional basins, deltas, accretionary prisms and during slope failure, the major expressions of shear deformation are the fault systems which develop.

The possible importance of diffuse strains in accretionary prism toes is assessed in section 6.3.4. Movement on faults controls the state of shear stress and the geometry of the sediment body, causing surface subsidence due to lateral extension and reduction in slope angles. Shear deformation serves to maintain equilibrium between the imposed forces, the strength of the sediments and their geometry. Active faults will deform at the critical state, protecting the intervening host materials from shear failure. Under these circumstances, strain within the sediment body will become strongly polarized, with gravitational loading causing pervasive consolidation leading to a reduction in pore volume and/or the generation of excess pore pressures, while shear deformation remains localized on the fault planes. This is akin to the separation of deformation paths observed in the chalk deformation experiments described in section 2.2.5.

Although tectonic compression generally involves deformation of stronger geological materials (fully lithified sediments and crystalline rocks), in some situations it is also an important cause of sediment deformation. The strain in weak porous sediments in the deeper parts of accretionary complexes at convergent plate margins will be laterally compressive (section 6.3). Similarly, during continent–continent and arc–continent collision, accretionary material, late pre-orogenic and synorogenic sediments will become deformed. Current orogenic events, such as the arc–continent collision occurring in Taiwan (Brunsden and Lin 1991; Lin 1991), are rapid, and high pore pressures, generated because of shear deformation (Yassir 1989b), lateral consolidation and increased surface elevation, will not dissipate readily. Sediments associated with plate convergence and orogenesis will tend to be overpressured and either preserve large pore volumes and behave in an underconsolidated manner, or, if they were previously consolidated, will become overconsolidated owing to secondary pore-pressure generation. Materials in either of these states are prone to shear deformation under low deviatoric stresses, although the strain response will depend on both lithology and previous history. Under these circumstances shear deformation at the critical state will occur. This may be localized along thrust faults or, if more pervasive, may facilitate the development of shear zones, diapirs and some expressions of folding.

It is important to note that although deformation of sediments during collision orogenesis appears to take place in the presence of low effective stresses, it is not the case that the total stresses need be small. This is important because although high pore fluid pressures may protect the material from mechanical compaction due to consolidation, and maintain a low shear strength, it is not protected from pressure-dependent phase changes. Most pressure-dependent phase changes during diagenesis and low-grade metamorphism are driven simply by increase in the total lithostatic pressure and are independent of effective stress. A sediment subject to large total stresses and pore pressures may therefore acquire the mineralogical signature of high-pressure diagenesis or incipient metamorphism while retaining the ability to deform readily in the presence of low shear stresses.

Shale and mud diapirism is a common feature of both deltas and accretionary prisms (O'Brian 1968; Chapman 1974; Dailly 1976; Mascle, Bonhold and Renard 1976; Barber, Tjokrosapoetra and Charlton 1986; sections 4.3.2 and 7.1). It is known, for example, in the Mississippi Delta (Figure 4.10), in accretionary complexes such as the Oyo Complex, Nias, Indonesia and the Joes River Formation, Barbados; and in the active orogenesis of the western foothills region of the Central Mountains of Taiwan. Shales of this type may have moved upwards for several kilometres in the crust, and when these diapirs pierce the earth's surface they often produce mud volcanoes, which appear to dissipate excess water and gas pressures from their interiors (Yassir 1989a). Shale diapirism is one of the major natural expressions of pervasive shear deformation in sediments. Diapiric shales generally have a texture consisting of an anastomosing array of polished, curved surfaces with strong mineral grain alignments, between which the fabric is far less ordered. The origins of this scaly fabric (section 9.4.3) have been the subject of considerable attention, although most authors agree that shear deformation is the major factor (J.C. Moore et al. 1986). Similar fabrics can be produced during experimental shear deformation of both overconsolidated (Skempton 1966; Skempton and Petley 1967) and weakly cemented shales (Leddra, Petley and Jones 1992; Petley et al. 1992). The fabric also occurs in non-diapiric shales that have been subject to shear deformation, such as in the Carboniferous of the southern North Sea (Petley et al. 1992) and non-diapiric mélange deposits in Sarawak (Clenell 1992). The presence of a scaly fabric in an argillaceous rock may simply be an indicator that the sediment has experienced shear strain. If this is the case, it is not surprising that the fabric is normally pronounced in diapiric shales.

Shale and mud diapirism is of particular interest because within the diapir pervasive shear deformation must have occurred under steady-state or strain-softening conditions at the critical state. From the experimental data and discussion given in section 2.2.5, it would appear that such pervasive deformation requires very specific conditions, with failure of the intact, bonded fabric of the shale occurring close to the critical state, but beyond the intersection of the Mohr-Coulomb and critical state failure lines. (If failure had occurred at lower consolidation stresses, shear displacements would have become localized on to fault surfaces, whereas in the presence of larger consolidation stresses, shear failure would have been accompanied by hardening.) The materials most prone to diapiric activity will be overpressured shales in which bonding has preserved large pore volumes. Such materials are stiff while the bonding is intact, and will regain stiffness after shear deformation if the shear stress falls below the critical state line. This fits with field observations from Trinidad, where oil wells drilled through diapiric shales have encountered materials that are stiff and deform into uncased wells in a blocky manner, showing a high density of discrete shears (Yassir and others, personal communication 1986). This fabric is clearly seen in core samples of the diapiric shale and is similar to the deformation of the chalk sample illustrated in Figure 2.30b.

Shale diapirism, therefore, is not an indicator that the sediment has behaved in a fluid manner. This is an important lesson – ductility (plasticity) is not equivalent to fluidity. Sediments behave in a ductile manner when deforming in shear at the critical state, but this behaviour pertains to the prevailing stress conditions. Thus shear deformation to large strains can occur in any shale if appropriate stress conditions are attained.

2.3.4 Origin of overpressures in sediments

Pore fluid pressures, particularly fluid pressures in excess of the equilibrium hydrostatic pressure, exert a fundamental influence on the mechanical behaviour of sediments, and sedimentary rocks. The timing of overpressuring within the deformation history of the sediment is also important in so far that consolidation state, and the extent to which a sediment is under- or overconsolidated, strongly influences subsequent deformation. Its origin in sediments and sedimentary rocks must therefore be considered in any discussion of natural stress and deformation states in sediments.

Overpressures are an almost ubiquitous feature of sedimentary basins, deltas, and accretionary complexes, and are often sustained for very considerable periods. Their origin is often attributed to the combined effects of low permeability and rapid burial. Overpressures of this type are well known from deltaic sequences (section 2.3.3), but are generally restricted to mudrocks, or other lithologies that have become sealed by low-permeability mudrocks and then rapidly buried (Magara 1971; Chapman 1972, 1981; Carstens and Dypvik 1981). However, this is not the only cause of overpressure development; other mechanisms are also important. This is particularly the case in sediment sequences where overpressures have been maintained for appreciable periods (Bredehoeft and Hanshaw 1968; Hanshaw and Bredehoeft 1968).

Few sediments are actually impermeable, so the persistence of overpressures generated by sediment burial over appreciable periods of time seems unlikely. Measured coefficients of consolidation from highly overpressured Eocene shales recovered from depth in North Sea oil wells suggest that the overpressures would dissipate in 1–5 million years, even through 3 km of overburden (unpublished data). These shales also have compacted fabrics and a pronounced bedding-parallel fabric. This suggests that the overpressures either owe their genesis to an event which occurred later in the history of the sediment, following significant compaction, or that the mechanism responsible for the overpressures has been active over a considerable period of time. Watts (1983) argued that overpressures in the Cretaceous and Tertiary sequence in the North Sea central graben were sustained by the high capillary pressures required for oil to migrate into the chalk and for natural gas to migrate into the overlying shales. In Taiwan, Miocene deltaic sediments, folded along the western margin of the currently active tectonic belt contain hydrocarbons with extreme overpressures. These sediments have not been subject to rapid, deep or sustained burial but have been tectonized and now lie at the margin of the 3 km-high mountain chain. The large excess pore pressures are in part due to the nature of the pore fluid (a hydrocarbon gas), but the recent tectonic deformation, thermal history and geomorphological setting will also be contributing to their development (Chan 1964).

Yassir (1989a) has reviewed a number of possible causes for overpressure generation which are not related to the trapping of fluids during burial. These are as follows: tectonic deformation of the sediment; artesian conditions; dehydration of hydrous minerals during burial and invasion by fluids released during dehydration of more deeply buried rocks (Bruce 1973, 1984; Magara 1975a; Pittman and Reynolds, 1989); geothermal heating (Barker 1972; Magara 1975b; Barker and Horsfield 1982; Daines 1982); capillary pressures due to the migration and gravitational segregation of multiphase fluids, generally gas–oil–brine systems (Fertl 1973; Hedberg 1974; Archer and Wall 1986); and osmotic effects (Hanshaw and Zen 1965; Young and Low 1965).

The origins of overpressures in sediments and sedimentary rocks are thus diverse and may occur at different stages during the history of the sediment. The presence of overpressures in a naturally occurring sediment are thus not necessarily evidence for rapid burial, low permeability and underconsolidation. Many, maybe the majority, of overpressured sediments owe the presence of the excess pressure to a mechanism of secondary pore-pressure generation. In such sediments, the state of consolidation will depend upon the magnitude and timing of secondary pore pressure generation with respect to the development of intergranular bonds.

2.4 CONCLUSIONS

This chapter has been concerned primarily with the mechanics of how sediments deform, with stress systems, the role of pore-fluid pressures and the influence that deformation may have on the properties of the material. Although the principles have been presented with reference to theory and laboratory tests, the final part of the chapter has given some glimpses of the relevance of the ideas to the explanation of natural sediment deformation phenomena. These latter aspects are explored in the following chapters. Much remains to be learnt about the mechanical behaviour of large sediment bodies before the generalizations presented above can be quantified. The careful and informed application of modern geotechnical theories and methods applied to such problems will provide a basis through which this quantification can be achieved. The principles, concepts and ideas presented in this chapter lead to the following broad conclusions.

Sediments are amongst the most deformable of geological materials, although this depends crucially on their previous deformation history, and the pore fluid pressure. They tend to deform by a combination of volume change, due to expulsion of pore fluids in response to an increasing isotropic component of the applied stress system, and shear due to imbalance in the applied stresses. Compaction is generally pervasive, normally driven by the gravitational components of stress (the passive stress system) and often constrained to be one-dimensional owing to the presence of adjacent compacting rocks and the generation of earth pressure. Shear deformation is favoured by increasing deviatoric stress, and at shear failure at the critical state will cause large strain deformations at constant pore volume and constant deviatoric stress. Shear deformation is also favoured by large excess pore fluid pressures, which prevent consolidation. Both shear and consolidation are strongly influenced by the lithology and fabric of the sediment, and by its previous stress and pore-pressure history.

Sedimentary rocks are the result of lithification of sediments, and exhibit significant elasticity at low stresses and small strains. With increasing stress this elasticity is progressively destroyed and the rock undergoes a change of state, its behaviour reverting to that of the equivalent particulate sediment. The behaviour of the rock at and beyond yield depends on the stress path and on previous consolidation history. At low stresses, or in the presence of high pore pressures, yield at the critical state is accompanied by pronounced bifurcation of the deformation path. Segments of the rock undergo disaggregation and become ductile shear zones, or faults deforming in an ideal critical state manner, while the shear stresses decay in the remainder of the rock which then behaves in a passive manner. Similar bifurcation of the deformation path appears in landslips and large sediment bodies such as deltas or extensional basins. In the presence of larger consolidation stresses, shear deformation is more pervasive and leads to a form of ideal plasticity, although dilation may also cause work hardening.

Deformation of sediments and sedimentary rocks is thus entirely dependent upon the nature of the effective stress systems affecting the sediment, and therefore on the pore-pressure history. Large pore pressures preserve the pore volume but facilitate shear deformation in sediments. If the pore pressures are primary, the sediment will be underconsolidated and respond in a plastic manner, whereas if they are secondary, a stiffer, overconsolidated, type of shear response will occur. For all these reasons, the role of fluids and pore pressures in deforming sediments is mentioned frequently in the chapters that follow.

CHAPTER 3

Glacial deformation

TAVI MURRAY

3.1 INTRODUCTION

The world is an icy place. Some 10% of the earth's land surface is currently covered by glacier ice; at glacial maxima this coverage has been much greater. Moving glacier ice is a powerful geomorphological agent and many parts of the world have been distinctively altered by the action of glacier ice. It is at the base of the glacier or ice sheet, where ice and substrate are in intimate contact, that geomorphological activity and change is greatest: it is here that most glacial sediment deformation occurs. Modern research has shown that the theoretical principles outlined in the previous chapters are relevant to both the movement of the ice sheets and to the understanding of the structures that are produced. Thus glacial movement provides a fine example of how the mechanical principles of sediment deformation are applicable in a particular geological environment. This is the subject of the present chapter.

Early glaciological observation was concentrated in the alpine regions, where basal conditions of ice overriding hard, lithified material were inferred from the accessible margins of mainly small, mountain glaciers (Clarke 1987a). Deglaciated rock with striated and polished surfaces provided further evidence of the impressive erosive power of glacier ice in intimate contact with bedrock. Detailed theories for glacier motion by internal ice deformation and sliding over a rigid bedrock surface were developed. Thus, until recently, the glaciologists' view of this basal region was of a clean interface between ice and rough, lithified bedrock. However, evidence has mounted that this picture is unlikely to be accurate for a significant proportion of current ice masses and, further, that many palaeo-ice masses overlay unlithified sedimentary material rather than bedrock. Because access to the subglacial environment is difficult, much of this evidence comes from the now exposed beds of past glaciations. Retreat of ice following the 'Little Ice Age' maximum is continuously exposing sediments that were covered by ice until very recently. Together with the sedimentary cover of the mid-latitude regions that lay beneath Quaternary ice, these materials provide evidence of the basal conditions within the sediments beneath these ice masses. If ice were to advance again in Europe and North America to occupy positions of previous maxima, it would do so over an extensive sediment cover. Sediments are estimated to have underlain some 70–80% of the mid-latitude ice sheets outside of their central regions (Boulton and Hindmarsh 1987).

It is not known what proportion of modern glaciers overlies unlithified sediments; the basal environment lies hidden beneath tens to hundreds of metres of glacier ice. However, evidence for the widespread presence of sediments beneath modern ice masses has accumulated in recent years. The first indications came from direct observation of active sediment deformation beneath the margin of a modern glacier in Iceland (Boulton and Jones 1979). These observations were followed by the collection of high-resolution seismic results from Ice Stream B, West Antarctica, which revealed what has been interpreted as a metres-thick layer of saturated sediment beneath this active ice stream. Subsequently, unprecedented interest has arisen in sedimentary glacier beds and in the mechanics of the deformation processes within them.

The Geological Deformation of Sediments Edited by Alex Maltman Published in 1994 by Chapman & Hall ISBN 0 412 40590 3

Despite the logistic difficulties of gaining access to glacier beds, sediment samples have been retrieved from under the ice, and direct *in situ* measurements of deformation have been performed. Initially, these measurements were made close to the glacier margins, where tunnels were used to access basal sediments, allowing derivation of the first flow law for subglacial sediments (Boulton and Jones 1979; Boulton and Hindmarsh 1987). Further upglacier, where basal conditions are probably more characteristic of the glacier bed as a whole, borehole access has allowed the collection of sediment samples (Engelhardt *et al.* 1990; Blake, Clarke and Gerin 1992) and the emplacement of sensors to measure directly the deformation of basal sediments and the sliding of ice over this unlithified material (Fahnestock and Humphrey 1988; Blake and Clarke 1989; Kohler and Proksch 1991; Blake, Clarke and Gerin 1992; Humphrey *et al.* 1993).

Where ice overlies predominantly soft and water-saturated sediments, deformation may occur in response to gravitationally driven forces. Both compressive and shear stresses operate on the sediment. Compression results from the vertical component of stress due to the ice overburden, and can cause dewatering and consolidation. The process is in many ways analogous to the burial pressure that arises through sediment accumulation (section 1.2.2), and has recently been analysed by Boulton and Dobbie (1993). Shear stresses are driven by changes in ice-surface and sediment-surface elevation, and may result in shear deformation within the basal material. This deformation can contribute directly to ice-surface motion. Unlithified sediments undergoing active deformation have been identified beneath a number of modern ice masses, including: Ice Stream B, Antarctica (Alley *et al.* 1986; Blankenship *et al.* 1986; Engelhardt *et al.* 1990); Columbia Glacier, Alaska (Fahnestock and Humphrey 1988; Meier 1989); Trapridge Glacier, Yukon Territory (Clarke, Collins and Thompson 1984; Blake and Clarke 1989); Urumqi Glacier No. 1, China (Echelmeyer and Wang Zhongxiang 1987); Breidamerkurjökull, southeast Iceland (Boulton and Jones 1979; Boulton and Hindmarsh 1987); and Storglaciären, Sweden (Kohler and Proksch 1991).

In addition to subglacial deformation, sediments within the glacial environment may deform in a number of other distinct settings, such as supra- or proglacially in debris flows, or englacially in layers of debris or debris-rich basal ice. Furthermore, the wide range of glacial settings, both geographical and thermal, results in the possibility of both glacial and glacio-aqueous deformation occurring either before or after deposition. The dynamic nature of the proglacial environment – where sediment is commonly reworked after deposition and structure is created and destroyed by marginal processes such as dessication, dewatering and overriding – results in complex and potentially hard-to-interpret sediments in which processes may be only subtly recorded (Clarke 1987b). No-one who has spent time in the forefield of a glacier will dispute either the activity or complexity of the environment.

This review is concerned largely with the subglacial deformation of terrigenous sediments. Such glaciotectonic deformation is defined as the direct result of glacier motion or loading (INQUA Group on Glacial Tectonics; see also section 1.1.1). Marginal, depositional and proglacial processes will be discussed only briefly, and largely in the context of the preservation of features developed subglacially. Subglacial sediment deformation and the resulting sedimentary structures will be developed from the perspective of sediment properties, the physics of till mechanics (Clarke 1987b) and the processes operating at the base of the ice.

3.2 SUBGLACIAL CONDITIONS

3.2.1 General

Basal conditions beneath a real ice mass overlying an unlithified bed are complex (Figure 3.1). The sediment is inhomogeneous, rheologically complex, and parameters such as fluid pressure may vary both spatially and temporally, potentially within a brief time-scale (e.g. Clarke, Meldrum and Collins 1986). This section discusses some idealized scenarios in order to indicate the range of conditions that may occur beneath ice masses, and it defines the terminology used in the rest of this review.

3.2.2 Bed type

The bed underlying an ice mass may consist of sedimentary material or bedrock, but the nature

Subglacial conditions 75

Figure 3.1 The nature of the ice–sediment interface near the margin of a temperate glacier. Note the wide size distribution of the sediment, the presence of very coarse material and the high water content of the sediment. (a) Clean ice–sediment interface. (b) Glacier sole covered with loose, wet sedimentary material.

of the material underlying the ice may not be the same at all locations and also may vary temporally. The definitions used here are as follows.

Hard bed: the glacier bed consists solely of bedrock material.

Deformable bed/unlithified bed: a glacier bed consisting of unlithified material, which has the potential for active deformation under the appropriate subglacial conditions and driving stresses but is not necessarily actively deforming at a given time.

Active deformable bed: an unlithified sedimentary bed that is underoing current deformation, driven by forces resultant from the overlying glacier ice and/or the glacial hydrological system, and resulting in a contribution to ice-surface motion.

3.2.3 Thermal regime

In general, the role of temperature in sediment

deformation is to affect modestly the stress configuration (section 6.2.3), but in glacial deformation the influence is major. The thermal conditions govern the extent to which ice rather than water is present in the sediment pores. A sedimentary bed underlying a glacier may be ice-infiltrated or ice-free, a factor that will be determined largely by the basal temperature and basal melt rate. Temperatures recorded beneath glacier ice range from -13 to $-18°C$ (Hansen and Langway 1966; Paterson 1976) beneath cold-based ice masses, to the pressure melting point beneath warm, wet-based ice masses. In this context three scenarios may be envisaged.

1. **Frozen bed:** water contained in the bed is below the pressure melting point and present largely as ice. Such a frozen bed was observed beneath Urumqi Glacier No. 1, China (Echel-meyer and Wang Zhongxiang 1987), where a frozen debris layer has been observed to have a viscosity less than glacier ice and to undergo enhanced deformation. There is, however, no clear distinction between sediment-rich ice and ice-rich sediment; that is, between basal ice and a frozen sedimentary bed.

2. **Unfrozen bed:** water contained in the sediments is at, or above, the pressure melting point and is present in the liquid phase. Such a bed would behave in a characteristically different way from a frozen bed, and its viscosity would be much lower. Such an unfrozen bed is believed to exist under Ice Stream B (Blankenship *et al.* 1986; Engelhardt *et al.* 1990). The sediments comprising such a bed are normally assumed to be water saturated.

3. **Partially frozen bed:** water at, and within, the bed is present in both the solid and liquid phases. Such a bed may result from ice infiltration into sediments, such as would be expected to occur if the pore water pressure within the bed is low (Shoemaker 1986; Boulton and Hindmarsh 1987). It is likely that on a larger scale most beds will vary spatially in thermal regime and, therefore, will be partially frozen. Furthermore, where debris-rich basal ice or frozen sediments overlie unfrozen sediments, spatial and temporal variation of the freezing front within the sediment may occur. Migration of the interface between frozen and unfrozen sediment may then cause deposition and erosion of sediment by melt-out or freezing-on of sediment (Menzies 1981).

3.2.4 The ice–bed interface

It is largely the nature of the interface between the glacier ice and the underlying sedimentary bed that will determine to what extent the ice slides over the bed or deforms the sediments beneath it. Theoretically, three situations can be envisaged.

1. **Complete decoupling** of the ice and sediments: the ice and sediment are buffered from one another and motion will be by slip alone (Figure 3.2a). Decoupling could result from the presence of a water film or cavities at the interface (Shoemaker 1986; Lingle and Brown 1987).

2. **Complete coupling** of the ice and deformable bed: no slip occurs between the sedimentary material and overlying ice (Figure 3.2b); in this situation the ice and sediments must move as a continuum. Such coupling could result from the infiltration of ice into the bed (Shoemaker 1986). If no sliding of the ice occurs, then ice will infiltrate into the underlying sediments by the processes of Darcian flow and ice regelation (Boulton and Hindmarsh 1987) unless basal melt rates equal or exceed the infiltration rate.

3. **Partial decoupling** of the bed and sediments: the bed is incompletely coupled to the overlying ice and both sliding and deformation occur (Figure 3.2c). Partial decoupling also covers active sliding, such as the ploughing of sediments by clasts entrained in the ice (N.E. Brown, Hallet and Booth 1987), and active grinding and glacier motion due to deformation within a thin coarse layer of sediment at the base of mountain glaciers (Robin 1989). Measurements beneath glaciers suggest that partial decoupling does occur. At Trapridge Glacier both sliding and bed deformation have been measured *in situ*, with the former accounting for some 40% of measured surface motion (Blake, Clarke and Gerin 1992). At Breidamerkurjökull, ice infiltration has been observed, although basal slip accounted for ~10% of the glacier motion (Boulton and Jones 1979; Boulton and Hindmarsh 1987).

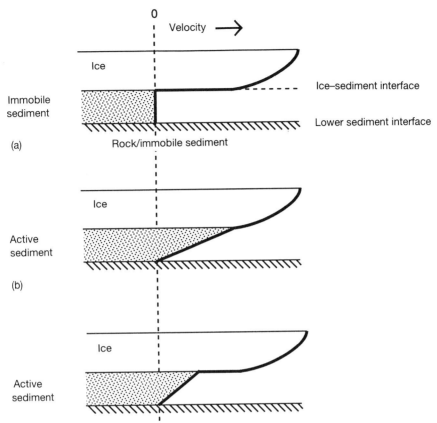

Figure 3.2 Schematic diagrams of the influence of the ice–bed interface on surface velocity: (a) complete decoupling of the ice and sediments (sliding only); (b) complete coupling of the ice and sediments (active deformation only); and (c) partial decoupling of the ice and sediments (sliding and active deformation). Diagram assumes no slip at the lower interface and a linear velocity–depth profile within the active sedimentary layer. (After Alley *et al.* 1986.)

The above discussion has assumed an idealized transition from glacier ice to basal material in the vertical direction. For a real glacier, the ice–bed interface is unlikely to be this clearly defined everywhere, and bands of sedimentary material will probably exist incorporated into the ice above both frozen and unfrozen beds. Such sedimentary bands were observed at Urumqi Glacier No. 1 and Shoestring Glacier, Washington (Brugman 1983; Echelmeyer and Wang Zhongxiang 1987).

3.2.5 Bed thickness

A sediment layer is **thick** with respect to the actively deforming layer if sediment deformation occurs above a layer of immobile sediments (Figure 3.3a). The resulting two-layer structure has been observed in Icelandic glaciers (Boulton, Dent and Morris 1974; Sharp 1984).

If the sediment layer is **thin** with respect to the active layer (Figure 3.3b), then deformation occurs down to bedrock and erosion of the lower bedrock surface is possible. This bed type is considered to exist beneath Ice Stream B, at the site of the Up Stream B field camp (Alley *et al.* 1987a).

3.2.6 Realistic basal conditions

It is possible for the base of the glacier to vary spatially or temporally between these bed types (Menzies 1989a). A spatial variation between a hard and soft bed would occur, for example, if there are bedrock protuberances that equal, or exceed, the sediment layer thickness, and the bed

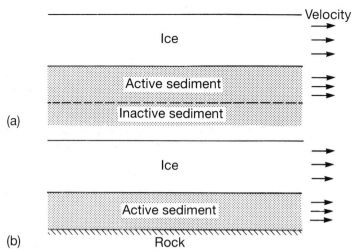

Figure 3.3 Schematic diagrams representing: (a) a thick, deformable bed where sediment thickness exceeds the active layer thickness; (b) a thin active bed where the sediment layer thickness equals the active layer thickness.

was therefore hard in some places and deformable in others. Equally, a surge-type glacier, such as Trapridge Glacier, which overlies a sedimentary bed (Clarke, Collins and Thompson 1984), probably experiences a dramatic temporal increase in the activity of its deformable bed when it passes from quiescent to active surge behaviour (Raymond 1987). On a shorter time-scale, evidence for such temporal switching arises from *in situ* strain measurements made within the basal sediments beneath Trapridge Glacier. Although these represent downglacier flow over longer time periods, they also show short periods of both zero and negative strain (Blake, Clarke and Gerin 1992).

3.3 STRESSES ARISING FROM OVERLYING ICE

3.3.1 General

Glacial deformation of the sediments underlying an ice mass results from stresses that arise from the overlying ice. As in other contexts of sediment mechanics, it is the effective stress, rather than the total stress, that controls deformation, consolidation and swelling of the sediment (e.g. Skempton 1970; sections 1.2.5 and 2.2.3). A change in pore-water pressure alone causes virtually no change in volume – as water is virtually incompressible – or in the intrinsic properties of the sediment, but it dramatically influences the response to stress.

3.3.2 Compressive stresses

Direct loading by glacier ice gives rise to compressive stresses, and potentially results in consolidation of sediments resulting from the loss of water and spatial rearrangement of solid particles (Mathews and MacKay 1960), in direct analogy with the submarine accumulation of a sedimentary pile (Boulton and Dobbie 1993). Such consolidation is normally accompanied by alteration of sediment properties, including increased shear strength, and decreased compressibility and permeability (sections 1.2.2 and 2.2.3). Loading a sediment causes the pore pressure within it to rise. Pore water then starts to be expelled from the sediment at a rate determined by the sediment permeability and volume compressibility, until the excess pore pressure is dissipated. As water is expelled the effective stress increases correspondingly.

Consolidation of a sediment can lead to dewatering structures (section 9.5.3), and potentially to overconsolidation if changes in pore pressure occur repeatedly. Such cyclic pore-pressure changes are common in the subglacial environment, and may drive alternating consolidation and swelling of the sediment in a non-reversible process that leads to a reduction in sediment porosity known as **ratchetting**. This process results from a reorganization of particles within the sediment body (Schofield and Wroth 1968; Clarke 1987b) and can be accompanied by the development of anisotropy. Large water-pressure variations are common within the basal hydraulic system, on a variety of time-scales, ranging from fluctuations on a scale of minutes to hours through diurnal to annual fluctuations (e.g. Clarke, Meldrum and Collins 1986). However, it is not clear how well the hydraulic system communicates with pore water within the basal sediments.

3.3.3 Shear stresses

The deviatoric component of the effective stress tensor (section 2.2.2) can result in the shear deformation of sediments and the formation of structures resulting from high strain. Because deformation results from the spatial rearrangement of sediment particles, properties such as shear strength and permeability will be affected. The response of sediments to applied shear stress will be dependent on basal conditions. Shear forces must be transmitted from ice to sediment; hence coupling across the ice–sediment interface must occur, at least locally. Coupling across the ice–sediment interface is unlikely to be uniform, and so the shear stress transmitted to a portion of the glacier bed may vary both spatially and temporally. The heterogeneity of the resulting deformation is probably more marked than in other geological environments. Deformation may be pervasive (e.g. Boulton and Hindmarsh 1987), and occur through a significant depth of basal material, or, alternatively, discrete deformation may occur (Kamb 1991), resulting in localized planes or regions of high strain with little strain occurring in surrounding sediment. Clasts entrained at the ice–sediment interface may plough through the underlying sediments if the sediment yield strength is exceeded only locally (N.E. Brown, Hallet and Booth 1987). Ploughing represents an intermediate state between sliding (no coupling of ice to sediment) and pervasive deformation (good coupling across the ice–sediment interface) (Alley 1989a).

Further effects of deformation on sediment and the structures that may be developed will be discussed in section 3.8.

3.4 SEDIMENT PROPERTIES

The grain-size distribution of sediments is a fundamental control on sediment deformation properties, and on the physics operating within sediments at the glacier bed. Sediments produced by the direct erosion of lithified material by glacier ice typically contain particles spanning a large size distribution, and are frequently characterized by high specific surfaces due to the presence of fine material. *In situ*, these sediments may have high water contents and porosities (Ronnert and Mickelson 1992), particularly where active deformation is occurring (Figure 3.1). Deformation results in high strains within basal sediments, often as the result of high pore-water and low effective pressures. Furthermore, sediment properties are altered by ice-basal processes, which are controlled by conditions at the bed. These basal conditions are, in turn, affected by changes in sediment properties. The subglacial sediment system is characterized by a subtle interplay and feedback between sediment properties and basal processes.

The bimodal size of rock fragments produced by the erosion processes of mechanical crushing and abrasion (Dreimanis and Vagners 1971) is reflected in the subglacial environment by the classic grain-size distribution of subglacial sediment – rich in both fine and coarse material. The fine component reflects the terminal mode, or modes, of the parent lithology (Dreimanis and Vagners 1971; Haldorsen 1981), typically 2–9 phi (0.25 mm to 2 μm), and the proportion of coarse to fine material reflects the travel distance of material. Increased travel distance results in a depletion of clasts with resultant progressive fining of sediments downglacier (Dreimanis and Vagners 1971).

This bimodal distribution means that subglacial sediment may be considered to consist of clastic material suspended in a matrix of saturated fines (Clarke 1987b). The coarse material is typically elongate and relatively rounded (Boulton 1978; Lawson 1979), reflecting the frequency of collision and abrasion events that result in comminution where deformation occurs. These processes of comminution result in the production of matrix material. The fine material that makes up the matrix may be relatively infrequently rich in clay-sized material (>9 phi, <2 μm), and especially clay minerals, unless an ice mass has advanced over marine-sorted sediments (Milligan 1976). Marine sedimentation tends to result in sediment sorting, with progressive fining of sediment material in ice-distal locations. This, accompanied by chemical breakdown of sediments in the marine environment, means that ice masses overriding material deposited below eustatic sea-level may have significant proportions of clay-sized particles and clay minerals in the matrix material. Ice that is not advancing over previously deposited sediments, but is producing a sedimentary bed by primary erosion processes will require very long erosion distances to produce significant clay-sized material, especially as the terminal modes for many minerals are silt-sized rather than clay-sized (Dreimanis and Vagners 1971). Clay minerals are likely to be uncommon, because chemical weathering rates resulting in the formation of clay minerals will be suppressed at the low temperatures and relatively low chemical activities associated with the subglacial environment; in contrast, physical abrasion processes will be relatively rapid. Particle sizes in the sediments at the margins of many modern glaciers: e.g. Matanuska Glacier, Alaska (Lawson 1979); Trapridge Glacier (Clarke 1987b); sediments from Icelandic and some Spitsbergen glaciers (Boulton 1976); sediments from beneath Ice Stream B (Engelhardt et al. 1990); and sediments underlying past ice masses: e.g. Norwegian sediments (Haldorsen 1983); sediments from central Finland (Haldorsen 1981; Virkkala 1969); those from the Canadian Shield (Scott 1976); and central and northern Sweden (Haldorsen 1981); indicate that many land-based glaciers and ice sheets overlie sediments that are silt- rather than clay-rich.

As indicated above, the proportion of clastic to matrix material will be fundamental in determining the deformation characteristics of the sediment. Where a continuous, or quasi-continuous skeleton of clasts exists, deformation will be dominated by physical interactions and frictional forces between this clastic material. The physical properties of the bulk sediment will be controlled by the large size and inert nature of the clasts. Deformation of such a sediment is controlled by particle size, shape, surface texture and size distribution. Furthermore, clast size may be such that one or more particles may bridge the actively deforming layer and support significant shear stress.

As the proportion of matrix material increases, clasts become effectively isolated from one another within the matrix and interactions between larger particles become rare. Evidence from soil mechanics suggests that approximately 30% of the solid particles by mass must be matrix material to effectively isolate clasts from one another (Smart 1985). The matrix material within subglacial sediment may be either silt- or clay-rich. The deformation behaviour of a sediment in which clasts are isolated will depend largely on the percentage of clay-sized material; as the amount varies, changes in the character of the forces dominating deformation occur. At the free surface of solid particles of any size, a potential field exists associated with unsatisfied intrasolid bonds. This potential field gives rise to surface forces that result in cohesion/adhesion bonding. Summed over the surface of a solid particle, the magnitude of these forces will be insignificant compared with other forces, such as weight, for silts, sands or larger particles. As the surface area to volume ratio increases with decreasing particle size, the importance of such bonding increases, typically becoming significant compared with other forces at a particle size of $\sim 1-2\,\mu$m (i.e. for clay-sized particles). Consequently two types of matrix can be envisaged: silt-rich or clay-rich. Within a silt-rich sediment, deformation will be dominated by physical frictional forces between particles, and the properties of individual particles controlling physical deformation will be similar to those controlling deformation of a clast-rich sediment. As the particle size decreases and the sediment becomes clay-rich, deformation will be dominated by surface effects and factors, such as mineralogy and pore-water impurities, that perturb conditions at particle surfaces.

The control of particle-size distribution on deformation results from: (i) the changing role of surface forces with particle size; and (ii) the effect of particle size on permeability (e.g. Freeze and Cherry 1979). The permeability/water-through-flow dependence will be discussed in section 3.7. Note that particle mineralogy also affects the magnitude of surface forces and, hence, values of the Mohr-Coulomb friction and cohesion coefficients ϕ and c, the residual strength (e.g. Mitchell

1976) and the porosity control on deformation (Terzaghi 1955). Because different mineralogies produce different sized characteristic abrasion and crushing products, physical comminution results in changes in mineralogy with particle size within a natural sediment. Sorting by size will, as a result, create mineralogical contrasts and further affect sediment properties.

Subglacial sediment is inhomogeneous on many scales. Firstly, its wide range of particle sizes gives rise to an inherent inhomogeneity – compare the size of a clay particle with a large glacial erratic. However, further heterogeneity is superimposed on the sediment by internal structure and the spatial grouping of particles by depositional and mobilization processes (Boulton 1976; Menzies 1989a). Not only is coarser material preferentially deposited at local surface perturbations, and particularly where other large particles have lodged, but the resulting clusters of coarse material are further resistant to remobilization or erosion by water. Finer material, in contrast, both couples readily to water, resulting in the flushing of fines (Clarke 1987b), and mobilizes more readily under many conditions away from the ice margin. Furthermore, many sediment properties, such as shear and residual strength, viscosity and permeability, are controlled, or affected by changes in particle-size distribution. Because deformation occurs preferentially at local sediment weaknesses (Alley 1989b), such inhomogeneities become self-perpetuating.

Deformation results from the spatial rearrangement of particles within the sediment. As this rearrangement occurs, continuing deformation will progressively affect sediment properties. Shear deformation in many sediments is likely to cause dilation and particle alignment accompanied by the development of anisotropy, although those that have undergone extremely high strains, such as may be common in the subglacial environment, may be characterized by homogenization and mixing (Hart and Boulton 1991).

Therefore anisotropy and inhomogeneity occur due to the deformation of sediment and its deposition, with structure developing at all scales. In a thick bed such changes in properties result in a two-layered structure, with mobile, deforming sediments lying above an immobile lower layer.

3.5 THE SEDIMENT TRANSPORT SYSTEM: PRODUCTION, ALTERATION AND LOSS

In the glacial environment sediment production, deformation and deposition cannot be separated conceptually. A brief outline follows of the sediment system, from initial production to final deposition.

Much as the basal hydraulic system transports water beneath a glacier or ice sheet, the basal system of an ice mass overlying an active deformable bed transports sediment. Measurements suggest that 88% of ice-surface motion may result from sediment deformation (Boulton and Hindmarsh 1987), so the sediment transport rate may be significant. At Ice Stream B, where the basal sediments are of the order of 6 m thick and the surface velocity is $\sim 500 \text{ m a}^{-1}$ (values at the Up Stream B field site; Alley et al. 1989b), sediment transport will occur at the rate of $\sim 600 \text{ m}^3 \text{ a}^{-1}$ per unit width of grounding line (Alley et al. 1989a).

For an ice mass in equilibrium, the rate of transport of sediment must equal the sediment production rate. Where a glacier is producing its own sedimentary bed by primary erosion of lithified material, the sediment transport system may be considered analogous to the glacier itself. Upglacier primary production of sediment occurs by direct erosion of bedrock beneath a thin bed, the sediment layer thickens, and sediment flow is extensive. Downglacier deposition occurs, the immobile layer thickens and sediment flow is compressive (Boulton 1987). Between these regions a conceptual sediment **equilibrium line** may be defined where no depositon or erosion occurs. This line may not be fixed in either time or space.

Sediment production for an ice mass overlying a soft bed is less efficient than for a hard bed. The primary production of sediment from underlying lithified material can occur only at a bedrock interface (either ice–bedrock or sediment–bedrock). A thick deformable bed has no access to a bedrock interface, and so no ability to produce sediment. An ice mass overlying such a bed, and producing this bed by erosion of underlying bedrock (rather than advancing over previously deposited material), must have a source region with either a hard or thin bed where intimate contact with bedrock occurs. Further downglacier the sediment layer will thicken until a region is reached where deposition occurs and active sediment deforms above previously deposited sediments – the region of thick bed.

As this sediment is transported it will undergo comminution by intrasediment processes. Sediment generation at the bed of an ice mass can occur by the processes of plucking, abrasion, bedrock crushing and water-aided erosion (Röthlisberger and Iken 1981; Drewry 1986; Iverson 1991). Typically, the primary processes of sediment production (plucking and rock crushing) will produce larger fragments and a coarse particle size distribution, whereas subsequent abrasion and erosion produce fine material. Thus, the particle distribution will tend to become finer down glacier (Dreimais and Vagners 1971). Accompanying this variation in particle-size distribution will be changes in bulk properties.

Sediment output from the basal system occurs in a region of deposition. Provided the bed remains unfrozen, sediment will be initially deposited from the base of the deforming layer. Such deposition occurs when the frictional forces resulting from the underlying bed are greater than local shear stresses. Lodgement will occur preferentially at surface roughnesses of underlying bedrock or sediment, where these frictional forces are largest, for example, where a large particle has previously been deposited. Consequent groupings of larger particles result and may form boulder clusters or boulder pavements within sediments (Boulton 1976). Sorting as a result of deposition tends to reinforce internal inhomogeneities within the sediment, and large-scale features, such as drumlins, can be initiated by this process (Boulton 1987).

Over the last 10 ka, basal sediments deposited at the grounding line of Ice Stream B have created a delta tens of metres thick and tens of kilometres long (Alley et al. 1989a). Smaller ice masses and outlet glaciers that overlie, or partially overlie, active deforming beds will also produce a zone of deposition at the margin. Meltback at the snout of such a glacier revealing thick sediments does not necessarily reflect basal conditions further up the glacier flowline, but may characterize marginal conditions only.

3.6 MODELS OF SEDIMENT PROPERTIES AND DEFORMATION

3.6.1 General

The deformation of sediment beneath a glacier results in changes of structure, particle arrangement and water content. If we are to be able to decode the resultant structure, and to reconstruct the physical processes and conditions that have produced it, we must develop the science of till mechanics (Clarke 1987b). This section summarizes some current models of sediment deformation and physical properties.

3.6.2 Deformation of a homogeneous sediment body

The simplest models of sediment beneath an ice mass consider it to be a homogeneous material with no internal structure. Typically the sediment is considered to obey a failure criterion, although some models set this to be zero. At applied stresses below failure no deformation is considered to occur. Once the criterion is exceeded, deformation begins, usually considered to be governed by a fluid-type viscous flow response.

Subglacial sediment overlies much of the mid-latitudes of Europe and North America, on which substantial development and building has occurred. As a result, the engineering properties of these soils and sediments have been tested extensively by geotechnical engineers. The resulting geotechnical theory is often used to predict the failure stress of subglacial sediment; frequently the Mohr-Coulomb criterion is used (e.g. Boulton and Jones 1979; Boulton and Hindmarsh 1987; Alley 1989b; section 1.2.3). The physical properties of glacial sediment reflect the mode of deposition, internal and sedimentary characteristics, such as particle-size distribution, and mineralogy. As a result, the values for parameters such as friction angle and cohesion quoted, or used in models, vary widely (e.g. Boulton and Dent 1974; Beget 1986; N.E. Brown, Hallet and Booth 1987; Clarke 1987b).

Other failure criteria are used in soil mechanics. The Hvorslev failure criterion defines porosity-dependent cohesion and friction functions (e.g. Das 1985) such that shear strength varies with water content. The friction angle typically varies only slightly with porosity (Gibson 1953), but the cohesion may be sensitively dependent on this parameter, and is often written:

$$c'(\eta) = c_0 \exp\left(-B \frac{\eta}{1-\eta}\right), \quad (3.1)$$

where c_0 is the cohesion at zero porosity, η is the porosity and B is a constant. This failure criterion is used by Clarke (1987b), with the dependence of the friction angle on porosity taken to be negligibly small. The Tresca and von Mises criteria use different combinations of the principal stresses to define failure conditions for soils. There is no evidence that these would prove superior to the well known Mohr-Coulomb criterion, and to my knowledge have not been used for subglacial sediments.

In fact, the traditional approach of soil mechanics may not turn out to be entirely appropriate for sediments under the conditions of high porosity, and often high strain, typical of ice-basal environments. Geotechnical engineers are primarily interested in non-deforming sediments, or in very small strains – deformation would typically result in some form of failure of a human-built structure, a clearly undesirable outcome! A clear-cut concept of failure stress has an obvious place in soil engineering, although creep, in the sense of a time-dependent strain that can occur below the failure condition, is acknowledged. It may be that on a glaciological time-scale, sediment deformation will occur at all applied stresses (Barnes and Walters 1985; Boulton and Hindmarsh 1987).

In this view, once deformation has begun, the sediment is typically considered to deform as a viscous fluid with a characteristic flow behaviour. Often a linear Newtonian or Bingham fluid is used (e.g. Boulton and Hindmarsh 1987; Alley et al. 1987b; Menzies 1989b) such that

$$\dot{\varepsilon} = \frac{1}{2\mu}(\sigma' - \tau_y), \quad (3.2)$$

where $\dot{\varepsilon}$ is the resultant strain rate and μ is the Newtonian sediment viscosity, assumed constant in this model (Figure 3.4). A Bingham fluid shows no variation of viscosity with either basal conditions or sediment strain rate. The measurements obtained from beneath the Lower Columbia Glacier, Alaska, by Humphrey et al. (1993) indicated that the till rheology had to be more complex

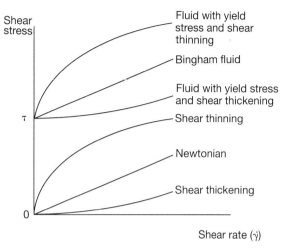

Figure 3.4 Schematic stress–shear rate curves for idealized fluid flow, illustrating the terminology used in the rheological approach to modelling sediment deformation.

than a Newtonian or plastic behaviour. Other fluid models are typically non-linear with strain rate, with the sediment displaying either shear thickening or shear thinning (Figure 3.4), or some more complex behaviour.

Evidence for non-linear flow behaviour in subglacial sediments arises from a number of sources. *In situ* measurements of strain rates beneath the margin of a glacier in Iceland resulted in the extrapolation of an effective-pressure-dependent flow law (Boulton and Hindmarsh 1987) of the form:

$$\dot{\varepsilon} = \frac{L_b}{(\sigma')^b}(\tau - \tau_y)^a \quad (3.3)$$

where a, b and L_b are constants, and τ represents the shear stress. However, marginal conditions are characterized by high shear and low normal stresses, and by what may be higher than average sediment water contents; basal conditions here may be unlike those further from the terminus. At Trapridge Glacier, away from the ice margin, *in situ* measurements of sediment deformation (made using tilt-cells inserted through boreholes in the glacier ice into basal sediments) show temporal variations of strain rate that appear poorly correlated with fluctuations in basal water pressure (Blake, Clarke and Gerin 1992). These temporal variations should result from variation in either applied shear stress, or effective pressure driven by pore-pressure variation, if the Boulton–Hindmarsh equation (3.3) is universally valid.

Unfortunately neither of these parameters has been measured except close to the ice margin (Boulton et al. 1979; Boulton and Hindmarsh 1987). Because pore-water pressures within the sediment have yet to be measured beneath a realistic thickness of glacier ice, basal water pressure measured just above the ice–sediment interface is often used as a surrogate variable. The shear stress is usually assumed to be constant and to be driven by ice and bed surface slopes alone. The validity of these assumptions is unknown.

Subaerial debris flows also display non-linear deformation behaviour (e.g. Iverson 1985; Phillips and Davies 1991). Such debris flows often consist of material with a wide particle-size distribution, although size distributions appear to be unimodal and denuded in both fines and very coarse material compared with tills (Johnson and Rodine 1984; Pierson and Scott 1985; Major and Pierson 1990). Porosities vary widely (Pierson and Scott 1985) and the sediment may or may not be saturated. For these reasons, direct extrapolation of flow laws derived for such material to sediment deforming in the subglacial environment may be inappropriate, especially since conditions of effective and pore pressure differ considerably.

Further evidence for non-linear variation of strain rate with applied stress comes from laboratory study. Sediment shear strength is dependent on the work done during deformation, and is made up of components related to friction and cohesion effects, particle rearrangement and dilation (Rowe 1962), and hence is deformation-mechanism dependent. Therefore, it is important that such laboratory studies should realistically reflect basal conditions, and that failure within a laboratory apparatus should occur by similar mechanisms to those causing deformation beneath a glacier. Laboratory experiments on concentrated suspensions, including suspensions of matrix material from subglacial sediments, deformed at high porosity suggest non-linear deformation behaviour dependent on strain rate, with both shear thickening and shear thinning occurring (Cheng and Richmond 1978; Beazley 1980; Murray 1990; Major 1993), although strain rates are orders of magnitude higher than those occurring subglacially. At lower porosities, measurements using subglacial sediment in a geotechnical apparatus may also result in a non-linear dependence of either failure strength or residual strength on strain rate (Skempton 1964, 1985; Kamb 1991). Purging coarse material, which is required for most small-scale laboratory experiments, may turn out to be a major problem when extrapolating results of such experiments to the subglacial environment.

3.6.3 Deformation of a structured sediment body – anisotropy and inhomogeneity

Sediment is not well represented by a homogeneous fluid; structure exists at many scales: (i) subglacial sediments are inhomogeneous. Such inhomogeneity is reflected in spatial changes in sediment internal structure, in particle-size distribution and in particle arrangement. (ii) Subglacial sediments are anisotropic. The anisotropy results from previous deformation and deposition processes.

Sediment properties are expected to be both inhomogeneous and anisotropic, with deformation occurring preferentially at weaknesses within the sediment. Localized perturbations in sediment properties can cause regions where the failure criterion is exceeded, although this is not the case for the entire bed. In these regions deformation will occur preferentially, and result in relative motion between parts of the bed.

The deformation of subglacial sediment, resulting in property changes and anisotropy, is discussed by both Clarke (1987b) and Menzies (1989b), who predict a hysteresis behaviour for the sediment. Three states are assumed: an immobile pre-deformation state; an active mobile state; and an immobile post-deformation state (Menzies 1989b). Sediment in the post-deformation state is not assumed to have the same properties as that in the pre-deformation state. This prediction is consistent with fabrics seen in naturally deformed sediments, which suggest that an initial assumed random structure becomes non-random after deformation, such that the particles are aligned parallel to ice flow (e.g. MacClintock and Dreimanis 1964). After deformation most flow parameters will be anisotropic in behaviour with respect to directions both parallel and perpendicular to the direction of active deformation.

While there is clearly feedback between the deformation, basal conditions and sediment properties, this is as yet unrecognized by models of sediment deformation. Interpretation of sediment structure arising from deformation must require consideration of both the anisotropic and inhomogeneous nature of subglacial sediments. After all, deformation of homogeneous sediment provides only minimal structural information!

3.7 BASAL PROCESSES AS A CONTROL ON DEFORMATION

3.7.1 Hydraulic processes

Sediment deformation is controlled by both porosity and effective pressure. These parameters are in turn controlled by the amount of water present at the bed of an ice mass, its distribution, throughput rate and throughput mechanisms. The configuration of the hydraulic system is a fundamental control on the distribution of water, and on sediment porosity, pore-water pressure and effective pressure, and thus on deformation.

An ice mass overlying a significant thickness of sediment may have a hydraulic system with a fundamentally different nature than an ice mass overlying lithified bedrock (Figure 3.5). Because bedrock is often highly impermeable, water flowing beneath a glacier with a hard bed does so largely at the interface between ice and bedrock, either in channels (Röthlisberger 1972; Nye 1976), linked cavities (Walder 1986; Kamb 1987), or as a thin sheet beneath the ice (Weertman 1972). In contrast, the sediments comprising a unlithified bed may be both relatively permeable and actively deforming. Although both sheet (Alley 1989b) and channelized flow (Shoemaker 1986; Alley 1989b) have been hypothesized to occur, water transport can also occur within the bed by advection with deforming sediment, by Darcian flow through the sediment, or preferentially through inhomogeneities within the sediment (Walder and Fowler 1989).

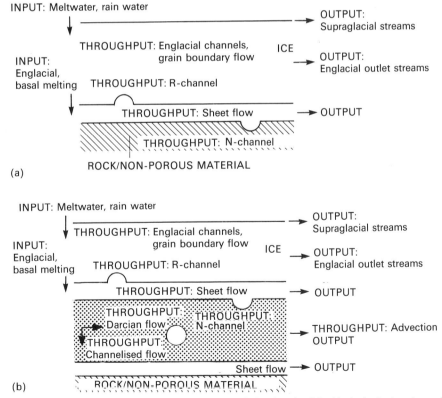

Figure 3.5 Schematic diagrams showing throughflow processes in hard and soft bed hydrological systems. (From Murray and Dowdeswell 1992.)

The basal hydraulic system evacuates water to the margin unless it is lost, for example, through an underlying aquifer (Alley 1989b). How the basal water is distributed between the flow mechanisms discussed above will strongly affect basal conditions, and the location of water at the bed. Water transported by channelized or sheet flow at the ice–sediment interface may not affect the underlying sediments greatly; an ice mass that evacuates its basal water largely by these mechanisms is analogous to a hard bed ice mass, and may have little communication with the underlying sediments. However, because sediments are permeable, water can flow into, and through, the bed. Because basal sediments are not homogeneous such water flow will be localized. Where sediments are anisotropic, permeabilities will also be anisotropic. The inhomogeneous nature of the basal sediments suggests that water flow and the resulting water distribution within the sediments will also be inhomogeneous. Channelized flow within basal sediments represents one such heterogeneous water-content distribution. Water flow may also occur through connected regions of high local permeability, which could result, for example, from changes in particle-size distribution or sediment dilation.

Open channels within basal sediments will tend to close by creep and viscous flow (Alley 1989b) and the flow of water through such channels may be unstable. In contrast, preferential flow through sediment is likely to be a stable process. Water flowing through non-uniform sediments will take the path of least resistance through the bed and will tend to follow paths of high hydraulic conductivity through connected regions of enhanced porosity and reduced tortuosity (Arch and Maltman 1990; Murray and Dowdeswell 1992). Water flow by processes that are essentially Darcian will be enhanced along sediment unconformities, shear planes and other lines of weakness. These water paths may tend to reinforce themselves in positive feedback, as the flow itself can remove fines thus increasing the local sediment permeability.

Heterogeneous water flow through channels, or along preferred paths through the sediment, will create inhomogeneities of porosity and, potentially, fluid pressure gradients between regions of the bed. Thus the hydraulic system itself has the potential to create conditions for heterogeneous deformation within the sediment.

Since deformation results from the spatial rearrangement of particles within the sediment, deformation alters sediment properties. Shear deformation may lead to dilatancy and thus to increased sediment permeability. Consolidation, in contrast, decreases porosity and hence permeability. These processes act in competition, controlling sediment permeability. Resulting local changes will affect both the rate and route of water movement through the sediment and may also affect sediment porosity and pore-water pressure. Variation in porosity has been shown to control sensitively both shear and residual strength (Henkel 1960; Maltman 1987), and to control viscosity (Beazley 1980; Murray 1990), with the resistance to deformation decreasing with increasing porosity. Thus, the distribution of water at the bed is a critical control for the deformation of basal sediments.

3.7.2 The consolidation–dilation competition

The processes of consolidation and dilation act in competition in a sediment beneath a glacier or ice mass. A lodged, inactive sediment experiences overburden that tends to consolidate it. After retreat of an ice mass, the sediment will be overconsolidated and may preserve dewatering structures and a horizontal fabric associated with consolidation. Shear deformation will, in contrast, tend to dilate sediment, increasing its water content. The material changes that accompany dilation resulting from shear deformation, including the increase in porosity, tend to weaken sediments. At Skálafellsjökull, southeast Iceland, the upper, dilated layer is reported to be much weaker than the underlying layer as measured by shear-vane tests (Sharp 1984). These sedimentological changes mean shear deformation will tend to be self-perpetuating. Consolidation also tends to be self-reinforcing; loss of water strengthens sediment and makes shear deformation less likely. The feedback loop is complicated, and made more subtle, by changes in the pore-water pressure and permeability that accompany these processes. As in other geological settings, as sediment consolidates, permeability is reduced, evacuation of water through sediments is retarded and pore-water pressures may rise. Deformation might be facilitated by this increase in pore-water pressure while consolidation is occurring. In contrast, shear deformation accompanied by dilation may increase permeability, facilitating the dissipation of pore pressure. These feedback systems are as yet incompletely understood in the glacial environment.

3.8 EFFECTS OF DEFORMATION

3.8.1 General

The deformation of sediment beneath a glacier results in the spatial rearrangement of particles within the sediment, and potentially in the development of three-dimensional form. Such form develops at the upper and lower interfaces of the sediment as deformation or deposition cause spatial changes in the thickness of the sediment layer and, hence, the formation of ice-parallel or ice-transverse features. Although this book is concerned with structures formed within sediments, in the glacial situation these structures are so closely associated with the surface form of the sediment mass that a brief outline of sediment form is given here. Together, the spatial arrangement of solid matter and the three-dimensional form compose the structure of the sediment (Figure 3.6). This structure develops in response to a complex set of driving processes and range of basal conditions. The structure developed reflects the deformation history of the sediment.

1988; Feeser 1988; Owen and Derbyshire 1988); at the macroscale through structures such as deformation zones and internal sediment sorting (e.g. N.E. Brown, Hallet and Booth 1987); and at the very large scale through mega-structures such as flutes, drumlins, moraines and tunnel valleys (Boulton 1987). Deformation structures in homogeneous sediments, or arising from homogeneous deformation, may be particularly difficult to decipher, whereas those in inhomogeneous sediments will be emphasized by contrasts in properties and characteristics that will better record the results of deformation. Where sediments are inhomogeneous, deformation will vary spatially. Resulting structures will reflect localized changes in deformation, and will be rendered visible firstly by variation in properties and secondly by changes superimposed by the deformation process.

In the following sections the record that will result within basal sediments from the processes of consolidation and shear deformation at all scales is described.

3.8.2 Features arising from consolidation

Sediment structures resulting from consolidation include those of dewatering, overconsolidation and jointing (Harrison 1958; Boulton 1976; Dowdeswell and Sharp 1986; Feeser 1988; Menzies and Maltman 1992). Dewatering may

Sediment parameters	+ Basal conditions	+ Processes	=> Sediment structure
Mineralogy	Pressure	Consolidation	
Particle size/shape	Temperature	Loading/unloading	
Pore water chemistry	Hydraulic system	Freezing/thawing	
		Deformation	
		Sediment generation	
		Abrasion/grinding	
		Water flow	
		Pressure, temperature and time effects	

Figure 3.6 Processes determining the structure of a subglacial sediment.

Most structures developed subglacially result from sediment inhomogeneity and its reinforcement by inhomogeneous deformation and strain-induced anisotropy. Structure occurs at all scales; at the microscale through structures in matrix material (e.g. Derbyshire 1978; Love and Derbyshire 1985; Arch, Maltman and Knipe

lead to the alignment and intrusion of fines, fluid injection features, layer rupturing and localized fluvial sorting (Lowe 1975; Derbyshire, Edge and Love 1985; Nocita 1988; Owen and Derbyshire 1988). Sediment mixing or the development of consolidation laminations may occur as pore water is expelled from the sediment. Particle alignment can result within the sediment and individual grains may become crushed as the solid skeleton readjusts to occupy a smaller volume (Owen and Derbyshire 1988), with grain size and grain-size distribution playing an important role in the features developed.

3.8.3 Features arising from shear deformation

Dilation

The shear deformation of sediment requires that particles move relative to one another, either sliding over, or rotating past one another (Schaeffer 1990). Such shear deformation may result in dilation accompanied by an increase in porosity (Reynolds 1885), particularly while it is being initiated. Evidence for dilation of actively deforming sediments beneath glaciers arises from the structure of Icelandic tills, and the high porosity inferred for sediment beneath Ice Stream B (Engelhardt *et al.* 1990). The effects of this dilation have been observed at the margin of Icelandic glaciers with thick sedimentary beds, where it results in a two-layered structure (Kozarski and Szupryczynski 1973; Sharp 1984; Boulton and Hindmarsh 1987). At Breidamerkurjökull, the upper layer that has undergone shear deformation is dilated, with a porosity reported to be ∼0.4. Below this lies a layer of largely undeformed sediment with porosity ∼0.2–0.3 (Boulton and Dent 1974) that appears to have deformed along discrete planes. Laboratory deformation experiments of sediments collected at the ice margin have also resulted in dilation (Figure 3.7a; Boulton, Dent and Morris 1974; Murray and Dowdeswell 1992), although it is suggested that such dilation might be suppressed by the occurrence of particle crushing at high normal stress (Boulton, Dent and Morris 1974). If such particle crushing occurs, then this feature, normally associated with high consolidation, could also be associated with shear deformation.

Particle alignment

Particle distribution is non-random in space and previous deformation is recorded as a preferred particle orientation (Figure 3.7b). Deformation is characterized by particle alignment (e.g. Love and Derbyshire 1985) and structures elongated by high strain in the direction of

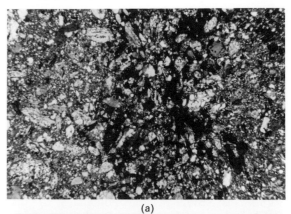

(a)

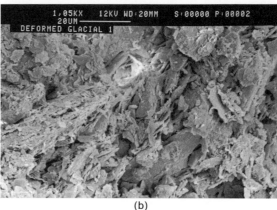

(b)

Figure 3.7 Some examples of the microscopic effects of deformation during the development of shear zones in direct shear tests. The sediments were collected from the margin of a Norwegian glacier and were then deformed in the laboratory. (a) Dilation of sediments. Long side of the photograph is 11.3 mm in length. Void space shows black. (b) Particle alignment in scanning electron micrograph. Scale bar at top is 20 µm in length. (From Murray and Dowdeswell 1992.)

shearing (Boulton and Hindmarsh 1987). Particle alignment results from the rotation of elongate particles to occupy positions of minimum torque, either parallel or perpendicular to deformation. Particle alignment occurs within both the clastic material (e.g. Dowdeswell and Sharp 1986) and the matrix (Sitler and Chapman 1955; Ostry and Deane 1963; Derbyshire 1978; Owen and Derbyshire 1988), although alignment within the matrix material will be heavily controlled by the presence of clasts. The direction of flow of matrix material will be deflected close to larger particles as the matrix flows relative to these clasts. Furthermore, where sediments are coarse, matrix motion may be dominated by clast dynamics, resulting in irregular matrix flow and consequent sediment mixing (Engelhardt, Harrison and Kamb 1978; Hart and Boulton 1991).

Deformation, and hence particle alignment, can occur either along discrete slip planes, resulting in the formation of shear zones (Sitler and Chapman 1955; Owen and Derbyshire 1988; section 9.4.1) in which intense particle alignment may occur, or as bulk deformation involving moderate thicknesses of sedimentary material. Shear zones may be slickensided (Boulton and Hindmarsh 1987; section 9.4.2); considerable strain can be accommodated by shear along such slip planes.

The alignment of particles, and in particular the alignment of matrix particles, will result in anisotropy within the basal sediments and in their deformation properties. Deformation, water throughflow and other processes will be facilitated in a preferred direction resulting from the deformation-derived anisotropy (Murray and Dowdeswell 1992).

Shear deformation of inhomogeneous material

The shear deformation of inhomogeneous material will commonly be localized. Deformation of low-viscosity material around more competent inclusions can result in streamlined features, analogous to drumlins, lending credence to the formation of at least some of these features by inhomogeneous deformation (Figure 3.8; Hart, Hindmarsh and Boulton 1990). At very high strains such inhomogeneities may become attenuated and preserved only in the horizontal direction, resulting in a secondary layering being formed within sediments (Hart, Hindmarsh and Boulton 1990). At extreme strains, even this layering may disappear as lamination thicknesses are attenuated to the average grain size within the layers, resulting in mixing (Boulton 1987).

3.8.4 Development of sediment form

Large-scale features of sediment form arise on scales from decimetres to the scale of landforms. They arise from large-scale changes in sediment properties, basal boundary conditions, or glacier forcings. Although there is controversy as to the genesis of many large-scale forms, features such as glacier flutings, tunnel valleys, Rogen moraines and drumlins have been suggested to be deformational in origin (Boulton 1987; Boulton and Hindmarsh 1987; Menzies 1989b).

Flutings (Figure 3.8b) are commonly observed where glaciers override deformable sediments. Such features are thought to be formed where glacier ice flows past boulders at the ice–sediment interface, forming a cavity at the downstream side. This cavity fills with soft sediment and forms a flute (Boulton 1976). If such flutings become attenuated and form streamlined features, drumlinoid features can result. A similar process of sediment intrusion is thought to produce crevasse-fill ridges, which form at the margin of surge-type glaciers (Figure 3.8c). Weak, water-saturated sediments are intruded into bottom crevasses within the ice forming features transverse to flow (Sharp 1985).

Drumlins are streamlined features formed at the base of ice masses. Sedimentary evidence for their mode of formation is equivocal; one set of theories suggests that they arise from spatial variation in subglacial deformation resulting from changes in sediment rheology or basal conditions (Smalley and Unwin 1968; Evenson 1971; Boulton 1979, 1987; Smalley and Piotrowski 1987). Alternative hypotheses suggest that the formation of these features results from: infilling of subglacial cavities; deposition from sediment-rich basal ice; or preferential lodgement.

If drumlins arise from spatial inhomogeneities within the deforming sediment there is no shortage of opportunities for their formation. Regions of inhomogeneity within the sediment can result from regions of changing sediment-size distribution, porosity, or basal stress. The most obvious cause of sediment inhomogeneity is the occurrence of a large boulder within a much finer matrix. In this case the drumlin core will be completely undeformable, although the boulder must be moving more slowly than the surrounding material in order to produce a streamlined form. Rock-cored drumlins have been reported by Hill (1973) and Minell (1973). Regions of high permeability resulting from a local coarsening of the sediment will potentially have lower pore pressures, and hence higher effective pressures. As this tends to increase the shear strength of the sediment, this region of sediment will have less tendency to deform than the surrounding sediment and may form a stiff inhomogeneity within the sediment. Drumlins formed in this manner

90 Glacial deformation

(a)

(b)

(c)

Figure 3.8 Some examples of the macroscopic effects of deformation. (a) Inhomogeneous deformation of material around a competent inclusion of material incorporated from the lower, undeforming, region. Some streamlining of the inclusion has occurred. Intense shearing of sediment material has resulted in the formation of a secondary, deformation-induced layering. Inhomogeneous deformation on a larger scale could have lead to the formation of subglacial mega-scale features such as drumlins. Feature is exposed at West Runton, East Anglia, UK. Pen for scale. (From Hart, Hindmarsh and Boulton (1990). Reproduced with permission from J.K. Hart.) (b) Fluting at the margin of Trapridge Glacier, Yukon Territory, a quiescent surge-type glacier. Flutes initiate at single boulders at the ice–sediment interface. (c) Crevasse-filled ridge at the margin of Trapridge Glacier.

will have a coarse-grained core. Drumlins cored by coarse grained material have been reported by Evenson (1971), Whittecar and Mickelson (1979) and Stanford and Mickelson (1985). Where a predominantly silt-rich bed exists with local fining, drainage may occur around rather than through clay-rich areas due to the lower permeability of the finer material, and hence fine material can also act as a drumlin core (J.T. Wilson 1938).

Contrasts in sediment rheology (Figure 3.8a), or in the stress that is driving deformation, results in relative motion between parts of the bed. Stationary, or slowly moving, portions of the bed form an obstacle to faster flowing sediment, resulting in erosion of material upstream of the obstacle, and deposition downstream; a streamlined form thus occurs.

In order for sediment inhomogeneity to be acting as a core for drumlin deposition it must be moving more slowly than the surrounding sedimentary bed. Although this is possible in the bulk sediment, due to the relative displacement of portions of the bed, the bed forms produced will be more likely to survive erosion and transport if the inhomogeneity occurs at or near the base of deformation (e.g. Menzies 1989a). Such inhomogeneities near the interface between deforming and undeforming material are more likely to initiate streamlined bed forms than those within the bulk sediment. Irregularities in this interface may also act as zones for drumlin initiation (Boulton, 1987; Smalley and Piotrowski 1987).

3.8.5 Sediments beneath surge-type glaciers

A surge-type glacier is an ice mass that experiences a cyclical flow regime. The cycle is characterized by a quiescent phase of ice stagnation or low surface velocity and an active phase of high surface velocity. The cycle is not related to climatic forcing or external parameters but is inherent within the glacier system (Meier and Post 1969). Where a surge-type glacier overlies deformable sediments, the cyclic flow regime should drive cycles of deformation within the sediments. These sediments may then provide a record of surge activity.

The active phase of surging is considered to result from either fast sliding or fast subglacial deformation rather than from fast creep within the glacier ice (Clarke 1987c). The mechanisms by which a surge is initiated are, however, not well understood, although surges are typically considered to result from the destruction of a well-developed drainage system, whether ice overlies a hard or deformable bed (Clarke, Collins and Thompson 1984; Kamb et al. 1985; Kamb 1987). The breakdown of the drainage system results in water accumulation at the bed, as water input to the basal system will exceed water output from it. Rapid velocities then occur as a result of either enhanced sliding on a water film, as obstacles are submerged (Weertman 1969), or the activation of a highly mobilized, water-saturated layer of deformable sediments (A.S. Jones 1979; Clarke, Collins and Thompson 1984).

Where a surge-type glacier overlies a deformable, unfrozen substrate, the stable, quiescent water drainage is likely to be by channelized or Darcian flow at sediment inhomogeneities. The destruction of these stable flow paths and the production of a high-porosity, low-viscosity 'slurry' at the glacier bed (A.S. Jones 1979; Clarke, Collins and Thompson 1984) could facilitate the high velocities of surging. Collapse of the drainage system must result from either channel closure or a large-scale reduction in sediment permeability. Clarke, Collins and Thompson (1984) suggest that such a change in permeability could result from consolidation or be caused by substrate deformation.

Certain sediment forms have been identified as characteristic of surge-type glaciers, such as crevasse-filled ridges (Sharp 1985, 1988), and both fluting and remoulding of sediments also occur, suggesting the presence of highly mobile sediments. These features may be destroyed or reworked by subsequent surges unless the surge is a net depositional process, or unless climatic amelioration results in successively decreasing ice advances. It is extremely unlikely that surface features would survive subsequent surges, although features lower in the sediment layer might do so. Net deposition of sediment is likely at the margin unless all the sediment transported by the basal system is removed, for example, by flushing out with basal water. Although the termination of the surge of Variegated Glacier, Alaska, was marked by a release from the bed of a pulse of extremely turbid water (Kamb et al. 1985), it is unlikely that flushing would remove all of the sediment transported by a soft-bed glacier.

3.9 PRESERVATION OF FEATURES

The glacial environment is poor for the preservation of features within sediment. Preservation requires a non-erosional environment, which will occur only near the ice margin (Boulton 1987). Features formed in the extensional region of net erosion have little chance of survival. Shear deformation leads to mixing of sediments (Kemmis 1981; Boulton 1987) and the very high strains imposed on subglacial sediments attenuate features rapidly. Hart et al. (1990) suggested that strains of greater than 700% may occur. Boulton (1987) estimated that at Breidamerkurjökull a feature within the basal sediments that started perpendicular would be rotated so that it was dipping 1–6° upglacier within 1 year.

Even if a feature survives to the glacier margin the proglacial environment is very active, features may commonly be reworked, and hence partially or completely destroyed. Marginal processes such as debris flows and mudslides are common (Lawson 1979). Such processes will be especially active in response to the high sediment-water contents that will arise when ice retreat causes blocks of ice to melt *in situ*, a process that was probably common at the margins of the great ice sheets at the ends of each of the ice ages. After deposition and subsequent ice retreat, events that are not necessarily contemporaneous, modification of the sediment continues (Boulton and Dent 1974). Dessication, dewatering and flushing of fines from sediment will occur both syn- and post-deposition. With all of these opportunities for alteration it is surprising that features are ever preserved!

Minimum requirements for preservation of features or structure within the sediment are: (i) a non-erosional environment, at least locally; (ii) inhomogeneity in sediment; and (iii) relatively stable marginal conditions. Luckily the first occurs close to the ice margin, and the second is superimposed by the deformation process itself, as changes in sediment properties caused by deformation are recorded within the sediment.

3.10 OTHER TYPES OF GLACIAL DEFORMATION

3.10.1 Proglacial deformation

The deformation of unfrozen sediments by the action of glacial ice is not limited to shear deformation of water-saturated sediments beneath an ice mass. Deformation of sediments also occurs subaerially in the proglacial environment. Sediment continuity implies that the shear forces applied to sediment, which are very high marginally, do not cease suddenly at the ice margin (Hart 1990) but rather continue into the glacier forefield. Compressive deformation of the sediment results. Typically push moraines are formed at the ice margin transverse to ice flow. Deformation in the proglacial environment is characterized by folding and thrusting, and in East Anglia has resulted in the folding of laminated sediments beyond the ice margin (Hart and Boulton 1991; Figure 3.9). Proglacial

Figure 3.9 Proglacial deformation of inhomogeneous material resulting in folding. Feature is exposed at Trimingham, East Anglia, UK. (From Hart (1990). Reproduced with permission from J.K. Hart.)

deformation has been reviewed by Hart and Boulton (1991).

3.10.2 Deformation of frozen-substrates–basal-ice

A glacier or ice mass may overlie frozen sediments. The lower portion of the glacier ice may be basal ice, which may be rich in debris and formed in close interaction with the bed. Such ice is often laminated. A frozen substrate is therefore not a distinctive situation, and a glacier may show a gradation from basal ice to frozen or unfrozen sediments without a sharp interface. Melt-out of both basal ice and frozen sediment may result in features within the unlithified material. Traditionally it was thought that glaciers frozen to their beds would show no basal slip and that their motion would be limited to internal deformation of the ice. However, where significant debris exists within ice, it appears that the ice is weakened and may deform more readily than the bulk of glacier ice (Mathews and MacKay 1960; Brugman 1983; Echelmeyer and Wang Zhongxiang 1987). Deformation rates hundreds of times greater than in clear glacier ice are quoted (Echelmeyer and Wang Zhongxiang 1987), although it is not obvious why high concentrations of debris weaken ice, especially as low concentrations of foreign material appear to strengthen it (Nickling and Bennett 1984). The deformation occurs as discrete slip along sediment bands with an extremely high debris content (Brugman 1983), or as homogeneous deformation (Holdsworth and Bull 1970), or both, and typically results in both particle alignment and banded debris.

Deformation of debris-rich basal ice or frozen substrate results in strong particle alignment that persists after deposition (Dowdeswell and Sharp 1986). During melt-out, formation of discontinuous lenses is likely to occur (Paul and Eyles 1990), which may produce weak stratification. Post-deposition sediments, thus, are poorly sorted, underconsolidated and have a strong particle alignment.

3.11 CONCLUSION

Deformation processes, and their effect on the structure, fabric and physical properties of the sediments, reveal subtle feedback systems between the sediment and subglacial conditions. Clearer identification of these feedback systems, and quantification of the effects of subglacial deformation on basal conditions will lead to a better understanding of both the conditions and the sediment structures, and of the response and interplay between the sediment and the deformation process.

CHAPTER 4

Sedimentary deformational structures

JOHN COLLINSON

4.1 INTRODUCTION

This chapter is concerned with those structures that develop as a result of deformation early in the burial history of a sediment. Historical aspects, including numerous literature references, are discussed by van Loon (1992). Some structures form very soon after or even during deposition and are sometimes referred to as **penecontemporaneous** structures. In those cases, the deforming stresses commonly relate to the same processes that deposited the sediment. Other structures come about when some compaction and/or diagenesis has been established. At what point sedimentological processes give way to those producing features of concern to structural geology, and hence which structures to include in the present chapter, is a somewhat arbitrary judgement. Mass movements are dealt with separately, in the following chapter. All of these processes take place in loose, highly porous materials and consequently, just as in other settings of sediment deformation, the role of the intergranular fluid is paramount.

Almost all sediments, with the possible exception of organically bound carbonates and deepwater oozes, are susceptible to disturbance after deposition. This may happen at or close to the depositional surface or following relatively shallow burial. Deeper burial may also lead to deformation, as is the case with salt diapirism and growth faulting, although the structures produced tend to be on a larger scale and be gradational with structures related to tectonic deformation. Early sedimentary deformation can occur in a wide range of sedimentary environments, ranging from aeolian dune sands to deepwater turbidites. This chapter deals only with those structures that develop as a result of inorganic, physical disturbance and with some products of early chemical processes in the sediments. Specifically excluded are structures caused by organic activity within the sediment or at the depositional surface, such as burrows, roots and animal surface tracks and trails. Also excluded from detailed discussion are those structures that develop as a direct result of sediment transport and deposition. Depositional lamination, whether due to fall-out of material from suspension or the sorting processes associated with the migration of bedforms, is not discussed in detail (for discussion of such features see Allen 1984, 1985; Collinson and Thompson 1989). However, some appreciation of the geometry of such structures is essential to the full interpretation of deformed structures, as it is commonly those same depositional structures that are deformed.

Early deformation, in many cases, has the effect of modifying the original structures and lamination that developed as the sediment was laid down. Many original structures may be still recognizable, and their growth and geometry may have exerted a close control on the deformation. In other cases, the deformation may modify or destroy depositional structures to the point where they are no longer recognizable. In certain cases deformation may create new structures in sediment that was depositionally featureless.

4.2 PRINCIPLES OF PHYSICAL DISTURBANCE

Before looking at the kinds of structures produced during early sedimentary deformation it is appropriate to consider briefly some aspects of

The Geological Deformation of Sediments Edited by Alex Maltman Published in 1994 by Chapman & Hall ISBN 0 412 40590 3

sediment deposition, and to recall some of the mechanical principles outlined in Chapter 1. Most sediments laid down by physical processes are delivered to the sediment surface either by fall-out from suspension or as the result of bed-load transport. Sediments from suspension are most commonly of finer grain sizes – silt and clay – though coarser sediment may be delivered from suspension by particularly powerful currents. Clay and silt-grade sediments commonly have a high proportion of platy or acicular mineral particles, which tend, on deposition, to have very loose packing and consequently high initial porosity. In addition, the presence of significant proportions of clay minerals means that the sediment will begin to develop cohesive strength soon after deposition. Coarser suspension deposits are likely to be rather poorly sorted and also to have been deposited rather rapidly. Both these features tend to give rise to loose packing and high initial porosity.

Bedload sediments are always silt grade or coarser, and tend to be better sorted and less rich in platy mineral grains. In the case of aeolian transport, grains are confined within a narrow range of sizes spanning fine to very coarse sand. Particles transported as bedload may be deposited in one of two ways. The first is by becoming lodged in a stable packing position on the bed over which they are rolling or bouncing. In that case the particle is most likely to lodge amongst neighbouring grains of similar size so that the accumulating layer is fairly well sorted and the packing stable. Fluctuations in current strength may lead to the accumulation of parallel layers of contrasting grain size. Progressive transport and deposition under these conditions, as on a sandy flat beach subjected to moderate wave swash and backwash, may lead to well-sorted sediment in a closely packed framework with low potential for instability. Higher depositional rates may lead to less good packing and higher initial porosity.

The second way in which bedload sediment may accumulate is in the lee of a bedform such as a ripple, dune or sandwave. There, deposition may result from grains falling directly to the bed having been thrown over the crest of the bedform in saltation or suspension. Deposition may also result from avalanche or grain flow down the lee side due to oversteepening of the upper slope by the addition of grains delivered mainly through rolling. Layers produced by grain fall and grain flow are likely to have contrasting packing, the former being somewhat more closely packed than the latter, which have high initial porosity.

Initial packing and porosity and the contrast in these properties between adjacent layers at a variety of scales are amongst the prime causes of potential instability in sediments and their susceptibility to deformation. Initial conditions of deposition are not, however, the only controls on sediment instability. After all, most of the sediments seen in the stratigraphical record appear to have maintained their initial depositional configurations without subsequent deformation other than normal burial compaction. For sediments to become deformed shortly after deposition, prevailing stresses must exceed sediment strength for a period of time long enough for stresses to achieve perceptible deformation. In most cases, the stresses are the result of gravity, either acting vertically on layers of contrasting density or as a shear component parallel with an inclined sediment surface. In less common cases, contractional stresses associated with volume loss and compressive stresses associated with crystal growth may apply.

The strength property involved is usually the shear strength of the sediment, although in some cases tensional strength may be relevant. As indicated in sections 1.2.3 and equation (1.5), shear strength is a function of particle cohesion and intergranular friction, and the pore fluid pressure (sections 1.2.5 and 2.2.3; equation 1.13). Therefore, shear strength can be reduced by a loss of cohesion or a reorganization of grain packing to reduce $\tan \phi$, or through an increase in pore fluid pressure. Cohesive properties of a sediment are largely a function of grain size and the proportion of fine-grained, commonly clay, particles. These are inherent properties of the sediment and are not readily changed. The frictional component is most readily reduced by an increase in pore fluid pressure, so that more of the support of overlying sediment falls upon the pore fluid and less upon grain contacts.

Excess pore fluid pressures commonly develop in near-surface sediment as a result of an initially high porosity being maintained during burial, through the sediment having too low a permeability to allow fluid to escape and thereby maintain the normal fluid, or hydrostatic, pressure (section 1.2.5). This is most likely in materials dominated by fine-grained sediment, silts and clays, particularly where they are rapidly deposited. Such successions are underconsolidated and the excess pore fluid pressures (overpressured conditions) are likely to be sustained for long periods of time. Deformation processes may be slow and long-lasting. The origins of overpressuring and the various ramifications for sediment deformation are discussed in sections 1.2.5 and 2.3.

In coarse-grained, loosely packed sediments, loss of strength is likely to be short-lived (G. Owen 1987). It may result from loose initial packing or be triggered by some form of shock, causing instantaneous loading of the pore fluid, and a tendency for the grain packing to readjust (Figure 4.1). Closer grain packing means excess fluid. The increase in pore pressure consequent upon the shock may result in loss of strength until the fluid has escaped through the permeable sediment, usually to the surface, or the triggering effect has subsided. During the relatively short interval of time between shock and loss of excess fluid, the elevated pore pressure may lead to a complete or partial loss of strength and the sediment–fluid mixture may behave rheologically as a fluid or a plastic. Further aspects of the rapid loading of sediments are mentioned in section 1.3.2.

Total loss of strength, the result of the pore fluid pressure reaching lithostatic values, is known as liquefaction (section 1.3.2). The process of fluid loss whereby grain contacts are re-established and frictional strength restored will be by intergranular flow to the sediment surface. If this is sufficiently vigorous the moving fluid may carry particles with it and support them in the fluid, the process known as fluidization. Fluidization, as a result of sediment dewatering is, of necessity, short-lived, as it depends upon the finite volumes of the excess pore fluid expelled. In natural examples it tends to be localized in narrow pipes and channels, within which particles are completely reworked from their original depositional packing. The grains may be significantly translated within the sediment, with elutriation of finer particles.

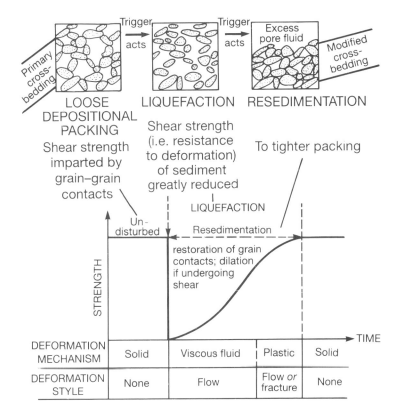

Figure 4.1 Model for the loss of shear strength in sands, its restoration through the establishment of tighter packing and the types of deformation that might prevail during these transformations. (After G. Owen 1987.)

Such sediment–water mixtures are commonly taken to behave rheologically as a Newtonian fluid (sections 1.2.4 and 3.6.2), in which case deformation will begin as soon as applied shear stresses exceed zero. The character of the deformation will therefore be controlled by the viscosity of the mixture. For liquefied and fluidized units, however, the dissipation of excess pore pressure and the draining of pore fluid will both give rise to eventual and commonly rapid re-establishment of frictional contact between grains and the return of strength to the sediment–fluid mixture.

Where the sediment is less well sorted and especially where there is a considerable content of clay particles, the sediment–water mixture will have a cohesive component to its strength. It may then behave as a rheological plastic (hydroplastic of Figure 4.3), which can sustain a certain value of shear stress before deformation begins. Once deformation does begin, it will continue until the applied shear falls below some critical value, probably the shear strength of the sediment–fluid mixture (Figure 4.1).

For both fluid and plastic states, the rapid re-establishment of shear strength is crucial to the preservation of most deformational structures. Such structures represent, in effect, snapshots taken at the point in the deformational process where strength was re-established.

In one group of structures, the sediment does not lose strength through the breakdown of intergranular friction, but rather it is disrupted very locally through the effects of tensional stresses. In some cases these act parallel with the depositional surface, such that the tensile strength of the near-surface material is exceeded and the sediment breaks up into a system of cracks. In order that the broken surface may be preserved it is also necessary that the sediment has significant cohesive strength, which commonly means a significant content of clay-grade material. In other cases, tensile strength is exceeded locally within some larger deformational package, with the result that extensional faults develop (Figure 4.2).

The stresses generated by crystal growth, especially of evaporite minerals at or close to the sediment surface, are sufficiently large to cause deformation of both the precipitating mineral layer and the host sediment in which precipitation is taking place.

DEFORMING FORCE \ DEFORMATION MECHANISM		Exceed strength		Reduced yield strength			Liquidize	
		Brittle	Plastic	High pore fluid pressure	Elevated temperature	Creep	Liquefied	Fluidized
Gravitational body force on slope		Slides	Slumps				Debris flows	Debris flows
Unequal confining load		Growth faults	Loaded ripples	Shale ridges Convolute lamination			Loaded ripples and sole marks	Loaded ripples and sole marks
								Clastic dykes Sand volcanoes
Gravitationally unstable density gradient	Continuous	Soft-sediment faults (mainly extensional)					Convolute lamination	
	Within a single layer	Soft-sediment faults (mainly extensional)			TECTONIC		Dish structures	Water-escape pipes and pillars
	Multiple layers not pierced	Soft-sediment faults (mainly extensional)					Load casts	Load casts
	Multiple layers pierced	Soft-sediment faults (mainly extensional)		Clay diapirs (mudlumps)		Salt domes	Ball and pillow/pseudonodules	Ball and pillow/pseudonodules
Shear stress	Current drag	Soft-sediment faults (mainly extensional)					Overturned cross-bedding	
	Vertical	Soft-sediment faults (mainly extensional)						Dish structures Pillar structures
Other	Physical	Desiccation and synaeresis cracks			TECTONIC			
	Chemical		Concretions Crystal growth					
	Biological		Bio-turbation					

Figure 4.2 A classification scheme for sediment deformation based on the interaction of loss of strength and the nature of the applied force. (Modified from G. Owen 1987.)

4.3 PHYSICAL DEFORMATION STRUCTURES

Structures resulting from post-depositional deformation may be classified in a variety of ways based on different geometric and genetic criteria (Figures 4.2 and 4.3). This account adopts essentially genetic criteria, which take into account both the degree to which sediment strength has been lost and the nature of the forces that produce the deformation during the period of reduced strength.

situation and reduction of shear strength of the layers can lead to foundering of the denser layer into the less dense layer, with the resultant formation of a set of structures that can all broadly be attributed to **loading.**

Load casts are most commonly seen as deformations of the interface between a sandy layer and an underlying layer of finer grain size, commonly mud or silt. Where the interface is seen, as on the base of a lithified sandstone bed, the sand is deformed downwards in a series of, commonly, rather equidimensional rounded lobes (Figure 4.4). Between the lobes are narrower

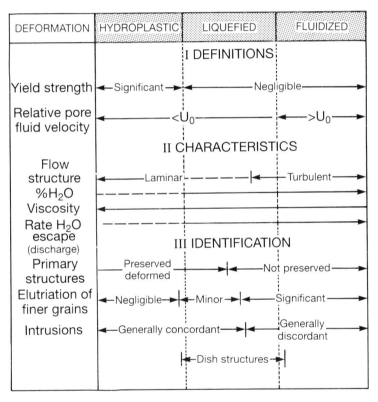

Figure 4.3 Processes of sediment deformation and the associated physical characteristics, behaviour and products. (After Lowe 1975.) 'Hydroplastic' here refers to the state of a sediment–water mixture with a significant yield strength, that strength being the result of cohesive or frictional forces. It is synonymous with 'rheological plastic' of the text.

4.3.1 Partial loss of strength and density inversion

Structures falling within this group are characterized by essentially vertical movements and are the results of gravitational body forces. They were introduced in section 1.3.2. The structures commonly result from the occurrence, within a sequence of sedimentary layers, of density inversion, so that a more dense layer overlies a less dense layer. This is a potentially unstable

grooves or channels, which represent zones of upward-moving mud (Figure 4.5). The sandy lobes are commonly referred to as load casts whereas the narrower muddy zones are **flame structures**, so called from their appearance in vertical section (see below). Although most commonly equidimensional in plan view, load casts may also be elongate, often with a preferred orientation. This may reflect a gradient to the sediment surface such that there is a component of downslope movement associated with the

Figure 4.4 Load casts on the base of a sandstone bed in an interbedded sandstone–mudstone sequence. The load casts do not appear to show any preferred elongation. Bude Formation, Upper Carboniferous, north Cornwall.

be parallel with the margin whereas the core is commonly more intensely disturbed, approaching convolute lamination (e.g. Brodzikowski and Haluszczak 1987; see below). In some cases it may be possible to discern the nature of the original lamination, as in the case of loaded ripples, where the cross-lamination can be seen. Where the sediments show lamination, flame structures show lamination parallel with the margin, with greater contortion in the centre.

In most cases, load casts occur on the base of a continuous bed of coarser grained sediment. In some cases, however, the process of vertical foundering leads to the total disruption of the coarse layer so that it occurs only as **isolated load balls** or **pseudonodules** floating in a matrix of finer grained sediment (Figure 4.6; Eschmann 1992).

Figure 4.5 Thin-bedded, fine-grained sandstones interbedded with dark mudstones. The upper sandstones have very small load casts on their bases, whereas the somewhat thicker beds in the lower half of the photograph show more extensive loading, almost developed to the point of detachment. Complex flames of mud extend up between the load balls, which have highly convoluted internal lamination. The mineralization of the small faults suggests that they are later tectonic features and unconnected with soft-sediment deformation. Bude Formation, Upper Carboniferous, north Cornwall.

loading. It may result from the loading process accentuating some earlier elongate feature of the layer interface, such as erosional sole marks or ripple forms in the upper layer.

In vertical section, load casts commonly show internal contorted lamination (Figure 4.5). Close to the edge of the structures, lamination tends to

Diameters of load balls, both attached and isolated, probably scale roughly with the thickness of the layer that has foundered, although in cases where loading has accentuated pre-existing structures, that relationship may not apply.

In some cases of loading, there is no obvious difference in grain size between the two layers,

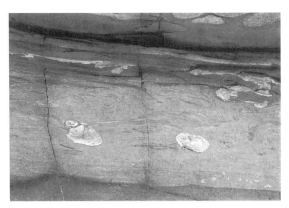

Figure 4.6 Isolated sandy load balls or pseudonodules floating in a unit of slurried siltstone that has undergone total liquefaction. Pseudonodules are around 5 cm across. Bude Formation, Upper Carboniferous, north Cornwall.

Figure 4.7 Convolute bedding in well-laminated sands. The sharp anticline and the more gentle synclines are typical of relatively simple convolute bedding and lamination. The core of the anticline is characterized by more intense deformation and the whole unit is truncated above by an erosive contact. Recent river terrace deposits, Tana Valley, Finnmark, Norway.

and sand appears to have sunk into similar sand. This occurs most commonly in thick-bedded, amalgamated turbidites. It appears that quite subtle differences in texture combined with the rapidity of the deposition lead to the instability. Sand-on-sand loading is also produced in intertidal sand bodies as the result of the entrapment of air during the rising tide. Overpressuring leading to loss of strength, as explained in section 4.2, is commonly thought of in terms of pore water, but in this situation compression of trapped air is thought to lead to strength loss and consequent density inversion (De Boer 1979). Such load structures, within a more or less uniform lithology, are gradational in character, with convolute lamination and bedding.

Convolute bedding and **convolute lamination** are structural styles that occur most commonly within a single bed, usually of silt- or sand-grade material. The terms are to some extent interchangeable, with a loosely defined size limit separating them. Deformation at the scale of decimetres or above would be 'bedding' and at the scale of centimetres 'lamination'. The structures are seen most commonly in vertical section, although they also appear in plan view.

They commonly involve original depositional lamination being folded and contorted. In some cases the folds may be upright and cuspate, with rounded synclines and sharper anticlines (Figure 4.7). In other cases a more chaotic style prevails. Where sharp anticlines are seen, these may reflect both an internal foundering within the deformed layer, akin to flame structures (see above), and also preferred routes for upward escape of excess pore water (Figure 4.8). In originally cross-bedded sands, it is sometimes the case that deformation is confined within a single set, with deformed laminae truncated by the overlying bounding surface. In other cases, the bounding surfaces themselves may be involved in the deformation, with multiple sets clearly having deformed together (Lang and Fielding 1991).

Some of the largest examples of convolute bedding occur in sands deposited as aeolian dunes and which were liquefied beneath the water table following burial (e.g. Doe and Dott 1980). These examples may involve sedimentary units up to several tens of metres thick (Figure 4.9). Deformation commonly crosses set-bounding surfaces and only rarely is truncated above by a bounding surface. More commonly, deformation dies out vertically both upwards and downwards. In some cases, deformed cross-bedding passes gradationally into more or less structureless sand within which it is possible to detect only very weakly defined, disturbed lamination. It appears that these structureless units represent the most extreme state of deformation, with the convolute cross-bedding being an intermediate state between structureless sand and undisturbed cross-bedding.

Aeolian sands tend to be very well sorted and to have high initial porosity, properties which

Sedimentary deformational structures

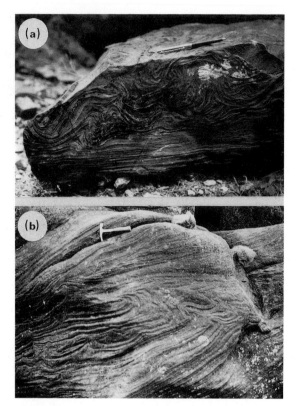

Figure 4.8 Convolute bedding within coarse-grained, cross-bedded channel sandstones. The sharp overturned anticline suggests water escape, perhaps coexisting with shear at the sediment surface. Roaches Grit, Upper Carboniferous, Staffordshire.

may make them susceptible to loss of strength when waterlogged (e.g. Doe and Dott 1980; Horowitz 1982; Collinson, Bevins and Clemmensen 1989). Total or partial liquefaction may be triggered by a shift in the water table, seismic shocking or rapid sediment loading. Exceptionally, convolute bedding in aeolian sands may result from a rapid rise of the water table through the dune sands as a result of rapid inundation of a desert area by the sea (e.g. Glennie and Buller 1983).

In some beds that display convolute lamination or bedding, the intensity of deformation increases upwards, often with undisturbed lamination towards the base. Although folds are often upright, overturning in a preferred direction also occurs. In plan view, the lamination is commonly seen to be in the form of basins and intervening ridges.

Convolute bedding and lamination records the internal foundering of liquefied sediment layers upon themselves, commonly in conjunction with active upward escape of pore water. Normally this appears to have been driven purely by gravitational forces normal to the deforming layer. Where anticlinal folds are overturned in a preferred direction, these forces may have operated in conjunction with a down-slope component of mass movement or with an applied shear from water flowing across the top of the layer (see also overturned cross-bedding).

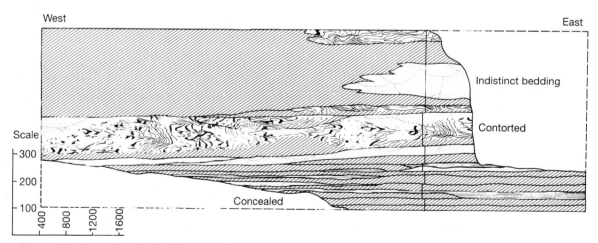

Figure 4.9 Somewhat simplified diagram of large-scale convolute bedding in aeolian dune sandstones. Note that contorted units pass upwards into units with indistinct bedding. An active water table was probably responsible for the loss of strength, the contorted bedding reflecting short-lived and partial loss, the indistinct layers total liquefaction with loss of lamination. (After Doe and Dott 1980.)

Convolute lamination is commonly rather irregularly distributed in a sediment unit, often localized within an otherwise intact bed. It appears that in many cases the liquefaction and loss of strength were spontaneous consequences of the early post-depositional state of the sediment layer (e.g. Allen 1977), although external triggering may be invoked where the disturbance is very widespread or occurs in an unusual setting. In certain turbidite sandstones, convolute lamination seems to be especially associated with the Bouma 'C' division of ripple cross-lamination and appears to have developed virtually synchronously with deposition.

The deformation of aeolian dune sands mentioned above is an example where introduction of water is the destabilizing agent. In other cases, heavy wave action on shallowly submerged aqueous bedforms may cyclically load the sediment to instability (Dalrymple 1979; section 1.3.2). Where convolute bedding is confined within single cross-bedded sets, it is likely that the loss of strength was a product of rapid deposition and the grain-size characteristics of the sediment. Where deformation involves several sets and their bounding surfaces it is more likely to have been triggered by an external agent, probably an earthquake (e.g. Reimnitz and Marshall 1965; Davenport and Ringrose 1987). Where seismic activity is thought to have been the triggering mechanism, the areal distribution of convolute lamination and bedding within sands of fluvial channel origin has been used as an indicator of which faults were syn-depositionally active (Allen 1986a; Leeder 1987).

The use of liquefaction structures for the characterization of palaeoseismicity is discussed further in section 9.5.2.

4.3.2 Structures due to progressive loading of cohesive sediment

Where fine-grained cohesive sediments are progressively but fairly rapidly buried by younger sediments they may develop overpressure, and lateral and vertical flowage may ensue. Gravitational instability may lead to the development of diapiric structures. The most common types of structure produced by this type of behaviour are the **mudlumps** or **mud diapirs** which rise through prodelta and mouth-bar sands of rapidly prograding deltas. These are particularly well seen off the mouths of the major distributary channels of the present-day Mississippi Delta, where muds from a burial depth of many tens of metres rise up through later sediments and, in many cases, emerge above water level as low temporary islands (Figure 4.10; Morgan, Coleman and Gagliano 1968). These are commonly eroded over a number of years by wave action and the muds dispersed for redeposition. Internally, the diapiric muds commonly show steeply dipping discrete shear surfaces and a characteristic small-scale brecciation. There are many analogues with the large-scale diapirs and diatremes found, for example, at convergent plate margins (section 7.1). In addition, large-scale mud diapirs occur on the lower parts of prograding continental slopes off the fronts of major muddy deltas such as the

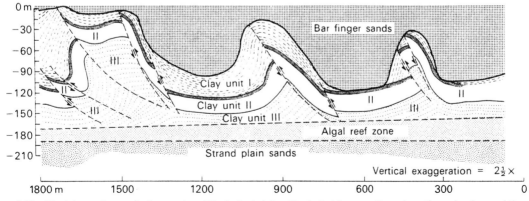

Figure 4.10 Diapiric mudlumps in the modern Mississippi delta, illustrated by a section along the axis of one of the distributary channels. Note the high-angle reverse faults associated with the crests of the diapirs and the marked thickening of the mouth-bar sands into the areas between the diapirs. Contrast the scale of these structures with the large diapirs seen on the lower parts of progradational continental slopes (Fig. 4.26) (after Morgan, Coleman and Gagliano 1968).

Mississippi and Niger. These are associated with the lateral flowage of deep, overpressured clays from beneath the deltaic depocentre towards the free surface (section 4.3.5).

The close association of mudlump diapirs with the site of most active and rapid sedimentation in the mouth bar, and the coincidence of mudlump emergence with periods of flood deposition, suggests that their growth is rapid and closely related to sediment load applied to the overpressured layer.

The emergent mudlump islands are thought to represent peaks or spines on the crests of shale folds or ridges, which are initiated by the basinward lateral flowage of clays in front of the advancing mouth bar. As folds are progressively overtaken by this progradation, the movements take on a more vertical component and, in a sense, the diapiric movements which then dominate are akin to those associated with sediment loading (see above). Ancient examples of mudlump diapirism within deltaic sediments are well known from the rock record (Figure 4.11), those of the Namurian of Country Clare being particularly well displayed, with both penetrative and non-penetrative types recognized (Pulham 1989).

As well as deforming both the diapiric muds and the overlying sediments, this type of behaviour also affects local depositional patterns, especially the thickness of overlying units. As diapirs move upwards, mud is withdrawn laterally from an area around the mudlump or shale ridge, and this leads to the development of local depocentres between and around the diapirs. Mouth-bar sediments associated with diapirs show patterns of spectacular thickness change as a result (Figure 4.10).

At a larger scale, diapiric deformation occurs in association with the burial of thick units of evaporitic halite, which deforms in a viscous fashion to form large salt domes, pillars and sheets. Associated features, such as crestal faults and rim synclines, are broadly similar to those of mud diapirs although on a larger scale. Deformation of this type takes place after deep burial and occurs so long after deposition that it can hardly be considered 'early'.

Figure 4.11 Small mud diapir which has burst through overlying mouth-bar sands in a deltaic sequence. The mud is derived from a unit around 10 m thick, in the base of the upwards coarsening unit and the whole deformation is confined within that progradational interval. Other diapirs within the same succession are non-penetrative. Figure (arrowed) for scale. Central Clare Group, Upper Carboniferous, County Clare, western Ireland.

At a smaller scale and in other settings, deformation of this type is probably rather rare. It has, however, been suggested for folding and convolution in Devonian siltstones underlying a laterally restricted set of large-scale cross-bedding, probably of aeolian dune origin (B.G. Jones 1972). The folding in the siltstone is most intense under the axial part of the overlying dune set and asymmetric folds verge away from the axis (Figure 4.12). It is suggested that the progressive advance of the aeolian dune loaded the cohesive silts and caused them to flow laterally before they were fully buried.

material moves. Slumps are distinguished from slides in that the moving material is itself subjected to internal plastic deformation as it moves. The relative movements of different masses of material within the moving layer give rise to a spectrum of deformational structures. These range from extensional structures with normal fault geometries in the upslope head region of the slump, to compressional folds and thrusts in the toe region. The patterns may be very complex and difficult to decipher in terms of processes. The topic is dealt with in Chapter 5, and numerous illus-

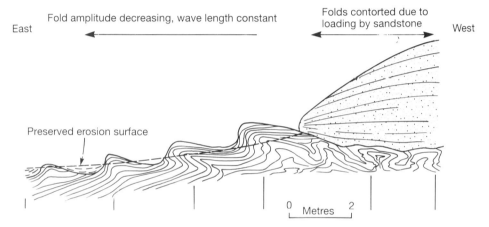

Figure 4.12 Schematic reconstruction of the development of contortion in a siltstone unit as the result of the advance of a large aeolian sand dune across its surface. Langra Formation, Upper Devonian, central Australia.

4.3.3 Partial loss of strength and applied shear

Structures described here are the product of a temporary loss of strength during a time when the weakened layer was subjected to a shear force, due either to the sediment resting on a slope (i.e. a downslope component of gravity body force) or to the effect of an overriding flow on the upper surface of the layer. The first situation gives rise to the phenomenon of mass movement, which was introduced in section 1.3.3 and is comprehensively reviewed in Chapter 5. Perhaps the most commonly seen consequences of the downslope movement of near-surface sediments are **slump folds**, the result of the movement of weakened sediment on a slope, in response to the downslope component of gravity. Within slumped masses, the greatest shear strain is commonly concentrated on the basal slip surface above which the slumping

trations are provided of the resulting structures.

Overturned cross-bedding is a common style of deformation in sandstones of shallow marine and fluvial origin where powerful currents were involved in the migration of dune-scale bedforms. Overturning of foreset laminae is confined within single cross-bedded sets, and is always in the direction of the foreset dip (i.e. down current) (Figure 4.13). In some examples, the overturning is restricted to the uppermost part of the set, whilst in others virtually the full thickness of the set is involved (Figure 4.14). In all cases the degree of steepening and overturning increases upwards through the set. Although most cases involve simple overturned folds, some examples show more complex folding and/or the loss of definition of lamination (Allen and Banks 1972; Hendry and Stauffer 1977; Doe and Dott 1980).

Sedimentary deformational structures

Figure 4.13 Overturned cross-bedding in coarse sandstones of probable shallow marine origin. Kap Holbæk Formation, Lower Cambrian, Danmarks Fjord, N.E. Greenland.

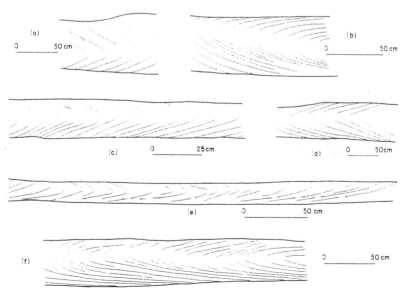

Figure 4.14 Variations in the geometry of overturned cross-bed foresets from a variety of settings. Note the variable position of the fold axis. (After Allen and Banks 1972.)

The structures result from the loss of strength of the cross-bedded set very soon after it was deposited, so that the overriding current, which produced the bedform in the first place, was still active. The boundary shear stress exerted by this current was able to shear the weakened sediment layer and cause essentially laminar (viscous) shear within the layer (Figure 4.15b). Re-establishment of shear strength within the deforming layer appears to have taken place from the bottom upwards as excess pore fluid escaped (Figure 4.15a). As a front of reconsolidation migrated upwards, higher levels in the set would be subjected to progressively longer durations of shearing, and hence the degree of overturning would increase upwards (Figure 4.15c; Allen and Banks 1972; Allen 1984). In Figure 4.15b, a tabular set of thickness h, which was instantaneously liquefied with a grain concentration, gc, is sheared by an overriding current such that the velocity of the surface layer is U. The instantaneous velocity (u) of sedimentary particles at a height y_0 above the bed is determined by the slope of the velocity profile, itself a function of the applied shear and the viscosity (μ) of the sediment dispersion. The total displacement (X) of a particle (Figure 4.15b) at a particular height above the base of the bed is an integral of its changing horizontal velocity (u) as the base of the liquefied layer moves upwards at velocity v. The resultant

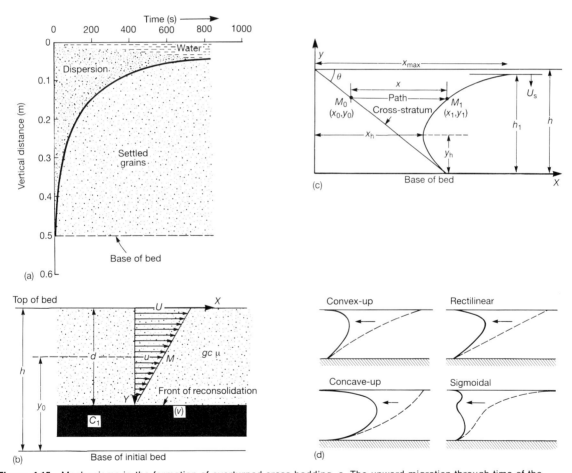

Figure 4.15 Mechanisms in the formation of overturned cross-bedding. a. The upward migration through time of the front of redeposition following initial loss of strength. b. Definition diagram of simple shear operating on the liquefied bed as the front of redeposition moves upwards. c. Deformed foreset trace of an initially planar foreset following simple shear and progressive base-upwards resettlement. d. Influence of original foreset shape on the form of the overturned foresets. (After Allen and Banks 1972; Allen 1977, 1985.) The symbols in (b) and (c) are discussed in the text; further details are given in Allen and Banks (1972).

values of X through the bed determine the shape of the overturned foresets. A suite of different curves may be generated depending on values attached to particle concentrations in the liquefied and consolidated layers, the particle settling velocities (U_s) within the liquefied layer and the shear rate. The curve generated on Figure 4.15c derives from an initially planar foreset. More complicated shapes arise from the deformation of initially more complex foresets (Figure 4.15d).

4.3.4 Structures related to upwards escape of pore water and sediment–water mixtures

As well as the upwards deflection of depositional lamination sometimes seen in association with convolute lamination and bedding, the escape of excess pore fluid to the sediment surface may also generate structures in its own right and sometimes in sediment which otherwise would be structureless.

Very rapid deposition of sandy sediment from suspension, as when a heavily loaded, powerful current decelerates rapidly, will commonly lead to the trapping of large volumes of pore water. This will tend to escape upwards as the sediment settles and the escape may take several forms.

Dish-and-pillar structures are subtle and rather uncommon features which reflect the local clay enrichment of small zones within sand (Lowe and Lopiccolo 1974; Lowe 1975). The dish structures are small, concave-upwards features which, in plan view, are roughly equidimensional dishes of a few centimetres diameter (Figure 4.16). Pillar structures have more or less vertically elongated zones which cross the dishes and often stem from their upturned edges.

The clay particles, whose enrichment defines the dishes, are thought to have been carried upwards with the escaping pore waters and to have been filtered out at slight inhomogeneities – possibly incipient lamination – within the sand. Once initiated, the reduced permeability of a clay-enriched layer would tend to filter out more clay and also force the escaping waters sideways. Vertical escape would be increasingly around the margins of the developing dishes, dragging them upwards, and concentrated in vertical conduits – the pillar structures. These may exhibit a slightly cleaner texture owing to the elutriation of some of the clay.

In late Quaternary glacial outwash deposits, Cheel and Rust (1986) described dish structures as the last of a series of liquefaction products. Their inferred development is illustrated in Figure 4.17. Initially, local fluidization of low permeability sediment at the base of the unit generated convolute lamination, with disrupted anticlines evolving into ascending vertical diapirs or coalescing flames. In the overlying, less weakened sediment, ball-and-pillow structures formed, owing to the penetration of the rising diapirs and the foundering of more dense material above. The lower burial pressures at the top of the layer allowed the fluid pressure to decrease, so that the rising material passed from

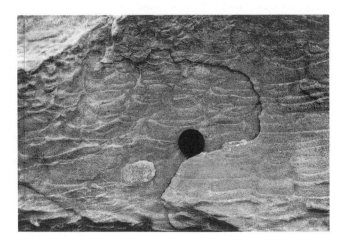

Figure 4.16 Dish structures in thick, massive turbidite sandstone bed. Note how the rather flat dishes in the lower part of the unit pass upwards into shorter, more curved dishes higher in the bed. The upturned ends of the dishes connect with pillar structures which acted as the main conduits of water escape. Eocene, San Sebastian, Spain.

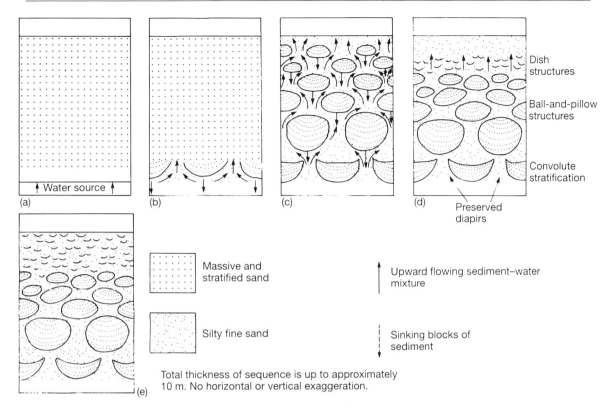

Figure 4.17 A model for the development of convolute bedding, ball-and-pillow structures and dish structures within glacial outwash deposits. The excess pore fluid pressure may have resulted from shocking or from the melting of buried ice within the sediment. The genetic model is based on an idealized composite model (e) for the distribution of the structures in the beds. (After Cheel and Rust 1986.)

a fluidized to a liquefied state. The final stages of upward movement of fluid in the liquefied sand generated dish structures at the top of the unit.

Lowe (1975) envisaged dish-and-pillar structures as developing optimally under conditions of moderate water escape (Figure 4.18). Under conditions of normal upward seepage, original structures are undisturbed, whereas at high rates of escape the liquefaction of the bed may be so great that major internal foundering of sediment masses (internal loading) occurs. Dish structures form in the middle parts of a bed while the more rapid water escape towards the top leads to liquefaction and to the development of convolute lamination (Figure 4.18).

In coarse-grained, poorly sorted sediments that appear to have been deposited very rapidly, water escape during a phase of liquefaction appears to have led to downward settling of coarser clasts. This apparently gave rise to pebble clusters and pocket structures, which are bowl-shaped concentrations of granules and pebbles a few centimetres wide (S.Y. Johnson 1986). In addition, some elongate pebbles may be reorientated subparallel with the direction of water escape.

Sheet dewatering structures are similar to pillar structures but are more linear in plan view and may occur independently of dish structures in otherwise structureless sands (Laird 1970). They are often characterized by cleaner sand than that of the host sediment and may be closely spaced within the bed. They may also coalesce vertically upwards, sometimes defining horizontal zones (Figure 4.19).

In the case of the above structures, the escaping pore waters locally modify the texture of the sediment by preferentially enriching or depleting it in particular components. The water escape does not involve the wholesale translation

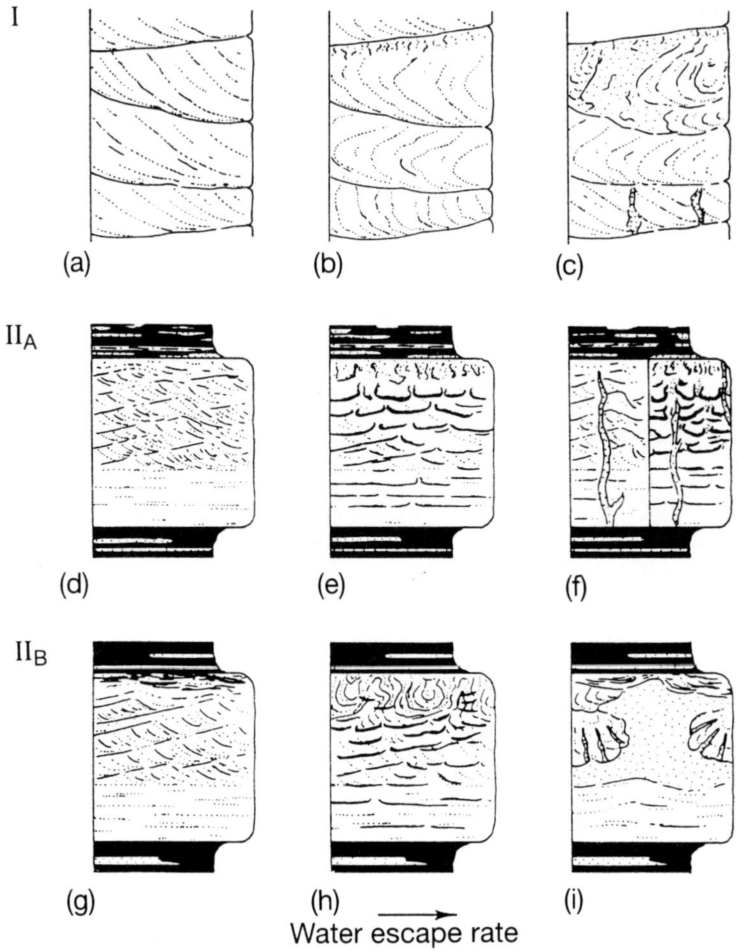

Figure 4.18 Schematic variation in water escape structures in response to rate of water escape and original depositional facies: I, medium to coarse-grained cross-stratified sandstones; IIA, turbidites with non-cohesive Bouma C and D divisions; IIB, turbidites with cohesive Bouma C and D divisions. Left-hand column indicates undisturbed depositional structures, preserved as a result of low water-escape rates. Middle column, at moderate escape rates, gives rise to dish structures and to small water-escape pipes in non-cohesive sediment with convolute lamination disturbing the upper parts of more cohesive units. Highest rates of water escape (right-hand column) give rise to more general liquefaction and loss of lamination and to more strongly penetrative, vertical water-escape pillars. (After Lowe 1975.)

of sediment, but such bulk movement does, however, occur. The resultant structures are described below.

Sediment injection structures are increasingly recognized in the rock record, mainly in the form of dykes, although sills and associated extrusive features are also sometimes found. Historically, they were amongst the first sediment deformation structures to be recognized and some early investigations are recalled in section 1.3.4. Sedimentary dykes occur in a range of depositional settings, from deep-water turbidite associations (e.g. Truswell 1972) to subaerial mass-flow deposits (e.g. Collinson, Bevins and Clemmensen 1989). They are also intruded into a range of host lithologies, whose main property must have been to have had strength at the time of injection. In practice, this means that the host must have at least some clay material in its overall composition in order to impart some cohesion to the material. Dykes vary in width up to several tens of centimetres and they may extend over many metres vertically (Figure 4.20). Both dykes and sills may wedge out rapidly. Dykes may become folded, sometimes into tight ptygmatic folds when the host has undergone significant compaction on burial (e.g. Truswell 1972). It is possible that sandstone dykes may be confused with sandy fills of desiccation and synaeresis cracks if seen only in vertical section (see below). As remarked in section 1.3.4, such passive fissure-fills are sometimes called neptunean dykes. In plan view, dykes are commonly random in orientation but some sets show a polygonal patterns whereas others may show a preferred orientation.

Both dykes and sills are commonly made up

Physical deformation structures 111

Figure 4.19 Sheet dewatering structures in a thick, massive turbidite sandstone bed. Note the upwards coalescence of the sheets, with merging taking place at discrete levels. Skipton Moor Grits, Upper Carboniferous, Yorkshire.

of rather structureless sandstone, although some show a marginal foliation parallel with the walls, reflecting shearing during intrusion. Where intrusions broke through to the contemporaneous sediment surface, sand volcanoes and/or extruded sand sheets may occur (see below).

Sand dykes and sills are the result of the forcible intrusion of liquefied sand into a cohesive host. The intrusion is probably most commonly from below, although it may in some cases have been sideways. In all cases the source of the liquefied sand must have been buried by less permeable cohesive sediment, in some cases for a considerable period of time prior to intrusion. Some sandstone dykes, which cut across and therefore post-date septarian carbonate nodules (see below) in the mudstone, must have been intruded quite a long time after deposition (e.g. Martill and Hudson 1989). In fact, there is no reason, in principle, why sedimentary dykes cannot arise in deeply buried sediments, provided the overpressuring is sufficient to overcome any strength increase due, for example, to cementation or compaction that occurred during the burial history. Section 9.5.2 mentions some examples that are thought to have formed

Figure 4.20 Vertical sandstone dyke cutting through poorly sorted conglomerate of mass-flow origin. Weak foliation parallel to the walls of the dyke suggest shearing of liquefied sand during emplacement. Moraenesø Formation, Precambrian, north Greenland.

at substantial burial depths in association with regional tectonism.

The process of liquefaction may have resulted from sudden shock, such as earthquakes (Reimnitz and Marshall 1965), or the sudden emplacement of the cohesive layer itself, as when They occur on the tops of both sandstone and more muddy beds that have been overlain by fine-grained sediment deposited from suspension (Gill and Kuenen 1958).

Sand volcanoes are broadly conical mounds, ranging in diameter from a few centimetres to a

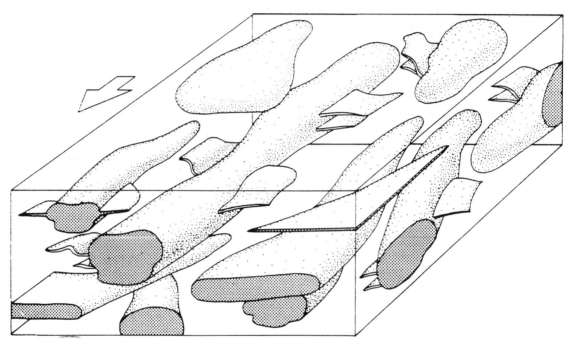

Figure 4.21 Schematic block diagram of the geometry of submarine sandy gulley fills in the Harelv Formation (Jurassic) of east Greenland. The gulleys were cut into slope mudstones and filled my mass-flow processes. The wing-like features on the flanks of the sandbodies are sand sheets intruded into the muds, probably as a result of overpressuring of the succession during burial. (After Surlyk 1987.)

subaerial mass flows are deposited on waterlogged sands (e.g. Collinson, Bevins and Clemmensen 1989). In other cases, the progressive burial of sands by fine-grained sediments may lead to increasing overpressure in the sands until they eventually break out into the surrounding sediments. This appears to have been the case in certain ancient channelized deepwater sands (e.g. Surlyk 1987; Newton and Flanagan 1993; Thomas, Collinson and Jones 1992; Figure 4.21).

Sand volcanoes and **extruded sand sheets** are relatively rare features in the geological record but are sometimes found on bedding surfaces. Sand volcanoes are preserved mostly in relatively deep-water sediments involving turbidites and slump deposits, whereas sheets are more widespread, occurring even on subaerial surfaces.

few metres (Figure 4.22). They commonly show many of the features of igneous volcanoes but on a much smaller scale, with central craters and flanking flow lobes. In vertical section they show an axial pipe analogous to an igneous vent, and inclined layers dipping radially away from the axis. In some cases it may be possible to trace the pipe downwards into the underlying layer and identify the source bed from which the sand had been expelled. In order for a sand volcano to form, a source bed must have become sufficiently overpressured for liquefied sediment to break out upwards to the depositional surface. Some type of shock may have been involved in the liquefaction process. As the sediment–water mixture was extruded at the surface, the excess water would rapidly escape and the flow lobes would become frozen in a

Physical deformation structures 113

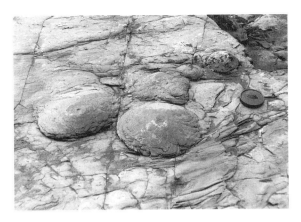

Figure 4.22 Sand volcanoes on the upper bedding surface of a sandstone bed. The sand of the volcanoes may have been sourced from the bed itself or from underlying layers. Note the axial vents and the preservation of delicate flow lobes on the flanks of the volcanoes, attesting to the quiet nature of the water at the site of the extrusion. Ross Formation, Upper Carboniferous, County Clare, western Ireland.

Figure 4.23 Folding in the transformation of a sandstone dyke (bottom right) into an extruded sandstone sheet (top left). The sand is intruded into poorly sorted conglomerate of mass-flow origin which, itself, must have been in a very mobile state at the time of intrusion–extrusion. Moraenesø Formation, Precambrian, north Greenland.

way somewhat analogous to the sieve lobes that are so commonly found in the stockpiles of sand-washing plants in quarries (e.g. Carter 1975). Very quiet conditions must have ensued in the basin for such delicate features to have survived until they were blanketed by fine sediment from suspension.

Extruded sand sheets are sometimes indistinguishable from beds deposited as part of the normal sequence. Their true identity commonly depends on observation of a transition from an intrusive dyke to a concordant bed (Figure 4.23). In some cases the distinction may not be totally clear, as some extruded sheets may be confused with sills. In plan view, extruded sheets commonly show signs of internal shearing in the form of weakly developed and folded foliation (Figure 4.24) and their upper surfaces may be modified by regular hydrodynamic bedforms, such as current and wave ripples. Extruded sand, in the form of a mobile slurry probably lost mobility very rapidly as the excess pore fluid drained to the surface, causing the extruded flow to 'freeze' (Hesse and Reading 1978; Collinson, Bevins and Clemmensen 1989). Modern examples are documented from areas subjected to intense seismic shocking (e.g. Swanson 1964; Reimnitz and Marshall 1965).

4.3.5 Synsedimentary faults

Faulting of very early post-depositional origin is quite common, occurring both in otherwise undeformed sequenes as well as in association with other deformation features. The results are not always readily distinguishable from later faults, although a useful rule of thumb is that early faults are seldom mineralized and commonly show smearing of sediment along the fault surface. This matter is discussed further in section 9.6.3.

Both extensional and contractional faults occur. Where the features occur in otherwise undeformed strata, they are most likely to be of extensional origin, that is, normal faults. Throws range up to a few metres, although more commonly they are at the scale of millimetres or centimetres. The extensional stress field may reflect contemporaneous tectonic activity or it may result from mass-movement of sediment on a slope. The head zones of slump and slide sheets are commonly characterized by such extensional movements.

Within larger, more complex deformation, such as convoluted layers or mud diapirs, both extensional and contractional faults may occur, reflecting the local stress field within the larger feature, probably at a late stage in its development when shear strength was being re-established.

At the larger scale, syndepositional growth

Figure 4.24 Irregularly folded foliation in an extruded sandstone sheet, suggesting internal flowage within the sheet prior to its consolidation. Moraenesø Formation, Precambrian, north Greenland.

faulting occurs, driven by gravity acting over a prograding delta slope and possibly driven by mass flowage at depth. Such faults, at the smaller end of their size range, may be confined to single progradational cycles, as in the Namurian of County Clare, western Ireland (Rider 1978; Pulham 1989), or to a small number of cycles, as in the Triassic of Svalbard (Edwards 1976; Figure 4.25). The largest occur at the scale of the continental slope and involve packages of progradational cycles, as in the Niger Delta (Figure 4.26) and the Gulf Coast of the southern USA (e.g. Weber and Daukoru 1975; Winker and Edwards 1983). All such faults have listric geometries and complex associations of synthetic and antithetic faults. Sequences thicken markedly into the hangingwall areas and there is commonly significant roll-over into the fault.

4.3.6 Structures due to sediment shrinkage

Structures dealt with above are all the product of loss of shear strength as a result of increased pore pressure. Volume loss of sediment may also lead to deformation through the development of extensional stresses. Layers of muddy sediment at the depositional surface may sometimes suffer a volume loss that cannot be fully accommodated by a loss of thickness. The result is that a tensional stress field is established within the shrinking layer. Once the tensile strength of the layer is exceeded it will begin to crack. For a horizontal layer, the tensional stress is likely to be homogeneous and the resultant crack pattern will be a system of broadly similar polygons, commonly hexagons. The diameters of the polygons seem to scale roughly with the thickness of the cracked layer and some surfaces may show coexisting large and small crack patterns. When the surface is sloping or where there are heterogeneities within the layer, the crack pattern may be influenced by these, with rectangular and elongated crack patterns developing (Donovan and Archer 1975).

Subsequent deposition of contrasting, commonly coarser grained, sediment above the

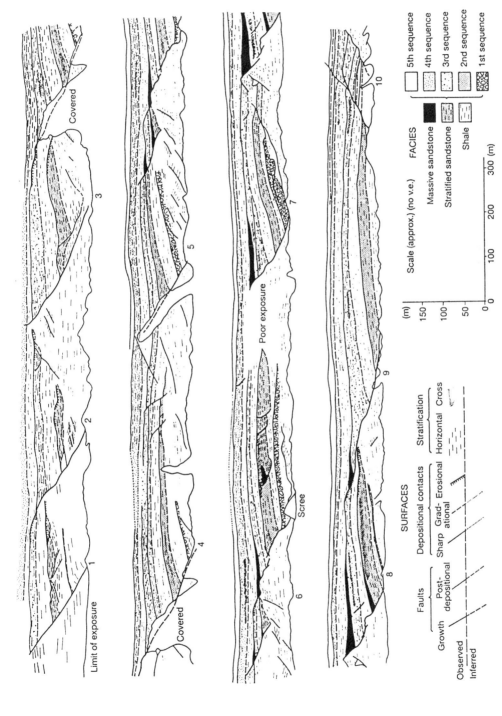

Figure 4.25 Succession of small-scale growth faults developed in deltaic sediments. There is slight overlap between the profiles from top to bottom. The faults stayed active during several phases of progradation, which was from left to right. Note the listric geometries of the faults and the thickening and roll-over in the hangingwall areas. The wedges of massive sand are infills of depressions that developed owing to continued movement on the faults after the last progradation. The apparent antithetic fault (10) may reflect a cuspate fault trace in plan so that faults 9 and 10 are the same, with a component of displacement out of the face. Triassic, Edgeøya, Svalbard. (After Edwards 1976.)

116 Sedimentary deformational structures

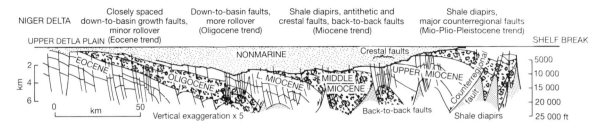

Figure 4.26 Large-scale cross-section through the Niger Delta complex showing extensive and deep-rooted growth faulting driven by basinward flowage of deeply buried, overpressured muds. Individual faults were active over long periods, involving many progradational cycles of the delta. The faults displace progressively younger sediments in a basinward direction. Near the free surface of the continental slope, major shale diapirs reflect the basinward flowage of the buried muds. (After Winker and Edwards 1983.)

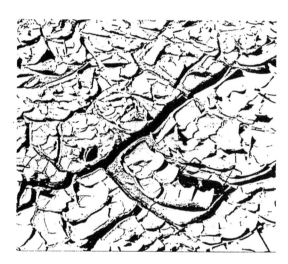

Figure 4.27 Desiccation cracks in present-day muds. Note the relationship between the diameter of the crack polygons and the depth of the cracking. The thin mud layer associated the smaller scale cracks is rolling up into potential mud clasts susceptible to future erosion. Sunglasses, bottom right for scale.

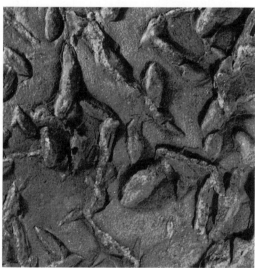

Figure 4.28 Casts of lenticular cracks on the base of a sandstone bed. Such cracks have been interpreted as being due to synaeresis, although some similar forms may result from evaporite pseudomorphs. Independence Fjord Group, Proterozoic, northeastern Greenland.

cracked surface will give sandy passive infills to the cracks. These commonly taper downwards but may be folded as a result of differential compaction of the host and crack-fill lithologies.

Shrinkage cracks occur in two main forms. **Desiccation cracks** result from sustained drying of a surface mud layer on an exposed river bed, lake floor or tidal flat (Figure 4.27). The cracks tend to be parallel sided and to have well-developed polygonal patterns. It is not uncommon for two or more scales of crack to coexist, smaller cracks reflecting the drying of a thin surface layer, and larger systems reflecting desiccation penetrating to a greater depth (Figure 4.27). Thin mud-cracked layers commonly roll up into convex upwards biscuits, which may then be subjected to erosion as mud clasts by the wind or by subsequent aqueous currents. These may become concentrated as mud-flake conglomerate layers above the cracked surface.

So-called **synaeresis cracks** are thought to develop in subaqueous settings where the contraction of the surface layer appears related to changes in the volume of clays, possibly as a result of changing water chemistry. Synaeresis cracks tend to be lenticular in plan and less inclined to develop into clear polygonal patterns (Donovan and Foster 1972; Figure 4.28). The details of the processes are not well known and the criteria for distinguishing cracks of this origin from those due to desiccation are somewhat ambiguous. Recently, Astin and Rogers (1991, 1993) have challenged the subaqueous origin for lenticular cracks of this type. They suggest that many, if not most, examples are the result of subaerial desiccation, with the lenticular forms being pseudomorphs of evaporitic gypsum crystals. Trewin (1992) and Barclay, Glover and Mendum (1993) have questioned this reinterpretation and suggested that cracks in organic-rich lacustrine mudstones are hardly likely to be other than subaqueous.

A more localized type of crack pattern, possibly of similar origin to synaeresis cracks, is that associated with certain carbonate concretions developed in mudstones. These concretions, called **septarian nodules**, are characterized by a pattern of irregular but somewhat radial cracks in their core. They are further described in section 4.3.9, together with explanations of their origin which do not involve sediment shrinkage.

4.3.7 Structures due to sediment wetting

These structures are features of some aeolian dune sands, especially those in coastal dune fields. When winds are actively supplying sand to the lee side of a dune, the grains are non-cohesive and accumulate by grain fall, grain flow or as the result of ripple migration. During inactive periods, the sediments of the lee slope may become moistened by rainfall, dew or salt spray, and, as a result, the sand of the surface layer may develop some cohesion and tensile strength. Resting on a steep slope, this material may begin to move down slope and, in the process, buckle and/or brecciate (McKee 1979; McKee and Bigarella 1972). The result is the development of lenses of deformed sand within the steeply dipping foresets that normally typify the lee side of a dune (Figure 4.29).

4.3.8 Deformation related to compaction

In terms of structures produced, compaction plays a variable role during burial, depending upon lithology. In coarser sediments, compaction is of relatively minor importance and might be detected only through slight reductions in the inclination of dipping depositional surfaces such as cross-bed foresets.

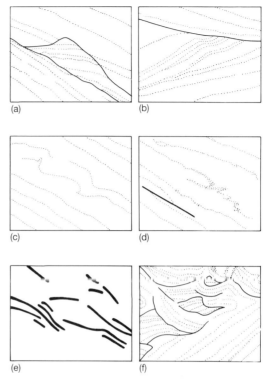

Figure 4.29 Styles of deformation in the foresets of aeolian dunes, including rotation, folding and brecciation. The establishment of tensile and cohesive strength in the sands, probably as a result of wetting, allowed the sands to behave in this way rather than as the more usual non-cohesive flow. (After McKee 1979.)

In finer grained and organic-rich sediments, compaction is much more significant and is commonly characterized by a general flattening of depositional features normal to bedding or lamination. There may be significant thickness change, up to an order of magnitude in the case of very organic materials such as peat. Not only is the thickness of bedding and lamination reduced but also fissility may be induced by the re-orientation of platy particles. Plant fragments and thin-shelled animal fossils may be reduced to bedding surface impressions. Deformation is perhaps more apparent in vertical section when there are inhomogeneities in the sediment, between which differential compaction has occurred. Such inhomogeneities, which may occur across a range of scales, may be depositional in origin, as where sandy ripple lenses float in a matrix of silt or clay, or they may be the result of early organic disturbance, as when sand-filled burrows occur in a finer matrix. In such cases, the

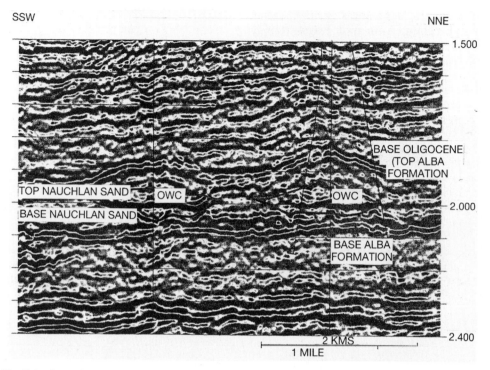

Figure 4.30 Seismic section across a major deep-water channel-fill sandbody. The top of the channel bodies are more conspicuous because later compactional deformation has caused convex upwards folding of both the top surface of the sand and the bedding in the overlying mudstones. Alba Formation, Alba Field, North Sea.

coarser sediment compacts less than the matrix, with the result that laminae in the muds are deformed around the coarser body. The burrow or ripple lens itself may also be subjected to some change of shape as a result of the compaction. Originally circular burrows may be flattened to elliptical cross-sections, and ripple lenses that were depositionally plano-convex may become convex both upwards and downwards. Section 4.3.9 outlines how deformation arises from differential compaction around diagenetic nodules.

At a larger scale, sand-filled channels, which originally had flat tops and convex-downwards bases, may develop distinctly convex-upwards tops as a result of differential compaction. Such changes of shape are particularly apparent in seismic sections (e.g. Newton and Flanagan 1993; Thomas, Collinson and Jones 1992; Figure 4.30). Inhomogeneities related to chemical precipitation and the differential compaction associated with them are dealt with in the following section.

4.3.9 Deformation related to early chemical precipitation

One of the most obvious ways in which sediments are modified after deposition is by the suite of chemical processes known as diagenesis (section 1.1.1). These processes often proceed without leading to any very obvious deformation, although, as has been made clear from the outset of this book, there is probably a close interlinking with mechanical processes, a topic which is complex and poorly understood.

Some mineral precipitation may be associated with overpressuring, such as that inferred for the curious bedding-parallel calcite veins spectacularly developed in Jurassic rocks of southern England, and known as 'beef' structure (Stoneley 1983).

Where fine-grained sediments have become locally cemented as concretions or nodules early in their burial history (e.g. Mozley and Bums 1993), these hardened patches may resist compaction and preserve bedding or lamination close to its depositional state. Compaction of surrounding sediments commonly leads to deformation of lamination around the concretion (Figure 4.31). Some measure of compaction may be deduced from comparison of lamination thickness within and outside the concretions. In turn this may give an indication of the relative burial depths at which the cementation processes began. For example, Craig (1985) compared the separation of laminations as they bow around septarian nodules in Lower Palaeozoic rocks south of Aberystwyth, Wales. Typically, the greatest difference in separation was in the range 52–58%. Assuming that this represents porosity loss due to compaction, the implication of such large values is that the concretionary growth must have begun within no more than a few metres of the sediment surface. Conversely, those nodules that show little deflection of the laminations are inferred to have initiated after greater burial and volume loss. The idea is summarized in Figure 4.32.

As well as leading to physical deformation through controlling differential compaction, concretions may be regarded as a type of sediment deformation in their own right. The mineralogy of the concretion-forming cement may vary considerably. Carbonate minerals (e.g. calcite, siderite), sulphides (pyrite), sulphates (e.g. barite,

Figure 4.31 Chert concretions with some compactional draping of bedding in the surrounding limestone. Portfjeld Formation, Lower Cambrian, north Greenland.

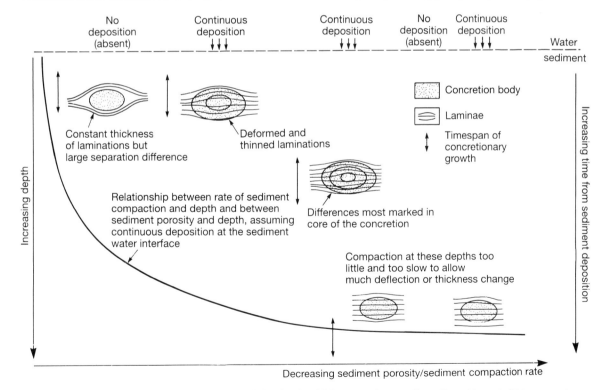

Figure 4.32 Schematic depiction of the changes in lamination thickness and separation adjacent to and within concretions formed at different burial depths. (From Craig (1985). Used with permission of Jon Craig.)

gypsum and anhydrite) and silica (flints and chert) are all common concretion-forming minerals. They may be precipitated within the host sediment simply to locally fill the original pore space, or the concretion may grow either replacively, preserving many of the host's original features in a changed mineralogy, or displacively, pushing the host sediment aside and thereby deforming it.

Some concretions follow and to some degree pseudomorph earlier features of the host sediment. They may follow burrows or root traces or may nucleate around a fossil fragment. In some cases the concretions may infill a preexisting cavity, caused by dissolution of some original soluble material. In the case of concretions formed as a result of pedogenic processes, the concretions may be distributed to define a distinct profile, reflecting a period of sediment starvation during which the near-surface sediments were subjected to the groundwater regime of the soil. In semi-arid settings, the concretions are most likely to be of calcite (caliche or calcrete soils) (e.g. Allen 1986b; Figure 4.33). In humid, water-logged soils, siderite is the more likely concretionary mineral, as in coal measure seat-earths (e.g. Besly and Fielding 1989).

The growth of certain concretions, particularly those of calcite within mudstones, may establish secondary stress fields, which, in turn, lead to further local deformation within and around

Figure 4.33 Calcite concretions in siltstone in fluvial overbank setting characteristic of a caliche palaeosol. The progressive downwards reduction in the intensity and interconnection of the nodules from a discrete surface is typical of a mature profile resulting from an extended period of sediment starvation. Lower Old Red Sandstone, Llanstephan, Dyfed, Wales.

the nodules. The extent to which developing concretions can cause deformation in the host sediment by exerting their own growth force has long been debated (Maliva and Siever 1988). Deformation within concretions is exemplified by the features known as **septarian nodules**. These are characterized by roughly radial or polygonal cracks within the concretion (Figure 4.34). The cracks are typically filled with calcite, although dolomite, ankerite, quartz, barite, celestite and pyrite are found, in various degrees of mutual replacement. The isotopic composition of the minerals may provide a record of the palaeohydrology (Desrochers and Al-Aasm 1993; Wilkinson 1993).

Astin and Scotchman (1988) invoked overpressuring during shallow burial to explain the septarian fractures and the textures of their mineral infill, but other accounts have emphasized the role of nucleii of decaying organic matter (Allison 1988). The alkaline, reducing microenvironments that are created around the nucleii favour the precipitation of calcium from the sea water. Recent explanations of the process have involved intermediate soapy materials (Duck 1990), the calcium stearates being derived from the saponification of fatty acids (Berner 1968; Hesselbo and Palmer 1992). This not only explains the localization of the concretions but also the origin of the septarian structure, as the subsequent conversion of the calcium soap to calcite involves a volume reduction and leads to shrinkage and cracking of the cemented volume.

Another striking structure within some concretions is known as **cone-in-cone**. The feature typically consists of opposing hemispheres of nested cones, with their apices directed inwards (Figure 4.35). The origin has long been debated, with either a gravitational consolidation stress or an outward force of concretionary growth commonly being invoked to explain the arrangement of the cones. More recent views have emphasized the mutual interference of bundles of growing calcite fibres, growing with a preferred crystallographic orientation normal to the sedimentary laminations (Marshall 1982). Many workers have surmised that the apical angle of the cones reflects the plasticity of the host sediment at the time of growth. The relatively minor

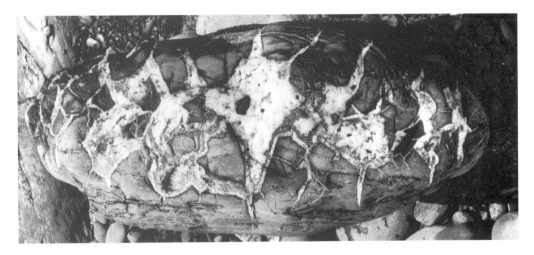

Figure 4.34 Septarian nodules (concretions). Bedding-normal appearance of septarian vein systems in eroded concretions. Silurian, Traeth-yr-ynys Lochtyn, west Wales. (Photographs provided by Jon Craig.)

effect of cone-in-cone nodules in west Wales on the laminations in the adjacent rocks, together with their incorporation of diagenetic chlorite, was interpreted by Craig (1985) to indicate initiation of the structure at greater burial depths than the other concretions in the district (Figure 4.36).

Significant stresses are also set up within near-surface sediments as a result of the growth of displacive evaporite crystals, concretions and layers. As precipitation proceeds, a surface layer of evaporite mineral may be pushed up into pressure ridges with a broadly polygonal plan pattern, perhaps originating from precursor desiccation cracks (Eugster and Hardie 1978; Figure 4.37). Displacive growth of gypsum and anhydrite nodules within earlier sediment, as in present-day sabkhas, pushes aside the host sediment and in extreme cases this process may lead to the development of **chicken-wire texture** (Figure 4.38). Layered anhydrite may develop quite intense folding as a result of the horizontal compressive stresses associated with crystal growth. This so-called **enterolithic** structure may cause extensive small-scale folding confined within a single layer (Figure 4.39). Such deformation may

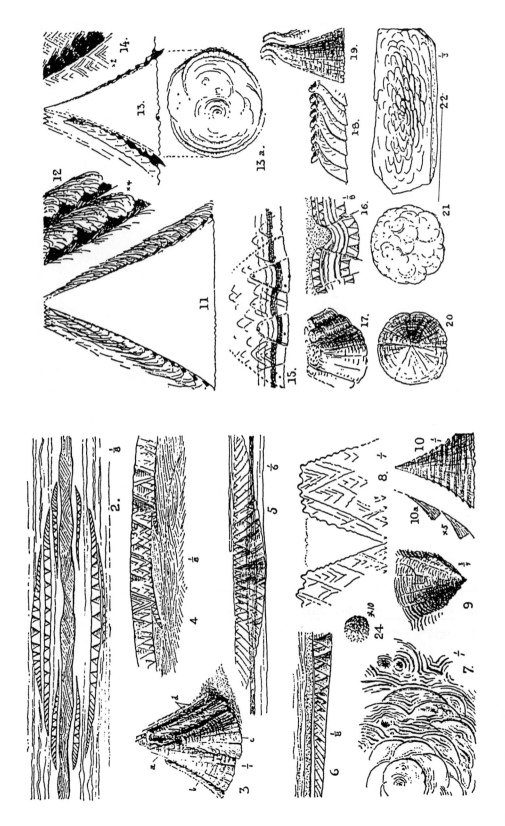

Figure 4.35 Cone-in-cone nodules (concretions), as portrayed by Gresley (1894). Drawings 3, 9 and 17 are perspective views showing the nested arrangement of the cones; the remainder are either plan or section views.

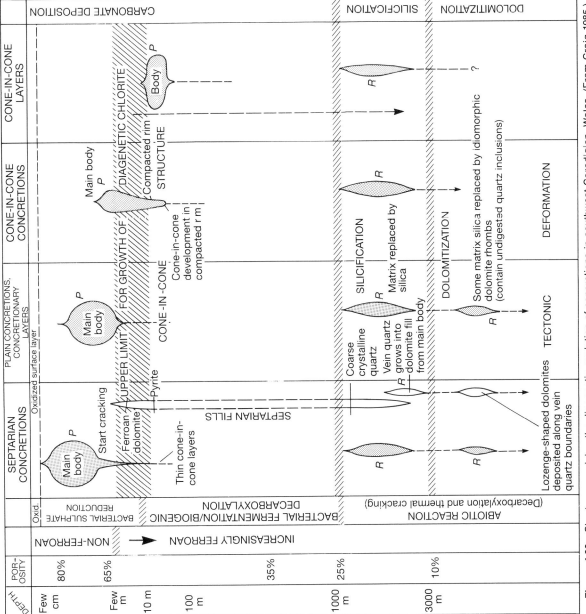

Figure 4.36 Chart summarizing the diagenetic evolution of concretions in southwest Ceredigion, Wales. (From Craig 1985.) Depth of burial axis is not to scale and all numbers are approximate. (Used with permission from Jon Craig.)

124 Sedimentary deformational structures

Figure 4.37 Polygonal pressure ridges developed on the surface of a playa lake as the result of the precipitation of evaporitic halite. Death Valley, California. (After Hunt et al. 1966.)

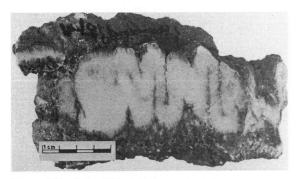

Figure 4.39 Enterolithic folds in a gypsum layer, probably a pseudomorph after evaporitic anhydrite. Lower Purbeck (Jurassic), Worbarrow Trout, Dorset. (Reproduced from Shearman (1980), with permission from Editions Technip.)

also be detected in pseudomorphs after precursor evaporites, the transformation occurring at quite an early stage. Chert and quartz pseudomorphs after anhydrite nodules (e.g. Chowns and Elkins 1974) and after complex sodium silicate evaporites in the soda lakes in East Africa (e.g. Eugster 1969) are examples of this.

4.4 CONCLUSION

The structures that are developed or modified early in the post-depositional history of sediments are widely but irregularly developed

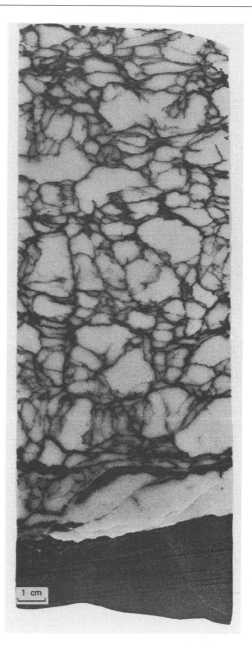

Figure 4.38 Chicken-wire texture developed by the growth of displacive anhydrite nodules, which have pushed aside the host sediment into thin sheets between nodules. Lower Lias (Jurassic), Aquitaine Basin, southwest France. (Reproduced from Bouroullec (1980), with permission from Editions Technip.)

throughout the stratigraphical record. Some types of sequence, particularly those which involved high depositional rates or which are located in areas of high seismic activity, are perhaps more susceptible to deformation but those relationships are far from clear-cut. That being the case, it is essential for the scope, scale and style of deformation structures to be constantly borne in mind when trying to unravel the structural history of an area or of a particular succession. This is particularly important in trying to identify and work with strain indicators. Some potential strain indicators, such as burrows or bedforms, may have undergone significant modification through early processes of alteration prior to becoming involved in any later disturbance. Equally, some deformation structures, particularly more regular and simple ones, such as desiccation cracks, could be used with care as strain indicators in their own right.

CHAPTER 5

Mass movements

OLE MARTINSEN

5.1 INTRODUCTION

5.1.1 General

Mass movements are significant geological processes which have an important impact on human life. They are environmental hazards, both on land and in the sea, and justify intensive attention. Much research has been related to human or industrial loss and welfare, and the importance of further efforts cannot be underestimated. The effects on modern communities of avalanches, major rock falls and mudslides are well known from recent catastrophies (Bolt *et al.* 1975; Press and Siever 1978; Voight 1978a; Brunsden and Prior 1984), and mass movements can seriously affect the stability of offshore installations (Prior & Coleman 1982). The importance in the context of this book is that mass movements present a very widespread situation in which sediments are subject to deformation.

For the geologist, gravitational mass movements span a wide range of processes, both subaerially and subaqueously (Brunsden 1971; Schumm and Mosley 1973; Voight 1978a,b; Saxov and Nieuwenhuis 1982; Brunsden and Prior 1984; Allen 1985; Morton 1993). At what point in this spectrum the processes are regarded as deformational is a very subjective judgement. All the products provide important information on the depositional setting, and in the ancient record are powerful tools, both for sedimentological and stratigraphical analysis, yet all could be argued to involve deformation of one kind or another. The processes involved in slumps and slides are deformational in any view, and the resulting products of relevance to structural geology. This chapter therefore presents an overview of the spectrum of mass movement processes and products, but emphasizes slumps and slides. Both modern and ancient examples are used.

The theoretical background to how such weak sediments deform has been outlined in Chapter 2. The present review concentrates on the qualitative macroscopic and morphological aspects of mass movements, as seen in nature. For the reasons mentioned above, those movements that are mainly depositional in nature, such as falls and flows, are treated only briefly, and more attention is given to movements that modify sediments already deposited, as these are unarguably deformation processes. Even these cannot be treated here in detail. Rather, this general summary of the important aspects of each process or group of processes is designed to provide the reader with a sense of the wide variability that exists.

5.1.2 Classification schemes

Mass movements have been classified on the basis of a wide variety of factors, including process and rheology, product, climate, type of material moved, local geology and triggering mechanisms (Ladd 1935; Ward 1945; Dott 1963; Crozier 1973; Middleton and Hampton 1976; Nardin *et al.* 1979; Hansen 1984; Pierson and Costa 1987). Many of these schemes are complex (Varnes 1978), and frequently difficult to use, particularly in the field. Some schemes are concerned only with subaqueous gravity flows (Lowe 1979), and do not include slope failures such as slides and slumps. Others aim to classify all subaqueous processes, whether gravity-driven or not (Pickering *et al.* 1986). Geologists are often concerned that classification schemes should be

The Geological Deformation of Sediments Edited by Alex Maltman Published in 1994 by Chapman & Hall ISBN 0 412 40590 3

simple, that they concentrate on descriptive and morphological factors, and that they be easily applicable. A useful scheme for mass movements should also point the user towards the genesis of the particular unit observed.

The scheme suggested by Kruit *et al.* (1975), and further developed by Rupke (1978) and Stow (1986), has probably come the furthest in developing a classification scheme within which process and product are considered in a simple and easily applicable way. The scheme, based on mass movement rheology, was simplified by Nemec (1990), who divided the movements into six categories (Figure 5.1a). This classification is simple and logical, both for subaqueous and subaerial processes, because it shows a range from slow movement of coherent masses (creep), with little or no relative movement of individual grains ('quasi-static' grain contacts), through increasingly turbulent movements, to rapid mass movement of grains which move almost to entirely independently of other grains (e.g. falls of debris).

Although the various mass movement processes are presented under individual headings, it is important to note that they represent parts of a process continuum (Figure 5.1b), and in many instances one may evolve into another with time, or the depositional effects of one type may trigger other processes. Thus, no one mass movement process should be considered entirely independent of the others.

Another useful and more detailed way of classifying mass flows was suggested by Pierson and Costa (1987) based on the mean velocity of the flow and sediment concentration in per cent (Figure 5.2). This scheme also emphasizes flow or movement rheology, and from the products important quantitative estimates of flow velocity and sediment concentration can be made.

5.1.3 Basic theory

To develop an understanding of mass movements it is necessary to recall the rheological principles mentioned in section 1.2.4. Newtonian fluids, like water, have no shear strength and deform immediately once a shear stress is applied. The amount of flow depends on the viscosity of the material, which may decrease (shear thinning) or increase (shear thickening) with increasing rate (section 3.6.2). The mass movements described

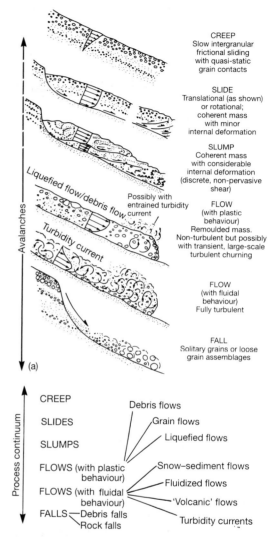

Figure 5.1 Classification scheme of mass movements: (a) details six categories, based on movement rheologies; (b) emphasizes that the categories comprise a process continuum. (From Nemec 1990.)

here as fluidal flows (section 5.3) are defined as showing approximately Newtonian behaviour.

Distinguished from fluidal flows are those mass movements that require the applied shear stress to have exceeded a yield point before initiating the irreversible deformation. At their simplest, where the shear rate increase with increasing shear stress is linear, these materials are showing Bingham behaviour (Figure 3.4), but they are treated here as being generally plastic materials and termed plastic flows (section 5.4).

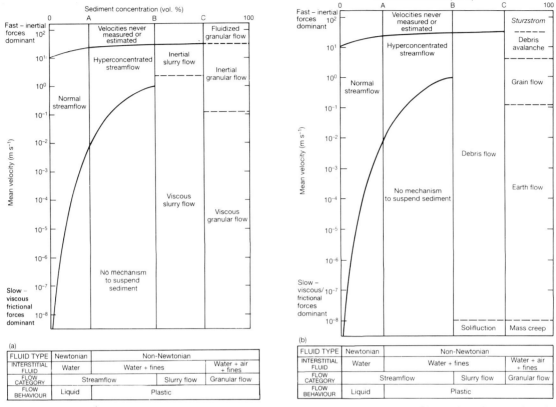

Figure 5.2 Classification scheme based on flow velocity and sediment concentration. (a) Rheological classification of sediment–water flows. The vertical boundaries A, B and C are rheological thresholds and functions of grain-size distributions and sediment concentration. Boundary A marks the initiation of importance of yield strength, boundary B marks the sudden rapid increase in yield strength and onset of liquefaction behaviour, and boundary C marks the end of liquefaction behaviour. (b) Application of existing flow nomenclature to (a). (From Pierson and Costa 1987.)

Generation of mass movement requires a slope, which may vary in inclination from less than 0.1°, such as on modern delta fronts (Prior and Coleman 1978b), to vertical and overhanging surfaces, where rock falls may occur. Mass movements arise where the shear stress acting on the sediment or rock exceeds the material's shear strength or shearing resistance. Following the formulation of Terzaghi (1962), the shearing resistance is a function of the cohesion, stress-dependent internal friction and pore-fluid pressure of the material, as given by the Mohr-Coulomb relation (equation 1.13). This equation is fundamental to understanding the initiation of mass movements.

From the Mohr-Coulomb relation, a decrease in cohesion, friction or normal stress at the potential slip plane will decrease the shearing resistance of a rock or sediment, and vice versa. Most importantly, overpressuring (section 1.3.3) is a very common process of reducing shearing resistance and causing slope failures. Slope oversteepening, e.g. by high sedimentation rates, undercutting or retrogressive sliding, is another common process of reducing shearing resistance.

In the following six sections, each major group of mass movement is dealt with, in order of decreasing amount of solitary grain movement and turbulence. Thus, falls are summarized first, then gravity flows, starting with those with fluidal behaviour, and moving on to flows with plastic behaviour. As mentioned above, slumps, slides and sediment creep are reviewed in more detail, emphasizing their deformational aspects.

5.2 FALLS

5.2.1 Introduction

The term falls refers to downslope movement of solitary grains or relatively loose assemblages of grains where each particle moves more or less independently of others. In subaerial settings the process is constrained to rock falls from rocky headwalls, a very common process is mountainous terrain. In subaqueous settings, falls occur in two modes: rock falls essentially similar to those in subaerial settings; and debris falls, which are produced by similar processes to rock falls, but are formed from previously deposited sediment.

5.2.2 Rock falls

Process

Rock fall occurs when slabs or pieces of bedrock come loose from their original position and fall down slopes or mountain sides (Figure 5.3). The volume of the rock fall may vary from single small pieces to cascades of solid rock, which may cause extensive damage. In addition to steep slopes, three factors are important for the formation of rock falls: (i) bedrock type; (ii) climate (subaerially); and (iii) bedrock fracturing and structural deformation (see extensive review by Whalley 1984). Carson and Kirby (1972) recognized several types of rock fall, their formation largely depending on the scale of the falling fragments (Figure 5.3). Rock falls are probably more common in subaerial settings than in subaqueous, largely because of the relative lack of significant slopes in the latter. However, in steep fjords (Prior and Bornhold 1990) and on frontal margins of carbonate platforms (James and Ginsburg 1979; Mullins and Cook 1986), rock fall may be a significant process.

Bedrock type Massive, homogeneous rocks, such as some igneous rocks, are probably less prone to producing rock falls than layered and heterogeneous rocks, such as metamorphic and most sedimentary rock packages, unless considerably influenced by other factors. Rocks that are layered are more susceptible to breaking apart, particularly along bedding planes, and some lithologies are more easily weathered than others. However, there is no simple correlation between rock strength and probability of rock fall formation (Abele 1972).

Fine-grained lithologies will tend to produce rock falls with smaller fragments than coarser and more massive lithologies (Whalley 1984).

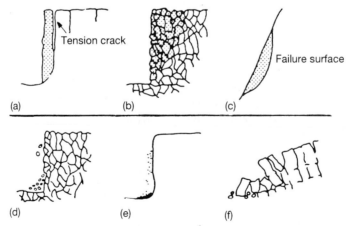

Figure 5.3 Types of rock fall, dependent on size and number of falling fragments. (Based on Carson and Kirby 1972.) Type (a) is characterized by a large fragment bounded by a well-defined, vertical fracture. Type (b) is characterized by numerous smaller fragments, also bounded below by a relatively well-defined surface. Type (c) is characterized by a sliding rock mass above a well-defined, inclined but non-vertical basal shear surface. It generally compares with (a) and (b), but contrasts with (a) in the nature of the basal contact and with (b) in the number of fragments involved. The characteristics of type (d) are freely falling, larger fragments and no extensively developed basal contact. Type (e) is similar to (d), but only very small fragments are involved. Type (f) is termed 'toppling failure' (de Freitas and Watters 1973), where the fragments are large and only move a short distance, with the maximum movement occurring at the top of the fragment. It has some resemblance to (a) and (d). (From Whalley 1984.)

Furthermore, the rock falls formed from massive rocks tend to have much larger volumes than rock falls formed from fine-grained rocks, although there is probably no one-to-one relationship. This is largely because fine-grained rocks generally weather more easily and form rock falls continuously, whereas massive rocks tend to break off in larger volumes and generate more voluminous rock falls.

Climate Climate is a critical factor for rock-fall generation in subaerial settings. Temperate and humid climates are probably the ideal conditions for rock-fall formation. Rain-water percolating continuously in cracks and advancing weathering, lubricates potential slip planes in the rocks and significantly diminishes the frictional resistance to slippage. Periods of high rainfall are therefore particularly critical periods for rock-fall formation.

Temperatures fluctuating around freezing also promote formation of rock falls. Repeated freezing and thawing of percolating water causes rapid volume changes in cracks and joints (Martini 1967; S.E. White 1976). During freezing periods the water expands in volume, causing an expansion of the cracks, whereas the subsequent thaw causes volume decrease. These rapidly alternating stress conditions lead to rapid propagation of the cracks and frequent slippage and rock-fall formation.

Fracturing and structural deformation Pre-existing joints and fractures will enhance rock-fall generation (e.g. Genç 1993), particularly if their strike orientation parallels that of the rock wall. This is a common feature in granitic terranes (Selby 1977). Furthermore, if the dip of the rock wall exceeds that of the fractures, particularly large and frequent rock falls may occur, because the bedrock may be inherently unstable and prone to break-off.

Products
Rock falls most commonly produce a debris cone or talus cone at the foot of the slope (Figure 5.4). Such cones can have a wide variety of shapes, largely depending on the extent of the area from which the rock falls are produced. Thus, a spectrum from narrow, wedge-shaped debris masses to wide, rectangular masses is to be expected. The debris is generally poorly sorted and has an openwork, clast-supported texture.

A first-look, qualitative assessment of the risk, frequency and common quantity of subaerial rock falls in a particular area can often be made by viewing the expression of the debris cone. A vegetated, major tongue of debris at or beyond the base-of-slope point suggests that rock falls are infrequent but of major scale when they occur. Conversely, a non-vegetated talus built up on to the rock wall suggests small but frequent rock-fall events (Figure 5.5).

Rock-fall products range in scale from cones less than 10 cm wide, which can be observed on any rocky slope, to rock falls of more than 10^6 m^3 in volume. There seems to be an inverse relationship between the scale of the rock fall and frequency, so that the daily rock falls generally are small, whereas infrequent ones are of larger scale (Rapp 1960; Gardner 1970). There are numerous examples of historically recorded rock falls that have caused extensive damage (Jørstad 1968), either because of the impact of the rock fall itself, or as a result of associated events such as catastrophic waves when the rocks fell into the sea.

Rock falls are also a dominant process along carbonate platform margins and constitute an important component of facies models from these areas. Erosional slopes around the Bahamas and off the Belize coast in the Caribbean are dominated by rock falls (James and Ginsburg 1979; Schlager and Ginsburg 1981).

5.2.3 Debris falls

Process
Debris falls (a term introduced by Holmes 1965) are straightforward downslope movements of dispersed debris, with single, freely moving grains responding to the downslope pull of gravity (Nemec 1990). The momentum within these falls is transferred by 'streaming' (in contrast to flowing) (Campbell 1989), which is entirely controlled by the gravity pull on the particles. Grain collisions are only of subordinate importance. Debris falls are transitional to **cohesionless debris flows** and may change into such flows if slowed, for example on a gentler slope (Nemec, 1990; see also below).

Figure 5.4 Rock fall and snow avalanche cones from near Moraine Lake, close to Lake Louise, Alberta, Canada. Several cones overlap, each originating from re-entrants in the rocky headwall. Avalanche tracks, produced from snow–sediment avalanches, are developed on several cones. The mountainside is around 300 m high. The cone on the right has more big boulders on it, suggesting a greater influence from rock-fall processes.

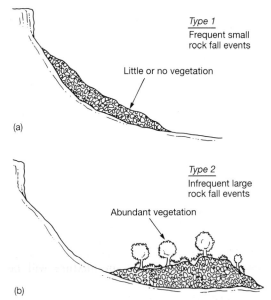

Figure 5.5 End-members of talus cones: (a) shows a non-vegetated talus cone along the rock wall, reflecting frequent but relatively small rock-fall events; (b) shows a vegetated talus cone at the foot of the rock wall slope, which reflects infrequent but large rock-fall events.

Debris falls may form if slope oversteepening occurs due to rapid sedimentation, retrogressive sliding or slumping, or wave undercutting. This can destabilize an unconsolidated, cohesionless sediment that is either at rest or slowly moving. Failure will then occur, causing rapid avalanches controlled by debris fall processes (Nemec 1990).

Because the movement of the individual particles is controlled mainly by gravity, the larger particles tend to move the farthest, causing a distinctive downslope coarsening of debris fall deposits (Figure 5.6).

Products

Debris falls tend to produce elongate tongues of coarse sediment which are coarsest at their downslope end (Figure 5.6). Furthermore, deposits of successive debris falls may coarsen up, because the increased bed roughness and frictional slow-down, where former debris falls were deposited, enhance deposition of coarse debris in succeeding debris falls (Nemec 1990). This is provided that no deposition of fines and consequent smoothing of the depositional surface occurs.

Texturally, debris fall deposits tend to be openwork gravels which may show normal grading

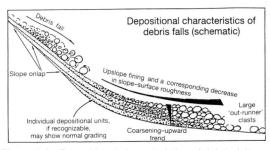

Figure 5.6 Depositional characteristics of debris falls. (From Nemec 1990.)

(Figure 5.6) due to decreasing flow competence with time as the largest clasts are deposited first and possibly overridden by the finer tail of the event (Nemec 1990). Debris fall clasts may also occur as isolated clasts or groups of outsized pebbles, for instance within or between turbidites deposited basinwards of the main debris fall deposits (Figure 5.6). The clasts may be 'outrunner' clasts rolling into otherwise sandy deposits.

5.3 FLUIDAL FLOWS

5.3.1 Introduction

Fluidal flows comprise mass-movement processes that are fully or dominantly turbulent during flow. They show a Newtonian behaviour. Some flows, such as high-density turbidity currents, may attain a non-Newtonian behaviour due to very high particle concentrations, and this is commented on below. Three main categories of flow will be described: turbidity currents, some flows related to volcanic eruptions, and snow and ice flows (avalanching and entraining clastic sediment).

5.3.2 Turbidity currents

Process

Turbidity currents are generated from sediment suspensions with an excess gravitational potential to flow down slopes. The main grain support mechanism, particularly in low-density turbidity currents, is the fluid turbulence. During steady state of the flow, a state of **autosuspension** (Bagnold 1962; Pantin 1979) may exist. The suspended sediment is put into motion due to the gravitational pull downslope, the motion causes turbulence because of shearing along the bed, and this further causes suspension. In this manner, the flow of the turbidity current is largely maintained by its own motion.

Fundamental work on turbidity current flow mechanisms has suggested that these flows should be treated as several grain-size populations (Bagnold 1954; Middleton 1967; Walker 1977, 1978; Lowe 1982). There are four models for classification of turbidites, depending on the grain size involved (Figure 5.7). For the three finer grained models (Bouma 1962; Piper 1978;

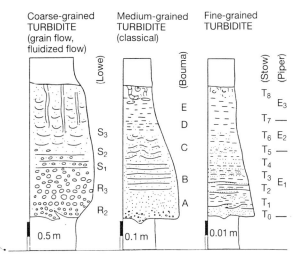

Figure 5.7 Models of turbidity current deposits based on grain size. (Modified from Stow 1986; based on Bouma (1962), Piper (1978), Stow (1979) and Lowe (1982).)

Stow 1979; Stow and Shanmugam 1980), fluid turbulence and associated effects such as boundary layer shearing are the dominant transporting and grain-supporting mechanisms.

For the coarse-grained model, other processes are also important. Lowe (1982) divided the grain sizes transported by turbidity currents into three categories, of which the second and third are important for the coarse-grained model.

1. Clay to medium-grained sand that can largely be suspended by flow turbulence alone;
2. Coarse sand to small pebble-sized gravel, which is suspended by flow turbulence, with settling hindered due to high particle concentrations and by buoyancy created by the interstitial mixture of fine-grained sediment and water;
3. Pebble- and cobble-sized clasts constituting more than 15% of the grain mixture will be supported by the combined effects of fluid turbulence, hindered settling, matrix buoyant lift and dispersive pressure resulting from grain collisions.

Generally, particle concentrations around 20–30% have been considered to divide low-density from high-density flows (Bagnold 1954; Middleton 1967; Wallis 1969). Lowe (1982), however, considered this division to be somewhat arbitrary, because for turbidity currents made up of fine-grained sediment (category 1) flows appear to be stable over the whole range of possible grain-size concentrations. Coarse-grained flows

134 *Mass movements*

with particle concentrations below 20%, however, may be unstable and collapse rapidly.

Products

Turbidity currents mostly produce sharp-based beds, which generally fine upward, and within which there are successions of sedimentary structures suggesting waning flow conditions with time. Deviations may occur (see also below). The most well-known product of deposition from turbidity currents is the Bouma sequence. The B (parallel laminated) and C (cross laminated) divisions of the Bouma sequence (Figure 5.7; Bouma 1962) are formed by traction, which has also been discussed extensively, for example by Walker (1965), Middleton (1967, 1970), Allen (1970) and Middleton and Hampton (1973, 1976). The overlying D and E divisions of the Bouma sequence are formed from further slow-down of the flow, which initiates deposition from suspension of the finest grains. (Figures 5.7). Some traction may occur during deposition of the D division, which produces the fine lamination caused by textural sorting related to near-bed shear (Walker 1965; Stow and Shanmugam 1980; Hesse and Chough 1981).

In very mud-rich turbidity currents, waning flow will cause a fining-upward succession essentially similar to the C, D and E divisions of the Bouma sequence (Figure 5.7; Piper 1978; Stow and Shanmugam 1980). However, this succession is dominated by repetitive sharp-based laminae, the lowest of which may have very low-angle, climbing, fading ripples (Figure 5.7, divisions T_{0-2}). Upwards, the laminae fade into wispy, convolute ones, and at the top into graded mud (Figure 5.7, divisions T_{3-8}; Stow and Shanmugam 1980). The repetitive laminae are thought to form within the same muddy turbidity current as a result of multiple phases of boundary layer shear causing textural sorting (Stow and Bowen 1980).

Coarse-grained turbidity currents go through three stages of deposition, according to Lowe (1982): (i) a tractional stage; (ii) a traction-carpet stage; and (iii) a suspension sedimentation stage (Figure 5.8).

Tractional stage A depositing turbidity current can form bed forms such as dunes and plane beds, owing to flow interaction with the surface on to which it is depositing (Govier and Aziz

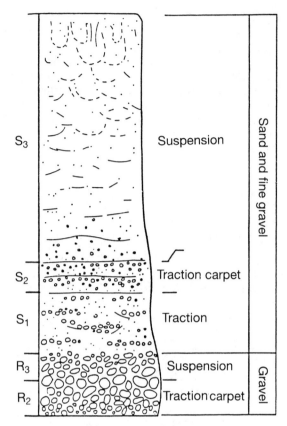

Figure 5.8 Idealized vertical section through a coarse-grained turbidite, showing the various divisions and their depositional mechanisms. (From Lowe 1982.) Lowe's symbols for the divisions are given to the left of the section.

1972; Mutti and Ricci Lucchi 1972; Walker 1978). The dunes will usually not be very organized, owing to the relatively short-lived nature of the flow and its unsteadiness (Lowe 1982).

Lowe (1982) suggested that cross-beds in turbidites were formed from high-density flows. This is controversial, because tractional movement of sediment would most easily occur from low-density flows (W. Nemec, personal communication 1992). For example, Hiscott and Middleton (1979) documented cross-bedding on the top of massive to stratified turbidites in the Ordovician Tourelle Formation on Gaspé Peninsula, eastern Canada. These authors suggested that the cross-bedding was formed from reworking by the low-density tail of an initially high-density turbidity current.

Traction-carpet stage As the turbidity current continues to deposit sediment, the increasing concentration of particularly coarse-grained particles near the base of the flow leads to an increasing amount of grain collisions (Bagnold 1956; Shook and Daniel 1965). This causes development of a layer near the bed maintained by dispersive pressure and fed from the settling sediment above. Turbulence will be suppressed in this layer, eventually leading to inverse grading (Figure 5.8a and b). In many traction-carpet deposits, several superimposed, inversely graded layers are observed, suggesting that as one carpet froze and was deposited, a new one formed rapidly above (e.g. Hiscott and Middleton 1979). Traction carpets probably do not develop in sand finer than coarse grained, owing to the insignificant dispersive pressure between grains of these size grades (Lowe 1982).

Suspension sedimentation stage The final stage of the deposition is direct sedimentation from suspension, which occurs when sediment fall-out rate is so high that there is no time for formation of traction structures or a traction carpet. The suspension deposits correspond to the Bouma A division, and can be graded (Figure 5.8a) or massive. The grading is either a **distribution-grading**, if late-stage turbulence has slowed down the deposition of the fine material, or a **coarse-tail** grading, if the sediment settled mainly as a non-turbulent cloud (Middleton 1967). Arnott and Hand (1989) challenged the view that the massive Bouma A division results from suspension sedimentation. Based on experimental work they suggested that the massive character resulted from deposition during upper flow regime traction and suppression of parallel lamination due to a very high rate of grain fall-out from suspension.

Lowe (1982) termed these three sedimentation stages S_1–S_3, for coarse sand and gravel, the S_3 being equivalent to the Bouma A division, and R_1–R_3, for pebbly and cobbly flows. A tractional stage is not likely to occur in the coarsest flows, because grain collisions and formation of a dispersive pressure will dominate (R.V. Fisher 1971; Walker 1975). Therefore, cross-bedding will probably not form at the bases of such flows (Figure 5.8a), and only R_2 and R_3 divisions are usually found.

In several studies it is observed that outsized clasts up to an order of magnitude larger than the normal grain size can occur in turbidites, also within normally graded beds (Winn and Dott 1978; Hein 1982; Clifton 1984). Postma, Nemec and Kleinspehn (1988) explained this feature, based on an experimental study, by large clasts being transported along a rheological boundary between two distinctly different parts of the flow (Figure 5.9). The lower and basal part of the flow is a pseudolaminar, inertia layer, along which the clasts 'glide' and are partly submerged. The clasts are driven downslope by faster moving, turbulent shear stresses set up by the top layer of the flow, which is fully turbulent (Figure 5.9; Postma, Nemec and Kleinspehn 1988).

5.3.3 Flows related to volcanic eruptions

Introduction

During volcanic eruptions, subaqueous and, particularly, subaerial catastrophic mass flow processes can occur, with a tremendous impact on surrounding areas (Fisher and Schmincke 1984). The dramatic effects of the Mount St Helens eruption in Washington, USA, on 18 May 1980 are easily recalled. Also, the pre-historic destruction of the city of Pompeii due to an eruption of Vesuvius was a dramatic event, caused at least in part by mass flow processes related to a volcanic eruption. Three processes are summarized here which appear to be the most common mass flow processes related to volcanic eruptions: (i) debris avalanches; (ii) pyroclastic flows (or nueés ardentes); and (iii) lahars. A detailed account of volcaniclastic mass movements is given by Fisher and Schmincke (1984), and case studies are described by Ballance and Hayward (1983) and J.D.L. White and Busby-Spera (1987).

Many of the flows probably attain a turbulent to quasi-turbulent state, due to the admixture of hot air, water, volcanic material and bed rock, although some, particularly debris avalanches, are prone to be transitional into rock falls, slides and debris flows. Siliciclastic debris flows are also common in relation to volcanic eruptions (Korosec, Rigby and Stoffel 1980), and are considered below under 'Debris flows'.

Debris avalanches

Debris avalanches occur when large portions of a volcanic caldera are displaced downslope under gravitational forces, as a result of caldera fragmentation (S.G. Evans and Brooks 1991;

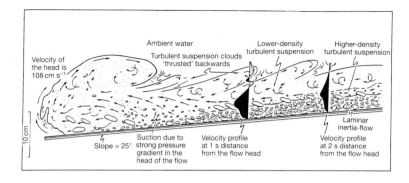

Figure 5.9 Drawing of a high-density turbulent flow showing development of a basal laminar flow zone able to carry larger clasts on top and within the middle of the entire flow. This model explains the occurrence of 'outsized' clasts in relatively fine-grained turbidites. (Based on laboratory experiment by G. Postma; from Postma, Nemec and Kleinspehn 1988.)

Smith 1991; Palmer and Neall 1991). Numerous such events are recorded historically (Siebert 1984). Although debris avalanches generally may be large, single-event mass movements, this process can also be cyclic, with recurrence intervals around 150–200 years (Beget and Kienle 1992).

Debris avalanches can exceed 1 km³ in volume and leave large, amphitheatre re-entrants in the volcanic cones (Figures 5.10 and 5.11). There is often a close correspondence in volume between the debris avalanche deposits and the missing parts of the volcanic calderas, suggesting that gravitational mass movement is the main process of removing material from the calderas (Siebert 1984). The morphologies of the debris avalanche

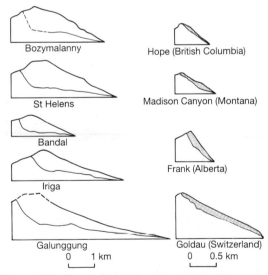

Figure 5.10 Comparisons of profiles of volcanic mountains affected by debris avalanching (left) and non-volcanic mountain sides affected by rock slides (right). Note the concave-upward lower boundary of the debris avalanches and their larger volumes (note difference in scale). (From Siebert 1984.)

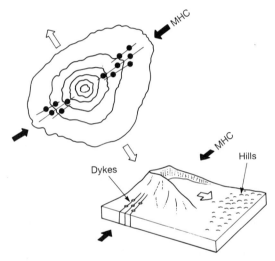

Figure 5.11 Dyke orientation in relation to generation and transport direction of debris avalanches. Most dykes are oriented parallel to the areal maximum horizontal compression (MHC) direction, and perpendicular to transport direction. Downslope from the fragmented cone, the debris avalanche forms numerous small hills (see text for discussion). (From Siebert 1984.)

craters are asymmetrical, concave-up forms that resemble slump and slide scars (Figures 5.10 and 5.11), and which differ from the more concentric depressions of collapse calderas.

Debris avalanches seem to have a systematic orientation in relation to principal stress directions in a volcanic terrane. Dykes tend to parallel the maximum compressional stress orientation, and the movement of the debris avalanches is generally at right angles to this (Figure 5.11; Moriya 1980; Siebert 1984). Therefore, the debris avalanches are probably controlled by structural weaknesses along dykes and related fractures, and are triggered during particularly violent eruptive events.

The avalanches can attain speeds up to and probably exceeding 100 km hr^{-1}. They have great mobility, and can be transported large distances. The mean ratio between vertical drop and travel distance of numerous examples of debris avalanches was calculated by Siebert (1984) to be 0.11, illustrating this point. It is likely that their mode of transport, at least temporarily and initially, can approach that of debris falls (see above). During continued transport they may be able to achieve a quasi-turbulent state depending on the content of hot air and water, which can enhance turbulence, although water is not an essential component for explaining the long runout distances of some debris avalanches (McEwen 1989). Direct observation of the Mount St Helens debris avalanche suggested that this was initiated as a slide and evolved into a flow (Voight *et al.* 1983). The frequent preservation of undeformed, primary and laminated fragments of the volcanic cone in some debris avalanches has been interpreted to suggest laminar plug flow (Voight *et al.* 1981). Thus there is still a great deal of uncertainty connected to flow processes in debris avalanches.

The deposits of debris avalanches are generally very hummocky or hilly masses of unsorted and brecciated volcanic debris. However, it is common that the size of the material and the size of the hummocks decrease away from the source of the avalanche (Siebert 1984). The hummocks often seem to occur in a predictable fashion within a debris avalanche deposit. The largest hummocks or hills occur near the axis of the deposit, while the hills may be slightly elongated, and their long axes radiate away from the volcanic cone in the transport direction of the avalanche (e.g. Aramaki 1963; Glicken 1982). The hills may occur in clusters, separated by flat areas. The inter-hill areas may be dominated by lahar deposits and reworked material (Ui 1983). The hills are thought to form by flow segregation around large fragments in the flow (Crandell 1971).

Pyroclastic flows (nuées ardentes)

Pyroclastic material is volcanic material ejected into the air during violent eruptions. Water and gas are natural components of magma, and when released from the volcano together with significant amounts of pyroclastic material, clouds of excess density in relation to the surrounding air can be set up which will respond to the gravitational force along the sides of the volcano. The result is a nueé ardente, or pyroclastic flow surge, which flows down the volcano side with tremendous speed and can cause incredible damage. The flows are fully turbulent, keeping the volcanic material fully suspended, and the flows may entrain material that they flow over on their path. Thus, they may be the closest subaerial analogue of subaqueous turbidity currents, although the turbulent medium is not only water but an admixture of water, gas and hot air.

The deposits of pyroclastic flows are unsorted masses of volcanic, angular fragments, which have a distinctive appearance of being welded together. The reason for this is that a substantial component of the flows is glass, and when cooled it gives the welded appearance. Pyroclastic flow deposits are usually called **welded tuffs** or **ignimbrites**, and can attain thicknesses up to 100 m and cover large areas (Press and Siever 1978). Some parts of the deposits may not be welded, and are characterized by chaotic masses of volcanic debris (Figure 5.12).

Figure 5.12 Unwelded part of pyroclastic flow deposit of andesitic composition from Milos, Greece, in the Aegean Sea. Note the chaotic texture of boulders of volcanic debris set in a finer grained tuff matrix. (Photograph courtesy of Harald Furnes.)

Pyroclastic flows can cause tremendous damage. In 1902, on the island of Martinique, a nueé ardente flowed down from the volcano Mont Pelée into the town of St Pierre at a speed of 100 km h^{-1}. The flow was estimated to have had an internal temperature of 700–1000°C, and it killed 30 000 people (Bolt *et al.* 1975).

Lahars
Lahars form when pyroclastic flows or debris avalanches flow into a standing body of water such as a river or lake. The entrainment of fluid into the pyroclastic flow or avalanche results in a transformation from a fully turbulent flow to a cohesive, laminar flow, which in most ways compares with 'normal' debris flows (see below). Lahars have great competence, and may carry large boulders several kilometres.

Lahars may also form when lakes close to volcanic craters breach, and intermixed volcanic material and water flow down adjacent slopes. In Java in 1919, when the volcano Kelut erupted, a lahar formed when the crater lake was breached, flowing down the volcano sides and killing 5500 people (Press and Siever 1978). In glacial regions, lahars form when pyroclastic flows or lava melt surrounding ice during eruptions.

The deposition and cohesive freezing of lahars is probably a result both of temperature reduction and other factors that reduce the applied shear stress as slope angle reduces. Thus, although lahars possess many of the same characteristics as 'normal' debris flows, their flow potential is to a large degree controlled by the ability of the flow to retain its temperature.

5.3.4 Snow- and ice-generated (avalanching) flows

Description and formation
Mass movements of snow and ice are important not only because they move snow and ice, but also because these movements induce mass movements of clastic sediments, and it is this respect that will be concentrated on here. (For a thorough review of mass movements of pure snow and ice see Perla (1978) and Mellor (1978).)

Two factors are essential for the initiation of avalanches: (i) climate; and (ii) steep slopes.

Climate Regions with high amounts of precipitation and frequent temperature fluctuations around freezing during the winter months are more likely to produce avalanches than arid and stable cold regions. Regions with frequent temperature changes are generally also windy and therefore local accumulations of snow may be built up in failure-prone areas near or on slopes. Therefore, coastal mountainous regions in temperate to arctic regions may be some of the highest risk areas for avalanches.

Shifting conditions between wet and dry snow decrease the stability of the snow column. The boundaries between different snow types are generally very sharp, and such boundaries may eventually form slip planes, for example, if wet, new-fallen snow accumulates above earlier frozen snow (Mellor, 1978).

Steep mountain slopes The inclination of mountain slopes is also a critical factor for producing avalanches. Generally, a 30° slope is needed to produce avalanches (Perla 1978, figure 7), unless extreme conditions are present, for example where the underlying sediment fails as a result of heavy rainfall. Therefore, relatively steep slopes are needed to trigger avalanches, although they seem to be able to flow on much less inclined slopes once generated. Slush flows, which are entirely saturated with water (up to 35–40%; Mellor 1978), may initiate on much less steep slopes than dry snow avalanches. These flows behave as mud flows, and slopes as low as 10–12° are sufficient to cause flowage (Rapp 1960; Mellor 1978). Slush flows also may develop from more turbulent drier snow flows if they flow into stream channels and become saturated with water.

Two mechanisms primarily control the incorporation of clastic material into the avalanches: avalanche turbulence and basal shear. Avalanches of relatively dry snow are probably more turbulent but have less basal shear than slush flows and avalanches of wet snow and ice. Therefore, the latter may be more likely to entrain larger pieces of rock and vegetation during movement. Mixed snow–sediment avalanches may be fully turbulent because of lubrication by snow between entrained sediment.

Mixed mass movements of snow and sediment are mostly relatively unsorted and correspond in several ways to subaerial debris flows (see below). They can behave as surging flows, with one event being dominated by several surges (Blikra and Nesje 1991).

Products

At their upslope end, snow–sediment avalanches tend to leave a concave-downslope depression very similar to slide and slump scars (Perla 1978; Blikra and Nesje 1991). This depression tends to narrow downslope into a rather narrow **avalanche track**, which is mainly a bypass zone of sediment, snow and ice (Figure 5.13a). At the downslope end, the avalanche track widens on to a debris cone or scree cone (Figure 5.13a and b), or an alluvial fan if significant alluvial processes also take place. In the case that the scree cone occurs at the land–water interface, a scree-apron delta may develop, with a pronounced fan shape if developed from a point. If there are several closely spaced avalanche tracks, a 'ramp' of debris may occur of some lateral extent.

The deposits of individual avalanche events are mostly unsorted masses of debris, with large tree trunks and boulders to fine sand and mud being entrained into the avalanche (Rapp 1960; Luckman 1971). The deposits of mixed flows after snow melt may resemble those of subaerial debris flows, but the presence of large tree trunks and a largely openwork, clast-supported framework (Figure 5.13b) of a very heterogeneous clast-type mixture may suggest deposition from a mixed snow–sediment flow. In talus screes, deposits of snow avalanches can be rather easily differentiated from those of rock falls. Whereas rock-fall deposits are rather well-sorted and characteristically coarsen downslope, the avalanche deposits are unsorted masses of debris, often with a significant portion of finer material at the downslope end (see Figure 5.13b). Debris tails may also occur (Rapp 1960), which extend unslope and downslope from large boulders embedded in the scree (Luckman 1971).

The preservation potential of such flows seems relatively small because of the generally high-relief regions that they tend to occur in. Most of the deposits will probably be redeposited after reworking by valley rivers, renewed avalanching or other mass flow types, or be resedimented in lakes or in the sea either by gravity flows or wave reworking.

Mountain regions of Norway provide many excellent examples of avalanches of mixed snow and sediment. The Sunnmøre region, western Norway, is a particularly high-risk area for such mass movements because of high mountain slopes and abundant precipitation and fluctuating

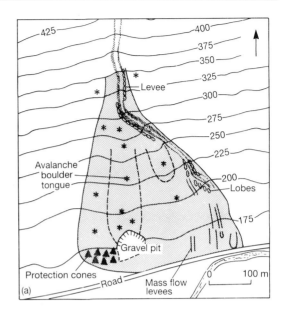

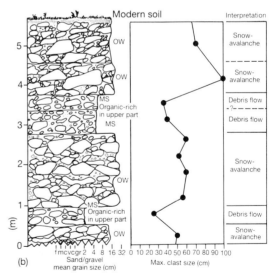

Figure. 5.13 (a) Plan-view sketch of the Bøndergjerde fan in Skorgedalen, Norway. The fan is built by deposition of debris flows and mixed sediment–snow flows. Note the narrow avalanche track extending from the top of the diagram on to the western part of the fan, and the relatively narrow avalanche boulder tongues. (From Blikra and Nesje 1991). (b) Vertical section from the gravel pit on the lower part of the Bøndergjerde fan. Note the interbedding between debris flow deposits and mixed sediment–snow avalanche deposits. The former are generally coarser grained, less sorted, and possess an openwork framework of clasts. (From Blikra and Nesje 1991.)

temperatures around freezing during the winter months. An excellent example is provided by avalanche deposits on the Bøndergjerde fan in the Skorgedalen Valley (Figure 5.13a; Blikra and Nesje 1991). There, elongate avalanche boulder tongues up to 80 m wide and 250 m long were deposited by mixed snow–sediment avalanches. The individual flow deposits are 1–1.8 m thick and contain openwork boulder and cobble beds interbedded with true subaerial debris flows, which are matrix-supported and generally finer grained than the avalanche deposits (Figure 5.13b).

5.3.5 Fluidized flows

Fluidized flows are sediment flows where the moving grains are entirely supported by pore fluid moving upward (Lowe 1976a, 1979). Deposits of fluidized flows have seldom been observed, mainly because this flow mode is unstable and probably represents only one stage of a broader evolution during flow development or flow deposition. Lowe (1982) states that fluidized flows probably decelerate to become liquefied flows or accelerate and become fully turbulent turbidity currents. Therefore, fluidized flows are not treated further here.

5.4 FLOWS WITH PLASTIC BEHAVIOUR

5.4.1 Introduction

Flows with a plastic behaviour include debris flows, liquefied flows and grain flows. There is a close linkage between fluidal and plastic flows, and particular flows may show characteristics of both flow behaviours (Nemec 1990). Specifically, fluidal flows may develop into plastic flows upon deposition (Lowe 1979, 1982), or plastic flows may develop into fluidal flows when accelerating (Nemec 1990). Below, end members only are summarized, but the transitional aspect is important.

5.4.2 Debris flow

Process

Debris flows are cohesive to non-cohesive masses of relatively unsorted debris that can flow on very low slopes, depending on the grain support mechanism (see below). Mud flows are analogous to debris flows processwise, the main difference being that they do not carry large volumes of debris. In the definition of A.M. Johnson and Rodine (1984), debris flows are a process where 'granular solids, in general only admixed by minor amounts of clay, entrained water and air, move readily on low slopes'.

There is a critical thickness T_c for initiating or stopping debris flows, assuming Coulomb behaviour of the debris (A.M. Johnson and Rodine 1984):

$$T_c = \frac{(c/UW \sin \theta)}{(1 - \tan \phi / \tan \theta)} \quad (5.1)$$

where c is the cohesion, ϕ is the angle of internal friction, UW is the unit weight of the debris and θ is slope angle of the surface (and base) of the debris flow. Thus, for instance, more cohesive debris flows can attain greater thicknesses, whereas greater slope angles will favour thinner flows (Figure 5.14; Van Steijn and Coutard 1989).

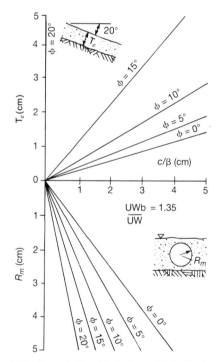

Figure 5.14 The relationship between T_c (critical thickness), R_m (radius of the largest clast), c (cohesion), uw (unit weight of the debris), UW_b (unit weight of boulder or large clast) and ϕ (angle of internal friction) of debris flows. See also equation 5.1.

The ratio between the critical thickness and surface slope angle is important for the internal shear characteristics of the flow. When the surface slope angle is at the critical value, all the internal debris will behave rigidly (Figure 5.15a). When the surface slope angle is greater than the critical value, the thickness of the flow will be less than the critical thickness. In this case, there will be a laminar shear zone at the base of the flow, with a rigid plug above (Figure 5.15b). In debris flow channels the same relationships will apply as shown in Figure 5.15a and b, not only in the depth profile (Figure 5.15c and d) but also in plan view (Figure 5.15e).

Subaerial debris flows, in particular, tend to move down channels and therefore have a very elongate plan-view expression. Their downslope margin is often lobate, and they have a well-developed, relatively steep **snout** at their downslope end (Figure 5.16; A.M. Johnson and Rodine 1984). Most often, the flows have superimposed 'waves', which tend to move at higher velocities than the main body of the flow (Figure 5.16). Subaqueous debris flows often may also be sheet-like, since the entrainment and rapid mixing of the sediment with water may cause lateral flow expansion and sheet development.

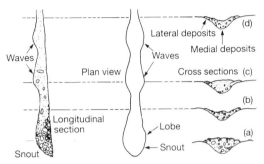

Figure 5.16 Schematic view of a debris flow, showing waves and deposits formed by successive waves of debris. (From Johnson and Rodine 1984.)

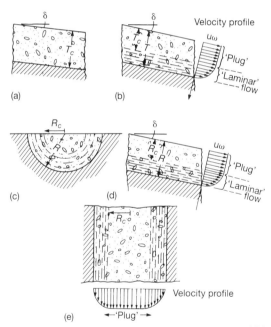

Figure 5.15 Ideal flow of debris in channels: (a) and (b) show flow in infinitely wide channels, whereas (c), (d) and (e) show flow in semi-circular channels. (a) The surface slope angle is at critical value so that the thickness of the flow is equal to critical thickness. All the debris is rigid. (b) The surface slope angle is greater so that the flow thickness is less than the critical thickness. There is a 'rigid' plug of debris above, and a zone of shear or laminar flow below. The velocity profile is shown. (c) Transverse section of semi-circular channel filled with debris. The critical radius R_c is less than the radius of the channel. (d) Longitudinal section of (c). (e) Plan view of (c) showing velocity distribution expected on debris flow surface. (From Johnson and Rodine 1984.)

Most debris flows tend to show a matrix-supported texture, with the largest clasts being positioned towards the top of the flows. This is caused by dispersive pressure between the clasts in a buoyant matrix with some cohesive strength, causing the largest clasts to move towards the top of the flow. There is usually a density difference between the debris and the dense matrix, which also gives the largest clasts buoyancy (Rodine and Johnson 1976). Hampton (1979) showed that the debris flow buoyancy is caused by two factors: (i) high density of the matrix, and (ii) loading of the pore fluid by clasts or matrix causing overpressure that buoys up the clasts. Hampton further showed that even in grain-matrix mixtures with a grain percentage up to 90, the largest clasts could be supported. Rodine and Johnson (1976) further showed that in poorly sorted debris flows, mobility was sustained when matrix was as low as 5%.

During flow, debris flows tend to sweep clean their pathway and thus effectively entrain large clasts (A.M. Johnson and Rodine 1984). The coarsest debris is generally carried in the snout of the flow, so that an upslope fining is commonly observed (Figure 5.16). Therefore, the finer and more fluid upslope debris sometimes remobilizes underlying coarser debris. Finer debris flows tend to travel further than coarse, but because fine and fluid flows readily incorporate coarser material, a simple coarse to fine gradation in a particular depositional setting (e.g. alluvial fans) should not always be expected (A.M. Johnson and Rodine 1984).

Flowslides are a particular kind of violent and very rapid debris flow (Rouse 1984). These are mixtures of sediment, water and air which may originate from very large landslides or rockfall events and in which the excess pore pressures are very high so that the sediment strength is greatly reduced. The flows can travel at enormous speeds, up to 500 km h^{-1}. It is believed that the flows travel upon cushions of fluid or air (Shreve 1968), which greatly reduce the basal friction. These flows can also travel uphill, and are known to have become airborne (Plafker and Ericksen 1978) when overtopping hills. In 1962, two catastrophic avalanches in the high Peruvian Andes caused initiation of very rapid debris flows or flowslides. These flowed for 16 km and attained speeds up to 280 km h^{-1}. In two towns, Yungay and Ranrahirca in the Rio Santa valley in Peru, 22 000 people were killed by the flows (Plafker and Ericksen 1978). The Huascarán flow, destroying Yungay on 31 May 1970, travelled at a speed of 480 km h^{-1}. It bounced off the gorge walls, causing superelevation at each impact point, overtopping a 150 m-high hill and becoming airborne before descending on Yungay (Plafker and Ericksen 1978; Rouse 1984).

Products

The textural difference between subaerial and subaqueous debris flow deposits is often pronounced and is important for interpretation of depositional environment (Gloppen and Steel 1981). **Subaerial** debris flows tend to be more clast-rich and less muddy (Figure 5.17a) than **subaqueous** flows (Figure 5.17b) due to the general lack of incorporation of water and subaqueous muds during flow (Gloppen and Steel 1981; see also section 5.3.4). Subaerial flows are generally also coarser, and may display a clast-supported framework. The bed thickness to maximum particle size ratio is relatively low, and generally less than 3 (Figure 5.18a). The fabric may be unordered, or elongate clasts may lie subparallel or parallel to bed boundaries. Grading is rare, although the lower part of the bed can show inverse grading. Subaqueous flows appear as lower-concentration flows. They are usually finer grained due to the lower competence caused by water incorporation (Figure 5.18b). Grading is more common and tends to be inverse to inverse-to-normal. Clast imbrication is also more common in subaqueous than subaerial flows (Gloppen and Steel 1981).

Debris flows are common on many modern, coarse-grained delta slopes, on continental slopes (e.g. Masson, Huggett and Brunsden 1993) and were also important on many ancient slopes (Nemec and Steel 1988, and references therein). Fjord deltas in British Columbia, Canada, are generally associated with relatively high slope gradients, and debris flows are readily formed, both on the fjord head deltas (Figure 5.19), and on the very steep underwater deltas along the fjord margins (Bornhold and Prior 1990; Prior and Bornhold 1990).

5.4.3 Liquefied flows

Process

Liquefied flows are sediment flows where the moving grains are partially supported by upward-displaced pore fluid generated by grain settling (Lowe 1976a, 1979). They are conveniently discussed in the present section although they are strictly not a rheological plastic. Two ways of generating liquefied flows are by sediment slumping or by spontaneous liquefaction (Terzaghi 1947) on slopes that exceed 3–4° (Lowe 1976a). Liquefied flows may move as, and subsequently deposit sediment from, laminar suspensions. However, if they accelerate, the liquefied flows can become turbulent and readily evolve into high-density turbidity currents (Inman 1963; Lowe 1976a).

Turbidity currents can, in their very latest stages of flow, become laminar and attain characteristics of liquefied flows. This is because flow deceleration causes particle settling and hyperconcentration, preventing turbulence and displacing pore fluid upward, which in turn causes temporary and incomplete grain support.

Liquefied flows may be unable to travel very long distances because the grain-support mechanism is insufficient to maintain prolonged movement. Experimental work suggests that the travel distance of liquefied flows depends on grain size, velocity of the head of the flow and bed thickness

Figure 5.17 (a) Subaerial debris flow deposit from the Devonian Hornelen basin, western Norway. Note the high clast-to-matrix ratio. Lens cap for scale. (Photograph courtesy of Ron Steel.) (b) Subaqueous debris flow deposit from the Eocene St Llorenç del Munt fan-delta complex, Catalan Coastal Range, Spain. Note the steeply imbricated clasts (*a*(p)*a*(i)-fabric), suggesting band shearing in the latest stages of flow (Nemec 1990, figure 33), and the lower clast-to-matrix ratio than in (a).

(Lowe 1976a). The finest and thickest flows are able to travel the farthest, up to an estimated maximum travel distance of 10 km (Figure 5.20; Lowe 1976a).

Products

Deposits of liquefied flows tend to be massive beds, generally without any evidence for traction-produced structures (Figure 5.21). However, in some cases, the top zone may be cross-laminated, suggesting late-stage reworking from the tail of the depositing flow or, alternatively, shear between the flow and the ambient water (Lowe 1982; Martinsen 1987). Grading may be present, as normal, coarse-tail grading (Lowe 1982). In contrast to massive turbidites, liquefied flow deposits show no sole marks, such as flutes, suggesting non-turbulent conditions during deposition. In several cases, water escape structures, such as dish and pillars, are present (Lowe and LoPiccolo 1974; Lowe 1975; section 4.3.4).

5.4.4 Grain flows

Process

Grain flows are flows of granular solids on a slope where the grain support mechanism is, by dispersive pressure, caused by grain collisions, and in which the interstitial fluid is the same as the ambient fluid (Lowe 1976b). **Modified** grain flows are flows where excess density of the interstitial fluid, ambient fluid current-drag or escaping pore fluid contributes to grain

144 Mass movements

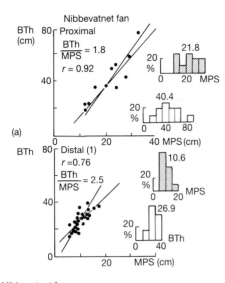

Figure 5.18 Bed thickness (BTh) to maximum particle size (MPS) ratio of (a) subaerial and (b) subaqueous debris flows from the Devonian Nibbevatnet alluvial fan in the Hornelen Basin, western Norway. Note the much lower BTh/MPS ratio for subaerial flows. Note also the difference in scale for the subaqueous flow diagram. (From Gloppen and Steel 1981.)

collisions in sustaining the flow (Lowe 1976b).

True grain flows cannot exceed 5 cm in thickness, largely because dispersive pressure alone becomes insufficient in maintaining movement and support of the grain dispersion. Thicker beds must have had one or several additional grain-support mechanisms. Lowe (1976b) suggested that the term grain flow should not include flows where other grain-support mechanisms operate. Other authors have also discussed usage of the term grain flow (Stauffer 1967; Middleton and Hampton 1973).

Density modified grain flows are cohesionless flows of granular solids in which the excess mass of large grains is supported by the excess density of the matrix (Lowe 1976b, 1982). This usage of the term grain flow is confusing, as these flows can be better termed **cohesionless debris flows**, as the only difference between them and true cohesive debris flows is the nature of the matrix. Therefore, these types of flows are included under debris flows.

Products

True grain flows are common both in subaerial and subaqueous settings. They are limited to slip-faces of sandy dunes and bar forms of largely any scale, or to inclined, destabilized surfaces, for example in slump scars that approach the angle of repose. On the lee sides of aeolian dunes, grain flows occur as relatively narrow, elongate tongues of sand, and they probably have the same geometry on subaqueous dunes. Grain flows form when the slip-face inclination increases above the angle of repose, either due to increased deposition or from erosion. Their downslope extent depends on slope inclination relative to angle of repose. Where the inclination becomes less than the angle of repose, grain flows will freeze and deposit, and vice versa.

In cross-section, grain flows are characteristically lenticular, a few centimetres thick and on slip-faces of bar forms are most often coarser than the surrounding sediment (Figure 5.22). This is because they are supplied from higher up on the lee face by mass movement and are not subject to size-sorting processes imposed on the normal saltational or suspended load, which also is deposited on the lee face.

5.5 SLUMPS

5.5.1 Introduction

Slumps are downslope movements of sediments above a basal shear surface where there is significant internal distortion of the bedding (Stow 1986). Nevertheless, the bedding should be recognizable. There is a continuous transition between slides (see next section), slumps and plastic flows, and some units may show characteristics of all three modes of transport (Bakken 1987; see also below). Therefore, careful analysis is required to fully understand the movement behaviour of the deformed unit, and to categorize it satisfactorily.

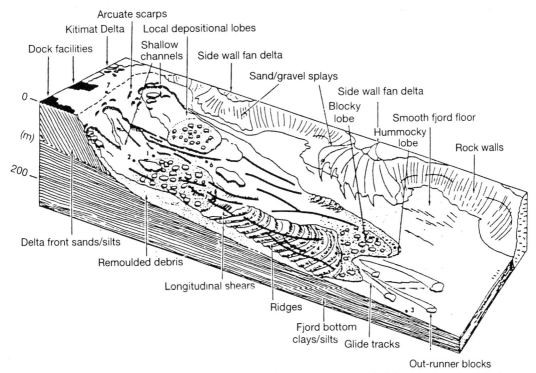

Figure 5.19 Conceptual drawing of a submarine debris flow in Kitimat Arm, British Columbia, Canada. Note the heterogeneous nature of the flow and the lobate pressure ridges in the toe region. (From Prior, Bornhold and Johns 1984).

5.5.2 Process

Slumping is a common process, particularly on subaqueous slopes, and especially where there is a significant input of fine-grained sediments. The slumps form above a basal shear surface (Figure 5.23), the depth to which is decided mainly by the pressure gradient in the sediment. Where the pore pressure approaches or balances the normal stress induced by the weight of the overburden, the shear strength is sufficiently reduced to allow slippage along a basal shear surface, given a sufficiently high shear stress. The magnitude of the shear stress acting on a slope can be viewed as:

$$\tau = \rho g s h \tan \theta, \qquad (5.2)$$

where ρ is the sediment density, g the acceleration due to gravity, s the solidity (the complement of porosity), h is sediment height (thickness) and θ is the slope inclination (see also equation 1.15). Failure at the basal surface occurs when the shear stress exceeds its strength, which is given by equation (1.5). The above equations were discussed in the context of mass movements by Middleton and Southard (1978) and Hampton (1979).

Once initiated, the shear surface will propagate upslope in a radial fashion from its nucleation point (G.D. Williams and Chapman 1983; Farrell 1984), leading to the formation of a scoop-shaped, concave-downslope depression or slump scar, often with an irregular outline (Martinsen 1989). The shear surface is probably initiated as a slope-parallel feature, but at some point steepens to intersect the sediment surface. The transition may occur at lithofacies boundaries (Figure 5.24), or at sites of pore-pressure jumps where material strength contrasts are present (Crans, Mandl and Harembourne 1980).

The moving slump deforms intensely internally, and produces a wide variety of deformational structures. Folds, boudins, microfaults, internal shear surfaces and faults are all common structures in slumps (Figure 5.23). The occurrence of these structures suggest that the slumps go through a main phase of plastic/ductile deformation whereby folds and boudins are formed.

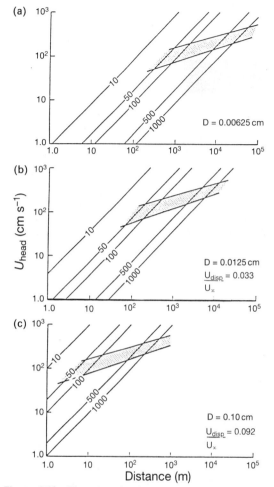

Figure 5.20 Diagrams showing the relationship between flow distance and velocity (*U*) of the head of liquefied flows of quartz spheres for silt to very fine sand (a), very fine to fine sand (b) and coarse to very coarse sand (c) in flow depths of 10, 50, 100, 500 and 1000 cm. *D* is the thickness of flow head. The shaded parts represent approximate ranges of natural liquefied flows. (From Lowe 1976a, which contains further details.)

The ductile phase is followed by a very late brittle phase where the faults form. It is common to see strain overprinting, where early formed folds are truncated by late faults (Figure 5.25; Martinsen 1989; Martinsen and Bakken 1990).

The slump folds are mainly sheath folds, formed by simple shear (Figure 5.26; Martinsen, 1989), although buckle folds can also occur (Woodcock 1976). Faults can be both extensional and contractional (Figure 5.25), and may either occur randomly within the slumps or be related to local obstacles or shear surface irregularities. Idealized models of slumps and slides (Lewis 1971; Allen 1985) show the deformed units to have a well-defined upper extensional zone and a downslope contractional zone. However, it is quite likely that a significant amount of lateral compaction will occur when slumps, particularly those that are fine-grained, come to a halt, thus preventing development of downslope or toe contractional zones (Crans, Mandl and Harembourne 1980; Garfunkel 1984).

Farrell (1984) suggested that if slumps halt first at their downslope margin, a contractional strain wave will propagate upslope through the slump, overprinting any earlier formed structures by contractional structures. In contrast, if the slumps halt first at their upslope margin (for instance due to initial pore-water escape there), an extensional strain wave will propagate downslope through the slumps, causing extensional structures to overprint earlier formed structures.

Slumps form on low slopes, as little as 0.1° or less (e.g. Prior and Coleman 1978b). Very shallow slopes are most common when slumps are formed from sediment finer than sand. This is because the internal deformation in slumps is generally ductile or plastic, a condition which is promoted by the interstitial water being more easily retained in the pore spaces of less permeable, finer sediments.

Slumps can range in thickness from 0.5 m (Martinsen 1987) to several hundreds of metres, for example, some major slumps on continental slopes (Dingle 1977; Jansen *et al.* 1987). There is a clear contrast in the scale of slumps observed on modern continental margins to those observed in ancient successions (Woodcock 1979a). The continental-margin examples are several orders of magnitude larger in cross-sectional area than ancient examples. This scale difference may have several causes. One explanation is that the size of outcrops of ancient slumps may be far too small to detect the extremely large slumps. Another explanation is that most ancient slump examples come from deltaic successions, whereas continental-margin successions, which may preserve the exceptionally large slumps, have poor preservation potential. The deformed units observed in most modern delta areas are generally less than 40–50 m in thickness (for example, in the Mississippi Delta; Prior and Coleman 1978a,b), which corresponds to the scale of deformed units in some ancient successions (Martinsen 1989).

Figure 5.21 Probable liquefied flow deposits from the Gull Island Formation (Upper Carboniferous), County Clare, western Ireland. Note the internally massive character and rippled top, particularly of the lowest bed, and flat base of this bed and the bed above, suggesting laminar flow conditions, at least in the final stages of flow. Measuring tape is 10 cm long.

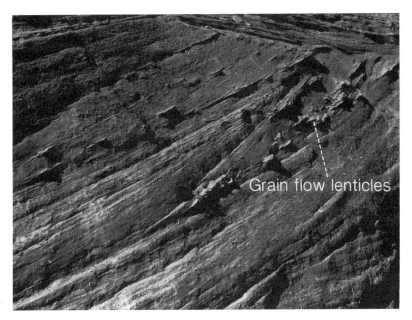

Figure 5.22 Ancient grain-flow lenticles in the aeolian Entrada Sandstone (Middle Jurassic), near Gallup, New Mexico, USA. Note the difference in geometry between the **grain fall** deposits, which thin to the left, the onlapping wind-ripple lamination, and the lenticular, coarser grained **grain flow** deposits. Lens cap for scale.

Slumps (and slides) may be triggered by a variety of processes. Seismic triggering (Seed 1968; Leeder 1987), cyclic wave-loading (Henkel 1970; Suhayda et al. 1976), high sedimentation rates and methane generation causing overpressuring (Whelan et al. 1976; Prior and Coleman 1978b) and slope oversteepening (Martinsen 1989) are all possible initiators of gravitational sliding. It has also been suggested that deposition of turbidites upon a poorly consolidated sediment may trigger failure, because of the induced shock (Martinsen 1987, 1989).

5.5.3 Products

Slumping produces lenticular units of deformed sediment of varying scale, bounded below by a distinct shear surface. Slumps tend to be characterized by a somewhat chaotic appearance, where primary bedding is distorted into isolated rafts or folded beds. Folds tend to occur throughout the

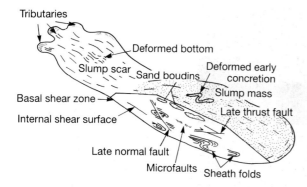

Figure 5.23 Idealized slump model showing the most common location of structures associated with slumps. Based on field observations in the Upper Carboniferous Gull Island Formation (western Ireland). (From Martinsen 1989.)

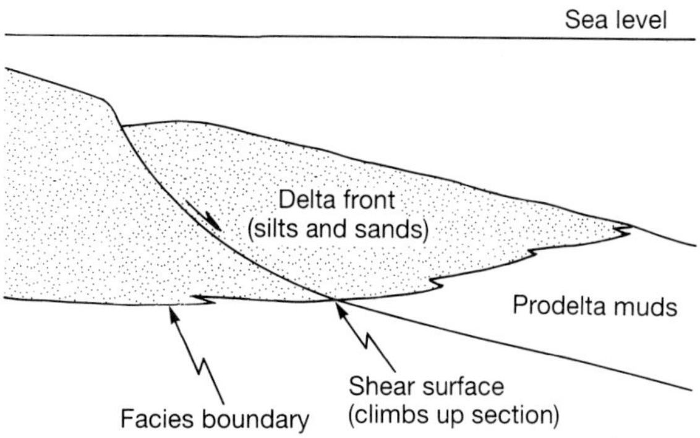

Figure 5.24 Basal shear surfaces may start to climb upsection at facies boundaries, here depicted in an idealized delta-front setting. (Based on Crans, Mandl and Harembourne 1980, figure 9a.) No scale implied.

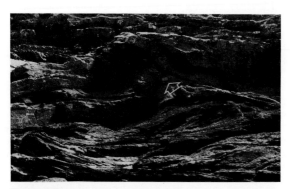

Figure 5.25 Late thrust fault truncating earlier slump fold in the Fisherstreet slump, Gull Island Formation (Upper Carboniferous), County Clare, western Ireland. This relationship of faults truncating folds is commonly observed in slumps (see text for discussion). Notebook, 20 cm long (arrowed), for scale.

units, although they are most common in the basal and middle parts because of the increased shear there. The slumps generally have very sharp lateral margins, which in effect are strike-slip or oblique-slip faults, depending on whether the slump margins maintain parallelism to the movement direction or are skewed. The upslope slump scar may have tributaries or show evidence of retrogressive failure, as a result of footwall unloading during initial slump movement (Martinsen 1987, 1989). The slump scars are generally filled by fine-grained sediment, or expanded by current erosion and filled by material such as coarser grained, turbidity-current deposits. In the latter case it may be impossible to differentiate the slump scar from an erosional channel.

Slumps form hummocky sediment masses in plan view, particularly at their downslope ends if toe zones are developed. The topography may be organized into lobate, convex-downslope ridges ('pressure ridges'; Figure 5.19; Prior, Bornhold and Johns 1984) which probably are the surface manifestation of toe-zone thrust faults (Martinsen and Bakken 1990).

Numerous authors have described slumps in the literature (e.g. Helwig 1970; Lewis 1971; Farrell 1984; Hill 1984; Postma 1984; Hein 1985; Farrell and Eaton 1987, 1988; Martinsen and Bakken 1990; Collinson et al. 1991). Some representative examples are described below, both from modern sediments and the ancient record.

The Fisherstreet slump (Namurian), County Clare, Ireland

The Kinderscoutian (Namurian, Upper Carboniferous) Gull Island Formation exposed in County Clare, western Ireland (Figure 5.27), represents slope deposits laid down on an inherently unstable basin slope (Martinsen 1987; Collinson et al. 1991). More than 75% of the 550 m-thick formation was deformed by syndepositional processes, and most of the deformation was by slumping. One extremely well-exposed slump occurs at Fisherstreet in north Clare (Figure 5.27).

The slump is approximately 20 m thick and it retains this thickness for about 4 km of outcrop. It consisted mainly of mud but included some sand beds toward the top. Plan-view exposures show a very organized style of folding (Figure 5.28a and b). The folds are non-cylindrical, in-

Slumps **149**

Figure 5.26 Eleven stacked, recumbent sheath fold hinges in the Gull Island Formation (Upper Carboniferous) in a slump at Killard, central County Clare, western Ireland. In some slumps, primary bedding may be shortened to less than 15% of the original due to superimposed sheath folds. Note the changing orientation of the fold axes. Compass (circled) for scale (10 cm long).

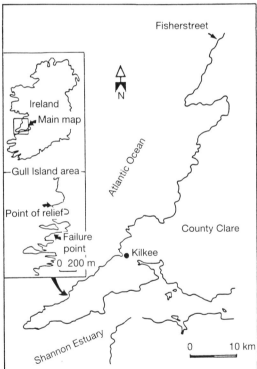

Figure 5.27 Map of locations in County Clare, western Ireland, described in the text.

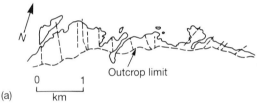

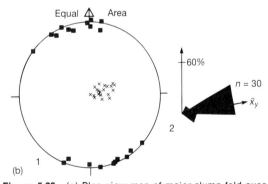

Figure 5.28 (a) Plan-view map of major slump fold axes (dashed lines) in the Fisherstreet slump to show regularity. The change in orientation toward the east may have been caused by increased lateral shear in this region.
(b) 1 = stereonet plots of fold axes ($\boxtimes$) and axial plane poles (x) of the Fisherstreet slump ($n=23$);
2 = palaeocurrent rose diagram for overlying turbidites.
Note the similarity in inferred palaeoslope attitude (toward the ENE), suggesting that there was relatively little rotation of fold axes. This is consistent with relatively little movement and/or internal deformation of the slump.

clined to recumbent, and tight to isoclinal with clearly thickened hinge zones (Figure 5.29). The folds resemble early development stages of large-scale sheath folds. The fold axes, allowing for the non-cylindricity, are consistently oriented NNW–SSE, normal to regional palaeocurrent measurements from turbidites (Figure 5.28b, diagram 2).

The axial planes dip to the WSW (Figure 5.28b, diagram 1).

Normal and reverse faults are also common, and these always truncate the folds where superimposed (Figure 5.29). The faults occur mainly as

Figure 5.29 Recumbent, isoclinal slump fold formed by simple shear and cut by later extensional fault (arrowed). Fisherstreet slump, Gull Island Formation (Upper Carboniferous), County Clare, western Ireland.

single faults, i.e. they do not seem to be grouped into 'families'; they sole out into a common décollement. Extensional faults and contractional faults occur in groups, but show no systematic areal distribution. Nevertheless, both the extensional and contractional faults strike systematically NNW–SSE (parallel to fold axes from slumps); the extensional faults always dip to the ENE, and the reverse faults always dip to the WSW.

Sandy turbidites drape the top of the slump, fill in lows, and thin over the preserved highs. In a few places, sand volcanoes occur on top of the turbidites. These were probably formed from water escape from the slump triggered by the deposition of turbidites (Martinsen 1989).

The Fisherstreet slump is probably a relatively organized slump, because little variation of the orientation data occurs. This may suggest that the slump moved only a short distance, or that it retained its internal coherence to a large degree. The latter may be the result of the main deformation and shear taking place along the basal shear plane, perhaps as a result of a high internal strength. Many slumps are more chaotic and show a large variability in the orientation data over a small area (Martinsen 1987). Nevertheless, the example from Fisherstreet serves to show the fold style and strain overprinting that is common in many slumps (Farrell 1984).

The Storegga slumps, Norwegian continental margin

A large slope failure area on the continental slope west of central Norway was described by Jansen *et al.* (1987), based on seismic and shallow-core data. Three periods of massive slope failure were recognized (Figure 5.30), involving a total volume of 5580 km³ of Cenozoic sediments. The first failure, which occurred before 30 000 years BP was the largest, displacing 3880 km³ of primarily young, unconsolidated sediments. The second and third failures occurred in the Holocene between 8000 and 5000 years BP. The last two events displaced a total of 1700 km³ of sediments, and cut into sediments as old as the Palaeogene. The second failure occurred in 3500 m of water depth and affected sediments over a distance of 800 km.

These slope failures were described as slides by Jansen *et al.* (1987), but are probably more correctly termed slumps because of the high degree of internal deformation in cored sediments. Most of the sediments are transitional from slump to debris flow deposits, and there are also associated turbidites (Jansen *et al.* 1987). The turbidites may have been carried as turbulent suspensions on top of the displacing sediment masses, or, alternatively, could have been triggered by it. Numerous fine-grained turbidites and creep deposits occur in the head region of the slump scar. These were probably triggered by both the instability and increased slope angles created by the downslope movement of the failed sediments (Jansen *et al.* 1987).

The slump scar is elongate, striking SE–NW (Figure 5.30). The first slump encompassed the largest area, and the two succeeding failures are superimposed on this (Figure 5.30), with the last event being the smallest in terms of area. This nested pattern suggests clearly that the triggering of the two last events was related to the morphology created by the large first event.

This example serves to show the enormous volumes involved in recent and subrecent mass movements on continental slopes. Other such examples include the slope failures off Nova Scotia (Heezen and Drake 1964), Brazil (T.C.

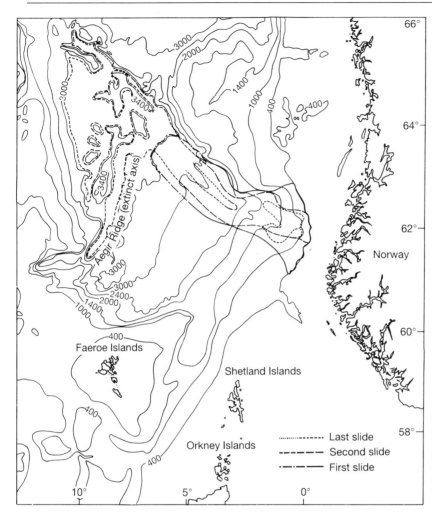

Figure 5.30 Map of the Storegga slumps, offshore central Norway. Note the nested pattern of the latest and smallest slump superimposed on the second slump, which is superimposed on the largest and earliest slump. (From Jansen et al. 1987.)

Moore et al. 1970), southern Africa (Dingle 1977, 1980), northwestern Africa (Jacobi 1976), the Canary Islands (Embly and Hayes 1974) and New Zealand (Barnes and Lewis 1991). A review of large submarine slides and slumps is given by D.G. Moore (1978).

Mississippi delta-front slumps

There is a wide variety of deformational features described from the delta-front area of the Mississippi River (Coleman and Garrison 1977; Prior and Coleman 1978a,b; Roberts, Suhayda and Coleman 1980). Some of these features do not involve lateral translation (for instance the collapse depressions and the mud diapirs, section 4.3.2), but most other features show downslope movement of sediment. A wide variety of terminology has been applied to these features but, in essence, they range from growth faults bounding slides with little internal deformation to mudflows with penetrative internal deformation of the sediment (Figure 5.31; Roberts, Suhayda and Coleman 1980). Slumps (*sensu stricto*) are common, and comprise the 'bottleneck' failures, and probably also some of the less deformed 'mudflows' of, for example, Roberts, Suhayda and Coleman (1980) (Figure 5.31).

All the slumps have a characteristic morphology, with an upslope head region that is lobate and concave downslope. In some cases, the head region narrows downslope into a 'bottleneck' from which the slump area widens into a toe zone, where the slump may overflow its margins on to the adjacent undisturbed

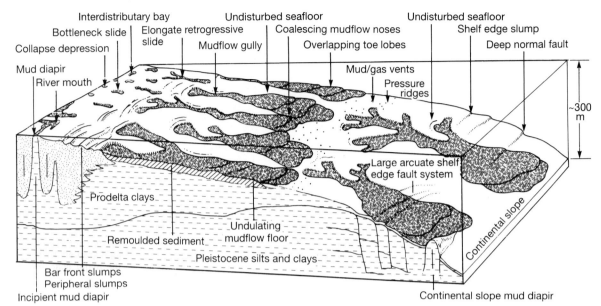

Figure 5.31 Overview sketch of instabilities on the Mississippi delta-front. Most instabilities range from slumps to flows of muds. (From Prior and Coleman 1982.)

sediment (numerous examples appear in the conceptual model in Figure 5.31). In other cases, the head region of the slump is dominated by several 'tributaries', which may be highly skewed in orientation to the main slump direction (Figure 5.32). In ancient successions this pattern is important when palaeoslope directions are measured. Tributaries are probably related to retrogressive failure, in which the upslope scarp formed by the initial slump movement becomes a 'free' end, and causes unloading of the adjacent footwall, leading to renewed failure (see Figure 1.9). A local slope then exists between the adjacent sediment and the slump scar, irrespective of the original orientation of the main delta slope.

The slumps leave a chaotic deposit characterized by floating blocks in a deformed sediment mass (Roberts, Suhayda and Coleman 1980). Whether the deposits should be termed slumps or flows is largely semantic, but to some degree depends on the movement distance of the deformed sediment. The further the movement, the more likely it is to turn into a flow.

Overpressure in the sediment caused by high sedimentation rates (hindering normal porewater escape), together with methane (CH_4) from degradation of the high amount of organic material in the sediment, are probably the primary mechanisms for causing instability (Roberts, Cratsley and Whelan 1976; Whelan et al. 1976). The slope of the Mississippi delta-front is generally very low (less than 2°, e.g. Prior and Coleman 1978b (see Figure 1.8)), so a significant reduction of the sediment shear strength is needed for instability to occur.

This example shows that slumps can occur on very low slopes, provided that the conditions are favourable. It also shows that a range of slump scar morphologies is to be expected when analysing ancient deposits.

5.6 SLIDES

5.6.1 Introduction

Slides are downslope displacements of sediments above a distinct shear surface where there is little or no internal deformation of the transported material (Figure 5.33; *sensu* Stow 1986). Slides are fully transitional from slumps, and in cases the differentiation may be semantic. However, an arbitrary boundary between slides and slumps may be at the point where the original bedding is folded and/or disrupted. Thus, in slides, the original bedding can be slightly rotated along fault planes, such as into hangingwall anticlines, but not deformed as a direct response to simple shear or buckling. A special case may occur where layer-parallel slip has caused intra-

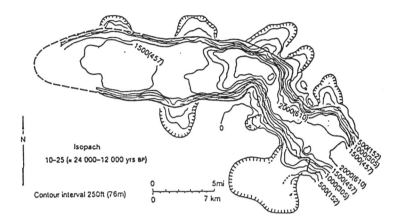

Figure 5.32 Plan view of major slump scar on the Mississippi delta-front. Note the highly skewed orientation of the scar tributaries in relation to the main scar. This pattern is probably caused by a local excessive slope gradient being set up as a result of destabilization and footwall unloading following removal of the sediment in the major scar. (From Coleman, Prior and Lindsay 1983.)

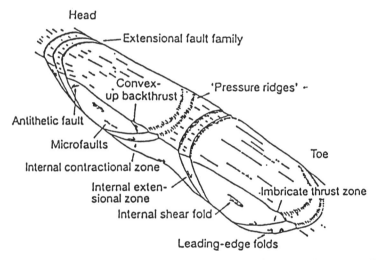

Figure 5.33 Idealized model of a slide showing the most common location of associated structures. (From Martinsen 1989; based on field observations in the Gull Island Formation (Upper Carboniferous), County Clare, western Ireland.)

stratal sheath folds to form. These failed sediment masses are still termed slides because the main part of the bedding is retained in its original configuration.

The following discussion encompasses sediments which at their upslope ends and bases are bounded by clearly defined fault planes. Therefore, sediments moved for only short distances compared with their thickness are included (**rotational** slides; Allen 1985), together with sediments moved for relatively long distances compared with their thickness (**trans- lational** slides; Allen 1985) (Figure 5.34). This means that sediments displaced by growth faults and shelf-edge fault systems are termed slides, because of the focus on their bounding discontinuities.

Slides span a great range of slope instabilities, including such diverse features as bank collapse features in river channels, subaerial mudslides, delta-front growth faults, shelf-edge faults and submarine glide-blocks and olistoliths. The following general outline of slide theory is discussed only briefly as they have much in common with

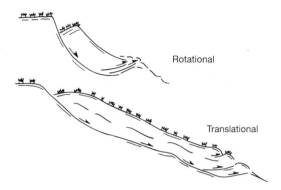

Figure 5.34 Schematic illustration of the difference between rotational and translational slides. Rotational slides are only moved for a very short distance relative to their thickness, thus the sliding motion is mainly rotational. Translational slides are moved for relatively long distances, or affect longer areas of the slope compared with their thickness, so that a translational mode of movement is dominant (see also Allen 1985).

slumps. The main difference between these two mass movements is simply the degree of internal deformation.

5.6.2 Process

In their most simple form, slides are spoon-shaped features with a three-part morphology of upslope head region, middle 'rigid' zone and downslope toe zone (Figure 5.35; Brunsden 1984; Gawthorpe and Clemmey 1985; Martinsen 1989). The upslope head region is concave downslope and dominated by extensional deformation, whereas the middle region is mainly translational and may behave in a rigid fashion and not show any particular strain signature above the basal

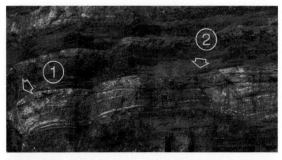

Figure 5.35 A family of two growth faults (labelled 1 and 2) at the upslope margin of a major slide complex at Edgeøya, Svalbard, Norwegian Arctic. Note the listric shape of the faults and the growth of several beds across (2). The section shown is 150 m thick.

shear surface. The downslope toe region is usually dominated by contractional deformation and has a convex downslope and characteristically lobate form.

The **head** region is usually dominated by extension along a group or 'family' of listric faults (Crans, Mandl and Harembourne 1980) which tend to sole out at a common level or basal décollement. In some large slides the fault families occur at several orders of magnitude, so that one small family may be entirely enclosed in the hangingwall of a larger order fault family (Figure 5.35). Quite commonly, antithetic extensional faults occur, which may be either listric or planar. Antithetic faults show downthrow in the opposite sense to the master faults, which can significantly confuse measurements of palaeoslope (Martinsen 1987). The existence of fault families in slides is an essential difference from slumps, where most faults are single faults (see above).

The central part of slides generally shows little evidence of the sliding itself, and the basal shear zone can be extremely difficult to detect, particularly in fine-grained sediments. In places, there may be evidence for internal slip between beds in the form of sheath folds or microfaults. These structures are important to detect and provide important information on the translated nature of the sediment when other features are not present.

The margins of the slides are dominated by strike-slip deformation, but the width variability of the slide scar can cause both transpressional and transtensional movement if the slide scar narrows or widens. Any evidence for strike-slip motion is important to detect, because it shows that the slide margin observed is a lateral margin, and not a head or toe region.

Contractional deformation dominates the **toe** region, and this is most commonly expressed in the form of thrust faults. The thrusts form classic duplex and imbricate zone geometries (Figure 5.36a and b; Lewis 1971; Dingle 1977; Martinsen and Bakken 1990), which often may be very difficult to differentiate from the familiar post-lithification structures. In plan view, the morphological expression of the contractional toe-zone is sometimes that of downslope lobate ridges, or 'pressure ridges' (Figure 5.19; Roberts, Suhayda and Coleman 1980; Prior, Bornhold and Johns 1984). Martinsen and Bakken (1990) suggested that three different types of contractional zones occur in slides (Figure 5.37). The

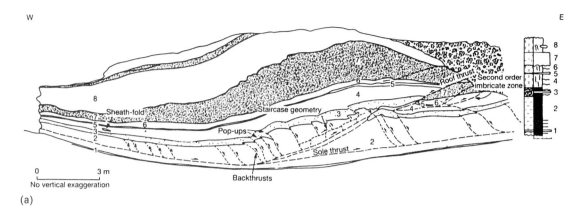

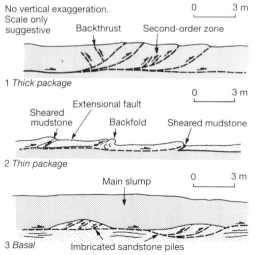

Figure 5.36 (a) Contractional slide zone in the Upper Carboniferous Gull Island Formation at Failure Point, Gull Island area, County Clare, western Ireland (see Figure 5.27 for location). (From Martinsen and Bakken 1990; based on Martinsen 1987.) (b) Picture of right-hand part of (a). The ramping thrust (marked 'roof thrust' in (a)) and the sole thrust are indicated. Compare with (a) for location of other structures. Backpack for scale.

Figure 5.37 Models for contractional zones in slides. (From Martinsen and Bakken, 1990; based on Martinsen 1987.)

thick package zone is the most complex and is most similar to mountain-belt thrust zones, although the scale is several orders of magnitude smaller. Thin package zones seem to develop at the frontal margin of heterogeneous slides, particularly where sand and mud is closely interbedded (Figure 5.37; Martinsen and Bakken 1990). Basal contractional zones are not related to toe zones but rather to obstacles, such as ramps, at the base of slides. Such zones can be mistakenly interpreted as toe zones and their context can be understood only through careful analysis.

In several instances, slides do not seem to have clearly defined toe zones, and some slides are 'open-ended'. This is probably caused by a high degree of lateral compaction so that the strain is taken up by porosity reduction rather than by thrust-zone formation (Crans, Mandl and Harembourne 1980; Mandl and Crans 1981;

Garfunkel 1984; Martinsen 1987; Barnes and Lewis 1991). This is probably particularly common for slides in muds and clays.

Toe zones of slides may be confused with contractional zones formed by **gravity spreading**. Gravity spreading forms as a result of loading a substratum by an overlying medium (Figure 5.38; for example progradation of sandy delta-front sediments over muddy prodelta deposits). During spreading of the load, a lateral stress component forms which causes the loaded medium to be pushed sideways and upwards along thrust faults (Bucher 1956; Galloway 1986; Pedersen 1987). In progradational settings, such as deltas, the lateral stress component will be oriented in the direction of progradation due to the progressive loading in that direction. Therefore, the thrust faults will show propagation in an offshore direction, parallel to the generally expected sliding direction of slides and slumps down the delta slope. Pedersen (1987) listed criteria to distinguish gravity sliding from gravity spreading (Figure 5.38), but in areas of insufficient exposure, the two may be inseparable.

Gravity spreading occurs on a variety of scales, from minor deltas to plate-convergent zones (see Pedersen (1987) for a discussion). Galloway (1986) also discussed gravity sliding and gravity spreading in Cenozoic sediments of the Gulf of Mexico. The Niger Delta shows excellent examples of gravity spreading caused by loading of the deltaic sediments on offshore fine-grained deposits (Weber and Daukoru 1975).

Because both slides and slumps can occur on very low slopes, piling up of deformed material can easily generate local slopes of orientation highly skewed to the regional slope. Therefore, both extensional and contractional structures with very variable orientations should be expected to occur on, for example, delta slopes. These can occur particularly where main slide scars are fed by tributaries (Coleman et al. 1983).

Significant departures can occur from the idealized model discussed above. Commonly, the slide will be a heterogeneous mass that moves above a heterogeneous substratum. Therefore, local stress regimes can be set up within the slide mass, and cause both local extensional as well as contractional zones to form. Consequently, extensional zones are not constrained only to the head region of a slide, neither are contractional zones confined to the toe region of a slide (Martinsen 1989; Martinsen and Bakken 1990).

A particular kind of slide occurs when exotic rock units slide downslope. These slide blocks are termed **olistoliths** and are particularly common in carbonate terranes (Conaghan, Mountjoy and Edgecomb 1976; Bosellini 1984), and can be of gigantic size, up to several cubic kilometes in volume. Mélange terranes are other settings where slide blocks are common (Swarbrick and Naylor 1980; Naylor 1982), but they can also be found in purely clastic marine settings (e.g. Ineson 1985). Naylor (1982) listed characteristics of exotic blocks displaced by gravity sliding, focusing on field relations and orientations of the blocks, external morphology and host-sediment

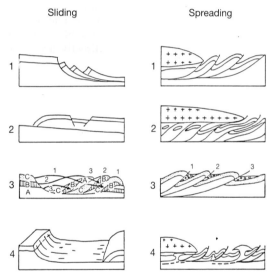

Figure 5.38 Conceptual diagram to show the difference between gravity sliding and spreading. The criteria for distinguishing gravity sliding are listed in the left column and include: 1, listric normal faults at the trailing end of deformed sheet; 2, flat-lying thrust sheet disturbed by extensional deformation; 3, **diverticulation**, whereby the upper stratigraphical units are displaced further than the lower; 4, exposure of a 'peel-off' region in the rear end of the thrust fault region. Four criteria for distinguishing gravity spreading are listed in the right column and comprise: 1, an imbricate fan formed by listric, splay, thrust-faults in front of the overlying spreading mass; 2, a duplex of imbricate thrust sheets formed in the deformed and overthrust sediments – because of loading, boudins may be formed beneath the spreading unit; 3, in the frontal part of the gravity-spreading deforming system, syndeformational basins are formed with progressively younger sediments away from the front of the spreading mass; 4, due to increasing overpressure in the décollement, water-escape structures and mud-diapirs occur in the frontal region of the gravity spreading system. (From Pedersen 1987.)

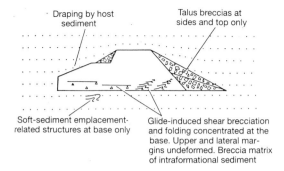

Figure 5.39 Schematic illustration of principal features of submarine slide blocks. (From Ineson 1985; based on Naylor 1982.)

deformation. Ineson (1985) pointed out that in complex tectonic terranes it may be difficult to differentiate between gravity slides and thrusting. In particular, the internal deformation of the gravity-slid block may be very similar to tectonically generated structures (Figure 5.39). However, gravity sliding may be indicated by increasing internal disruption towards the base of the block, and by the style of deformation, which is mainly by simple shear under low overburden (Figure 5.39; Ineson 1985).

5.6.3 Products

Gela submarine slide, Sicily foredeep

The Gela Basin, southwest of Sicily in the Mediterranean (Figure 5.40), is filled with up to 2500 m of marine sediments that shallow upward, and are of Pliocene to Quaternary age. The basin is situated at the front of the Gela nappe, and is a foredeep basin to the Maghrebian fold–thrust belt (Trincardi and Argnani 1991).

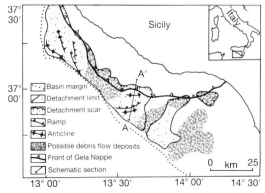

Figure 5.40 Location of the Gela slide, Sicily foredeep. (From Trincardi and Argnani 1991.)

The Gela slide itself occurs immediately above the tip of the Gela nappe. It covers 1500 km², and has a strike-parallel, very elongate shape. In cross-section (Figure 5.41) the slide appears as an undisturbed mass, where the original bedding is well-preserved, as shown by the continuous and parallel reflectors in seismic sections.

Trincardi and Argnani (1991) recognized four main parts of the slide and related features, namely slide head, slide toe, lateral ramps and a chaotic unit which probably filled the slide scar (Figure 5.41). The slide head is dominated by one extensional fault that slopes 15° at the most towards the basin, but it is a listric fault and soles out underneath the slide. The slide scar, the surface expression of the detachment, is continuous along strike for at least 120 km. The relatively long extent along strike is probably related to the fact that the slide occurs directly above the Gela nappe, which seems to extend for a similar distance (Figures 5.40 and 5.41). Only one extensional fault is found in the head region. This also may be related to its position immediately above the nappe.

Immediately downslope from the head, the slide appears undeformed. This central **propagational** or **translational** zone is typical of many slides, and represents the 'rigid' central zone where little or no deformation occurs above the basal detachment. In several instances, it is impossible to detect evidence for sliding within this zone because deformation is confined to the basal shear zone, which itself may only be a few millimetres thick.

Further downslope, the slide toe region shows well-developed contractional deformation with thrusts, folds and imbrication zones. The thrusts caused thickening of the slide body in this region (Figure 5.41), producing a positive feature on the sea bottom (Trincardi and Argnani 1991).

Lateral ramps formed along the sides of the slide. Local detachments have formed where abrupt changes take place in their distance from the basal slip surface. The slid material is deformed and folded above the lateral ramps to accommodate the geometry of the basal slip plane.

Above the head-zone roll-over, a **chaotic unit** fills in the depression created by the downslope-translated material. Trincardi and Argnani (1991) interpreted this unit as a debris flow deposit, possibly formed by the locally increased

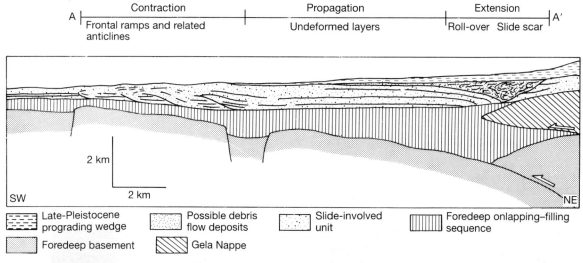

Figure 5.41 Cross-section (A–A' in Figure 5.40) of the Gela Slide. See text for discussion. (From Trincardi and Argnani 1991.)

slope-angle due to the formation of the slide scar. This relationship of slumps or debris flows filling slide scars is also observed in ancient slide deposits (see below).

The Gela slide is an example of a simple slide where the tripartite division of extension, translation and contraction is well displayed. There is no evidence for strain overprinting in the internal extensional and contractional zones. Strain overprinting may occur if heterogeneities are encountered (for example, uneven or ramping décollements), if the slide stopped abruptly so that a compressional strain wave propagated upslope through the slide, or if the sliding motion was hindered in the head zone, causing an extensional strain wave to propagate downslope through the slide (Farrell 1984).

Point of Relief slide, County Clare, Ireland

The 2100-m-thick Namurian (Carboniferous) succession of County Clare, western Ireland, records the fill of a deep, symmetrical ENE–WSW elongate trough which formed above the Iapetus Suture as a response to N–S extension (Rider 1974; Collinson *et al.* 1991). The fill is a shallowing upward succession, from deep basinal shales, through turbidites and a slope succession, into deltaic cyclothems. The slope succession, the Gull Island Formation, is 550 m thick, of which 75% is deformed as a result of sediment deformation processes (Martinsen 1989). The deformation style is either slumping, sliding, water escape or combinations of these. Slides are especially common in the upper part of the slope succession, which is particularly muddy and fine-grained.

One of the slides, informally named the Point of Relief mudslide, is well exposed in a cliff-section along the Atlantic Coast (Figure 5.27). The slide disturbs more than 35 m of mudstone, and occurs approximately 100 m below delta-front sediments of the overlying deltaic cycle. The depositional setting was therefore probably a distal prodelta environment.

The largest deformation features are two westerly dipping, normal master-faults (EF and WF; Figure 5.42a). The sediments east of EF are undeformed, thus the two faults probably represent the head region of a slide. Displacement can only be determined along WF, and is around 12 m. Both faults are concave-up and listric, but no décollement is seen as the faults extend below sea-level. Between the faults, several minor scale faults occur, both with synthetic and antithetic displacement (Figure 5.42a). Some of the bedding is clearly rotated, but the displacement on these faults does not exceed a few metres. There is normal drag of the bedding in the hangingwall of EF, but this is not so clearly developed along WF. The pre-lithification nature of the faults is clearly displayed by the truncation of WF by the overlying beds (Figure 5.42a).

Other deformation features occur in the cliff, which are related indirectly to the sliding. A very contorted zone occurs near the base of the cliff,

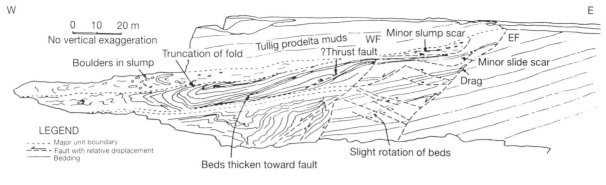

Figure 5.42 (a) Drawing of the Point of Relief (for location, see Figure 5.27) slide in the Gull Island Formation (Upper Carboniferous), Gull Island region, County Clare, western Ireland. (From Martinsen and Bakken 1990; based on Martinsen 1987.) (b) Photograph of areas labelled WF and EF on (a). Note fill in hangingwall. EF is seen in the lower right-hand corner. The cliff is 50 m high.

and this is cut by WF and therefore pre-dates the slide. Above this, a wedge-shaped bedset occurs which records infill of the depression in front of WF (Figure 5.42a and b). The beds above this wedge are slumped and folded into an isoclinal, recumbent fold which covers the entire thickness of the slump, but is truncated on the top. It is probable that the slump moved in the slide scar and was banked up against the slide-scar wedge. The same may be true for the overlying slump, which also thins towards WF.

The Point of Relief slide shows the importance of interaction and superimposition of sliding, slumping and *in situ* deformation (the contorted zone at the base of the cliff). The similarity to the Gela slide (Trincardi and Argnani 1991; see above), where debris flows filled the depression in front of the master fault, is significant. The temporal evolution of increased remoulding with time (sliding followed by slumping and/or debris flows) suggests that sliding is an important process in forming local excess slopes. These in turn prompt further failure and other more intense types of deformation.

Kidnappers slide, New Zealand margin

The Kidnappers slide occurs off the eastern margin of New Zealand's North Island. This is a convergent plate margin with the sea floor sloping around 1–5° into the Pacific Ocean, where the Pacific plate is being subducted underneath the Indo-Australian plate (Figure 5.43). In the early Holocene, the lowstand sediments of the last glacial age failed on the upper continental slope seaward of Cape Kidnappers. The slide was originally investigated using seismic data by

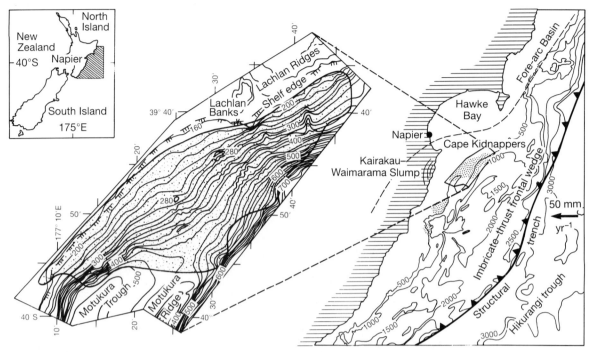

Figure 5.43 Location map of the Kidnappers slide area, offshore North Island, New Zealand. (From Barnes and Lewis 1991.)

Lewis (1971), but recently has been re-examined by Barnes and Lewis (1991).

The Kidnappers slide is a composite unit of several slides and slumps dominated by extensional, rotational failures (Figure 5.44a). The minor deformed units range in thickness from 20 to 140 m, and slope failure is thought to have occurred in several phases. The main failure probably took place in early Holocene time, but failure occurred more or less continuously from mid-glacial time up to the present. The entire slide area is around 720 km^2 in area, and the total displaced sediment volume is around 33 km^3.

Three phases of deformation can be discerned. The earliest phase comprised formation of now-buried slumps, bounded upslope by normal faults. The middle phase was the most voluminous, and the slide area can be divided into two areas, northern and central (Figure 5.44b). In addition, deep rotational slumping occurred. The central slide block was the largest, covering approximately 500 km^2, and can be described as a large sheet failure affecting 20–70 m of sediment, where a lower basal slide surface can be recognized. The northern slab is affected mainly by deep-seated normal faults, and no basal slide surface can be detected (Figure 5.44b). The differences between these two areas and the deep, rotational slumps illustrate important along-slope variation in slide style (Figure 5.44b). The latest phase of deformation encompasses translational sliding in the central slide area, which has deformed some of the earlier-formed normal faults.

Only extensional zones can be documented within the entire slide area, and no evidence is found for contraction. Therefore, only a series of normal fault scarps is observed, which produces a very uneven sea-bottom topography. The absence of contractional faults may be explained by the underlying large-scale tectonics. Imbricate thrust faults occur within the sediment package, producing growth anticlines (Figure 5.44a and b). The upslope anticline appears to grow the fastest, causing an increase of slope inclination toward the basin. This has probably induced a general state of extension within the entire sediment package, so that all the stress is resolved by the normal faults. In addition, lateral compaction may have contributed to the prevention of compressional structures.

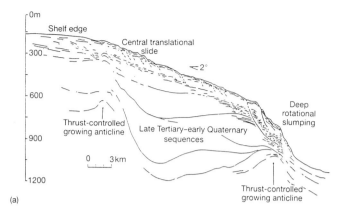
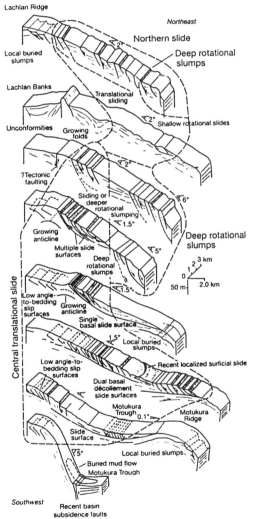

Figure 5.44 (a) Cross-section of part of the Kidnappers slide, showing the relation of extensional zones to thrust-controlled growing anticlines of the offshore imbricate-thrust wedge. (b) Various cross-sections along-slope of the Kidnappers slide. Note the along-slope variation in structural style. (From Barnes and Lewis 1991.)

The Kidnappers slide is important because it shows that 'open-ended' slides may be quite common. In addition, the variation in structural style along the slope is an important contribution, indicating that the slide is in fact a complex and composite unit of several deformed units and several deformational styles.

5.7 CREEP

5.7.1 Introduction

Creep, used here in the conventional geomorphological sense (see section 1.2.4), is the slowest-moving mass movement and involves the slow translation of sediment or soil down a slope (Radbruch-Hall 1978). Creep takes place at a rate that cannot be observed directly with the naked eye (McKean *et al.* 1993). The moving material may move only a few centimetres a year or even less, but rates up to $20\,\text{cm}\,\text{day}^{-1}$ have been measured (Radbruch-Hall 1978). Nevertheless, it is an important process both subaerially and subaqueously because it can lead to substantial land waste, and, in addition, creep often triggers other and more violent mass movements, such as landslides (Heim 1932; Müller 1964).

5.7.2 Process

Subaerial creep

Subaerial creep can be divided into two classes (Carson and Kirby 1972): **rheological creep** and **physical/organic creep**. The former relates directly to soil clay minerals and is caused by continuous breaking and re-establishment of clay mineral bonds under the influence of gravity. Each individual bond break is insignificant, but the cumulative effect on a soil mass is important in causing downslope movement.

Physical creep is more important than rheological creep and occurs on a seasonal or diurnal basis in soils. Three mechanisms can be distinguished (Allen 1985): (i) heating and cooling of particles; (ii) wetting and drying; and (iii) freezing and thawing of interstitial water in soils. These mechanisms all produce contraction and expansion of the soil. During expansion or heave, the particle moves along the resultant vector between the directions of expansion (directed upwards normal to the surface) and the particle weight force (Figure 5.45a). During contraction,

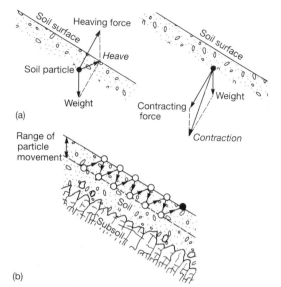

Figure 5.45 Diagram to explain the origin of slope-induced physical creep. See text for full explanation. (From Allen 1985.)

the same happens, only the contracting force is oriented in the opposite direction to the expansion force. The resultant downslope motion thus defines a zig-zag pattern as a consequence of repeated expansion and contraction (Figure 5.45b; Allen 1985). Organic activity by animals and plants also contributes to creep.

Thus, this kind of creep is controlled mainly by climate (Allen 1985; Feda 1992), and the controlling mechanisms differ between climatic regions. In tropical and subtropical areas, wetting and drying, heating and cooling, and organic activity control creep, whereas in mountain areas and arctic regions, creep is mainly controlled by freezing and thawing. **Solifluction** is a special and relatively sporadic (but also relatively rapid) form of creep which combines the effects of freezing and thawing with rheological creep, because of internal slippage during thaw due to wetting and soil saturation.

Rates of creep depend to some degree on climate and vary between a maximum of $0.3\,\text{m}\,\text{a}^{-1}$ ($0.01-0.1\,\text{m}\,\text{a}^{-1}$ is more usual) in solifluction-dominated arctic and mountainous areas, $0.002-0.01\,\text{m}\,\text{a}^{-1}$ in temperate continental areas, a maximum of $0.002\,\text{m}\,\text{a}^{-1}$ in temperate maritime climates and up to $0.005\,\text{m}\,\text{a}^{-1}$ in tropical rainforests (Allen 1985; see also Brunsden 1979).

Soil creep decreases below the surface, and can reach depths up to 10 m with solifluction. More commonly, the maximum depth is less than 1 m. In temperate and warmer regions, creep rarely exceeds depths of a few tens of centimetres (Allen 1985). Vertical profiles of creep rate were published by Fleming and Johnson (1975) and are shown in Figure 5.46. Convex-up, rectilinear and inflected profiles occur, but the last two may be most common (Allen 1985).

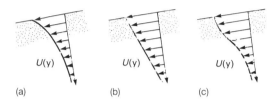

Figure 5.46 Variation in velocity profiles in creeping soils. U_y represents the creep rate in the y direction, parallel to the land surface. (From Allen 1985; based on Fleming and Johnson 1975.)

Periglacial environments are particularly prone to mass movement of weathered debris because of seasonal freeze and thaw, water-saturated sediment from meltwater and frozen ground at depth, which prevents deep percolation of water (Brunsden 1979). A particular kind of creep occurs in periglacial environments and is called **gelifluction** (*sensu* Baulig 1956). This occurs where weathered, thawed debris saturated with meltwater flows over frozen subsurface beds. Gelifluction can be more rapid than creep or solifluction due to freeze–thaw cycles, but in places the two processes can operate at the same time (Brunsden 1979).

Rock glaciers are also a periglacial phenomenon and occur as lobes of angular boulders below cliffs. Their shape is similar to glaciers and they occur preferentially on northern slopes near the snow line in mountain areas (Flint 1971). Rock glaciers form and move as a result of interstitial ice and snow 'lubricating' grain contacts under the influence of gravity (Wahrhaftig and Cox 1959). This results in downslope creep, which can reach up to 150 cm a^{-1}. Ridges, lobes and crevasses form that are analogous to those formed in glaciers (Flint 1971).

Rock creep is a common process on slopes where loose or solid rock is influenced by gravitational pull (a detailed account was given by Radbruch-Hall 1978). This generally causes faults and fractures to form, which most commonly are oriented parallel to valley walls (Figure 5.47; e.g. Ferguson 1967). The joints at the top of the valley slope are commonly more developed than those at the base of the slope. A gouge or mylonite-like zone can be developed at the base of the valley slope, suggesting that the entire valley slope is moving outwards, towards the valley centre (e.g. Ferguson 1967; US Corps of Engineers 1973).

Rock creep can range from less than 2 cm a^{-1} (Huffman, Scott and Lorens 1969) to more than 20 cm day^{-1} (Müller 1968). It can affect the bedrock up to depths of 300 m (Nemčok 1972).

Subaqueous creep

On subaqueous slopes, contraction and expansion of sediment cannot explain creep generation because continuous or semi-continuous mechanisms producing this behaviour do not exist. Rather, it is thought that slow, intergranular frictional sliding of non-cohesive sediment better explains creep (Nemec 1990). The strain rate must be low, preventing the development of well-defined slip planes.

Creep on subaqueous slopes has not been studied in detail. It may be an important process on steep, coarse-grained shorefaces or deltas where the slope inclination may be up to 35° (W. Nemec, personal communication). In these settings, creep can be important in stabilizing the slope, as the slow sediment movement reduces the gradient, which may reduce the risk of more massive slope failure. It is unlikely that creep is an important process on low-angle delta slopes of fine-grained sediment. The sediment is unlikely to behave in an intergranular frictional manner, and slow mass movement is probably rather by sliding.

One way of initiating subaqueous creep is perhaps by 'gravitational winnowing' or expulsion (Nemec *et al.* 1984; Postma 1984), in which the interstitial fines in heterolithic mass-flow deposits liquefy and are expelled from the steep snout of the flows. This renders the coarse, remaining fraction more susceptible to creeping because the overall frictional resistance will be reduced when the pore spaces are emptied of matrix. Finally, flow of pore water downslope will exert a stress or drag on grains which can lead to creep. This may be important on steep slopes.

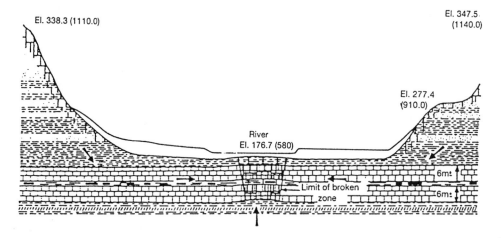

Figure 5.47 Cross-sectional view of a valley in the Alleghany Plateau region, eastern USA, showing vertical jointing of valley sides and bulging and fracturing of the valley bottom as a result of rock creep. (From Radbruch-Hall 1978; based on Ferguson 1967.)

5.7.3 Products

Subaerial soil creep produces characteristic deposits (Cotton and Te Punga 1955; Flint 1971; Feda 1992) and morphologies, which can have important environmental effects. Creep can locally increase the slope inclination, causing mudslides and debris flows to be generated. More commonly, creep results in such well-known effects as the bending of trees, moving of fences and formation of creep tongues (Figure 5.48). The tongues lack sorting, and are mainly fine-grained. Sometimes, clasts up to boulder size are included. The larger clasts may be imbricated, and are characteristically angular and of local provenance (Flint 1971).

The creep tongues are generally less than 1 m thick, but on flatter areas they may attain greater thicknesses, particularly where several tongues are superimposed. The tongues are smooth, especially where the soil is relatively permeable, allowing downward percolation of rain-water and prevention of tongue destruction from rain-water runoff (Flint 1971). Gelifluction produces similar deposits to solifluction, but is constrained to treeless areas.

Rock creep produces blocks and disrupted bedding (Figures 5.49), and is a prime initiator of rock fall. The rocks deform mainly by buckling in a convex-upward fashion. This causes the rocks below buckles to be steepened, commonly prompting the rocks upslope to penetrate underneath the downslope part of the buckle (Figure

Figure 5.48 Solifluction tongues on a grassy, cultivated slope, near Ulven in Os, western Norway. Note the high number of tongues, which generally are 10–30 cm across.

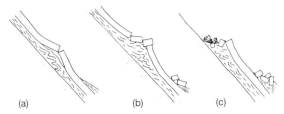

Figure 5.49 Conceptual sketches to show steps in the buckling of a sandstone bed above mudstone due to rock creep. (From Radbruch-Hall 1978.)

5.49). Apparent thrust faults may therefore form (Radbruch-Hall 1978).

Rock creep may initiate major catastrophic mass movement events. In 1806, at Goldau, Switzerland, a block of Tertiary conglomerate slid down a 20° slope (Heim 1932) and 457 people in the village were killed. Ter-Stepanian (1969), citing Zay (1807), pointed out that creep had occurred for 20 years at the site before the slide. In addition, animals were restless for several hours before the slide, suggesting early movement, perhaps by enhanced creep (Heim 1932).

At Monte Toc, in the Italian Alps, a mass of limestones slid into the Vaiont Reservoir in 1963. The slide was initiated by creep, and caused a flood wave 100 m high to form, which overtopped the local dam and flowed into the Adige Valley. Almost 2000 people were killed in the town of Longarone (Müller 1964). These two examples show the importance of creep in generating catastrophic mass movement. Therefore, monitoring creep can be an important practice in risk analysis of mass movements.

Although subaqueous creep is known to occur (Hill, Moran and Blasco 1982), it is not clear to what extent it produces characteristic deposits in the way that, for example, turbidity currents or debris flows do. The effects of subaqueous creep may be difficult to recognize but more research needs to be carried out to assess the characteristics and overall importance of this process.

CHAPTER 6

Tectonic deformation: stress paths and strain histories

DAN KARIG and JULIE MORGAN

6.1 INTRODUCTION

The traditional endeavours of structural geologists are now being extended, as indicated in section 1.3.5, to shallow levels of the earth's crust and to situations where incompletely lithified sediments have been subject to tectonic stresses. Such stresses, although by their nature of deep-seated origin, can extend to the highest levels of sediment piles, and, as explained in section 2.2.7, the particulate deformation that characterizes sediments can persist to substantial depths of burial. Much of the sediment that is affected by tectonic stress is in the broad and ill-defined area of partial lithification. Here, as this chapter illustrates, the mechanical effects of diagenesis become important, as do the elastic responses of the sediment.

One objective in structural geology is the interpretation of deformation histories of geological bodies and of the stress paths responsible for those histories. This task, incorporating the evolution of volumetric strains, is particularly relevant to particulate sediments and their equivalents in the geological record. The principles were outlined in section 2.2.5. This chapter focuses on the deformation that sediments are likely to encounter in settings where tectonic stresses operate, and it discusses the subject largely in terms of deformation paths. Such stress and strain histories are important not only for a proper understanding of sediment behaviour but because, for example, the roles of pore fluids and the resulting mechanical behaviour influence patterns of fluid flow and hence mineralization and fault behaviour. There are also very practical reasons for understanding the evolution of stress in the shallow crust, and especially in sediments. These include assessments of the stability of boreholes, the behaviour and manipulation of subsurface reservoirs, and the suitability of sediments for the foundations of structures.

Geomechanical models are now becoming more realistic and reliable, but they require adequate knowledge of stresses at specific mechanical states, such as at brittle failure. However, because stresses are not preserved in rocks and sediments, they cannot be determined directly but must be deduced from the sequences of structures and strain fabrics that happen to have been preserved. Such interpretive analyses are very difficult for several reasons. First, there is still only a poor understanding of applicable rheologies that couple strain to stress along geological deformation paths. Second, deformation paths involve both elastic and inelastic strains, and only the inelastic components are preserved in the rock fabric. Third, there may prove to be non-unique couplings of stress paths to deformation histories. In addition, changes in the stress tensor, rotations of the rock element in the stress field, and changing physical properties along the deformation path greatly complicate the deduction of a stress history.

The understanding of the nature of stress and of stress paths that produce deformation of sediments has advanced remarkably over the past decade, both from laboratory experimentation and from *in situ* measurements. Both approaches

The Geological Deformation of Sediments Edited by Alex Maltman Published in 1994 by Chapman & Hall ISBN 0 412 40590 3

are valuable, but both have serious limitations and present only partial solutions.

Mechanical experiments on sediments in the laboratory are relatively easy to perform, with modern servo-controlled systems interfaced to computers. Such equipment is fairly reliable and the experiments are inexpensive in comparison with *in situ* measurements. Laboratory experiments also permit uniform sediment to be subjected to varying conditions to better isolate functional dependencies. In these experiments, close control of stress and strain can be maintained over the range of stress applicable to purely mechanical deformation of sediments. Examples have arisen earlier in this book, especially in section 2.2.5.

There are, however, serious problems with the experimental approach to mechanical behaviour. Experimental strain rates are many orders of magnitude faster than natural strain rates, which precludes adequate measurement of creep or viscous (time-dependent) effects. Diagenetic processes are not easily mimicked in mechanical experiments, and these will be shown to have very pronounced effects on behaviour. A most serious limitation is that only simple stress paths can be applied during experimental deformation. Only a very few machines permit plane strain or true triaxial tests.

In situ measurements are, in effect, the monitoring of natural experiments, which would seem to be an approach preferable to laboratory experimentation, but most of these measurements are difficult, time-consuming and presently are very expensive. Techniques such as hydrofracturing, although rapidly improving, are still subject to problems of reliability and accuracy. Technical problems with *in situ* measurements can be severe, particularly in environments with higher differential stresses. Where data from *in situ* techniques are taken to be the results of natural experiments they must be analysed with caution, because boundary and initial conditions are often difficult to document. For example, rates of lateral strain are often unknown. The effect of lithological variations among the data can be a problem because, in most cases, variatioos in a spatial field must serve as substitutes for temporal changes along a stress path. At this time and probably far into the future, both approaches will have to be pursued concurrently. Moreover, the two approaches will have to be co-ordinated and integrated far better than has so far been the case.

Data from both these techniques can be applied to two structural regimes that fall near the ends of a spectrum of common deformation paths most likely to affect sediments. These paths are the uniaxial strain that accompanies deposition in a stable basin and the plane strain in accretionary prisms that form at some convergent plate margins. Moreover, these deformation paths are those most familiar to the present authors. They will also serve as a framework upon which to base concepts that apply to the general mechanical behaviour of sediments. This chapter reviews and discusses data from laboratory based experiments and *in situ* measurements from these two regimes of tectonic sediment deformation. The two contrasting paths also provide a framework within which to illustrate some of the more general aspects of sediment deformation.

6.2 STRESS PATHS DURING BURIAL AND UPLIFT OF SEDIMENTS IN BASINS

6.2.1 General

Perhaps the simplest deformation path in geology is the loading of sediments in basins and in other tectonically quiescent environments by subsequent deposition. It was introduced in section 2.3.2. This path might be considered non-tectonic but, because literally all sediments are deformed in some way during burial before being tectonically deformed, the burial path constitutes an essential part of the total deformation. Moreover, even the most stable basins are subjected to tectonic stresses of lithospheric origin (Zoback and Zoback 1989) and the deformation from these stresses cannot be adequately understood except against a basis of simple burial loading. Burial approximates a case of uniaxial (vertical) strain in that there is very little lateral strain in these settings, although this assumption will be addressed later in some detail. Deviatoric stresses are usually low in basins, but yet can lead to failure: jointing and normal faulting being good examples of extensional failure in this setting.

Despite this apparent simplicity, the state of stress in basins can vary widely as a function of such variables as stress history, lithology, thermal conditions and pore-fluid pressure (Evans and Engelder 1989). The stress history of a sediment is crucial because the loading path (consolidation) is dominantly an inelastic process, whereas unloading is more nearly elastic, and the two have quite different stress–strain relationships. Unloading can result from erosion, cooling, increased pore-fluid pressure, or extensional tectonic strain. Variability of each of these and other factors leads to a very broad range of stress states in basins and thus to uncertainty concerning the interpretation of stress measurements even in that environment. Nevertheless, simple uniaxial consolidation in a basin with hydrostatic pore pressures and a normal thermal gradient can be viewed as a fundamental or reference stress path, from which much could be learned about the intrinsic mechanical behaviour of sediments.

As outlined in section 2.1, the uniaxial consolidation of sediments entails the plastic yielding and volume reduction of the grain framework by the effective stress due to the overburden. This volume reduction requires the expulsion of pore fluid, the rate of which is controlled by the permeability of the sediments in the system. Consolidation is a compactive process generating deformation that is largely irreversible. Although sediments undergoing consolidation are defined by inelastic stress–strain relationships, they are capable of elastic deformation when effective stresses are reduced and the stress path lies within the yield envelope. Only this component of strain is recoverable upon unloading, although it will be shown that such elastic strain is not always linear with respect to stress. Elastic and inelastic responses are related, in that elastic parameters, such as the Young's modulus (E) (section 1.2.4), depend on the state of consolidation through reduction of porosity. A summary of the symbolic notation used in this chapter is provided in Table 6.1.

Vertical stresses responsible for both elastic and inelastic strain are governed by the effective overburden stress because the system has a free upper surface. The compactive vertical strain induced during uniaxial consolidation of soils has been related empirically to this stress by:

$$e = L \log \sigma'_v \text{ (e.g. Wood 1990),} \quad (6.1)$$

where e is void ratio and L is a constant depending on factors such as lithology. Geologists are more familiar with an approximate empirical relationship between porosity (η) and depth (z) in basins:

$$\eta = \eta_0 \exp(-Bz) \text{ (Athy 1930),} \quad (6.2)$$

where η_0 is the porosity at the surface and B is a lithology dependent constant.

Consolidation in natural sediment sections would be better related to σ'_v than to z (Figure 6.1), because σ'_v is not related simply to depth, owing to non-linear vertical variations in bulk density and in pore fluid pressures. Moreover, there are significant departures from idealized porosity versus σ'_v relationships in most basins, reflecting variations in lithology, cementation and stress ratios, as will be discussed later.

If the vertical effective stress is changed as a sediment follows a strain path that is elastic, for example by erosion or by pore-pressure increase, the vertical strain (ε_v) can be calculated for an isotropic sediment using the uniaxial strain equation:

$$\varepsilon_v = \frac{\Delta \sigma'_v - 2\nu \Delta \sigma'_h}{E}, \quad (6.3)$$

where ν is Poisson's ratio (section 1.2.4), E is the Young's modulus, and $\Delta \sigma'_h$ or $\Delta \sigma'_v$ represents the change in horizontal effective or vertical effective stress. Thus ε_v depends on $\Delta \sigma'_h$ as well as on several elastic parameters, which will be shown to vary widely under the range of geological conditions.

Horizontal stresses during uniaxial consolidation are dependent upon lithology and σ'_v, but a large amount of uncertainty surrounds this relationship. For the relatively low stresses and short time periods characterizing the consolidation of soils under a construction load, geotechnical engineers have concluded that the ratio σ'_h/σ'_v, termed K_0 (section 2.2.6), is constant for a given sediment. Complications regarding the magnitude of K_0 in nature were discussed in section 2.2.6. Whether K_0 remains constant over the stress range applicable to mechanically consolidated basinal sediments or whether it is constant over geological time spans are also points of considerable debate.

During elastic deformation, changes of horizontal stress are dependent on σ'_v as well as on Poisson's ratio (ν). For isotropic materials:

$$\sigma'_h = \sigma'_{h0} + \frac{\Delta \sigma'_v \nu}{1 - \nu}, \quad (6.4)$$

Table 6.1 Symbolic notation used in formulae

Stress

σ_1, σ_3	Maximum and minimum principal stresses
q	Differential stress: $\sigma_1 - \sigma_3$
σ_v	Vertical stress
σ_h, σ_H	Horizontal stress (H subscript denotes maximum and h subscript denotes minimum horizontal stress where horizontal stress is not uniform)
p'	Mean effective stress
σ'_c	Effective consolidation stress: maximum vertical effective stress to which the sediment has been subjected.

Note: primed superscript denotes effective stress; un-primed denotes total stress

Elastic or related parameters

ε	Strain, with subscripts as for stress
E_h, E_v	Young's moduli $\left(\dfrac{\sigma}{\varepsilon}\right)$ in vertical and horizontal directions
E_c	Constrained Young's modulus
G	Shear modulus
$\nu_{hv}, \nu_{vh}, \nu_{hh}$	Poisson's ratios, where the first subscript denotes the direction of the induced strain and the second is the direction of the applied stress (and strain) e.g. $\nu_{hv} = -\varepsilon_h/\varepsilon_v$
β	Bulk compressibility; $\beta = \dfrac{\partial V}{\partial P}$, where V is volume and P is pressure
β_g	Compressibility of mineral grains
B	Constant, relating porosity and depth
α	Thermal compressibility; $\alpha = \dfrac{\partial V}{\partial T}$, where T is temperature in °C.
L	Constant, relating void ratio and compactive strain
P_{H_2O}	Pore fluid pressure
P_b	Breakout pressure: borehole fluid pressure necessary to cause horizontal tensional failure in unfractured impermeable rock
S_T	Tensile strength

Note: Other subscripts used are: g (property of sediment grains) and w (property of water)

Other parameters, largely of soil mechanics origin

η	Porosity
e	Void ratio
ρ_b	Wet bulk density
p	'Pressure' – vertical effective stress in $e - \log p$ tests
K_0	Stress ratio, σ_h to σ_v, for no lateral strain
M	Ratio of $\dfrac{\Delta \sigma}{p'}$ at critical state failure
T	Temperature

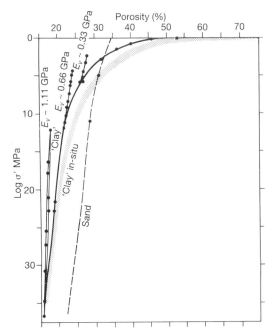

Figure 6.1 Porosity versus effective vertical stress (σ'_v) for clay-rich sediments. The solid curve is from an experimentally consolidated silty clay (Karig and Hou 1992), including elastic unload–reload cycles, from which values of vertical Young's moduli (E_v) can be calculated. For comparison, typical relationships for naturally consolidated clay (e.g. Bryant et al. 1981) and for experimentally consolidated sand (Karig and Hou 1992) are shown.

where $\Delta\sigma'_v$ denotes the change in vertical stress from some initial state and σ'_{h0} is the horizontal stress at that initial state. The effect of any viscous secondary consolidation, or creep, would be to increase σ'_h. The calculation of σ_h with an elastic relationship is sometimes used for sediments undergoing consolidation (e.g. McGarr 1988; Engelder and Lacazette 1990). This is not only incorrect conceptually but also produces numerically different results.

The properties of most sediments are anisotropic to some extent, in both elastic and inelastic responses. On the scale of laboratory samples, these anisotropies reflect the properties of both the grain framework geometry and of the grains themselves (e.g. Moon and Hurst 1984). Clay-rich sediments are more highly anisotropic than sands but there are remarkably few data quantifying these anisotropies and relating them to the mechanical state of deformation histories. Basinal sediments are often assumed to be laterally or horizontally isotropic, largely to simplify analyses, but the observed azimuthal variations in horizontal stresses suggest that the anisotropy might better be described as orthotropic (having three principal orthogonal anisotropies). Even lateral isotropy introduces a directionality to elastic moduli and to inelastic parameters such as compressibility, which significantly affect the calculated stress paths and deformation of sediments. Here basinal sediments will be assumed to behave as laterally isotropic; derivations leading to equations relating principal stresses and strains for this condition are found in Nye (1957), Jaeger and Cook (1979) and Pickering (1970).

Consolidation is responsible for some porosity reduction in almost all sediments, but its use as a model is limited to pressures and temperatures below which pressure-solution-dependent (diffusion dominated) processes are not significant. This boundary is not well defined and probably varies with a number of environmental factors, but certainly is lithology dependent (e.g. Rutter 1983). Carbonate sediments show intense pressure-solution effects such as stylolitization, at a differential stress equivalent to burial depths of a kilometre or so (Engelder and Marshak 1985). Although minor pressure solution has been reported in near-surface sands (Palmer and Barton 1987), empirical relationships from outcrop and well samples suggest that mechanical consolidation dominates the porosity reduction in sands and clays in the upper several kilometres or until porosities drop to about 15% (Sprunt and Nur 1976; Rutter 1983). Because of the problems with pressure solution, which are difficult at best to emulate in laboratory based mechanical testing, this review is limited to sand and clay-rich sediments.

The approach taken here will be to review laboratory studies on experimentally and naturally consolidated sediments and then to discuss the parameters that affect *in situ* sediments, but which are not adequately treated during most laboratory tests (temperature, pore pressure, lateral strain and time). With the combined information, theoretical stress paths can be developed for a variety of consolidation and unloading (elastic) histories, but which assume that no time-dependent stress relaxation occurs. That assumption will be addressed by comparing available *in situ* stress measurements in basinal settings with predicted stress states.

Finally, several common structural features observed in basinal settings will be discussed in the light of theoretical and observed stresses.

6.2.2 Laboratory studies of consolidation

The consolidation of sediments has been characterized by several exponential or logarithmic relationships between σ'_v and some parameter related to volume change. The applicability of these relationships can be tested by consolidation tests on 'synthetic' sediments or disaggregated natural sediments, as well as from porosity versus depth measurements in consolidating basins.

Consolidation of a 'synthetic', three-component silty clay (Karig and Hou 1992) to values of σ'_v from 0.1 MPa to almost 40 MPa (equivalent to a depth of several kilometres) shows a progressively poorer fit to a linear e versus $\log \sigma'_v$ relationship with increasing σ'_v (Figure 6.2). If, instead, porosity (η) from the tests on the synthetic silty clay is used as a measure of consolidation and plotted against $\log \sigma'_v$, the data show a very nearly linear relationship over the entire range of σ'_v (Figure 6.3). Thus, over a geological stress range, the volume change during consolidation of a clay-rich sediment seems better characterized by a linear relationship of porosity rather than a void ratio against $\log \sigma'_v$.

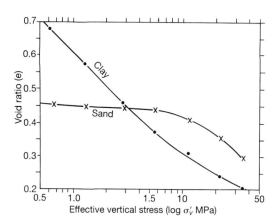

Figure 6.2 Plot of void ratio (e) versus log effective vertical stress ($\log \sigma'_v$) for experimentally consolidated silty clay and sand. (From Karig and Hou 1992.) The slope for the clay is approximately linear over the stress range below about 1 MPa but decreases gradually at higher stresses. The slope for the sand is also approximately linear at low stress but increases very sharply at stresses greater than about 2 MPa. Compare with the idealized curves in Figures 2.13 and 7.4.

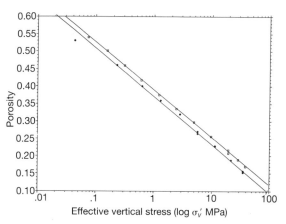

Figure 6.3 Porosity versus log effective vertical stress ($\log \sigma'_v$) for two samples of similar silty clays, experimentally consolidated over a range of stress encompassing both geotechnical and basinal sediment conditions. In contrast to the e versus $\log \sigma'_v$ plot (Figure 6.2), these curves remain linear over the entire stress range.

Similar consolidation tests on a fine-grained sand (Karig and Hou 1992) showed an approximately linear relationship between e and $\log \sigma'_v$ to a σ'_v of 5 MPa. Above that stress there was a marked increase in the absolute value of the constant L (equation 6.1) to the highest stress applied (36 MPa). This sharp increase in compressibility probably reflects some cataclasis and resultant enhancement of packing (Chilingarian and Wolf 1975; Wong 1990; Rutter and Hadizadeh 1991). The increase in compressibility gives rise to nearly linear porosity–depth profiles for both experimentally consolidated sand and natural sands in basins (e.g. Magara 1980).

The experimentally consolidated clay-rich sediments display a much higher initial porosity than do the sands but they are also much more compressible at that state, which leads to the intersection of the two η–σ'_v curves near 3 MPa (Figure 6.1), and thus to lower porosities for clays than for sands at higher stresses. This inversion is seen in basins (e.g. Plumley 1980) but at depths corresponding to somewhat higher stresses. The difference can be attributed to the greater diagenesis in natural sediments, as outlined in the following section.

The ratio of σ'_h to σ'_v during consolidation is another relationship considered in the geotechnical literature to be constant over the low σ'_v range. Although this ratio is termed K_0 for all uniaxial strain paths in soil mechanics literature, it is here restricted to consolidation paths. K_0 does vary with lithology, from as low as 0.3 for coarse sand to 0.8 for some clays (from data in Lambe and Whitman (1979); see section 2.2.6). The extent to which this stress ratio remains constant as σ'_v increases has received only little attention (e.g. Brooker and Ireland 1965). A series of high-stress consolidation experiments (Karig and Hou 1992) showed that this ratio remains remarkably constant for a silty clay, at a value near 0.63 up to $\sigma'_v = 35$ MPa (Figure 6.4). Similar tests on fine sand resulted in a lower ratio, but one that increased moderately with stress from 0.44 to 0.53 (Figure 6.4). This increase in K_0 is again attributed to grain crushing, which is interpreted as leading to a reduction in q. These results confirm that stronger sediments, such as sands, are capable of supporting higher differential stresses but are counter-intuitive in that the stress ratio for sand increased with stress, whereas that for the weaker clay did not.

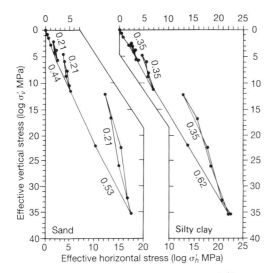

Figure 6.4 Plot of effective horizontal stress (σ'_h) versus vertical stress (σ'_v) for experimentally consolidated silty clay and sand under uniaxial strain conditions. (From Karig and Hou 1992.) For each lithology, the elastic stress cycles have indistinguishable slopes at widely varying porosities and consolidation stresses. This observation implies that Poisson's ratio is constant for a given lithology.

Uniaxial reconsolidation–consolidation tests are often used to estimate the maximum effective vertical stress to which a natural sediment, especially a clay-rich sediment, has been subjected. This approach relies on the assumption that the maximum σ'_v, termed the consolidation stress (σ'_c), occurs at the break in slope of the $\sigma'_v - \varepsilon_v$ curve between the elastic reconsolidation strain and the plastic first-time consolidation strain. In other words, the yield stress for a sediment loaded in a manner identical to that it underwent naturally should be σ'_c.

Reconsolidation–consolidation tests produce agreement between σ'_c and the *in situ* σ'_v for some shallow sediments, but show poor agreement for others, especially more consolidated sediments. Commonly, σ'_c is greater than the *in situ* σ'_v. In some cases this can be shown to be the result of erosion of overlying material, which reduces σ'_v from its maximum value, but more often the inequality occurs in continuously sedimenting and consolidating basinal sequences where erosion has not occurred. The general explanation is that almost all naturally consolidated sediments are, to some degree, 'cemented' (see also section 2.2.7). Here 'cementation' refers to various diagenetic processes, such as secondary mineralization and electrochemical effects, as well as deposition of pore material, and it is equivalent to the term 'structure' as used in the geotechnical literature (e.g. Burland 1990; section 1.1.1). Cementation imparts a component of strength in addition to that developed by mechanical consolidation, which explains the higher porosities in natural sediments than in those experimentally consolidated to the same σ'_v (Figure 6.1).

Typical consolidation tests, which measure only σ'_v and ε_v, do not clearly discriminate between the effects of stress and cement; this can be done much better during tests in which σ'_h is also determined. Such tests show very large stress ratios over the stress path just past the yield or consolidation stress (σ'_c), with a return to a K_0 stress ratio at higher stress (Figure 6.5). Such behaviour represents the breakdown of cement as σ'_c is exceeded (Lerouiel and Vaughan 1990; Karig 1993). This effect is even more graphically illustrated by plots of $\Delta\sigma$ versus p', where $\Delta\sigma$ rises well above the K_0 line just beyond the yield point before returning to the K_0 line (Figure 6.6). Clearly the *in situ* value of σ'_v is less than σ'_c, but it is not clear how σ'_v could be deduced from the test data. In general, variations in cementation probably explain much of the departure of observed porosity–depth curves in basins from the ideal exponential relationships.

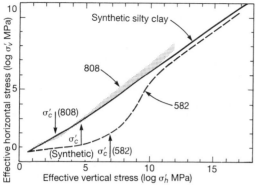

Figure 6.5 Comparative plots of effective horizontal stress (σ'_h) versus vertical stress (σ'_v) for silty clays from: DSDP Site 582, in the Nankai Trough; ODP Site 808, beneath the prism décollement; and from the experimental consolidation of a synthetic silty clay (from Karig 1993). These curves illustrate that the sample from beneath the décollement, in which the cementation (bonding) has been destroyed, behaves much like the experimentally consolidated analogue during the elastic and post-yield stress paths. In contrast, the sample from the Trough, which is strongly cemented, shows a large apparent overconsolidation and a pronounced phase of post-yield cement destruction. All samples have very similar K_0 values after deformation has destroyed the cementation. σ'_c is the effective consolidation stress.

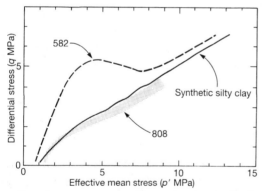

Figure 6.6 Comparative plots of differential stress (q) versus mean stress (p') for silty clays from: DSDP Site 582, in the Nankai Trough; from ODP Site 808, beneath the prism décollement; and from the experimental consolidation of a synthetic silty clay (from Karig 1993). These curves show that the sediment from Site 582 has a large component of strength from cementation, whereas the sediment from beneath the décollement at Site 808 has virtually no cement (bonding) induced strength, similar to the experimentally consolidated sample.

The destruction of cement can occur from stress oscillations within the elastic field as well as by plastic strains. Soils may be 'destructured' (Burland 1990) by rough handling or by natural means, such as oscillation of p' by pore-pressure

fluctuations. Destruction of cement reduces σ'_c and the size of the yield envelope back toward that of the uncemented sediment. Such destructuring could also lead to resumption of consolidation.

The elastic behaviour of sediments within the yield envelope is governed by mechanical parameters such as Young's modulus and Poisson's ratio, which are still ill-defined functions of porosity and cementation. This state of affairs is a result of the paucity of experiments addressing comparisons of parameters as functions of cementation. Nevertheless, a few data, and a broad view of sediment property compilations, offer an idea of trends.

Young's modulus (E) is one of the most important and often measured of the mechanical parameters, but is often of dubious accuracy. The initial slope of a uniaxial stress–strain curve usually provides a minimum value of E, as shown by subsequent stress–strain cycles (Wilhelmi and Somerton 1967). Most likely this reflects the initial compliance of the test apparatus as well as microcrack closing at low pressure. The most reliable measurement is obtained by stress–strain cycles at stress conditions above that necessary for crack closing, but uniaxial stress cycles are seldom undertaken. Slopes of initial loading curves are more useful for qualitative comparisons of uncemented and cemented sediments, as are the unload–reload curves from uniaxial strain tests. The slope of the elastic stress–strain curve during these uniaxial strain tests is termed the **constrained modulus** (E_c).

Carefully monitored unload–reload experiments on silty clay and sand illustrate the relationship between E (or E_c) and consolidation stress or porosity. As expected, E for sand is greater than E for clay at the same σ'_c, with the difference increasing with stress (Karig and Hou 1992), but the rate of increase of this difference decreases with increasing stress.

Natural sediments, with variable degrees of cementation (diagenesis), clearly have much higher values of E than their uncemented, experimentally consolidated analogues with the same σ'_c, but without a better data base no quantitative comparison can be made. It appears from compendia of sediment properties that cementation can increase E by several fold when the comparison is made on the basis of similar porosities (Karig 1993) and greater if the comparison is based on equal σ'_c (Karig and Hou 1992). Because porosity decreases and diagenesis generally increases with burial of sediments, rapid increases in E can be expected with depth. Very much less information is available concerning the **shear modulus** (G), a proportionality constant that relates the amount of shear strain to shear stress, but it appears to have a similar, if not even more pronounced response to changes in σ'_c and cement than does E.

The dependency of Poisson's ratio (v) on consolidation and diagenesis seems to be very different from that of E. Many workers have noted an increase of v with inceasing pressure (e.g. Price 1974), but this effect is limited to the stress range over which microcracks close. Above that stress range, v changes very little with stress (Jaeger and Cook 1979). For the sand in our tests, which we assumed to be nearly isotropic, v was 0.17 over the entire stress range explored. This ratio is similar to that given for a highly lithified quartzite (Bishop and Hight 1977) and fused quartz (Birch 1966). This suggests that v has little dependence on σ'_c or diagenesis, but is intrinsic to the sediment mineralogy.

Poisson's ratios for clays are more difficult to understand. Most clay-rich rocks are anisotropic during elastic strains. Not only are Poisson's ratios dependent on the directions of the applied and induced strains but it would also seem logical that this anisotropy in v would increase as σ'_c increased. Little of such behaviour was observed during our tests on silty clay (Karig and Hou 1992), which was attributed to relatively little strain and intrinsic anisotropy. More information on clay anisotropy and its relationships to other parameters is badly needed.

The point to be emphasized from this discussion is that, because the mechanical behaviour of a sediment differs markedly during elastic and inelastic strains, it is critical to understand correctly the nature of the deformation. Sediment consolidation cannot be modelled with elastic strain equations and elastic strains must be calcuated from appropriate initial stress states usc-ing realistic elastic moduli.

6.2.3 Effects of geological processes not duplicated in laboratory experiments

The consolidation behaviour of sediments in natural basins differs from consolidation in the laboratory, not only because of the effects of diagenesis but also because natural sediments are subjected to processes that are difficult or impossible to duplicate in the laboratory. Such processes operate both during the consolidation phase and during elastic deformation, often producing very different effects over the two paths. For example, natural sediments are subjected to changing temperatures, higher pore pressures, finite horizontal strain and effects due to the very much greater time over which the stresses act. These are reviewed in turn below.

Temperature changes

Temperature changes affect the mechanical behaviour of sediments through **thermal compressibility** (α), where $\alpha = \varepsilon_{ij}/\Delta T$. This parameter is a tensor quantity, reflecting the thermal anisotropy of minerals (Nye 1957). Few data for the relevant minerals exist. The thermal compressibility of single quartz crystals is known to be moderately anisotropic (Nye 1957), and that of the platey clay minerals is probably significantly more so. The random orientation of quartz crystals in most sands leads to a grossly isotropic response, and the lack of adequate numerical data for the thermal behaviour of clays has led to a general treatment of α as being isotropic, as it is in the following discussions.

Sediments are heated during burial as they move down the geothermal gradient, and they cool during erosion and exhumation. Although the magnitude of the temperature change over the two paths may be the same, the thermal effects during the inelastic consolidation associated with burial are quite different from those during elastic unloading.

Temperature effects during consolidation have been explored in soils, for which increasing temperature leads to modest increases in K_0 and compressibility (e.g. Mitchell 1976). These effects apparently reflect the addition of a thermal consolidation stress. In general, increased temperature engenders consolidation and creates a more nearly isotropic state of stress.

The effects of thermal strain on elastic stress paths can be calculated by equating the horizontal thermal strain to the horizontal elastic strain, as the two components must cancel under the condition where there is only vertical strain (and no change in vertical stress). This leads to the relationship:

$$(\Delta\sigma_h)_T = \frac{E_h \alpha \Delta T}{1 - \nu_{hh}}. \quad (6.5)$$

Here $(\Delta\sigma_h)_T$ is the change in horizontal stress due to temperature effects alone. If ν_{hh} and α are relatively constant over the linear elastic stress range, variations in thermal stress for a given ΔT depend primarily and directly on changes in the value of E. Thermal stress can be a major component of the total $\Delta\sigma_h$ during typical erosional unloading of sediment columns (e.g. Haxby and Turcotte 1976).

Pore fluid pressure

Although the range of pore pressures expected during the tectonic deformation of sediments in the upper several kilometres of the crust could be adequately duplicated in laboratory experiments they seldom have been. Because effective stress is thought to be the important variable in governing mechanical response, most tests are run over the appropriate effective stress range with pore pressures high enough only to drive any gas into solution. However, *in situ* pore pressures, especially in ocean-floor sediments, can be of the order of 10 000 p.s.i. (50 MPa). For several aspects of mechanical behaviour this difference in pore pressure has pronounced effects. Moreover, even the simple effective stress equation ($\sigma' = \sigma - P_{H_2O}$) is often misinterpreted.

It is logical to begin with a consideration of this misinterpretion, which is a result of the general inference that during an increase in pore pressure, total stresses remain constant. Although total horizontal stresses commonly are kept constant during laboratory triaxial tests, they seldom remain so in nature. In basins, the total vertical stress may remain constant during a change in porefluid pressure (P_{H_2O}), reflecting a fixed overburden, but in a laterally constrained system, where the fluid pressure varies gradually over space and time, the total horizontal stress will increase with an increase in P_{H_2O} (Mandl and Harkness 1987; Karig and Hou 1987). In the case of lateral confinement and uniaxial (vertical) strain, and assuming the simple effective stress law, the only reduction on σ'_h will be a result of the elastic Poisson effect caused by reduction in σ'_v:

$$\Delta\sigma'_h = \frac{\nu_{vh}}{(1 - \nu_{hh})} \Delta\sigma'_v, \quad (6.6)$$

where $\Delta\sigma'_v = -\Delta P_{H_2O}$.

The second aspect of importance is the compression of mineral grains by the pore fluid pressure, which modifies the effective stress law to:

$$\sigma' = \sigma - \frac{\beta - \beta_g}{\beta} P_{H_2O} \quad (6.7)$$

where β is the bulk compressibility and β_g is the compressibility of the mineral grains (Skempton 1970; Nur and Byerlee 1971). Although β_g is relatively small and its effects usually negligible, in laterally confined, cemented sediments of low porosity, where β approaches β_g, it can lead to significant changes in lateral stresses:

$$(\Delta\sigma'_h)_p = -\left[\frac{E_h \beta_{hg}}{3(1 - \nu_{hh})}\right] \Delta P_{H_2O} \quad (6.8)$$

or using relationships among elastic moduli:

$$(\Delta\sigma'_h)_p = -\frac{E_h}{E_{hg}} \frac{1 - \nu_{hh} - \nu_{vh}}{1 - \nu_{hh}} \Delta P_{H_2O}, \quad (6.9)$$

where $(\Delta\sigma'_h)_p$ is the additional change in stress when grain compression is considered. For example, natural sandstones or shale with porosities of 0.10 or less will have a ratio of E_h/E_{hg} of about 1/3, which will lead to a value of $(\Delta\sigma'_h)_p$ that will reduce σ'_h by about $1/4 P_{H_2O}$ without affecting σ'_v.

Lateral strain

Experimental sediment deformation is usually constrained to be axisymmetric. During uniaxial consolidation tests the axisymmetric horizontal strain is kept as close to zero as possible, in order to approximate the K_0 condition, but in natural basins the horizontal strain is neither axisymmetrical nor zero. Horizontal strains during sedimentary loading are probably non-zero because of tectonic stresses, earth sphericity and basin curvature. Several authors (e.g. Price 1974; Haxby and Turcotte 1976) have attempted to estimate the magnitude of horizontal strains in basinal sediments resulting from such geometries, but the natural conditions seem seldom to be so uniform or predictable. Here horizontal strain is treated as an independent variable, and variations in σ'_h and mode of failure with respect to horizontal strain are explored.

Determination of the effect of horizontal strain requires the estimation of the plausible range of the deviations in σ'_h from that of the uniaxial strain model. During inelastic consolidation, the minimum σ'_h possible is that developed during critical state failure with $\sigma'_v = \sigma_1$. With the simple assumption that, at the critical state, q/p' is a constant that varies from 0.9 to 1.2 (M.E. Jones and Addis 1986), σ'_h could reach a minimum value of 0.33 σ'_v. Similarly, consolidation associated with lateral shortening could lead to σ'_h as high as seven times σ'_v. Ratios anywhere near this high are extremely unlikely, however, as no unmetamorphosed basinal sediment has ever been reported for which the grain fabric displays a horizontal pole maximum. It is likely that the range in stress ratios during basinal consolidation is $0.4 < \sigma'_h/\sigma'_v < 1.0$. Thus, at a depth where σ'_v is 30 MPa (~ 2.5 km), the horizontal stress due to horizontal strain might vary between 12 MPa and 30 MPa. These values differ only moderately from the 15–20 MPa range expected for uniaxially consolidated sand and clay.

Geologically plausible elastic horizontal strains, on the other hand, can produce much larger changes in horizontal stress. From the elastic horizontal strain equation, the changes in horizontal stress from a horizontal strain, assuming $\Delta \sigma'_v = 0$ (a constant overburden stress) is:

$$\Delta \sigma_h = \frac{\varepsilon_h E_h}{1 - v_{hh}} \quad (6.10)$$

The magnitude of $\Delta \sigma_h$ depends on both ε_h and E_h, as v_{hh} has been concluded to vary little with stress over the range of linear elastic response. A lateral extensional strain of 0.01 in a typical naturally cemented sandstone ($E = 15$–20 GPa) would result in a stress reduction, $\Delta \sigma_h$, of near 200 MPa. Clearly, similar horizontal strains have a much greater effect on σ'_h over elastic stress paths than during consolidation.

Time

The duration of stress and the rate of application of stress may have significantly different effects on sediments experimentally consolidated over periods of a few hours to a few weeks and sediments naturally consolidated much more slowly, over periods of 10^5 to 10^6 years. During experimental consolidation of soils, the vertical strain has been shown to have a time-dependent component other than the primary phase, related to the rate of expulsion of pore water (e.g. Jamiolkowski el al. 1985; section 2.2.4). This secondary consolidation, or creep, decays with an approximately exponential relationship to time and there is some evidence that it decreases with decreasing porosity (Hou 1988). Secondary consolidation has been attributed to time-dependent reorganization of grains and to destructuring (Jamiolkowski et al. 1985).

The effect of secondary consolidation or creep on the stress ratio is uncertain, both for soils and for more consolidated sediments. A review of the evidence from geotechnical studies on soils (Jamiolkowski et al. 1985) concludes that the stress ratio changes little due to creep over engineering time spans.

Some workers feel that over geological time intervals there may be a lithology-dependent relaxation of differential stress, with clays tending toward an isotropic stress state more easily than sand (e.g. Swolfs 1984; Warpinski, Branagan and Wilmer 1985). On the other hand, *in situ* measurements of horizontal stress in clay-rich sediment sections indicate that these lithologies can support significant differential stresses over geological time (e.g. Bell 1990).

This problem of creep or elastico-viscous secondary consolidation is very difficult to address other than by *in situ* stress measurements in a basinal setting where a uniaxial strain condition is likely. The sparse data that might be applicable to this problem will be reviewed in a later section, after the generation of theoretical stress states and paths based on the sediment responses that have been discussed to this point. These theoretical mechanical responses assume there to be no time-dependent effects for sediments in the geological realm, and thus serve as limiting paths for the interpretation of observed *in situ* stresses.

6.2.4 Theoretical stress paths during sediment deposition and unloading

The individual effects of σ'_v, T, P_{H_2O} and ε_h, as outlined above for both consolidative and elastic stress paths, act in various combinations in natural basins. Some combinations undoubtedly occur more often than others, but there are still too few constraints from *in situ* measurements to permit construction of general models. Here several simple end-members are examined and some implications concerning sediment deformation are explored. Different stress paths will shape the character of deformation, whether by consolidation or by failure through shear or extensional strain. Section 7.6 discusses similar matters in the context of permeability. In the following section, theoretical stress paths associated with sediment burial and with various causes of elastic unloading are discussed. Elastic strain paths leading to brittle failure are given particular attention because of the interest in brittle structural features such as faults and jointing.

Sediment burial

During burial, sediments consolidate as the effective vertical stress increases due to additional sedimentation. The vertical effective stress is also affected by pore-fluid pressure, which rises with increasing depth along a hydrostatic or steeper gradient. Uniaxial consolidation can proceed, although more slowly, even with pore pressures much higher than hydrostatic. It is only necessary that the increase in σ_v be greater than the increase in P_{H_2O} in a sediment, so that σ'_v increases.

Several behavioural characteristics of consolidation during burial are affected by diagenesis and temperature. Some aspects were discussed in section 2.2.6. Diagenesis clearly inhibits volume reduction at a given σ'_v, but it is far from clear whether, or how, it affects K_0. Results of experiments designed to generate and maintain cementation during consolidation could not be found in the literature and *in situ* data are insufficient not only to gauge the effect of diagenesis but even to determine K_0 adequately. Reconsolidation of several natural mudstones (Karig 1993) showed that K_0 at the assumed *in situ* σ'_v was significantly less than that for uncemented, experimentally consolidated mudstone. On the contrary, reconsolidation of the experimentally consolidated mudstone to σ'_c did result in the original value of K_0 (Karig and Hou 1992). These data suggest that diagenesis tends to reduce the *in situ* stress ratio during uniaxial consolidation, but they are incapable of probing the effect of time on K_0.

Because increasing temperatures enhance consolidation and increase K_0, the effects of downward increases in diagenesis and temperature act in opposite directions and may have a small resultant. Thus in basins where lateral strain is approximately zero, σ'_h/σ'_v may not differ much from the laboratory values. Nor, as was shown, will the common range of lateral strains in basins change this ratio greatly during consolidation, whether such strains are induced by tectonism or by effects of deflections on a spherical earth. The point to be made is that, except for the possible effects of geological time, the horizontal stresses associated with natural consolidation probably vary relatively little, whereas stresses over elastic paths will be shown to vary much more widely.

Elastic deformation

Elastic deformation of a sediment ensues when some combination of reduction in differential and effective mean stress occurs that creates a stress path within the yield envelope of that sediment. Such paths may lead to brittle failure with increasing strain, either by shear or extension, depending on the magnitude of q. For basinal sediments, elastic strain is usually accomplished by some combination of erosional unloading, lateral extension and increase in pore-fluid pressure.

Unloading during erosion The simple case of uniaxial erosional unloading through a fixed geothermal gradient and with a reduction in pore pressure can be described by linearly combining the elastic stresses due to vertical stress reduction, temperature change and pore-pressure reduction into a pair of equations for $\Delta\sigma'_v$ and $\Delta\sigma'_h$:

$$\Delta\sigma'_v = \rho_b g \Delta z - \Delta P_{H_2O}, \qquad (6.11)$$

where Δz is the thickness of sediment eroded, and:

$$\Delta\sigma'_h = \frac{v_{vh}}{1-v_{hh}}\Delta\sigma'_v + \frac{E_h \alpha \Delta T}{1-v_{hh}} - \frac{E_h}{E_{hg}}\frac{1-v_{hh}-v_{vh}}{1-v_{hh}}\Delta P_{H_2O}. \qquad (6.12)$$

Both these equations can be divided through by a depth increment, Δz, to provide stress gradients ($\Delta\sigma/\Delta z$) with assumed values of $\Delta T/\Delta z$ and $\Delta P/\Delta z$. The stress gradient form is more common in geological literature, but it implicitly ignores variations in density with depth. Both forms describe the change in stress over only the elastic unloading segment of the stress path, which here is assumed to begin at the state of maximum consolidation.

In this discussion, a typical temperature gradient of 25°C km^{-1} and a hydrostatic pressure gradient of 10.67 MPa km^{-1} are used, but a comment on the latter is required. Reduction of pore pressures from those at maximum consolidation will be a function of the sediment permeability as well as the compressibilities α and β of the fluid versus those of the grain framework. For conditions of pressure and temperature in the upper few kilometres of the crust, the thermal expansion of the pore water is the dominant factor ($\alpha_w \sim 5.2 \times 10^{-4}\,°C^{-1}$ versus α_g and α of 0.1 to $0.3 \times 10^{-4}\,°C^{-1}$). The change in pore pressure during erosion and uplift of an impermeable sediment can be solved by equating the change in volume of the sediment framework with that of the pore water plus that of the grains, which in expanded, rearranged form is:

$$\beta_w \Delta P_{H_2O} - \alpha_w \Delta T = (\beta - \beta_g)\Delta \sigma'_m - (\alpha_g - \alpha)\Delta T. \quad (6.13)$$

A solution of ΔP_{H_2O} per kilometre of sediment eroded, assuming a temperature gradient of 25°C km^{-1}, isotropic elastic material response and the simple effective stress law, leads to a pore-pressure reduction of 30 MPa km^{-1} of erosion, which is far greater than the hydrostatic gradient.

Pore-pressure reductions to much less than hydrostatic, or even to negative values are well known in rapidly unloaded clay soils (e.g. A.W. Bishop, Kumapley and El-Rawayih 1975). Pore pressures significantly less than hydrostatic have been reported from oil and gas fields in several regions, all of which have recently been or are being uplifted (Hunt 1990; Powley 1990). Some of these underpressured sediments result from very low general porosities, whereas others are within 'sealed' compartments. Sealed compartments can also be the locus of abnormally high pore-fluid pressures (Hunt 1990; Powley 1990), but such seals generally develop at depths of 3 km or more, are attributed to cementation fronts and lie below the environment of concern in this review.

Because the values of many of the mechanical parameters are not known precisely, and vary with lithology, porosity and cementation, only a general calculation of horizontal stresses during uniaxial unloading is possible. For example, a quartz sand, consolidated to σ'_v of 30 MPa, representing burial to about 3 km, and assumed to be isotropic, might have $\eta = 0.2$, $E = 15$ GPa and $v = 0.15$. Substituted into equations (6.11) and (6.12), these values produce a stress ratio gradient, $\Delta \sigma'_h / \Delta \sigma'_v$, somewhat greater than that of consolidation and would lead to a state of horizontal tension at depths of less than 1 km (Figure 6.7). More highly cemented or less porous sandstones, with higher values of E, would require less unloading before reaching a state of horizontal tension, primarily because of the larger thermal contraction.

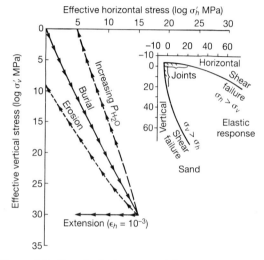

Figure 6.7 Hypothetical stress path for sand during burial to stresses equivalent to about 3 km and for several modes of stress reduction. 'Erosion' represents the purely elastic response to exhumation through a 25° C km^{-1} temperature gradient at hydrostatic fluid pressures. Initial erosion removes surficial sediment with an assumed ρb of 1.8, resulting in a curved unloading path. 'Increasing P_{H_2O}' represents the effects of increased pore fluid pressure alone. The curve labelled 'extension' shows the effects of a lateral strain of 10^{-3} acting alone. The inset illustrates the stress conditions that would lead to failure by shear (faulting) and by jointing.

Estimation of stress during the uniaxial unloading of shale is even less constrained because of its elastic anisotropy. In addition, far less information is available concerning shale and its constituent minerals. With plausible values of the necessary parameters, it appears that, during unloading, $\Delta \sigma'_h / \Delta \sigma'_v$ would be less for shale than for sandstone (cf. Figure 6.7 and 6.8), and because shales generate a higher $\Delta \sigma'_h / \Delta \sigma'_v$ during consolidation, it would be difficult for stresses in shale to become tensile at significant depths during simple erosional unloading.

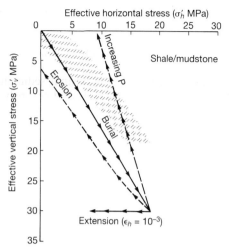

Figure 6.8 Hypothetical stress path for silty clay during burial to about 3 km and for several modes of stress reduction. Labelling as for Figure 6.7. The stippled band shows the ratios of σ'_h to σ'_v determined from hydrofracturing experiments in several stable basins undergoing uplift (Data from McGarr 1980; Hickman, Healy and Zoback 1985; Warpinski and Teufel 1987; Evans, Engelder and Plumb 1989.)

A conclusion of this exercise is that if a sediment is uniaxially unloaded by erosion, the tendency for σ'_h in that sediment to become tensile is directly related to its value of E. Only the least porous and most cemented sediments are likely to become horizontally tensile at depths much greater than 1 km.

Effect of pore fluid pressure on deformation Pore pressures do not, of course, necessarily remain hydrostatic during sediment consolidation or unloading. Subhydrostatic pore pressures generated during unloading were discussed in the previous section. Pore pressures that are higher than hydrostatic are common in basins with rapid sedimentation and low permeability strata, and are the topic of extensive discussions, primarily in the petroleum geological literature (e.g. Gretener 1981; Harrison and Summa 1991). Abnormally high pressures in basins have been attributed to downward decreasing permeabilities, increase in temperature, tectonic stresses, fluid influx and hydrocarbon generation. Rather than discussing the causes of abnormal pore-fluid pressures (which are covered in sections 2.3.4 and 7.3.6), only some effects of pore fluid pressure fluctuations on the deformation of sediments are considered here.

As derived earlier, the elastic stresses induced by a change in P_{H_2O} can be viewed in two parts: that due to reduction in σ'_v by the simple effective stress relation, and that caused by grain compression. During an increase in P_{H_2O}, the first of these has usually been considered alone to be capable of causing reduction of p' and a decrease in σ'_h/σ'_v. However, as shown by equation (6.6), this component of the pore pressure effect alone would lead instead to an increase in σ'_h/σ'_v and toward a horizontal σ'_1 (see also figure 4 of Mandl and Harkness 1987).

The total effect of increased P_{H_2O} (equation 6.6 plus 6.9), however, can enhance reduction of σ'_h, and under some conditions can result in horizontal tensional stresses. Substitution into equations 6.6 and 6.9 of the values for elastic moduli of the sandstone used earlier as an example, results in $\Delta\sigma'_h/\Delta\sigma'_v = 0.4$. This ratio is larger than those determined using the simple effective stress law but still not large enough to result in horizontal tensional stresses before σ'_v goes to zero. However, only modestly greater values of E will sharply increase $\Delta\sigma'_h/\Delta\sigma'_v$ and could generate an unloading stress path leading to horizontal tension. Such values of E (3×10^4 MPa or more) would be characteristic of well-cemented, low-porosity clastic sediments, which implies that an increase in pore pressure alone should lead to horizontal tensional stresses only in sediments that have been buried to depths of several kilometres or more. Even greater depths of burial would be required for sediments in which consolidation has been retarded by high values of P_{H_2O}.

In the uncompartmentalized upper levels of basins, with permeabilities typical of sediments without strongly cemented seals, excess pore pressures will dissipate after sedimentation ceases and erosion begins, unless dynamically maintained by tectonic strain (Cello and Nur 1988), or by fluid input to the system (Mudford 1990). Thus, the generation of horizontal tensional stresses, and natural hydrofracturing, seems most likely to occur when sediments are near their maximum depth of burial.

Deformation resulting from horizontal elastic strain Non-zero horizontal strains during elastic deformation of sediments may be one of the most significant factors in controlling sediment stress paths. It was shown earlier that horizontal strains have only a modest effect on horizontal stress during consolidation, but that even very small strains can produce very large changes in stress during elastic unloading. Horizontal elastic strains of as little as 10^{-3}, combined with the elastic effects of erosion and temperature reduction are easily capable of leading to horizontal tensional stresses and to failure in lithified sediments.

Horizontal extensional strains of 10^{-3} are very modest in a crustal context, and might be expected during many tectonic events, even during compressional ones where strike-parallel extension can be generated by the curvature of the deformational belt. The problem in appealing to elastic strain to affect sediment stress paths is that the deformational features preserved in a sediment are the results of inelastic strain only. The magnitude of the *in situ* elastic strain can be estimated only while the associated stress is still applied, and there are few data with which to address this problem.

General considerations of theoretical stress paths

Theoretical stress paths provide only simple and limiting cases, but they can be considered as a starting point and a basis of comparison with *in situ* observations. Moreover, they constrain the range of conditions under which various stress states can develop. For example, the limited circumstances under which horizontal tensional stresses can be produced have critical implications for the development of regional vertical joints, which are so common in basinal settings.

Processes leading to elastic unloading most likely operate in time-varying combinations, but again, these combinations are not arbitrary. High pore pressures are not easily maintained during uplift and cooling unless fluids are introduced into the system. Uplift combined with horizontal extension would result in very large values of $\Delta\sigma'_h/\Delta\sigma'_v$ such that the differential stress would be very high when σ'_h became tensile. This would favour the development of normal faults rather than development of vertical joints.

6.2.5 *In situ* measurements of stress in sediments

The geological applicability of simple theoretical stress paths could be tested by identifying geological settings in which the deformational history is known and is simple, and by comparing the theoretical stress paths with *in situ* stresses. This approach is currently being avidly pursued, but results are still too sparse and inconclusive to provide adequate comparisons.

Theoretical stress paths for basinal sediments, based on laboratory experiments and soil behaviour are developed relatively easily, but natural stress paths have only begun to be studied. The most straightforward approach to natural stress paths would be to measure contemporary stresses in different settings representing points along burial and unloading paths. Unfortunately there are only few data with which to attempt this approach, and these few data have not correlated well with theoretical paths.

Compilations of *in situ* stress data seldom do more than separate the values into those for 'hard' and 'soft' rock (e.g. McGarr 1980). To date there has been little attempt to take the deformational history of the rock into account so that differences in stress state between consolidating and unloading sediments can be explored. Analyses of *in situ* data are hampered further by the reporting of data as total rather than effective stresses, in part because pore pressures are seldom known adequately.

The validity of interpretations of observed stress data is dependent upon the assumed stress path leading to the observed stress. Many interpretations implicitly assume that sediments respond elastically from an initial isotropic unstressed state by the use of equations employing total stress rather than stress differences from the stress state at the end of consolidation (e.g. Warpinski 1986). Other analyses recognize the different stress relationships during consolidation (e.g. Engelder 1985), but still use an elastic response. Clearly, the assumptions made concerning the nature of the stress path are a critical factor in the validity of the interpretation of observed data.

In situ stress measurements in sediments have generally been made in basins undergoing erosion and in continental margins where the sediment cover is under extension. In general, compilations of *in situ* stress in sediments display an increase in differential stress $(\sigma'_v - \sigma'_h)$ with increasing depth (e.g. Swolfs 1984) as well as an upward increase in stress ratio near the surface (Brown and Hoek 1978). However, there is a wide variation in the values of the stresses and stress ratios as a function of depth.

Observational techniques

The magnitudes of *in situ* horizontal stress in sedimentary sections can be estimated with both *in situ* and laboratory techniques, but the most reliable and commonly used method is based on the hydrofracturing of a formation in a borehole (J. S. Bell 1990, and references therein). Other techniques for the measurement of *in situ* stress magnitudes, developed largely for applications in soil and rock mechanics (e.g. pressuremeters and overcoring), have proved to be less reliable or are not yet adequately calibrated.

Stress determination by hydrofracturing assumes that the borehole fluid pressure can be increased to a level at which the sediment fails tensionally along a fracture oriented perpendicular to σ_3. In addition, it is assumed that one principal stress is vertical; moreover, if σ_3 is vertical, the true value of σ_h is often not acquired because the fracture closure pressure is controlled by σ_v (Evans, Engelder and Plumb 1989).

The relationship of the maximum principal stress (σ_H) and minimum princpal stress (σ_h) in the horizontal plane to pore pressure P_{H_2O} and to the borehole fluid pressure at fracturing P_b is:

$$P_b = 3\sigma_h - \sigma_H + S_T - P_{H_2O}, \qquad (6.14)$$

where σ_H is the maximum total horizontal stress, σ_h is the minimum total horizontal stress and S_T is the tensional strength of the sediment. In a careful, research-oriented hydrofracturing test, a small amount of fluid is injected into a packed-off section of borehole, producing a pressure–time curve with several discrete sections. If a new fracture is created, a sharp pressure peak is followed by a drop to a relatively steady state representing crack propagation and fluid inflow. Cessation of pumping results in a complex pressure drop curve that includes an inflection point representing the crack closure. If an open fracture already exists, σ_h is approximated by the borehole pressure associated with this crack closure.

Several oilfield procedures undertaken to control drilling parameters are forms of hydrofracturing, and can also be used to estimate stress, although less accurately. A good general review of the hydrofracturing methods is included in Ervine and Bell (1987) and a more detailed assessment of the various approaches is given in Evans, Engelder and Plumb (1989). It is recognized that, in addition to the problem of underestimating σ_h when it is greater than σ_v, the tendency is to overestimate σ_h when it is less than σ_v. It is also recognized that estimates of σ_H are much less reliable than those of σ_h.

Estimates of *in situ* stress have also been made using core samples. Such methods are far simpler and cheaper than hydrofracturing but are of questionable reliability. The most promising of these is anelastic strain recovery (ASR), in which a small, time-dependent viscoelastic strain is measured as it decays after the core has been removed from the ambient stress field (e.g. Teufel and Warpinski 1984). If this viscoelastic component of strain is proportional to the total elastic strain tensor and if the compliance matrix is known or can be assumed, an *in situ* stress tensor can be calculated. These assumptions are still being discussed and compliance matrices to date are modelled by very simple isotropic or laterally isotropic materials (Blanton 1983). In the simple case of isotropic sands in basinal settings, there is a fairly good correspondance between ASR and hydrofracturing results (Teufel and Warpinski 1984). Other stress measuring techniques on cores involve measurements of strain and seismic velocity anisotropies, which are assumed to reflect *in situ* stress conditions, but these have not been generally accepted as reliable.

Examples of in situ *stress in sediments*

In situ stress measurements have been made in sediments with both research and economic objectives, from very shallow depths to more than 6 km. The shallower measurements in highly porous, less consolidated sediments are far more questionable because they are based on pressuremeter or less reliable techniques; hydrofracturing is very difficult under those conditions because packers generally cannot be emplaced adequately. Measurements have been collected in a variety of basinal settings, from erosional to depositional. However, no data from depositional basins without synsedimentary normal faulting which might be assumed to have no lateral strain, have been found in the litcrature. Economically driven measurements are often not as well-controlled or as completely reported as research measurements.

USA Gulf Coast By far the most *in situ* stress data have been collected from oil fields of the USA Gulf Coast, but much of this information is of questionable accuracy. It was collected largely for immediate engineering application. In the synoptic study by Breckels and van Eekelen (1982), as in most studies, the assumption of pre-fracturing was made. This led to the conclusion that the stress ratio (σ'_h/σ'_v) increases from 0.3 or less near the surface toward 1.0 at depths near 6 km. On the other hand, because most of the measurements used for this compilation were made in shale that is consolidating and theoretically unfractured, as well as functionally impermeable, the pressures recorded may more closely reflect the stress concentration (P_b) near the borehole. In such a case, the effect of T, which probably increases to about 3 MPa at 3 km cannot be ignored. Abnormally high fluid pressures ($P_{H_2O_e}$) which begin at depths near 3 km in Gulf Coast wells, should tend to raise the stress ratio above the K_0 value.

A probable lower bound on the effective stress ratio can be generated using equation (6.14), by assuming that $\sigma_H \approx \sigma_h$, and by assuming that the borehole fluid pressures recorded during fracturing operations are close to P_b. With the Gulf Coast data compiled by Breckels and van Eekelen (1982), σ'_h/σ'_v at 2 km, where P_{H_2O} is hydrostatic, would be near 0.5. This exercise would lead to the conclusion that the effective stress ratio does not increase with depth and, if anything, would decrease as the material involved became denser and stronger. Because of the many active normal faults in the Gulf Coast area, lateral extensional strain during consolidation must be suspected, and may be responsible for stress ratios that are lower than the laboratory values. In summary, although the present Gulf Coast data have led to the belief that σ'_h/σ'_v increases with depth, implying creep and temporal reduction of q, support for that conclusion requires much more careful *in situ* test data.

Scotian shelf A similar data base has been generated for *in situ* stresses on the Scotian shelf (Ervine and Bell 1987). Their more specific analysis of stress in several industrial wells of the passive margin off eastern Canada (Ervine and Bell 1987; Bell 1990) shows that the ratio of least effective horizontal stress to effective vertical stress, σ'_h/σ'_v, is roughly constant for shale and mudstone at 0.5–0.6. This value holds from near the surface to a depth of slightly more than 4 km, where pore fluid pressures begin to rise well above hydrostatic. These observations support a conclusion that σ'_h/σ'_v can differ from 1 for clay-rich sediments over periods of 10^6 to 10^7 years, and even these ratios could remain constant over such periods. That the σ'_h/σ'_v values in these mudstones and shales are slightly less than those of their uniaxially consolidated equivalents could again be due to slight regional extension, documented by synsedimentary normal faults (Bell 1990) or because the natural shales are affected by diagenesis. A downward increase in σ'_h/σ'_v through the overpressured section may reflect elastic unloading. This may be caused by increased pressure from an introduction of fluid, in addition to that resulting from restricted drainage (Mudford 1990).

Piceance Basin, Colorado Good sets of *in situ* stress measurements are available from several interior basins in the USA. Data from these basins reflect stresses from both subsidence and subsequent uplift, or other sources, and must be interpreted with this more complicated stress history in mind. Research-oriented data from the Piceance Basin of Colorado and from the Alleghanian foreland of New York were acquired from a range of lithologies, which also revealed the effects of variations in sediment mechanical properties on the state of stress.

A suite of stress measurements from depths of 1.3 to 2.5 km in three closely spaced borehole cores in the Piceance Basin defines a greater stress ratio for sandstones and siltstones than for interbedded shales and mudstones (Warpinski and Teufel 1987; Warpinski 1989). For the sandstones, σ'_h/σ'_v was about 0.7, which is significantly greater than that expected from an ideal uniaxial consolidation path for this lithology. Both the observed overpressures (Warpinski and Teufel 1987) and the unloading due to erosion could cause an elastic post-consolidation increase in stress ratio and are qualitatively capable of explaining the observed stresses. Without better control of the maximum depth of burial and of consolidation stress, and without data on the elastic moduli of these moderately lithified sediments, no quantitative comparison can be made.

Minimum horizontal stresses in the shale and mudstone sections of the boreholes were approximately equal to the overburden stress ($\sigma'_h/\sigma'_v \approx 1$). This observation was interpreted by Warpinski (1989) as a greater visco-elastic creep toward an isotropic state of stress for clay-rich rocks than for sandstones. However, the conclusion that values of σ'_h greater than σ'_v are very seldom measured (Evans, Engelder and Plumb 1989) would suggest that the stress ratios in these clay-rich sediments are only minima. At the least, these data are not strong evidence for visco-elastic creep.

Stress measurements in clay-rich sediments at shallower depths of the basin (Bredehoeft et al. 1976) produced ratios similar to those at greater depths, except that at depths less than 120 m evidence was given for values of σ'_h/σ'_v greater than 1. In general this agrees with the increase in stress ratio with progressive uniaxial uplift.

The tensional fractures created by hydrofracturing in both these experiments were parallel to the dominant WNW trending regional joints and normal faults (Bredehoeft et al. 1976; Lorenz et al. 1988). It was also observed that the regional joints occurred only in the sandstone units (Lorenz et al. 1988). Thus, the stress orientation has apparently remained constant since the development of natural jointing and faulting, but the present magnitude of σ'_h is too high for jointing to occur and the stress difference ($\sigma'_v - \sigma'_h$) is too low for normal faulting.

Alleghany Plateau, New York A second set of *in situ* stress measurements and physical properties data is available from Devonian sediments of the Alleghany Plateau of New York. This sequence, primarily of sandstone and shale, has had several kilometres of overburden removed and is still undergoing erosion (Evans, Engelder and Plumb 1989). In a group of three closely spaced boreholes, stress measurements were made over a depth range of 200–1000 m. Above 700 m, values of σ_h in all lithologies were about equal to σ_v and were shown to be 'clipped' at the value of the minimum principal stress (Evans, Engelder and Plumb 1989). At greater depths, σ_h in the shale becomes increasingly less than σ_v, whereas σ_h in the sandstone and limestone remains clipped at σ_v and must have at least that magnitude. Thus, the shale in this setting is capable of supporting significant differential stress. Minimum effective stress ratios in this deeper zone, calculated from total stresses and assuming that pore pressures were locally hydrostatic (Evans, Engelder and Plumb 1989) are about 0.5, but probably increase upward.

The greater σ'_h and stress ratios in the sandstones and limestones than in the shale correlate with the much larger values of E in the former, and have been explained logically by an ambient horizontal shortening strain acting on rocks of differing stiffness (Evans, Oertel and Engelder 1989; Plumb, Evans and Engelder 1991). This is corroborated by the orientation of the induced hydrofractures, which parallel the present regional maximum stress vector. The general upward increase in stress ratio can be attributed to an unloading stress path, but the sharp upward decrease in $\Delta\sigma_h/\Delta z$ in the shale, which is not accompanied by recognized changes in physical or mechanical properties (Evans, Oertel and Engelder 1989), is more enigmatic. A study of strain in these shales, based on fabric orientation (Engelder and Oertel 1985), indicated a lower degree of compaction in the deeper strata with lower σ_h values. This was interpreted as a result of high pore pressure during the early compactional history (Evans, Oertel and Engelder 1989). Regardless of the details, it does seem likely that the early stress history has strongly affected the present stress magnitude, which implies that the shale has the ability of maintaining high stress differences for very long periods of time.

Stress measurements in another well in the same general region, but which penetrated Lower Palaeozoic strata (Hickman, Healey and Zoback 1985), demonstrate a similar correlation between effective stress ratio and sediment stiffness (Plumb, Evans and Engelder 1991). This observation further strengthens the conclusion that regional shortening after consolidation can explain the horizontal stress patterns in these sediments. One significant difference between the stresses in this well and those in the Devonian section is that the minimum effective horizontal stress extrapolates upward to zero at a depth of about 300 m (Hickman, Healey and Zoback 1985) and is probably tensional above that depth. This shallow horizontal tension is associated with the development of vertical jointing (Engelder 1987), because there is a joint set perpendicular to the present σ'_h. However, these joints are restricted to the upper few hundred metres of the borehole, and are much less well developed in outcrop than are the regional joint sets.

Conclusions and implications The results of *in situ* stress measurements in the well-defined basinal sections reviewed above lead to several summary conclusions and implications. Despite the lack of measurements designed specifically to explore the effect of time on the stress ratio in uniaxially consolidated sediments, the available measurements indicate that sediments, even clay-rich lithologies, can maintain significant differential stresses over geological time. The case can also be made that stress ratios in actively consolidating basins may even be close to the short-term values obtained by experimental consolidation.

Effective horizontal stresses calculated from almost all *in situ* measurements are compressional. The stress values in different lithologies are functions of the mechanical properties of the sediment as well as the loading history of those sediments. Uniaxial consolidation and subsequent elastic unloading lead to higher values of σ'_h in shales than in sandstones, as predicted by theoretical loading paths. If unloading is associated with a horizontal shortening strain, this stress difference can be reversed, but the actual stress values will also reflect the prior inelastic consolidation history. It is generally invalid to assume an isotropic stress state at the beginning of the elastic strain path. Finally, it appears that theoretical stress paths based on data from laboratory experiments can offer reasonable models for the measured stresses in sediments.

6.2.6 Geological implications from theoretical and measured stresses

With the advancing understanding of geological stress paths, structural analyses of deformational features observed in sediments should be undertaken in light of their stress histories. Examples of attempts to use this approach include the interpretation of jointing on the Alleghany Plateau (Engelder 1985), slip along the San Andreas Fault (e.g. Rice 1992) and of deformation related to low-angle normal faulting (Axen 1992).

Tectonically-related deformational features in 'basinal' settings other than inelastic consolidation are relatively simple and limited, including joints, normal faults and near-surface compressional structures. Of these only the near-surface compressional structures are easily explained, resulting from the horizontal compressional stresses generated during erosional unloading.

Joints are almost ubiquitous and apparently simple structures, but are still puzzling in that the state of horizontal tension required for their formation tends not to be observed by *in situ* stress measurements. Normal faults require horizontal extension but commonly form as syndepositional features in basins, particularly along continental margins. Normal faults are shear zones associated with brittle failure and might seem incompatible with compactively and ductile deforming sediments. The discussion in the following section concentrates on joints and synsedimentary normal faults because of the interest in and data about these features, and also because they are not as easily explained as is often assumed.

Joints

Almost all joints are mode 1 cracks (e.g. Engelder 1987) that require a rather restricted range of stress conditions for their formation. First, σ'_3 must be negative or tensional and oriented perpendicular to the joint plane. In fact, the tensional stress must exceed the tensional strength of the sediment. Equally important, q must be quite small so that the material will not fail in shear. Vertical joints, which are the most difficult to explain, and yet are extremely common, require that σ'_3 be horizontal. The problem is: what process or processes will produce the requisite stress conditions for joint formation, and where along the consolidation and/or unloading paths can those conditions develop? Clearly, the conditions responsible must be common and often repeated, and yet tensional effective horizontal stresses have very seldom been recorded in sedimentary sections.

The literature on joints and joint formation is extensive (e.g. Secor 1965; Price 1974; Engelder 1987; Pollard and Aydin 1988), but stress paths responsible for the formation of vertical joints do not seem to be satisfactorily resolved. Natural hydrofracturing resulting from abnormally high pore pressures has been suggested most commonly as the major cause of jointing (Secor 1965; Engelder 1985), but the conditions within which natural hydrofracturing is likely to occur have not been explored. As pointed out earlier, a simple increase in P_{H_2O} will lead to horizontal tensional failure, unless E_h/E_{hg} is of the order of 0.5 or more. Such a condition is obtained only in very stiff sediments, which implies low porosity and, generally, deep burial. Vertical jointing can also develop where steep pressure gradients occur, for example during artificially induced hydrofracturing or where fluids with very different properties are involved (e.g. magmas or hydrocarbons; Mandl and Harkness 1987). However, such conditions are difficult to perceive for the development of regional vertical joint sets in most basins with aqueous fluids, and where fluid pressure gradients are low.

The orientation of these vertical joints provides another constraint for their time of origin along a stress path with changing principal stress orientations. For example, Mid- to Upper Devonian strata of the Alleghany Plateau were deposited in a stress field related to west-directed Acadian thrusting, whereas the regional joints in these sediments reflect the Carboniferous–Permian Appalachian collison, with N-S shortening (Engelder and Geiser 1980). The present maximum horizontal stress trends ENE and is probably related to the opening of the Atlantic (Zoback and Zoback 1989), and may have existed since the Mesozoic. Correlation of the stress history with the deformational history of the sediments thus argues against the formation of jointing by hydrofracturing during basin filling or much after the unloading by erosion began.

Lateral elastic extension during unloading was shown to be a very effective process in reducing σ'_h, but must be associated with a marked reduction in σ'_v so that q will be small enough to result in tensional rather than shear failure. Sources of extension during an elastic unloading path could be deep crustal strain, gravitational 'spreading' of the sediment prisms along continental margins, and the lateral spreading of mildly deformed sediments in arcuate orogenic zones.

On the Alleghany Plateau the vertical cross-trend joint system developed after the initial pressure solution and intragranular deformation (Engelder 1979) and may have alternated with pulses of stylolitization along vertical planes perpendicular to the joints (Engelder and Geiser 1980). A mechanically plausible explanation for formation of these joints is regional extension parallel to structural trends, together with pulses of increased pore fluid pressure that may have favoured jointing over stylolitization. A quantitative solution is impossible for this strain-driven system in which the stresses generated during stylolitization are unknown.

Lateral strains induced by the increase in crustal curvature associated with uplift (Haxby and Turcotte 1976) are up to an order of 10^{-3} and probably much less than the lateral strain suggested for the Alleghany Plateau. Strains induced by basin uplift (Price 1974) depend on the curvature of the basin but would vary between the condition assumed by Haxby and Turcotte and compressional strains as basin curvature decreased.

Normal faults

Normal faults are common in many sedimentary basins, and are very often active during sedimentation. Examples due to differential consolidation, were mentioned in section 1.3.2. In some cases the extension defined by this faulting may reflect regional gravitational tectonics, whereas in other basins the displacements may be related to regional crustal extension. These faults are of economic interest in that in different settings they can either trap or channel fluid hydrocarbons (e.g. Hooper 1991). These faults are also of fundamental interest because they are brittle failure structures that develop along a generally ductile consolidation stress path. As brittle failure surfaces they should show dilative strain, which would favour increased permeability, but evidence has been presented for porosity reduction along some small fault zones (e.g. Knipe 1986a; Aydin and Johnson 1978). The stress paths associated with synsedimentary faults, which may well constrain fault characteristics, have received much less attention than have the fault characteristics themselves.

Synsedimentary normal faults are classically listric, resulting in antiformal or 'roll-over' structures in the hanging wall. The decrease in dip with increasing depth has been explained with fair success by rotation due to continuing consolidation at depths greater than the fault-line tip (M.E. Jones and Addis 1984). Geometric and kinematic analysis of hangingwall structures has been extensively treated by Xiao and Suppe (1992) and others but is outside the scope of this review.

Of interest here is the range of conditions whereby horizontal extension is accommodated by brittle faulting rather than by ductile flow within a generally consolidating basinal environment. The problem could be visualized as the response of the sediment to variations in the ratio of the rate of increase of vertical stress ($\partial \sigma'_v / \partial t$) to the rate of horizontal extensional strain ($\partial \varepsilon_h / \partial t$), both of which can be considered as independent variables. High rates of sedimentation relative to the extensional rate would clearly favour ductile flow, whereas high extensional rates relative to sedimentation will favour faulting. How the mechanical behaviour of extending, consolidating sedimentary sections is quantitatively related to the ratio of $\partial\varepsilon_h/\partial t$ to $\partial\sigma'_v/\partial t$, and how it is affected by differences in lithology and cementation, has not yet been probed by experimental deformation. The loading paths for experiments designed to explore synsedimentary faulting would require triaxial or, preferably, plane-strain paths. Hints as to what might occur can be drawn from triaxial deformation tests on sediments, reviewed in the following section on the deformation in accretionary prisms.

6.3 STRESS PATHS ASSOCIATED WITH DEFORMATION IN ACCRETIONARY PRISMS

6.3.1 General

The converging margins of lithospheric plates are the loci of intense deformation involving a wide spectrum of geological materials and mechanical responses. One common response is the offscraping of sediments that were deposited in an oceanic trench and basin into a prism with a triangular cross-section (termed an accretionary prism) at the leading edge of the upper plate (Figure 6.9). Accretionary prisms display relatively simple and well-known geometries and boundary conditions, the study of which helps understand more complicated settings, such as those presented in many continental fold-and-thrust belts. Yet, because of formidable difficulties in acquiring *in situ* measurements, studies of stress and deformation in accretionary prisms are still in their infancy. In this section we will summarize recent data regarding the stress and strain histories of accretionary prisms, but to suggest even simple models we must rely to a great extent upon extrapolation of data from soil mechanics studies and from a small suite of experimental data oriented toward the conditions in prisms.

Accretionary prisms are commonly composed of imbricate thrust sheets of offscraped sediments that accumulated in the trench and oceanic basin of the subducting plate. The prism is separated from the subducting plate by a sharply defined shear zone or décollement, the properties of which are critical to defining the mechanical state of the prism. Over a period of several million years, the gross geometry of the accretionary prism remains approximately constant and sediments can be treated as if they flow through a steady-state system (Figure 6.10). The along-strike

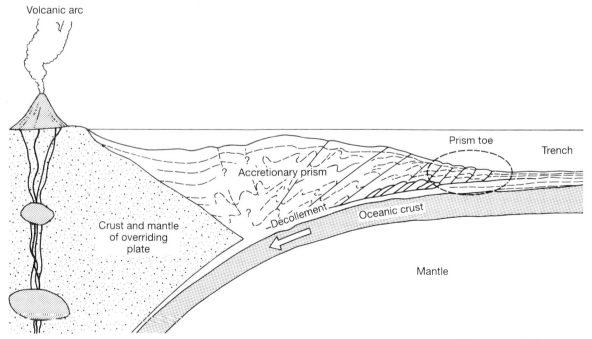

Figure 6.9 Generalized section of a convergent plate boundary with an accretionary prism. The section of the prism informally termed the toe is included within the dashed oval. (Modified from Langseth and Moore, 1990).

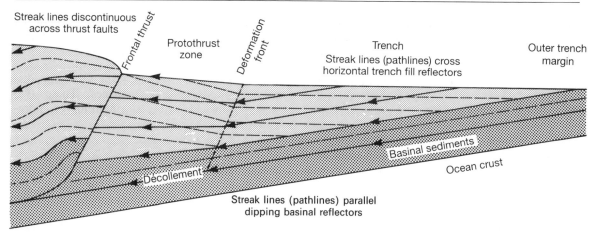

Figure 6.10 Steady-state model for deposition of sediment in a trench and accretion into a prism. Stratigraphical horizons, marked by dashed lines, terminate arcward against the sea floor and seaward against the underlying basinal strata. Sediment streak lines (or path lines in a steady-state system), marked by solid lines, are subparallel to the oceanic crust and the contact between the basinal and trench strata. Streak lines are oblique to bedding reflectors in the trench, which travel passively with the moving sediment element. Thrust faults represent moving discontinuities across which streak lines are displaced.

uniformity of many prisms also permits a simplifying plane-strain approximation.

The accretion of initially very porous sediments leads to changes in their physical properties and to a complex progression of deformational fabrics. To date, the nature of sediment deformation in accretionary prisms has been largely inferred from seismic profiles and velocity data, and from deformational structures and porosities measured in deep-sea drill cores. The observational data represent a spatial data set but, if adequate control on the distribution of strain is available, these spatial variations can be reformulated into temporal variations of individual units of sediment during their accretion. The progressive evolution of the properties of sediments as they flow through the prism can be compared with the results of deformation experiments and can provide at least some constraints on the mechanical behaviour and stress paths within the prism.

Unfortunately, structural features and physical properties observed in sediments from the prism toe convey complicated, even contradictory, implications for mechanical behaviour during deformation. Deep-sea drill cores document ductile strain, which is indicative of compactive deformation, through reduction in porosity and by the development of preferred orientation in minerals. The cores also demonstrate brittle, and presumably dilative, deformation, as evidenced by abundant faults and localized shear zones. Moreover, the combined structural evidence from seismic and core data indicates that most deformation in prisms occurs after the sediments have reached a state of failure, and under continuously changing stress conditions. This indicates a history of deformation which may include the sequential or simultaneous development of ductile and brittle strains during tectonic deformation. At present, the integration of observational, numerical and experimental studies provides only tentative answers, but these are valuable in showing what could be accomplished with better data sets.

In this discussion we will first present a general framework for the sediment deformation in prisms and draw comparisons among some of the better known examples of prisms. We will then review our state of knowledge of the nature of deformation in accretionary prism toes. The results of deformation experiments undertaken at stress conditions relevant to those in prism toes will then be presented, along with implications for the stress paths and deformational histories in accretionary prisms.

6.3.2 Lagrangian description of sediment accretion

Of necessity, the distribution of deformation in accretionary prisms is described in a spatial or Eulerian reference frame, because information is collected as a function of position within the prism. This compares with recording deformation data at various outcrops, and provides no information regarding the history of any deformed element. However, the approximately steady-state nature of accretionary prisms allows individual sediment paths through the Eulerian field to be predicted so that progressive stress paths and strain histories can be inferred (Karig 1990). This represents a material or Lagrangian description, using a reference frame fixed with respect to the particle, and is analogous to the structural analysis of progressive deformation in a single sample or outcrop.

Sediment flow paths can be described within an approximately steady-state system fixed to the toe of the prism (Figure 6.10). Material deposited at the sea floor in the trench or adjoining basin is buried by subsequent sedimentation as the subducting plate moves arcward. Sediment trajectories are subparallel to the subducting ocean crust, and thus nearly parallel to the dipping basinal strata, but oblique to bedding in the trench fill. At the deformation front, the particle trajectories diverge as sediments tectonically deform.

The presence of thrust faults, which represent moving discontinuities, causes a breakdown of true steady-state conditions. However, the finite strain accrued by a sediment element can be estimated by projecting the particle back to some undeformed configuration, defined at the sea floor. The plotting of the current positions of particles that passed through a single point at the sea floor defines a particle 'streak line'. Whereas a trajectory (path line) represents the path taken by a given particle from the initial point to its present position, a streak line simply identifies all particles that passed through a given point. By projecting sediment streak lines upon the Eulerian deformation field, the strain history of an element can thus be determined. The calculation of sediment streak lines and diffuse deformation (ductile and small-scale brittle) has now been accomplished for the Nankai prism using a numerical solution (Morgan, Karig and Maniatty in press).

6.3.3 Large-scale variations among prisms and mechanical implications

Significant variations in gross geometry and deformational style are observed among the better known examples of accretionary prisms and are assumed to indicate differences in mechanical behaviour. The Nankai accretionary prism off the southeastern coast of Japan and the Barbados (Lesser Antilles) prism are among the most studied prism settings and will serve for discussion in the following pages. These two examples provide contrasts in lithology, pore-pressure and strength which can be related to mechanical differences. Differences are even observed along strike within a single prism, as indicated by comparisons between two transects across the Nankai Trough and prism about 100 km apart. Further background on the Nankai prism, and discussion of its structures in a different context, are presented in section 8.2.

The most striking aspect of a prism is its cross-sectional shape, or the wedge taper defined by the angle between the surface and the basal décollement (Figure 6.11). This has been related to the contrast between the internal strength of the accreting material and the shear stress along the décollement (e.g. Davis, Suppe and Dahlen 1983). High internal strength or low basal shear strength contributes to a low taper of the prism toe, as along the northern Barbados prism (4°; Ladd et al. 1990) and the eastern Nankai prism (4–5°). The 7° taper on the western Nankai transect suggests a somewhat lower internal strength or higher basal strength in that toe.

These differences in taper angles could reflect either differences in the lithological composition within the prism or along the décollement, or differences in the distribution of pore pressure. The Barbados prism is clearly more clay-rich in composition than either parts of the Nankai prism, both of which contain a large fraction of sand-rich turbidites. Clay is intrinsically weaker than sand at the stress conditions in the prism toes and, furthermore, the much lower permeabilities of the clay-rich toe would engender higher pore pressures, further weakening it. The combination of effects would thus suggest that the Barbados prism should be weaker than either part of the Nankai prism. The reason for the similar tapers of the Barbados and eastern Nankai prisms, as well as the quite different tapers of the two Nankai prism examples must in large part reflect different conditions along their décollements.

The décollements of all three examples lie basically in clay-rich horizons, but are observed to differ in other respects. Several lines of

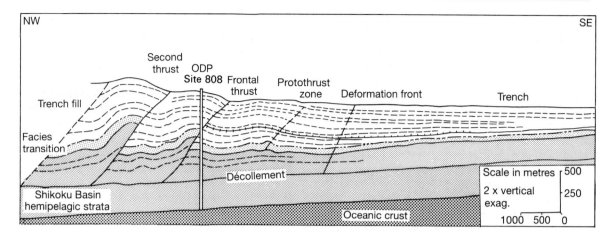

Figure 6.11 Line drawing of the prism toe from the eastern Nankai transect. (Based on seismic profile of G. F. Moore et al. 1990.) Diffuse strain within the protothrust zone passing is illustrated by diverging reflectors. The location of drilling at ODP Site 808, through the frontal thrust, décollement and subducting strata, is also indicated. (See also Figure 8.2).

evidence, including geochemical tracers (Gieskes, Vrolijk and Blanc 1990) and anomalously high temperatures (Fisher and Hounslow 1990) at the décollement of the Barbados prism, indicate that very highly pressured fluids are flowing oceanward along that zone, leading to very low basal shear stresses. There is no such evidence for fluid flow along the décollement of the eastern Nankai transect (Kastner et al. 1993), but the very sharp change in structural response across that zone (Taira, Hill and Firth 1991) also indicates a very low shear stress. This décollement, which lies within a thick mudstone section, also extends seaward beyond the deformation front (G.F. Moore et al. 1990), suggesting that it may initiate as something other than a simple shear zone. The décollement along the western Nankai transect initiates near the interface between the permeable turbidite-rich, trough-fill sediments and underlying basinal mudstones (Kagami et al. 1986). This location probably results in better drainage from the décollement into the overlying sediments, reducing pore pressures and thus accounting for a higher taper than along the eastern transect.

Another obvious difference among the three examples is in the development of the protothrust zone (PTZ), a zone of diffuse deformation lying seaward of the frontal thrust (Figure 6.11). The PTZ is over 5 km wide, with about 20% horizontal shortening along the western Nankai transect, but only 2 km wide and with less diffuse strain on the eastern transect. No PTZ at all was observed along the Barbados transect. Different stages in development of the outstepping frontal thrusts might explain part of the differences between the Nankai transects, but an incipient frontal thrust forming in the middle of the eastern PTZ precludes that solution. It is more likely that the effect of pore pressures within the PTZ on stress paths is responsible, a point to which we will return later after additional constraints are extracted from a review of other observations from ocean drilling and of deformation experiments.

6.3.4 Diffuse strains in prism toes

Although the stacking of sequential imbricate thrust sheets is the most apparent expression of deformation in accretionary prisms, several prisms exhibit distributed strain at the seismic scale through the progressive thickening of strata at the toe of the prism. This deformation could reflect the small-scale brittle and semi-brittle deformation structures recognized in drill cores (e.g. Lundberg and Moore 1986), as well as ductile strains suggested by porosity reduction along path lines. This distributed deformation is well illustrated by the two seismic profiles across the Nankai Trough and prism toe (G.F. Moore et al. 1990). These transects display zones of arcward stratal thickening in the PTZ (Figure 6.11). The stratal thickening is associated with an arcward increase in seismic velocities across this zone (Stoffa et al. 1992), which can be attributed to the dewatering of initially very porous sediments.

Although this stratal thickening is resolvable at the seismic scale, the finite diffuse strains cannot be quantified directly from the profiles because of the associated volume changes. However, volume changes can be estimated from the porosity distribution. A compact kinematic solution for estimating the diffuse strain field, as well as particle streak lines, based on the integration of porosities with changes in vertical spacing of continuous reflectors in the accreting sediments on seismic profiles, is offered by Morgan, Karig and Maniatty (in press). This method, applied to both the western transect across the Nankai toe (Morgan, Karig and Maniatty in press) and the more complex eastern transect (Morgan and Karig in press), indicates a heterogeneous distribution of strain within the toes of accretionary prisms.

The first-order results of these analyses indicate that strain generally accrues during pure shear. There is little evidence for shear strain immediately above the décollement along either transect, suggesting that the shear strains are almost completely restricted to the narrow décollement zone. This is consistent with the subhorizontal shortening axes observed in minor structures from drill cores (Karig and Lundberg 1990; Maltman et al. 1993b), and with the sharp drop in abundance of deformation structures in drill cores immediately below the décollement (Taira, Hill and Firth 1991). Higher order variations in strain include folding of strata in response to thrusting, which causes sedimentary elements to rotate within this flattening field; the total finite strain of these elements may reflect the non-coaxial superposition of strains and deformation structures.

The strain calculations also indicate a significant differentiation between volumetric and distortional strain, with deeper sediments accommodating horizontal shortening dominantly through vertical extension, and shallow, more porous sediments showing high-volume strain and little vertical extension. These results point to the importance for mechanical response of consolidation state of the sediment at the time of accretion. The magnitude of strain also appears to vary as a function of vertical position in the vicinity of the frontal thrusts. Along the eastern transect, sediments within the footwall of the thrust faults appear to have experienced up to 35% horizontal shortening, whereas sediments in the hangingwall show only 15–20% horizontal shortening (Morgan and Karig in press). These variations could arise from different stress paths associated with thrust loading of the footwall in contrast to uplift and erosion of the hangingwall.

Although these results suffice to establish a general strain field of accreting sediments at the scale of seismic observations, inferring stress histories of these sediments requires resolving the distributions of the ductile and brittle components. These components are best evaluated by examining porosities and deformation fabrics preserved in drill cores from active prisms, which are described below.

Brittle deformation

The brittle response within accretionary prism toes is the easier aspect of the deformation to document, as it is expressed by faults with large displacement imaged on seismic profiles, as well as by core-scale shear surfaces. Remarkable variations in the nature and abundance of brittle deformation are observed not only between the Nankai and Bahamas study areas, but also between the two Nankai transects.

Drill holes at two sites along a western transect across the Nankai accretionary prism partially penetrated the prism toe and provided information to a depth of 600 metres below the sea floor (mbsf). The most common structural features in these sediments were deformation bands, interpreted as brittle–ductile zones composed of Reidel R_1 zones, each of which was a kink-band (Karig and Lundberg 1990). These deformation bands were observed in clay-rich sediments at depths greater than 350 mbsf and at porosities of 45% or less. Detailed descriptions of these features have been presented elsewhere (Lundberg and Karig 1986; Karig and Lundberg 1990), but it is useful to note here that deformation bands result from compactive shear strain. These bands represent zones within which porosities are as much as 5% less than in the surrounding sediment. Moreover, the angle between conjugate bands, taken through σ_1, ranges from about 60° to more than 110°. This leads to the interpretation that the dihedral angles have been modifed by progressive horizontal ductile shortening (Karig and Lundberg 1990). Because the amount of ductile shortening calculated from the band rotation was far greater than the bulk shortening estimated for the prism toe at that site, some form of localized, intense, horizontal shortening strain was proposed. In retrospect there was a surprising lack of discrete core-scale shear surfaces in drill holes of the western transect; those originally identified (Kagami et al. 1986) were later shown to be a subset of brittle–ductile deformation bands (Karig and Lundberg 1990).

The complete penetration of the prism toe at ODP Site 808 on the eastern Nankai transect provided an extensively described suite of brittle structures (Byrne et al. 1993b; Lallemant et al. 1993; Maltman et al. 1993b; section 8.2), again most clearly observed in clay-rich strata. The only major faults cut at this site were the frontal thrust and the décollement zone. With the exception of an overturned fold having a wavelength of about 50 m and symmetrically disposed about the frontal thrust (Taira, Hill and Firth 1991), there was little evidence for tectonic folding.

Small-scale brittle shears, with geometries reflecting subhorizontal shortening, were observed below about 250 mbsf, at porosities of about 45% or less (Taira, Hill and Firth 1991). Because of poor core recovery between 100 and about 400 mbsf, it is difficult to determine adequately the frequency of brittle failure surfaces as a function of depth, but brittle structures related to accretion are clearly concentrated in, if not restricted to, the sediments above the décollement.

Deformation bands were also observed at Site 808, where the greater penetration showed them to be distributed over a zone symmetrically disposed (after correcting for the effects of incomplete core recovery) about the frontal thrust fault. The angles subtended by conjugate bands again range to more than 90° (Taira, Hill and Firth 1991), but suggest less ductile strain than in the western transect. The superposition of structures suggests that deformation bands developed prior to folding associated with the frontal thrust, whereas most brittle deformation appeared to develop after folding (Maltman et al. 1993b). The deformation bands thus appear to be the earliest structural features related to the development of the frontal fault zone.

The thickest zone of brittle shear penetrated in Hole 808 was the 20-m-thick basal décollement. The clay-rich sediments that comprise this zone occur as sheared phacoids several centimetres in length, evidence for the high shear strain in this zone. However, these phacoids are not significantly sheared internally, in contrast to the scaly clays recovered in the décollement of the Barbados example (D.J. Prior and Behrmann 1990b). Unlike the frontal thrust, deformation bands were not observed in the vicinity of the décollement, suggesting deformation histories differed between the décollement and the frontal thrust.

Despite the dominantly clay-rich lithologies in the Barbados accretionary prism, drilling in that setting revealed a far more extensive development of brittle deformation features than at either study area in the Nankai prism (Behrmann et al. 1988; Brown and Behrmann 1990). In general the deformation was much more complex and intense than in the Nankai prism, with far more large-scale faulting and folding, greatly reducing the seismic coherence of the prism. Deformation bands were absent, whereas brittle shear zones were observed at depths as little as a few tens of metres and at porosities as high as 65%. Again deformation was largely restricted to the region above the décollement, although more extensive development of mud-filled 'veins' was observed in the subducted sediments than in the Nankai sites. An arcward increase in intensity of deformation was documented in a transect of drill holes, in which greater bedding rotation and intense folding, axial planar surfaces and syndeformational calcite-filled veins were observed (Brown and Behrmann 1990).

Ductile deformation

Documenting and quantifying the ductile component of tectonic deformation in accretionary prism toes is more difficult than recognizing the brittle contribution. Subtle evidence for ductile strains includes changes in sediment porosity, preferred mineral orientation and the rotation of pre-existing structural features. Rotation of features was addressed in the preceding discussion; this section will concentrate on the expression and interpretation of porosities and mineral fabrics, which suggest a significant ductile contribution to strain in accretionary prism toes.

Porosities Porosity reduction reflects compactive volumetric strain and is associated with ductile deformation. Although this volumetric component of ductile strain is a function of stress path, it is also dependent on the ratio of q to p', which includes the effects of pore pressure, on lithological variations (e.g. Figure 6.1), and on the degree of cementation. As a consequence of all these factors, porosity–depth profiles in prism toes differ from those in basins and vary among themselves.

Porosities in accretionary prisms are generally observed to decrease both downward and landward. Along the western Nankai transect, drill cores obtained within the prism indicate slightly lower porosities than those at comparable depths in the adjacent trough, suggesting a component of tectonic dewatering (Bray and Karig 1986). Porosity contours estimated from seismic data (Stoffa et al. 1992) also converge landward, showing a moderate decrease in the average porosity of sediment columns within the prism. The porosity decrease along sediment paths (streak lines), by comparison, is quite high (up to 25% in the western PTZ), with the steepest gradient occurring quite close to the deformation front (Morgan, Karig and Maniatty in press).

Although the data set is less complete, similar trends in porosities and velocities are observed along the eastern Nankai prism (Taira, Hill and Firth 1991). Moreover, local variations in porosities can be identified that indicate greater complexity. Porosities increase sharply downward by 5% to 8% across the décollement, suggesting that this zone represents a barrier to fluid flow and/or separates different stress states. A similar, but lesser, downward porosity increase is observed across the frontal thrust, but can be explained readily by a relatively recent stratigraphical offset. More surprising, however, are the low porosities of sediment fragments within the décollement zone relative to the wall rock (Taira, Hill and Firth 1991). This phenomenon is inconsistent with the dilative stress path assumed for such brittle shear zones and requires an alternative or additional process.

Porosities within the Barbados prism are generally higher, and show significantly more variability than those within the Nankai prism (Mascle et al. 1988). This might be attributed to lower permeabilities of the clay-rich sediments along this transect, and possibly to greater structural complexity. In addition, porosity–depth profiles do not show a significant landward change (Mascle et al. 1988), a feature supported by lack of landward convergence of seismically derived porosities (Bangs et al. 1990). Porosity changes along sediment path lines are more significant, but path lines are not readily determined in this setting, due to intense thrusting. Vertical averages in porosities at points along the profiles of Bangs et al. (1990) show an arcward decrease, which requires at least some compactive path lines. Nevertheless, the less obvious reduction in porosity profiles along the Barbados transect may reflect a lower component of ductile strain, relative to brittle, than in the Nankai examples.

Fabrics The distortional component of ductile strain may be recorded in preferred mineral orientations of some sediments. Minerals with extreme dimensional ratios (e.g. phyllosilicates and magnetite) will tend to reorient during ductile flow, producing an anisotropic fabric; this strain can also be apparent through magnetic, acoustic and stiffness anistropies. Recent studies of these anisotropies have attempted to evaluate fabric evolution and quantify strains in accreting sediments.

Although the data on mineral fabrics are scarce, several studies have demonstrated a component of lateral fabric anisotropy in mineral sediments from accretionary prisms. Hounslow (1990) showed low degrees of magnetic susceptibility anisotropy in the Barbados prism, documenting some progressive lateral ductile shortening. Magnetic fabrics (Owens 1993) and clay-mineral fabrics (Morgan and Karig 1993) also document a larger ductile strain component of tectonic origin in the eastern Nankai prism toe. Using the simple March model (1932), which relates particle density and strain, Morgan and Karig (1993) record up to 10% lateral shortening in Hole 808, but with little variation with depth.

Such lateral fabric anisotropies are absent in sediments collected below the décollement surface at the Nankai prism (Morgan and Karig 1993), suggesting that these sediments have experienced no lateral strain. These observations are consistent with the lack of deformation structures in the subducting sediments and are evidence that these sediments remain at a state of vertical uniaxial consolidation.

Relationships between brittle and ductile strains

Accretionary prisms display differing degrees of ductile and brittle deformation, suggesting that stress paths during accretion can vary significantly within and among prisms. Along the eastern Nankai transect, lateral diffuse strains of 15% for the hangingwall sediments and 30% in the footwall in the vicinity of Site 808 contrast with a uniform 10% horizontal ductile strain based on clay-mineral fabrics. This points to between 5 and 20% brittle strain depending on position, suggesting that brittle strains dominate in this setting, at least at depth (Figure 6.12). The stronger convergence of seismically derived porosity curves along the western Nankai transect, as well as its higher taper angle and wider PTZ, suggest that ductile strains may be proportionally more important along this section than to the east. In contrast, the Barbados prism shows much less evidence for ductile strains in drill cores than in either Nankai example; it displays a small taper angle and lacks a diffusely strained PTZ. In this prism toe, it would appear that nearly all of the deformation occurs through brittle modes.

The relative timing of ductile and brittle deformation can be deduced from the superposition of structures where both modes are evident, as in the eastern Nankai toe. Structures identified in the vicinity of the frontal thrust at Site 808 suggest a sequence initiating with deformation bands, followed by folding and, ultimately, by faulting (Figure 6.12; see also section 8.2). High dihedral angles of conjugate deformation bands in cores from both the eastern and western Nankai transects suggest that some ductile straining may post-date the formation of the deformation bands, but this may be a reflection of local strains related to folding and thrusting. However, there is no evidence for post-rupture change in angle of conjugate faults. In general, ductile strains are assumed to occur quite early, and relatively close to the deformation front, as indicated by the steep gradient in porosity along streak lines. The

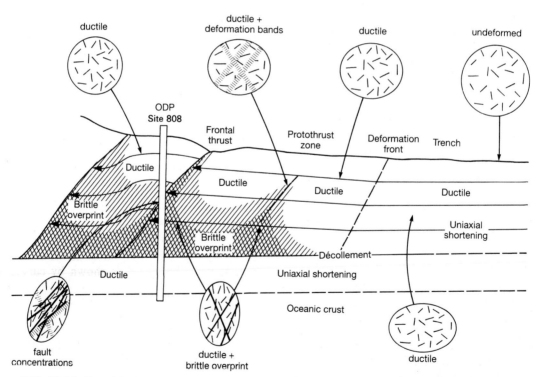

Figure 6.12 Distribution of ductile and brittle strains in an accretionary prism toe, inferred from porosities, clay-mineral fabrics and deformation structures observed in ODP drill cores, and from strain magnitudes estimated by Morgan and Karig (submitted). Ductile strains are indicated by white; brittle strains by diagonal lines. Superimposed grey lines represent streak lines. Uniaxial shortening (vertical consolidation) is assumed in the trench and in the subducting sediments. Tectonic strains occur landward of the deformation front; high porosity reduction near deformation front suggests that initial tectonic strains are ductile. Brittle shears are concentrated near thrust faults or incipient thrust faults, as well as at depth. Ductile–brittle deformation bands may develop during early stages of thrusting, before the formation of brittle shears.

apparent uniformity of lateral ductile strains indicated by clay-mineral fabrics suggests the early development in a strain driven system. This is consistent with an initial compactive stress path.

This proposed sequence and distribution of deformation in accretionary prisms raises questions regarding the stress histories leading to the observed strains. These are best addressed by examining the results of experimental sedimentary deformation of natural sediments and their synthetic analogues.

6.3.5 Experimental studies

General

The experimental deformation of sediments provides quantitative relationships between strain and deformation fabrics with known stress paths. Unfortunately, there is a remarkably small body of data that can be used to relate the strain fabric in porous sediments to the stress path responsible for that fabric, especially for geologically realistic stress paths. Moreover, much of that information is derived from the deformation of 'synthetic' and remoulded natural sediment, both of which lack the diagenesis of natural sediments. It has been demonstrated already that such differences lead to significant variations in the response of uniaxially strained mudstones. Such differences have been probed in our laboratory by two series of deformation experiments on synthetic and natural silty mudstones. These and other results provide a background on sediment mechanical behaviour for the discussion of deformation histories of sediments as they enter the prism toe and progressively lithify.

Experiments on synthetic sediments

Experimental deformation of synthetic sediments provides a good basis for the general behaviour of a given lithology because the material being deformed is uniform and reproducible. The use of synthetic sediments also overcomes the shortage of suitable natural material. For the series of tests undertaken in our laboratory, a three-component silty clay consisting of quartz, illite and montmorillinite was used (Karig and Hou 1992). This material was intended to mimic the natural silty clay that comprises a major fraction of the sediments in most accretionary prisms. Moreover, clay-rich sediments display much better developed strain features than coarser clastics and are generally much less disturbed during recovery from the drilling process. The basic objectives of the deformation experiments reviewed here were to extend our understanding of the mechanical behaviour from the low stresses of soil mechanics into the higher stress conditions of geological environments. Many of our tests involved large stress tensor rotations, which emulate the vertical consolidation subsequently subjected to horizontal shortening strains; such test paths are seldom undertaken in soil mechanics. In addition, our testing explored the post-failure behaviour of these sediments, again more complex in the geological realm than in soil mechanics.

One of the more important specific objectives of these initial experiments was to explore the conditions within which the sediment would fail brittly or ductiley. A common assumption among geologists (e.g. Knipe 1986b) is that the more porous a sediment, the more likely it is to deform by bulk ductile flow, yet soils with very high porosities have been shown to fail brittly when strongly overconsolidated, and highly lithified shales can deform ductiley when subjected to high p' values.

A general solution to the brittle–ductile problem has been provided by the critical state concept (e.g. Wood 1990; M.E. Jones and Addis 1986; section 2.2.5). The concept implies that a sediment, with a porosity greater than that of the critical state for a given p' will deform compactively and thus ductiley, whereas a sediment with a porosity less than that at critical state will deform elastically, with dilation leading to brittle failure. A corollary is that for a given p', the equilibrium porosity is a function of the q, with the porosity decreasing as q increases to the critical state. This effect is shown by curves for several stress paths on a porosity versus p' plot (Figure 6.13). The extrapolation of these curves from the better characterized conditions of soil mechanics into tectonic stress ranges is important because the effect of q on porosity change is strongly dependent on p'.

Results The mechanical behaviour of a synthetic silty clay, similar to that which dominates the core recovered from the Nankai prism, was explored to a p' of 30 MPa (Karig 1990) for a range of initial consolidation states. Most test paths had a constant p' and were run well into

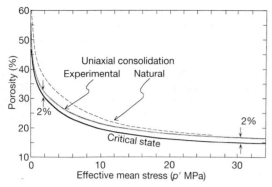

Figure 6.13 Porosity versus effective mean stress (p') for uniaxial consolidation and for critical state (ductile) failure from the experimental deformation of synthetic silty clay. The porosity difference, which is approximately the volumetric strain for small differences, remains constant at about 2% over the stress range explored. The dashed curve represents the relationship for naturally consolidated sediments, using data from Figure 6.1 and a K_0 value of 0.62.

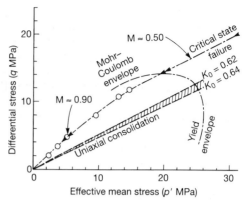

Figure 6.14 Effective mean stress (p') versus differential stress (q) plot for a synthetic silty clay experimentally deformed along uniaxial consolidation and triaxial strain paths leading to ductile failure. (Simplified from Karig 1990.) Tests in which σ_1 was either parallel (circles) or perpendicular (triangles) to bedding define a single critical state line with a slope that decreases as p' increases. The K_0 values for individual tests were constant over p' but varied slightly among tests. A hypothetical yield envelope, including both plastic and brittle sectors, is drawn for a single consolidation state based on the response of remoulded soils (e.g. Wood 1990).

the post-failure regime. All tests were run under drained (constant P_{H_2O}) conditions.

Both brittle and ductile responses were achieved. Initial brittle failure occurred only when p' in the sample was less than two-thirds of that responsible for consolidation. This response is predicted by tests on remoulded sediments, where it is equated with the stress path along which there is neither compaction nor dilation (Wood 1990). Tests run at higher p' resulted in ductile failure with a constant q, which was assumed to be a critical state despite incomplete data on volumetric strain. With this assumption, the tests show that, similar to the response seen in soil mechanics, the ratio, q/p', at critical state is nearly constant at 0.9 for less than 10 MPa. At higher p', however, this ratio clearly decreases, to about 0.5 at 30 MPa (Figure 6.14). Contradicting our results, undrained tests on generally similar clay-rich sediments resulted in an apparently constant ratio of 0.5 to 0.6 over a range of p' from 5 MPa to 60 MPa (Yassir 1990). If critical state and ductile failure conditions are similar, a decrease of pressure sensitivity with increasing pressure is expected from past rock mechanics experimentation, as in our results.

In the tests for which failure was ductile, the samples exhibited variable post-yield strain hardening. Although the test paths were not designed to probe this behaviour, it is clear that q associated with strain hardening increases with increasing p', as would be expected from the divergence of the K_0 line and the critical state line with increasing p' (Figure 6.14). The few data on strains during this phase do not show such a clear relationship, but there is no clear increase in the volumetric strain with increasing p'. The volumetric strain remained near 2% for constant p' tests over a range from 5 MPa to 31 MPa. These few data would extend into the tectonic stress range the assumption from soils mechanics studies that the volume–p' curves for uniaxial consolidation and for critical state are approximately parallel (Wood 1990). Axial strains associated with this stress path possibly decrease modestly with increasing p', ranging from about 9% at a p' of 5 MPa to about 5% at 31 MPa. However, because q approaches its critical state value asymptotically when plotted against ε_a, this strain is poorly defined.

Perhaps the most geologically significant observations concerned the nature of the post-failure deformation with changing p'. Specimens that were ductiley failing at critical state and subjected to increasing p' during continuing strain supported an increased q such that the stress path followed the critical state line. Although an increase in p' together with a very low applied axial strain rate could carry the stress path below the critical state line, samples from accretionary prisms are replete with structural features indicating continuing failure. This suggests that the ratio of strain rate to p' increase is sufficient to maintain a failure condition. Post-failure increases of p' in prisms and most other zones of crustal shortening are expected from tectonic thickening, which increases the vertical stress, unless offset by pore-pressure increases.

In contrast, reduction in p' after a specimen had reached a state of ductile failure led to the development of a discrete shear zone and to a stress path that basically moved down the critical state line (Karig 1990). This behaviour appears to involve the generation of a ductile instability that results from the localized weakness associated with the dilative deformation within the incipient shear zone. This behaviour can be viewed as brittle, but the sediment has satisfied the Coulomb criterion only at the condition where the Coulomb (brittle) and critical state (ductile) failure envelopes merge (Figure 6.14).

This load path has great geological significance, particularly in accretionary prisms, because sediments that are at ductile failure are suspected to be subjected to pulses of pore fluid, with pressures high enough to lower p'. Such pore-pressure fluctuations may result in the seemingly contradictory observations of brittle and ductile deformation in these sediments.

Experiments on natural sediments
Natural sediments differ from their synthetic analogues primarily in the development of cementation, which will be shown to introduce significantly different strain characteristics than those observed in synthetic sediments. Experiments on natural sediments have been run under stress conditions appropriate to deformation in prisms, but these usually have been oriented toward rock strength and the gross nature of deformation at failure (e.g. Hoshino 1981). Relatively less attention has been paid to the relationship among porosity, applied stress and resultant strain, as is done in soil mechanics.

Because of the similarity in lithology and testing schemes, the results of tests on a suite of silty mudstones from the Nankai Trough and the adjacent prism provide a direct comparison with the results from the tests on synthetic clays described previously. These natural samples were collected from DSDP Site 582 in the Trough and from ODP Site 808 in the toe of the prism. Samples from Site 808 came from both beneath and above the basal décollement.

Sediments from beneath the décollement are believed to have experienced only uniaxial strain, and test results on these have been described in the discussion of deformation in basinal sediments. Unfortunately, our test paths involved only uniaxial strains and no information is available for stress paths leading to either brittle or ductile failure. Samples from above the décollement were subjected to more relevant test paths, but these sediments were probably already at or near a state of ductile failure in the prism. Thus, in some respects the results of these tests cannot be substituted for those from tectonically undeformed analogues.

The principal objective of tests on samples from above the décollement was to estimate the *in situ* stress, with the rationale that the *in situ* stress was a failure state and close to the brittle–ductile transition. This required definition of the critical state line and the Mohr envelope for each sample, at least in the vicinity where the two criteria meet. An ancillary objective was the exploration of post-failure behaviour, especially the role of diagenesis on the strain fabric. This discussion will concentrate on the second objective.

Results With the limited number of samples available, only a very general outline of the yield and failure criteria could be obtained from each section of drill core, but as wide a range of stress paths as possible was run to characterize brittle and ductile failure criteria and to define the plastic yield envelope, especially in the vicinity of the critical state condition.

A single critical state line was defined by the combination of ductile failure states from all core samples, as well as by residual strength data from sediments strained to brittle failure and beyond. This critical state line closely matches that developed from the analogous synthetic mudstones and also passes through the few critical state points obtained from disaggregated and reconsolidated natural mudstone (Figure 6.15). This agreement is not surprising if the component of strength due to diagenesis is removed during the laboratory deformation to critical state.

Characterization of the yield envelope is important because its intersection with the critical

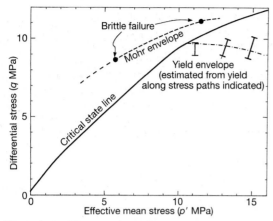

Figure 6.15 Effective mean stress (p') versus differential stress (q) plot for several samples of natural silty clay from a core at 844 mbsf in the Nankai prism toe showing sectors of the plastic yield and brittle failure (Mohr) envelopes near the critical state line. In contrast with the behaviour of the synthetic sediment of Figure 6.14, the two segments do not meet at the critical state line, which leads to complex brittle–ductile behaviour.

state line is one constraint on the brittle–ductile transition. For sediments that have been uniaxially consolidated, as in a basin, the yield envelopes are elliptical on a q–p' plane, with major axes oriented along the K_0 line. These envelopes are known to rotate if the principal stresses change significantly in ratio or direction during plastic deformation (Wood 1990). If a basinal sediment is subjected to a horizontal principal stress large enough to cause yield, the major axis of the yield ellipse can be expected to rotate toward the new stress condition in the q–p' plane, but the degree to which this is accomplished is unknown.

Using several uniaxial consolidation tests on other samples from the Nankai prism toe, Feeser, Moran and Bruckmann (1993) suggest that the yield envelope has rotated in phase with the changing stress orientation, and furthermore that, as in the case for the uniaxial strain, the stress state at the major axis represents the *in situ* conditions. Our data preclude such a large rotation and suggest that the major axis has rotated only partially, into near parallelism with the $q=0$ axis (Figure 6.15). This would argue against the method of Feeser, Moran and Bruckman (1993) to estimate *in situ* stress.

Samples subjected to triaxial deformation at relatively low p' failed brittly. Although there are not enough data to characterize the entire Mohr envelope, there is ample evidence that, for a given core section, the envelope has a significant curvature (Figure 6.16). The slope of the envelope estimated from Mohr stress circles at failure shows that $\tan \phi$ increases from about 0.2 near the brittle–ductile transition to at least 0.4 at p' about one-half the transition stress. The angles between σ_1 and the failure surface also reflect a decrease in $\tan \phi$ with increasing p', but in all tests these angles are less than predicted from the envelopes to the Mohr stress circles (Figure 6.17). These small fracture angles are most likely an effect of the fabric anisotropy during these tests, in which σ_1 was parallel to bedding.

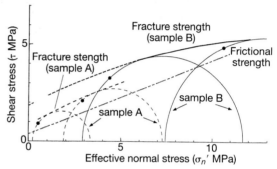

Figure 6.16 Plot of shear stress versus effective normal stress for two natural silty clay core samples from different depths in Hole 808C in the Nankai prism toe, showing peak (fracture) strength and residual (frictional) strength relationships. Sample A was from a depth of 685 mbsf with a porosity of 34% whereas sample B was from 844 mbsf with a porosity of 29%. The two Mohr envelopes are constrained only for short segments by two tests on each sample, but it is clear that the slope is lower for sample B, which was tested at stresses closer to its brittle—ductile transition. The Mohr envelope for sample B has been extrapolated to lower stresses using the data from sample A and assuming that Mohr envelopes for similar lithologies, but of different porosities, are self-similar. The filled small circles denote the stress state on the fracture surface, which always developed at a smaller angle to σ_1 than would be predicted from the Mohr envelope, probably reflecting sediment anisotropy. The Mohr envelope to stress states at residual failure on all tests was a straight line, coincident with the critical state line.

Both measures of $\tan \phi$ are significantly lower than generally assumed (e.g. Byerlee 1978) for the range of stresses expected in the prism toe. Such low values cannot be attributed to anomalous mineralogy because these cores contained only a small fraction of clays with very low $\tan \phi$ values, such as montmorillonite (section 1.2.4). The low values are better attributed to the fact that our

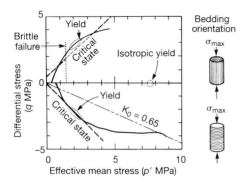

Figure 6.17 Synopsis of mechanical test data on a naturally deformed silty clay from 516 mbsf in Hole 808C through the toe of the Nankai prism. In this plot, differential stresses are defined as $\sigma'_h - \sigma'_v$. The curved stress paths are from uniaxial consolidation tests on samples with axes parallel and perpendicular to bedding. These, especially the sample cut perpendicular to bedding, show significant excess strength at yield (Figure 6.6), probably from cementation (bonding), which is destroyed at high strain where the path rejoins a K_0 line. Yield stresses (dotted ovals), although poorly defined in the uniaxial strain tests, are much less than for an isotropically consolidated sample (dotted circle). This indicates a yield envelope with a major axis closer to the isotropic stress line than to the K_0 line, contrary to the conclusions of Feeser, Moran and Bruckmann (1993).

tests were run at p' levels relatively close to the brittle–ductile transition, whereas many, if not most tests on lithified sediments are conducted at p' levels far below the transition, where the Mohr envelope has a steeper slope.

Frictional strengths for the Nankai mudstones were obtained by continuing the brittle failure tests to a state of residual strength. With stresses on the failure surface that were corrected for the changing area of that surface as displacement proceeds, the Mohr envelope tangent to all the Mohr stress circles defines a straight line with a slope of 0.4 (Figure 6.16). This slope, the coefficient of sliding friction, is again lower than generally assumed for lithified rocks but is typical of sediments of similar composition (Kenney 1967).

Tests run at values of p' close to the brittle–ductile transition show a complex combination of post-yield strain hardening followed by strain-softening and development of a shear zone. The most complete set of data for a single Nankai core sample requires that the intersection of the Mohr envelope with the critical state line lie above the intersection of the critical state line with the yield envelope (Figure 6.15). This is quite a different behaviour than that of a synthetic sediment, but is similar to that suggested for the few other natural porous sediments for which data are available (e.g. E.T. Brown and Yu 1988).

In brief, the range for brittle failure of natural sediments may extend somewhat closer to the p' of consolidation, but it is still difficult to understand how sediments deforming within prism toes could reach initial brittle failure without extensive erosional unloading or very large increases in pore pressure. Strongly compactive Lagrangian paths, as in the western Nankai toe, argue for initial ductile failure, but it may be possible that less compactive paths, as in the drilled transect of the Barbados prism, could produce a dual strain-hardening–strain-softening response.

The difference in response of natural and synthetic sediments most likely results from differences in diagenesis, but only general behavioural traits can now be suggested. Although the very different strain states and different test programmes for samples from Site 582 and above the décollement at Site 808 preclude direct comparisons, several lines of evidence indicate that sediments in the prism toe have undergone a decrease in cementation (Karig 1993, in preparation). On the other hand, some cement-related strength must remain or have been regenerated in these samples (e.g. Feeser, Moran and Bruckman 1993). For example, tests leading to ductile failure show large, continued, volume loss well after the stress path has reached critical state, implying a progressive breakdown of cement. This contrasts with only small lags in similar tests on synthetic sediments.

Some idea of the amount of volumetric strain resulting from cement breakdown during deformation from uniaxial consolidation to ductile failure is seen by comparison of curves of porosity versus σ'_v for natural sediments and their uncemented analogues. The porosity differences, which are approximate measures of volumetric strain between the two groups at the same stress are of the order of 10% at very low stresses but decrease sharply with increasing stress (Figure 6.1). This strongly suggests that, as basinal sediments are deformed to ductile failure, the cement-related component of volumetric strain should be greatest at the shallower depths and decrease downward.

Deformation-related bonding loss may also explain some examples of compactive shear zones.

In samples failed brittly and subsequently deformed under increasing p', strain is initially confined to within the shear zone, except for small elastic strains in the stronger material outside. Although not quantifiable, the data indicate that, as p' increases toward the brittle–ductile transition for the unbroken and still cemented sediment, the porosity in the shear zone decreases to less than that outside. This reflects the fact that the total strength inside the shear, which is approaching that outside, is borne through porosity, whereas the strength outside is partly supported by cement.

Although the previous discussion has been restricted to the mechanical behaviour of clay-rich sediments in the accretionary prism toe, that of sandy sediments is quite similar. Increasing p' leads to ductility (Rutter and Hadizadeh 1991), and strength reflects both porosity and diagenesis (e.g. Dunn, La Fountain and Jackson 1973). One significant difference is that cataclastic comminution of sand grains occurs in the post-yield, strain-hardening phase of deformation, with significant porosity reduction (Aydin and Johnson 1983; Wong, Szeto and Zhang 1992). In some respects, however, this is analogous to the effects of destruction of bonding in clays.

6.3.6 Implications for the deformational histories of accreted sediments from observation and experiment

The spatial and temporal distribution of the various structural features and related observations cannot by themselves define the deformational history of sediments in accretionary prism toes, but can be interpreted in the light of results of experimental deformation that constrain sediment behaviour. However, the paucity of data and limited observations will allow only a partial picture to be constructed, by necessity biased toward the three examples of prism toes from which data have been presented. The framework developed from this discussion is certainly speculative, but will be tested as new findings come to light.

Eastern Nankai

Interpretation of the observations from a prism toe in light of results from mechanical experiments on silty clays can be made most easily from the eastern Nankai transect, with its large data base. Sediments subducted beneath the décollement, as well as those within that zone, are silty clays directly comparable with the sediments used in our experiments. Although the strata accreted above the décollement include a large fraction of sandy sediments, the experimental data are applicable to the clay-rich fraction which represents the lower half of the accreted section as well as the best preserved part of the rest.

The evidence available indicates that the sediments subducted beneath the décollement remain at the stage of uniaxial strain developed in the basinal setting of the trench wedge or oceanic setting (Figure 6.18). This evidence consists largely of the lack of any significant biaxial strain superimposed on the uniaxial consolidation strain, even very close to the décollement. However, maintainance of a uniaxial strain state, with a vertical σ_1, does not preclude fluid flow or further vertical consolidation.

The low salinity of pore fluids observed below a décollement (Taira, Hill and Firth 1991) has been cited as evidence for basinward fluid flow from a deeper, more arcward source (Kastner *et al.* 1993). Such flow need not change the mechanical state of the sediments beneath the prism, but there is also evidence from reconsolidation tests on samples from beneath the décollement that the yield envelope has been collapsed nearly to that of the uncemented state (Karig 1993). This collapse would result in porosity reduction unless there was a compensating reduction in p', as through a pore pressure increase. The limited data suggest that both a porosity decrease and a pore pressure increase have occurred, but a more quantitative statement would require comparative data on the condition of the subducted sediments before they were affected by the décollement.

The décollement at the eastern Nankai transect appears to be a zone of nearly complete stress decoupling. The structural fabric immediately above that zone defines a subhorizontal σ_1 (Lallemant *et al.* 1993), requiring that the shear stress in the décollement be extremely low. Such low shear stresses cannot be achieved without very high pore pressures within the décollement because the coefficient of friction at a state of residual strength is about 0.4, far too high to account for the observed stress rotation or for

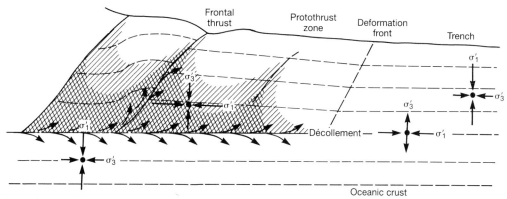

Figure 6.18 Synopsis of stress and strain relationships along various path lines in the eastern Nankai prism toe. Sediments beneath the décollement remain at a state of uniaxial strain with a vertical σ_1. Stresses along path lines above the décollement must rotate rapidly near the décollement tip to a state where σ_1 is nearly horizontal. Vertical stresses at the décollement tip probably become tensional in response to very high pore fluid pressures. The light, arrowed lines suggest fluid diffusion away from a high-pressure décollement, although diffusion rates may be relatively low. The distribution and mode of strain is as in Figure 6.12.

the very low taper (Davis, Suppe and Dahlen 1983).

Evidence for high pore fluid pressure within the décollement, as well as for an outwardly decreasing excess pressure field, is offered by the results of reconsolidation tests on samples below the décollement (Karig 1993) and from the dihedral angles between minor shear surfaces above the décollement (Lallemant *et al.* 1993). This interpretation of very high pore pressures within the décollement does not contradict the geochemical evidence for only limited fluid flow along that zone (Kastner *et al.* 1993); it points instead to a décollement with relatively high permeability, presumably along fractures, and inflated by a source of high-pressure fluid at depth within the prism (Figure 6.18). A lack of fluid egress from the shallow décollement, where it is contained by low-permeability sediments, would inhibit the escape of fluid from that zone and could account for the low fluid flow. Both the geochemical and mechanical data support a model of minor fluid diffusion upward and downward from the high-pressure décollement.

This high-pressure–low-flow model for the décollement also provides an explanation for its propagation, at least at the eastern Nankai transect, where the tip of the décollement extends basinward of the deformation front. High pore pressure within the décollement will create a halo of elevated pore pressure around the décollement tip. Within this halo the rising P_{H_2O} will cause a greater reduction of σ'_v than of σ'_h (see equation 6.6) until the sediment fails in tension along a horizontal surface (Figure 6.18). Thus the décollement would propagate as a tensional crack, creating a weak zone that subsequently becomes the locus of shear strain. This would explain not only the propagation of the décollement basinward of any horizontal shortening but also the lack of deformation bands preceding the brittle shear strain. A similar mode of décollement initiation was suggested earlier by Voight (1976), in a comment on a paper by Roberts (1972).

The sediments above the décollement, and accreted to the toe of the Nankai prism at the eastern transect, were clearly at a state of failure where sampled at Site 808. At what positions along their streak lines the sediment elements reached failure, be it brittle or ductile, remains an important question. The compactive strain observed in these sediments in a Lagrangian frame is the basis for assuming an initial ductile failure, at something approximating critical state. Estimation of the loci within the prism where failure is reached is important because the sediment mechanical response should change markedly at those loci. Beyond them, only increases in p' will result in further porosity reduction, and fluctuations in p' should become very effective in causing alternations between brittle and ductile failure models.

Volumetric strain and p'

Unfortunately, neither the amount of strain necessary for a natural sediment to deform from a state of uniaxial strain to a critical state nor the distribution of diffuse strain in the prism toe is well bracketed. However, even approximations provide useful insights and the approach outlined will clearly be more powerful as the data improve.

The volumetric strain of a sediment element deforming from a state of uniaxial consolidation to critical state can be estimated fairly well for highly porous, uncemented soils if the change in p' is known. The e-log p' curves for the two states are approximately linear and parallel over that low stress environment (e.g. Wood 1990), but it may not be valid to extrapolate these to stresses relevant to most of the prism toe. Moreover, if natural sediments have a significant component of strength due to cementation at their consolidation state, as has been claimed previously, and if some fraction of this strength component is lost during deformation to the critical state, as is indicated by results of tests in our laboratory, there will be an added component of volume strain due to decementation.

There are presently no experimental volumetric strain data for natural, uniaxially consolidated sediments that have been deformed from that state to critical state, but such tests on synthetic silty mudstones suggest compactive volume strains near 2% for constant p' paths over a wide range of p'. Stress paths with increasing p' would generate higher compactive volume strains. The argument was made earlier that the volume strain due to cement destruction could range from as high as 10% at near-surface stresses to perhaps several per cent at a depth of 1 km.

Volumetric strains in the prism toes of both the eastern and western Nankai transects have been estimated from seismic and drill data (Morgan, Karig and Manatty in press; Morgan and Karig in press) but in both areas a number of assumptions were made that reduce the quantitative reliability of the estimates. The most reliable constraint on volumetric strain, based on porosity changes along path lines, is that, at the rear of the PTZ on the western transect, the volume strain reaches about 7% near the décollement and increases upward. Although the eastern transect has no reference site, the 9% volume strain calculated from the porosity difference across the décollement should be a minimum value for a path line just above the décollement. The values of p' along path lines through PTZs are even less well known, but have been suggested to increase significantly along the western Nankai transect (Karig 1990), which would increase the volume strain to ductile failure. Smaller increases or even decreases in p' are suspected in the eastern Nankai transect, which could still result in ductile failure but with reduced strain. With the present poor constraints, the best that can be said is that the PTZ in the Nankai prism lies largely in the strain-hardening field between uniaxial consolidation and critical state.

The volume loss generated as the accreted sediments deform toward critical state can also explain the sharp upward decrease across the décollement in Hole 808C on the eastern transect. In other words, the much higher q in the accreted sediments than in the uniaxially consolidated sediments beneath the décollement results in greater volume loss. This explanation is more logical than an appeal to higher pore pressure beneath the décollement because there is evidence for very high pore pressure in the region just above the décollement as well (e.g. Lallemant et al. 1993).

More difficult to explain is the low porosity of the mudstone within the décollement at Site 808. Byrne et al. (1993) have appealed to porosity reduction during transient pulses of high-pressure fluid, but brittle shear alone should engender dilative strains. An alternative explanation is that these sediments have become more completely decemented than those above and thus must have a decreased porosity to balance the applied stress.

Eastern Nankai frontal thrust

The mechanical behaviour of the frontal thrust on the eastern transect is enigmatic, but clearly differs from that of the décollement. Evidence for fluid flow, or even high pore fluid pressure is lacking, but creation of such a brittle shear zone along a compactive deformation path is not consistent with sediment mechanical behaviour as presently understood. Evidence has been presented for ductile to transitional strain precursors to the development of the fault and of the concentration of diffuse brittle shears beneath the fault, but the cause for the transition is not obvious. A cryptic pulse of high-pressure fluid moving up the zone of folding and ductile shear would suffice mechanically and might not leave an observable geochemical trace. Alternatively, the greater reduction of strength through bond destruction in the nascent failure zone might create a zone of weakness that, for some reason, failed as a brittle rather than a ductile shear.

Eastern and western Nankai comparisons

The differences observed in the structural character between the eastern and western transects of the Nankai prism toe lead to a fairly consistent interpretation of a more ductile response in the west due to generally lower excess pore fluid pressures within that prism toe. Because of the similarity in the bulk lithologies of the prism toes of the two transects, the mechanical characteristics within the prisms, as well as those at the décollements, are assumed to be similar, which would require that the lower taper to the east be due to higher fluid pressure within that décollement. The higher pore pressure along the eastern décollement has already been attributed to the location of the décollement tip within a thick, low-permeability mudstone section, in contrast to the location of the western décollement tip along the contact between the mudstone and the more permeable trench-wedge turbidite section.

Another difference between the two transects that might be attributed to pore pressure differences is the much greater width and apparent ductile strain within the western PTZ. One possible line of reasoning based on sediment mechanical behaviour is that higher pore pressures at equivalent points along path lines in the eastern PTZ would lead to a reduction in p', a smaller increase in q and much lower compactive volume strain, resulting in a failure mode lying closer to the brittle–ductile transition. This suggestion, that the development of a PTZ reflects the stress paths of accreted sediments to a failure state, is attractive but awaits more information on the loci of failure within the prism toe.

Nankai and Barbados comparisons

A comparison between the Nankai and Barbados prism toes may further illuminate the effects of pore fluid pressure, but equally illustrates the lack of simple relationships. Despite the much higher clay content in the latter, the observed deformation is much more brittle than that in either Nankai example. Moreover, there appears to be a total lack of a PTZ in the Barbados prism. Both of these observations could reflect a more rapid increase in P_{H_2O} with respect to horizontal strain and lead to a more brittle, weaker prism. This must be balanced against the small taper, which requires a relatively weak décollement. As the toes of the Barbados and eastern Nankai examples have about the same gross geometries, the reasoning above would demand that the shear stress along the décollement of the former be even less than that along the eastern Nankai, which was argued to be very low.

6.4 CONCLUSIONS

The principal conclusion to be drawn from this review of tectonic deformation of sediments is that, despite the recognition of and interest in the mechanical behaviour of sediments, much more quantitative information is needed from *in situ* observations and from experimental deformation. Moreover, this information must come from focused and problem-oriented investigations. Particular attention should be paid to deformation paths of the more porous sediments, which display larger volume changes and a wide range of deformational modes. To reiterate a statement made at the outset, careful integration of observation, experimentation and modelling will be required.

Much has been made of the role of pore pressure in deformation, but very few pressure data are available from critical zones such as prism décollements. Such measurements are difficult, expensive and time consuming, but only perseverance will lead to cheaper and more efficient methodologies. Pore pressures and other hydrological parameters depend on *in situ* permeability data, but these are even rarer. The recent introduction of borehole flowmeters (Morin 1992) to acquire bulk permeability data represents just the sort of innovative approach required.

Measurement of *in situ* stress is advancing rapidly, but should become more clearly focused on the relationship between stress paths and deformation fabric. It would, for instance, appear most logical to address the problem of regional joint development with combined *in situ* stress studies and joint mapping in a carefully chosen range of settings.

Experimental studies of deformation in porous sediments is needed in many areas, but clarification of differences between synthetic or uncemented sediments with their natural, cemented analogues would seem of high priority. Better access to carefully preserved natural samples of a variety of lithologies from well-characterized settings would go far to ameliorate this problem. This in turn suggests that innovative interactions between those concerned with geological stress paths and the petroleum industry be developed, because opportunities from academic drilling projects will remain limited in number and scope for many years.

CHAPTER 7

Fluids in deforming sediments

KEVIN BROWN

7.1 INTRODUCTION

This chapter focuses on the hydrogeological processes that occur in deforming sediments. Muds and sands are given particular attention because of their abundance, especially in many active tectonic environments. In these materials the deformation is predominantly accommodated by grain-boundary sliding (section 1.1.1), although brittle fracture becomes progressively more important with burial. Although the general nature of this chapter means that some aspects cannot be discussed in as much detail as they warrant, its purpose is to illustrate some of the main processes through which tectonic and hydrogeological systems are coupled. Emphasis is given to how both pore fluid pressure and permeability interplay with tectonic and hydrogeological processes in a variety of tectonic environments.

In the deeper levels of hydrogeological systems, diagenetic processes and metamorphic mineral transformations can release significant quantities of fluids, which strongly influence both structural and geochemical processes as they are transported upwards. Although these fluids originate at depth they migrate towards the surface, where they mix with the fluids from shallow sources. In near-surface sediments such fluids play a vital role in deformation and in diagenesis. In fact, the addition or subtraction of chemical components in sediments is usually attributed to fluid movement.

Manifestations of such fluid flow are commonplace, and take on a variety of forms. Diffuse, upward movement of water and methane can produce substantial volumes of gas hydrates in both passive continental margins and accretionary prism environments (Kvenvolden 1985, 1988; Kvenvolden and Kastner 1990). Although such diffuse flow is important, structures such as faults can in some instances focus the fluid flow whereas in others they can act as barriers (Wang et al. 1990). In both cases the structures help determine the deposition of ores and the accumulation of hydrocarbon deposits. Regional fluid-flow systems tend to be directed out from subsiding basins and developing mountain belts, with flow being focused along permeable stratigraphical horizons and faults. This flow may carry both hydrocarbons or mineralizing agents, and it has been ascertained that certain types of ore deposit tend to cluster around faults near to the margins of these tectonic regions (Clendenin and Duane 1990; Leach and Rowan 1986).

In the subsurface, flow of fluid out of mountain belts and basins can be associated with significant regional and local temperature and pore-water geochemical variations (Smith and Chapman 1983; Hitchon 1984; Fisher and Houndslow 1990a,b; Gieskes et al. 1990; Gieskes, Vrolijk and Blanc 1990; Langseth and Moore 1990). In the marine environment, surface manifestations of the focused component of advective fluid flow include fluid seeps marked by biological communities. These communities tend to cluster along the surface traces of active faults, feeding chemosynthetically on the methane and hydrogen sulphide emanations. The effects are well seen in tectonically active systems such as accretionary prisms (e.g. Suess et al. 1985; Kulm et al. 1986; Moore, Orange and Kulm 1991) but many examples of fluid seeps have also been reported from passive continental margins and basins such as the North Sea (Hovland and Judd 1988). Substantial reef-size carbonate accumulations (bioherms) may develop (Hovland 1990)

The Geological Deformation of Sediments Edited by Alex Maltman Published in 1994 by Chapman & Hall ISBN 0 412 40590 3

if they form above the calcite compensation depth.

Other large-scale manifestations of the influence of fluids in the subsurface include mud diapirs and diatremes, which are associated with, among other factors, methane gas generation at depth (Hedberg 1974; K.M. Brown 1990). Diapirism involves both fluid flow through pores and the large-scale mobilization of sediments (Hovland and Judd 1988; K.M. Brown 1990; K.M. Brown and Orange 1993). Some diapirs reach the surface in large numbers to form substantial sized mud volcanoes and pockmarks on the sea bed. For example, K.M. Brown and Westbrook (1988) reported many hundreds of mud volcanoes on the Barbados accretionary prism. These commonly have diameters of 1–3 km, with maximum diameters of up to 10 km. Submersible investigations above potential diatremes in front of the Barbados prism indicate that they are the site of large-scale advection of warm, methane-rich fluids originating in and beneath the sediment wedge (Le Pichon et al. 1990). The substantial quantities of advecting fluid involved suggest considerable interconnectivity in the hydrogeological system, possibly associated with long-distance, lateral fluid-flow along major structural features, such as the basal décollement and/or permeable stratigraphical horizons in the under-thrust sediments (Westbrook and Smith 1983; Brown and Westbrook 1988). Mud volcanoes and diapirs are not the only hydrogeologically active diapiric features. Hydrated ultrabasic rock originating in the mantle wedge has risen into the Mariana fore-arc to form substantial serpentinite diapirs or volcanoes (Fryer, Ambos and Hussong 1985). These serpentinite diapirs have been found to act as advective conduits for deep-seated fluid sources (Fryer et al. 1990).

By determining patterns of pore-pressure fluctuation and diagenesis, the hydrogeological system exerts a strong control on the pattern of tectonic development on both small and large scales. The significance of the interaction between the hydrogeological and tectonic systems in both active continental and marine environments has long been appreciated, dating from the early ideas of Hubbert and Rubey (1959) on the significance of high fluid pressures for the movement of thrust systems, through to the concept of fault valving espoused by Sibson (1988). In accretionary prisms fault movement is dominated by aseismic creep on major faults, which also tend to be hydrogeologically active (Langseth and Moore 1990) and, thus, there is a strong potential for coupling between the hydrogeological and tectonic systems. Indeed, the predicted shape of accretionary prisms, when modelled as Coulomb wedges, is sensitive to this coupling (Davis, Suppe and Dahlen 1983; Dahlen 1990). Because of their sensitivity to hydrogeological activity, much active research is being focused on accretionary prisms and this work is used to illustrate certain important points throughout this chapter.

In other tectonic environments, numerous observations link fluctuations in pore fluid pressure and permeability to deformation during episodic faulting (Roeloffs 1988; Sibson 1981a; J.C. Moore et al. 1991; Tobin et al. 1993). For example, perturbations in hydrogeological activity persisted at decreasing levels for up to a year after the Kern County and Idaho earthquakes. This activity included high-pressure artesian activity, dramatic well-pressure increases, increased spring discharge, the formation of lakes and increased mine flooding (Sibson 1981b). Conversely, the initiation of seismic activity has been linked to changes in the hydrogeological environment. In the 1960s, the injection of fluid wastes into the fractured Precambrian crystalline basement beneath the Rocky Mountain Arsenal near Denver triggered a series of earthquakes with epicenters that progressively migrated away from the region of fluid injection with time. This was eventually attributed to the propagation of increased pore pressure into the fractured rocks around the boreholes (Hsieh and Brędehoeft 1981).

These examples illustrate how tectonic and hydrogeological systems are closely coupled through their respective effective stress–pore-pressure parameters. Further examples of the relationships are discussed elsewhere (e.g. sections 2.2.3 and 6.2.3). The overall importance of fluid pressures for sediment deformation is underlined by the fact that they arise in every chapter of this book! In the following section, the relevant basic hydrogeological parameters are explained and they are then discussed with reference to a number of real geological situations.

7.2 SOME BASIC HYDROGEOLOGICAL CONCEPTS

M.K. Hubbert indicated in 1940 that, like heat, fluid flow is driven by a **potential gradient** and that flow always occurs from regions with a high potential energy to regions with low potential energy. Hydrogeologists use **hydraulic head**, H, because this parameter takes into account the two main factors that drive fluid flow: (i) topographic effects (i.e. fluid flows down hill owing to differences in gravitational potential); and (ii) fluid flows in response to excess pressure gradients. The excess pressure component, commonly referred to as overpressuring (section 1.2.5) is the difference between the measured pressure and the expected hydrostatic pressure component (Figure 7.1a) resulting purely from the load of the column of fluid above. The excess pressure, therefore, essentially contains any 'tectonic component' of the fluid pressure. Other mechanisms by which overpressures can be generated are outlined in section 2.3.4. A formal evaluation of the relation of hydraulic head to potential energy is given in many textbooks on hydrogeology (e.g. Freeze and Cherry 1979), and it is not necessary to restate the details here. However, it is useful for the geologist interested in fluid flow to understand what is meant by a **head gradient**.

If a hollow tube or manometer is inserted into a saturated porous medium at a point, water will flow up the tube to a certain level (Figure 7.1b). The hydraulic head, H (subsequently called just **head**), is the sum of two components

$$H = Z + \Phi, \quad (7.1)$$

where Z is the elevation head of the point above an arbitrary datum and Φ is the pressure head (Figure 7.1b). The various heads H, Z and Φ all have dimensions of $[L]$. The pressure head, Φ, can also be written as $P_{H_2O}/\rho_w g$, where P_{H_2O} is the actual measured pore pressure at the base of the manometer and ρ_w and g are, respectively, the density of the fluid and the acceleration due to gravity. For comparison, the column of fluid that would correspond to the $P_{H_2O_e}$ component is also marked in Figure 7.1b. Fluid flow is driven by a difference in head $(H_1 - H_2)$ between two points in a saturated porous medium (Figure 7.2) or

$$\Delta H = H_1 - H_2 = (Z_1 - Z_2) + (\Phi_1 - \Phi_2). \quad (7.2)$$

In a hydrostatically pressured formation the fluid pressure is equal to the weight of the fluid

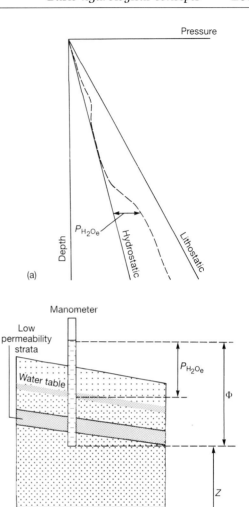

Figure 7.1 Schematic figure illustrating the meaning of (a) excess pore pressure, $P_{H_2O_e}$, and (b) hydraulic head. The total hydraulic head comprises an elevation head, Z, plus a pressure head, Φ, in units of length. Compare with Figure 1.4

column above, and the level of the fluid in the manometer would rise to be level with the water table. In Figure 7.2a two manometers are shown inserted into a hydrostatically pressured formation on a topographic slope. Although there is no excess pressure, water will still flow down slope because the topography produces a head difference (i.e. $H_1 > H_2$). Topographically driven regional groundwater flow can be significant in systems such as foreland-thrust belts or sedimentary basins. For example, Smith and Chapman

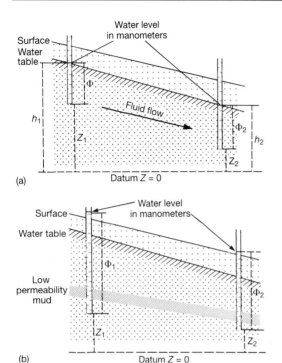

Figure 7.2 Schematic figure illustrating the meaning of (a) a topographically generated head gradient in a normally pressured formation and (b) a head gradient in an overpressured, confined aquifer.

(1983) modelled fluid flow in a region with a topography similar to that of the foothills of mountain ranges in both eastern and western North America. Their model was applied to a hydrostatically pressured system in which the water table is on average 1 km higher in the mountains than in the foreland. They showed that regional fluid flow was generally directed down out of the mountains and up through the foreland sequences. The regional flow pattern greatly affected the hydrogeological system and heat-flow profile down to depths of 5 km or more, reducing surface heat flow in the mountains and increasing it in the foreland. Topography alone can, therefore, exert a significant control on the nature of the regional fluid-flow patterns in subaerial systems, particularly in the upper, hydrostatically pressured part of the crust.

In contrast, in the marine environment, where the sediment deformation is completely submerged, the effects of sea-floor topography can often be ignored in the description of head because the top of the 'effective water table' is consistently at sea-level. It is not uncommon for workers modelling fluids in accretionary prisms, for example, to refer just to the excess pressure component (Shi and Wang 1988).

Fluid flow in response to a head gradient across a porous, saturated, sedimentary matrix can be approximated by Darcy's law (Darcy 1856). In its most basic form Darcy's law states

$$\frac{Q}{A} = K \frac{dH}{dl}, \qquad (7.3)$$

where Q/A is the rate of fluid flow per unit area or linear velocity, K is hydraulic conductivity and dH/dl is the gradient in hydraulic head.

The constant of proportionality in Darcy's law (equation 7.3), termed **hydraulic conductivity**, is a function of both the properties of the porous medium and the fluid (Hubbert 1956). Hydraulic conductivity is related to the intrinsic permeability of the medium alone, k, termed simply permeability in the following discussion, by

$$K = k \rho_w g \omega, \qquad (7.4)$$

where ρ_w and ω are the density and dynamic viscosity of water respectively, and g is the acceleration due to gravity (e.g. Freeze and Cherry 1979). Hydraulic conductivity, K, has units of length per time, typically m s^{-1}, and permeability, k, has units of m^2.

Hydraulic conductivity is affected by the physical properties of water, which vary with temperature and pressure. Figure 7.3 is taken from Smith and Chapman (1983) and shows the decrease in the dynamic viscosity, ω, and density of water, ρ_w, with increasing temperature. From Figure 7.3 we can see that the viscosity of water will increase by a factor of approximately 3.8 and the density

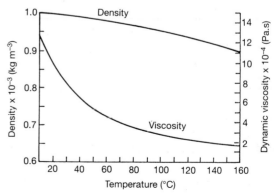

Figure 7.3 Variation in dynamic viscosity and density of water with temperature. (From Smith and Chapman 1983.)

by a factor of 1.05 as it moves from a temperature of 100°C deep in the earth to 20°C at shallower levels. If the permeability remains constant, this will result in a net decrease in the hydraulic conductivity by a factor of 3.6. Morin and Silva (1984) illustrated the effect of temperature on the hydraulic conductivity of an illite clay. When the data were corrected for viscosity changes due to temperature, the same intrinsic permeability was found for all temperatures.

Permeability, k, in rocks and sediments can range over many orders of magnitude. Therefore, permeability is typically at least one million times more variable than the other main physical factors affecting fluid flow, and hence its potential impact on deformation processes is substantial. For example, whereas topographic, tectonic and other effects readily produce fluid pressures greater than hydrostatic, the prolonged existence of large excess pressures in the subsurface requires the presence of low permeability units to retard free drainage to the surface. Indeed, as permeability becomes reduced, the required head or pressure gradient can rise considerably above hydrostatic levels, while maintaining only a low rate of fluid flux out of a buried formation (equation 7.3). Excess fluid pressures, or overpressures, are notated in this chapter as P_{H_2Oe} (e.g. Figure 7.1). The question of how excess fluid pressures are maintained over long periods of time is also addressed in section 2.3.4. The overpressure will make the fluid level in a manometer rise to well above the level of the water table (Figure 7.2b) and, by reducing effective stresses, can greatly affect how a porous material deforms. This fundamental effect of fluid pressure on the behaviour of a deforming sediment is emphasized in sections 1.2.5 and 2.2.3, and ramifications arise throughout this book.

7.3 FLUID SOURCES AND THE NATURE OF THE TECTONIC PROCESS DRIVING FLUID FLOW

7.3.1 General

Fluid sources or sinks are required to maintain the head differences that drive fluid flow. The fluid sources can lie outside or inside the sediment mass. Meteoric water falling on upland regions is an important external source of fluids for many subaerial hydrogeological systems. Head gradients in meteoric systems are commonly generated by topography (Figure 7.2a), which, in mountainous regions, can drive the fluid deep into the earth's crust (Smith and Chapman 1983). Tidal forces also may be significant in certain circumstances, particularly for shallow cyclic flow through the geochemically active regions near the sediment–water interface in marine environments, and for flow near the saline–fresh-water interface in coastal regions. Fluid density variations are significant driving forces in other environments. For example, convective hydrothermal systems in mid-oceanic systems are driven by temperature-related fluid density contrasts. In these convective systems, cool, dense sea water is drawn down into the sediment mass and then eventually expelled as less dense hot water, focussed in regions above the heat source. Salinity variations are responsible for other types of density driven flow, particularly around salt bodies in the subsurface, in semi-arid regions, and in lagoonal settings in regions such as the Bahamas (Whitaker and Smart 1990).

Internal fluid sources, mechanisms that commonly generate excess pressures, become significant at deeper levels in the earth. Fluids in the pore spaces of a sediment or rock sustain a certain proportion of the total overburden load. Even where topographic effects are not important, head gradients can still originate as a result of the spatial variation in the magnitude of loading of the overburden on the fluids (and, thus, fluid flow is intimately related to effective stress). A fluid source generally increases the proportion of the confining stress carried by the fluid (i.e. increases the excess pressure component), by increasing the volume of fluid in the pore space. Examples of this include hydrocarbon generation, smectite dehydration and aquathermal pressuring. In addition, the physical volume of the pore space available to contain the fluids may be reduced by processes of consolidation, cementation and pressure solution. In contrast, fluid sinks tend to reduce the load carried on the fluids either by increasing the physical volume of the pore space (examples of this are elastic swelling, dilatant deformation and dissolution of cements), or by reducing the fluid volume in the pore. Examples of the latter are cooling and contraction of pore fluids, and some mineral hydration reactions.

There can be both positive and negative deviations from a hydrostatic gradient in different regions in the subsurface as the productivity of the various sources and sinks varies. One control on the magnitude of excess pressure generated in the porous medium is rate of fluid production or uptake. Higher rates of fluid production or uptake will tend to lead to greater degrees of departure of the excess pore fluid pressures from hydrostatic values. The other important control on the magnitude of the excess fluid pressure fluctuations is the permeability of the porous medium, with substantial excess pressures normally being restricted to low-permeability formations. This aspect is discussed in detail below.

Various sources and sinks of fluid in the subsurface are considered in turn below. They are of three main types: mechanical, chemical and thermal. Some of the mechanical processes operating during deformation are transitory or cyclic in nature. These tend to redistribute fluids around the systems rather than cause large-scale advection out of the system. None the less, these transitory perturbations are important because they control the effective stress distribution and mechanical response during the period of deformation.

7.3.2 Consolidation and swelling

The principal mechanical source of fluids in the subsurface is the fluid expelled from consolidating porous sediments. On deposition, sediments may have porosities as high as 70–80%. As they are buried, effective stresses increase and consolidation commonly reduces porosities to below 30–40% at a few kilometres depth (Bray and Karig 1985). The rate of porosity reduction is generally rapid near the surface and decreases with depth. This reduction in pore volume is accommodated by the movement of fluids out of the porous sediment.

The volumetric response of poorly lithified sediments during burial can be shown as a stress path (Roscoe, Schofield and Wroth 1958; Schofield and Wroth 1968; Atkinson and Bransby 1978; and section 2.2). As the volumetric response of a sediment to burial along various stress paths is discussed in detail in sections 2.3.2 and 6.2, only a few concepts of direct hydrogeological interest will be outlined here. In an untectonized basin, sediments are typically subject to one-dimensional consolidation and as a sediment consolidates it will follow a path such as AB on Figure 7.4a. From the form of the void ratio, e (equation 1.2), and average effective stress, p' (equation 2.17), relationships, it can be seen

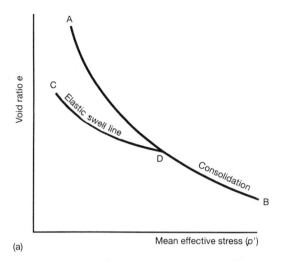

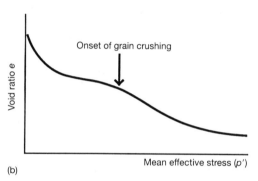

Figure 7.4 (a) The variation of void ratio (e) in sediments with mean effective stress (p'). The elastic swell (unloading) line and plastic one-dimensional consolidation path are shown. See also Figures 2.13 and 6.2. (b) The onset of grain crushing at high effective stresses causes a break in slope and an increased rate of volume reduction. Experimental studies, such as those of Zhang, Wong and Davis (1990), indicate that the minimum average effective stress for microcracking and grain crushing in medium- to fine-grained sands under hydrostatic conditions are affected by porosity and can be as high as 75–380 MPa (porosities of 35 and 21% respectively), although Zoback and Byerlee (1976) found that Ottawa sand (with a porosity of 31%) underwent grain crushing at effective hydrostatic confining pressures of as low as 50–75 MPa. In general, however, experiments show that the mean effective confining stress at which grain crushing and cataclasis are initiated is substantially reduced as the deviatoric stress increases, a significant factor for fault-zone permeability evolution in sandstone formations.

that sediments will consolidate at a decreasing rate as effective stresses increase during burial. Thus, fluid production from consolidation is most abundant early in the burial history, within the first 2–5 km of the sediment–water interface.

At high effective confining stresses the slope of the e versus p' relationship can change (Bishop 1966; Zoback and Byerlee 1976; Zhang, Wong and Davis 1990) with the onset of grain crushing in lithified sediments such as sandstones (figure 7.4b). Together with the increased activity of such processes as pressure solution, the onset of grain crushing could lead to a secondary phase of raised fluid production in deeply buried formations. Normally, however, visible signs of the large-scale crushing of grains is rare in the absence of increased shear stresses, and this mechanism may be only a local source of fluids, restricted to regions near faults.

Path AB on Figure 7.4a corresponds to sediments undergoing continuous consolidation under increasing average effective stresses. Where average effective stresses are reduced, by either the removal of overburden or raised pore fluid pressures, the sediments will swell and can exhibit an elastic volumetric response (Figures 7.4a and 2.13). In the elastic field the sediments are generally much stiffer and the slope of the swell line (CD) is lower, so that the volumetric response for a given change in effective stress is lower than for plastically deforming sediments (Figure 7.4). In the elastic field, reduced average effective stresses will cause the sediment to swell and take up fluid, whereas increased effective stresses will cause contraction of the void space and fluid expulsion.

Where the sedimentary sequence contains units with low permeabilities, fluid expulsion can be retarded, so that a large part of the overburden load is supported by the development of excess pore pressures. The raised fluid pressures retard further consolidation by reducing the average effective stress experienced by the sediment framework. Under these conditions the sediments are commonly said to be under consolidated and overpressured. Overpressuring extends the depth to which large volumes of fluid can be buried in an active tectonic system. Generally, significant overpressuring is associated with thick mud-rich sequences and rapid burial by sedimentary or tectonic processes. For example, in the Gulf Coast system, shales just under and above permeable sandstones consolidate normally, whereas away from the sandstone the muds become overpressured and undercompacted (Figure 7.5). These overpressured muds can act as pressure seals, because the high pore fluid pressures in them restrict vertical fluid

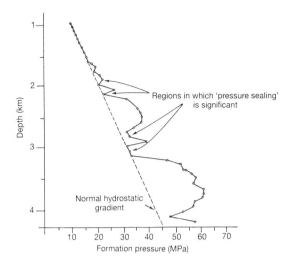

Figure 7.5 Overpressuring in muddy formations in the Gulf Coast and the formation of 'pressure seals' that restrict vertical flow out of less overpressured sands formations. In the sands much of the flow is constrained to be in a lateral direction. (From Magara 1981.)

movement so that fluids in the sands are forced to migrate laterally. Additional fluid production from a variety of chemical sources (e.g. smectite and hydrocarbon generation) associated with these muds has also contributed to the development of overpressures in the deepest regions of the basin (see below).

The geometrical relationships and tectonic processes at convergent margins differ from those in sedimentary basins, and this leads to differing patterns of fluid production. Modelling of consolidation-related fluid production within accretionary prisms (Karig 1990; Screaton, Wuthrich and Dreiss 1990; Bekins and Dreiss 1992) shows that it is most pronounced near the deformation front (Figure 7.6). Moreover, these models also indicate that a larger prism-taper angle (section 6.3.3) and a thin incoming sediment section lead to the expulsion of a greater volume of the fluid near the deformation front (Bekins and Dreiss 1992). This is because: (i) the large taper angle requires more rapid thickening near the toe of the prism than a small taper angle; (ii) a thin incoming sedimentary section tends to have higher porosity on average, and will therefore produce more fluid per unit volume than a thicker, less porous section. Mud-rich accretionary prisms are examples of systems in which rapid tectonic thickening commonly results in substantial overpressures.

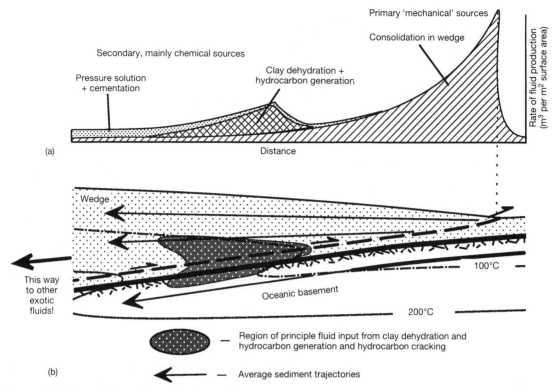

Figure 7.6 Schematic figure illustrating the relative contributions of the various fluid sources to the hydrogeological system in accretionary prisms. Although the contribution of the chemical sources of fluid in the deep regions of the accretionary system is volumetrically much smaller than the initial phases of consolidation at the accretionary toe, the low permeability at depth can promote significant overpressuring. See Figure 6.9 for general features of accretionary prisms and section 6.3.2 for further discussion of sediment trajectories.

In accretionary prisms, high plate-convergence rates allow burial rates of underthrust sediments (sediments beneath the basal décollement) to substantially exceed those of normal sedimentary basins, resulting in considerable overpressuring and undercompaction (Karig 1990). These underthrust sediments can be transferred to the prism at depth by duplex formation and underplating, potentially delivering large quantities of water and hydrocarbons to the base of the accretionary system. The contrast in kinematics, lithology and consolidation histories between overthrust and underthrust sediments therefore gives rise to very different fluid production histories. J.C. Moore and Vrolijk (1992) review current thinking on the production and interaction of fluids within accretionary prisms.

7.3.3 Transient fluid sources and sinks generated by faulting

Perhaps the most obvious coupling between hydrogeological and tectonic systems occurs during faulting events, when potentially significant transient pore-pressure fluctuations cause the local redistribution of fluids around active structures. Typically, such transient hydrogeological disturbances may affect the local pre-, syn- and post-faulting flow regime over a period of days to a few years (Roeloffs 1988; Roeloffs et al. 1989). For example, perturbations in hydrogeological activity persisted, at decreasing magnitudes, for up to a year after the Kern County and Idaho earthquakes (Sibson 1981b). The overall significance of the transient disturbances in flow regime does, however, differ between the hydrogeological and tectonic systems. The fault-related transient flow component may be relatively minor in terms of the total mass flux of fluids in a hydrogeological system because the recurrence intervals between fault events are commonly on time-scales of many tens to hundreds of years. However, it exerts a disproportionate control on the tectonic system, because it affects the transient effective-stress distribution and mechanical properties of the porous materials in and around the fault during the most significant periods of active faulting. The possibility of using *in situ* measurements of pore pressure for earthquake prediction continues to be evaluated.

The tectonic and hydrogeological coupling during a faulting episode can be very complex. Some of the parameters controlling the interaction are:

1. state of lithification
2. lithology
3. previous stress history
4. stress state at the initiation of faulting
5. strain rate
6. permeability–diffusivity of the deforming materials
7. temperature.

Much research has been published on this, as yet, still imperfectly understood subject and only a few relevant features are summarized here.

The rate of propagation of a fluid-pressure perturbation through a porous medium is controlled by its **diffusivity**, D. This is related to hydraulic conductivity, K, and specific storage, S_s, by the relationship:

$$D = \frac{K}{S_s}. \tag{7.5}$$

Specific storage, as used in studies of the response of simple aquifers to pumping at boreholes, is classically defined as the volume of water that a unit volume of porous material releases under a unit decline in head (Freeze and Cherry 1979) and is the sum of two terms:

$$S_s = \rho_w g (C_s + \eta C_f) \tag{7.6}$$

where C_s is the compressibility of the porous framework, C_f is the fluid compressibility, and η is porosity. Note that in a water saturated system $C_s \gg C_f$. Higher diffusivities allow perturbations to migrate more rapidly through a porous medium. From equation 7.5 it can be seen that the rapid transmission of pore-pressure fluctuations is assisted by high hydraulic conductivities and a relatively stiff pore-fluid and porous-framework response by the sediment.

Assigning a diffusivity to a deforming medium during a fault event is, however, complicated by the fact that both the hydraulic conductivity and specific storage will change with time as the fault goes through its pre-, syn- and post-failure stages (hydraulic conductivity changes will be dealt with later). These complications include the following factors. First, the pore pressure and stress changes can result in the sediment behaving elastically or plastically at different times, with the compressibilities generally being higher in the plastic field than in an elastic field (e.g. Figure 7.4a). Second, a sediment can both increase or decrease in volume depending on the variations in pore fluid pressure, stress state and the previous stress history. Third, even without a change in the pore pressure or average effective stress, variations in the stress field can also cause a volumetric response by causing the sediment to yield and fail. Due to these factors, marked differences may be expected in the pore-pressure response between the elastic walls, the damaged regions near the fault and in the failure region itself.

Rice and Clearly (1976), after Biot (1956), have developed constitutive and field equations for the coupled effects of deformation and diffusion associated with strains in a linear-elastic, porous medium near faults. These would be most applicable to the behaviour of the elastic wall regions, especially in significantly lithified sediments. One simple concept that becomes apparent from these studies is that if a poro-elastic material is subjected to an increase in average stress, the pore fluid pressures will rise and, conversely, if the average stresses fall so will the pore pressures. High rates of stress imposition and low diffusivities will increase the magnitude of this pore-pressure response but will allow only a limited volumetric change. Another basic concept is that the undrained response of a porous medium is stiffer than the drained response because the pore fluids are relatively incompressible (Figure 7.7). The types of pore-pressure response in the elastic region around the termination of a slip region, here termed the 'tip-line', are shown schematically in Figure 7.8a (Rice 1975; Rudnicki 1986; Rudnicki and Hsu 1988). Average stresses are commonly increased on one side of the fault and reduced on the other, generating, respectively, positive and negative pore-pressure perturbations. As the 'tip-line' propagates so will the pore-pressure response in the wall rock.

Roclofis and Rudnicki (1985) examined how the coupling of deformation and pressure diffusion

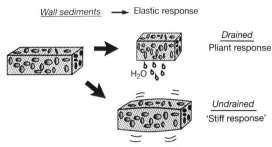

Figure 7.7 Schematic figure illustrating the differing physical responses to drained and undrained deformation. Because the fluids have time to migrate out of a porous medium it behaves in a significantly more pliant manner during drained deformation than during undrained deformation, when much of the stress is carried on the relatively incompressible pore fluids.

affected water-level changes due to the propagation of fault-creep events. They predicted that near the fault, pore-pressure fluctuations may reverse in sign as the creep dislocation passes (illustrated by the change in pore-pressure response in the elastic zone between sections ab and cd, Figure 7.8b).

7.3.4 Response of poorly-lithified sediments

The response of the porous sediments in the zones of yielding and failure are different between regions of shallow-level faulting in sediments and deep-level faulting in better lithified sediments. As detailed in Chapter 2, in the case of shallow-level faults, critical-state soil mechanics principles can be used to study the deformation behaviour of unlithified materials at low to moderate effective stresses (section 2.2.5). The degree of overconsolidation controls whether the sediment dilates or consolidates during yielding and failure and, thus, whether fault zones are net local contributors or sinks for fluids. Overconsolidated sediments have been subjected to higher than present average effective stresses. They are present in regions of substantial overburden removal or where the ambient fluid pressures have increased substantially over previous values. Heavily overconsolidated sediments tend to dilate and increase pore volume on yielding and failure (Figure 7.9), whereas slightly overconsolidated to normally consolidated sediments tend to lose pore volume on yielding and failure. Thus, the stress history of the sediment determines whether the failing materials are incipient local sinks or sources of fluid.

Because fluids have a very small compressibility and little chance to migrate in or out of the

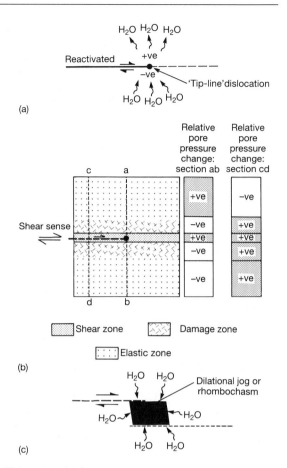

Figure 7.8 (a) Schematic figure illustrating the sign of the pore pressure change that is a response to the elastic strains around the tip-line dislocation of a propagating slip event on a fault. As the tip-line dislocation propagates past a point, the sign of the pore pressure response can change. (b) Schematic model illustrating the sign of the relative pore-pressure responses in the elastic zone, the damage zone and gouge zone of a major fault. The responses are shown along two sections (ab and cd) across the fault. (c) Dilational jog.

deforming porous sediments, only a limited volumetric response may occur during poorly drained events associated with high strain rates and low permeabilities. While muting the volumetric response, poor drainage results in significant fluctuations in pore pressures and, consequently, effective stresses. Pore pressures tend to be reduced during the faulting of overconsolidated sediments and tend to increase during the faulting of normally consolidated sediments. These pore-pressure variations exert a significant influence on the behaviour of the fault by respectively decreasing and increasing effective stresses

Fluid sources and tectonic processes

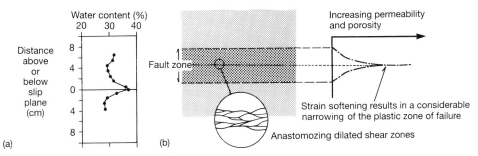

Figure 7.9 (a) Water-content profiles measured across the basal shear zone of a gravitational slide in overconsolidated London Clay showing the narrow strain-softened zone. (From Skempton 1964.) (b) Due to strain softened, the plastic dilation will result in a net permeability increase in the fault that will be limited to a thin zone.

during yield and failure of normally consolidated and overconsolidated sediments. For example, reduced effective stresses will tend to weaken a fault during the critical phase of movement, whereas increased effective stresses, associated with dilative behaviour, can stabilize a fault for a period of time. This latter effect has been termed dilational hardening (e.g. Rice and Rudnicki 1979). The rate of decay of the pore-pressure perturbation is controlled by the different diffusivities of the plastically failing fault-zone materials and the elastic wall rocks, with, for example, a high diffusivity reducing both the effectiveness and duration of the dilational hardening and stabilization process. Thus, in sediments the interplay between the tectonic strain rate and hydrogeological properties of the sediments determine the strength and behaviour of sediments during fault events, with a greater degree of interaction occurring during rapid undrained events than during slow-drained events. Therefore muddy sediments with a low permeability are more strongly affected by these processes than are relatively permeable sediments such as sands.

7.3.5 Response of well-lithified sediments

Even at shallow depths, the upward flow of chemically active fluids can result in the cementation of shallow sediments. A good example is the carbonate crust developed around the fault-related fluid conduits of the Cascadia Margin (Moore, Orange and Kulm 1991). In this case, the carbonates originate as a consequence of the oxidation of the methane (Ritger, Carson and Suess 1987) that is being expelled along the faults together with pore water. The localization of cementation and diagenesis around active tectonic structures changes the nature of the physical processes in these vital regions. Owing to their inherently low permeability, lithified sediments will tend to deform under poorly drained conditions, both during seismic and all but the slowest aseismic creep events. Commonly, lithified sediments dilate (Figure 7.10) as they yield owing to the development of intragranular extensional microcracks between grain contacts and in the cements (Brace, Paulding and Scholz 1966; Scholz and Kranz 1974; Zhang, Wong and Davis 1990). Dilation occurs whether the average confining stresses remain the same (Figure 7.11a), decrease (Figure 7.11b), or increase (Figure 7.11c). Thus, even though the overall average effective

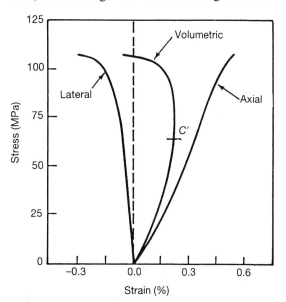

Figure 7.10 Stress and volumetric strain curve for Mesaverde Sandstone deformed in triaxial compression. Point c' denotes the onset of dilatancy. (After Lorenz, Teufel and Warpinski 1991.)

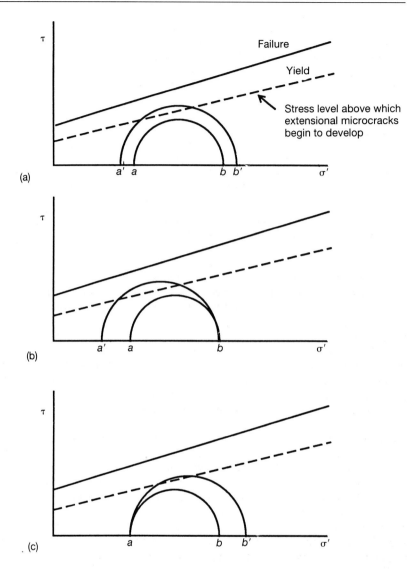

Figure 7.11 Extensional microcracking during yielding can occur under (a) constant, (b) decreasing and (c) increasing average stress conditions.

stress may be reduced and increased on different sides of the propagating 'tip-line' region (Figure 7.8a), a dilative response will always occur in materials undergoing yielding in the damage zone. Consequently, in a poorly drained system, significant negative pore-pressure perturbations will be generated during incipient yielding in the nucleation region, both immediately prior to a fault event and in the damage zone around the propagating 'tip-line' region during an event. Similar to the response in overconsolidated sediments, the drop in fluid pressure can lead to dilatant hardening and the partial stabilization of the fault (Rudnicki and Rice 1975; Rice and Rudnicki 1979). The local diffusivity of the fractured regions in and around the fault zones again exert a significant control on the rate at which the pore-pressure perturbations decay and the significance of the dilatant hardening process. As the negative pore-pressure perturbation decays the fault will weaken and fail. Some of the potential complexities in the pore-pressure responses of the elastic, damaged and failed regions in a dilative rock are illustrated in the simplified conceptual model shown in Figure 7.8. In Figure 7.8b the expected sign of the pore-pressure response across a propagating 'tip-line' region is shown along section ab.

A proportion of this fracture-related porosity, however, closes once the shear stresses are released (Scholz and Kranz 1974) and this will tend to increase pore pressures. The post-failure response in the damage zone (section cd, Figure 7.8b) will, therefore, be one of increased pore fluid pressures. It has been argued that at deep levels in the crust, dilatancy recovery due to the induced post-earthquake stress drop may cause a pore-pressure rise that acts as a trigger for aftershocks in the surrounding regions. Sibson (1981b) suggests that the cyclic dilation and closure of this fracture-related porosity in the periods immediately pre- and post-faulting could pump fluids into and out of the regions around fault zones. The width of the damage zone that may be involved in these events has not been determined but is probably <1–2 km in major faults such as the San Andreas (see Moony and Ginzburg 1986). Although, perhaps, a detailed discussion of such deep-seated processes is going beyond the remit of this book, many fundamental principals are transferable between deep- and shallow-level processes.

In many large tectonic faults cutting well-lithified sediments, the main slip zone is often associated with significant gouge development, particularly where deformation results in grain breakage. The stress level at which grain breakage becomes significant is relatively low; experimental and natural examples suggest that a reasonable transitional boundary between the independent particulate flow and cataclasis fields (Figure 1.1) might be as low as 1–5 MPa for medium to fine sandstones deformed with initial porosities between 31 and 41% (K.M. Brown and Orange 1993; Bishop 1966). Gouge thicknesses tend to increase with the magnitude of displacement (Figure 7.12), as local fault-wall irregularities and protrusions become sheared off and incorporated into the gouge body (Scholz 1987). At moderate depths mature gouges in major faults commonly come to be dominated by clay minerals (O'Neil 1985) owing to the hydration and breakdown of stress-damaged minerals such as feldspars, and they behave much like clay-rich sediments. Whereas mechanical deformation may cause dilation of the material, hydration of minerals will fill the voids with clays (rich in water). Thus, fault gouge, probably forms a limited local sink for fluids during its formation due to both mechanical and chemical effects. Eventually, a

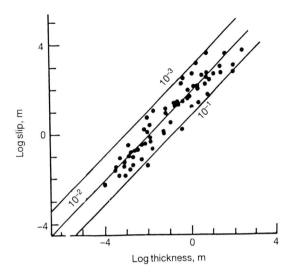

Figure 7.12 Log–log plot of gouge zone thickness, against total slip. (From Scholz 1987.)

'mature gouge' will undergo little further significant grain-size reduction, extensional crack development, or hydration. In addition, where faults follow lithologies already rich in clay, a response similar to an already mature fault gouge is likely. Mature, clay-rich gouges will behave rather like poorly lithified, low permeability muds (Morrow, Shi and Byerlee 1984). If the gouge is stressed slowly it will drain, consolidate and undergo strain hardening (Morrow, Shi and Byerlee 1982). If stressed rapidly, the gouge will respond in an undrained fashion owing its low permeability, and pore pressures will increase up to the failure point, weakening the fault (Figure 7.8b). It has also been suggested that frictional heating on faults may cause thermal expansion and pressurization of the pore fluids in the region of failure (Mase and Smith 1987). Where permeabilities and compressibilities are low it is possible that pore pressures can rise to near lithostatic values. These thermal effects will, however, be greatly reduced in weak, compressible, muddy gouges and fractured formations with high compressibilities and permeabilities.

The volumetric and pore-pressure response therefore can vary along an individual fault zone before, during and after a single fault event (Figure 7.8). The rate of migration and decay of these perturbations is controlled by the diffusivity of the deformed materials coupled to the rate of propagation of the 'tip-line' dislocations. Fractured, lithified sediments have higher compressibilities than undamaged sediments and this will tend to reduce the diffusivity (equations 7.5 and 7.6). The effect of higher compressibilities may, however, be offset by the significantly increased hydraulic conductivities of the fractures (see below). Propagating fault events may ultimately terminate at dilational jogs (Sibson 1985), which become volumetrically significant regions of substantially reduced fluid pressures (increased effective stresses) and major fluid sinks (Figure 7.8c). Because there are so many possibilities for the physical behaviour of deformed materials, and because major fault zones contain many anastomozing minor fault zones, each with an individual set of volumetric responses, it is not surprising that complex hydrogeological conditions are associated with perieods of active deformation.

7.3.6 Chemical sources of fluid: cementation, hydrocarbon generation and mineral dehydration

Chemical reactions occurring during burial-related diagenesis and metamorphism can release significant quantities of fluid. These fluids are liberated in two fundamental ways. First, cementation and the breakdown and redeposition of minerals can greatly reduce the volume of fluid-filled pore spaces in a rock. Second, fluids are released during the dehydration of minerals and thermal maturation of organic matter. Mechanical processes, such as consolidation (with or without grain breakage), can only reduce the intergranular porosity to a certain level. Ultimately, mineral transformations, pressure solution and cementation (e.g. Houseknecht 1987) can result in the almost total destruction of pore volume, reducing it from 20–40% at a few kilometres depth to below 1–2% porosity by the time sediments are buried to metamorphic depths. As this chemically induced porosity reduction proceeds, fluids must be expelled from the materials (Figure 7.6), which are themselves acquiring reduced permeabilites, with the corrolary that associated overpressures become more likely.

There are several mineral dehydration reactions that release fluids. Clay minerals are the most important water-bearing component of terrigenous clastic deposits and altered volcanic ash. Of the clays, smectites and mixed-layer clays contain the most significant quantities of water. In the presence of potassium ions, the kinetic reaction associated with the transformation of smectite to illite depends mostly on temperature and time (Pytte and Reynolds 1989; Elliott et al. 1991) with only secondary minor pressure dependence (Colten-Bradley 1987). The evolution from smectite to illite occurs through a progressive series of transformations between approximately 80 and 140°C depending on composition. For example, Bruce (1984) found that the smectite–illite transition occurs between 105 and 138°C for smectites in the Gulf Coast. In a subsiding basin, a smectite-bearing section will migrate through the 105–138°C region at several kilometres depth, generally when the initial consolidation-related, fluid-expulsion episode has waned (Figure 7.13). Ca-montmorillonite contains 10–20% by weight water (Keren and Shain-

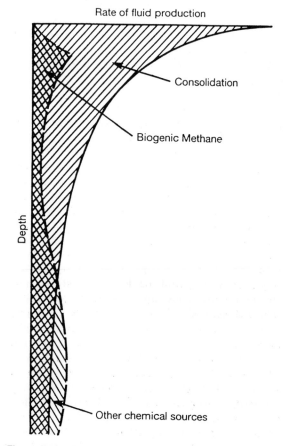

Figure 7.13 Schematic figure to show the variations of fluid production with depth.

berg 1975; Bird 1984), which is liberated when smectite is eventually transformed to illite. Overpressures can result because part of the rock matrix volume is converted to fluid volume and mud permeabilities at these depths are sufficiently low to inhibit drainage.

The dehydration during the mineral transformations from Opal-A to quartz in biogenic siliceous deposits can also be a locally important source of water in some formations (Keene 1976; Kastner 1981; Isaacs, Pisciotto and Gerrison 1983). As with clay dehydration reactions, these phase changes could cause a transmission of stress to the fluid and increase pore pressure in low-permeability formations until drainage associated with consolidation allowed the fluids and pressures to dissipate.

It has been estimated that at deep levels in the crust, secondary metamorphic dehydration reactions can release a further 5% (Walther and Orville 1982) to as much as 10% by weight (Cox, Etheridge and Wall 1986), from a typical shale host. The metamorphic dehydration reactions occur between 250 and 650°C (Peacock 1989). Bebout (1991a,b) suggested that breakdown of chlorite was a principal source of water in subduction zone mélanges heated to 400–600°C, and now exposed in the Catalina Schist. In their review of fluid sources at subduction zones, J.C. Moore and Vrolijk (1992) note that fractured and hydrothermally altered basaltic ocean crust, formed at oceanic spreading centres, constitutes another fluid source. The volume of water in hydrous minerals in the upper 1 km of oceanic crust is estimated at 8% by Kastner, Elderfield and Martin (1991), with perhaps about another 7% by volume of water trapped in pores and fractures in the upper kilometre of oceanic crust (Becker et al. 1990). This water may not be lost in a strong and incompressible rock until the onset of ductile deformation (J.C. Moore and Vrolijk 1992). Because the mantle is predominantly anhydrous, much of the water incorporated into the oceanic lithosphere must eventually be expelled at great depths in subduction systems. Indeed, isotopic evidence suggests that fluids can be taken down sufficiently deep to contribute to arc volcanism. A proportion of all these fluids that are liberated at depth will migrate upwards, perhaps supplementing the pore pressures in the overlying, unlithified sediments.

Oil and gas generation contributes both fluid and, in some regions, to the development of overpressures in the subsurface. Although the type and amount of organic matter found in sediments may not always be sufficient to form a source for economic oil and gas deposits, it can still be an important source of fluid in many tectonic environments. The widespread development of methane hydrates in passive continental margins and subduction zones (Kvenvolden 1985, 1988; K.M. Brown and Westbrook 1988; Kvenvolden and Kastener 1900) suggests that methane is the most volumetrically significant of the phases originating from the breakdown of organic matter (particularly in tectonic settings dominated by gas-prone terrigenous detritus). In the near-surface, below temperatures of approximately 50°C, methane is predominantly derived from bacterial action. As temperatures rise to 80–150°C thermogenic methane production becomes significant. J.C. Moore and Vrolijk (1992) note that in accretionary prisms, methane has been detected in variable concentrations in fluids from deep (45 km) (Bebout 1991a) to shallow levels (J.C. Moore et al. 1988; Fryer et al. 1990; Kvenvolden and Kastner 1990). Its concentration commonly exceeds its saturation level in water, so forming a separate phase at depth. These observations indicate that methane must constitute at least 1–5 mol% of the fluid (Hanor 1980), with few constraints on the upper limits of the amount of methane.

In low-permeability formations, methane and hydrocarbon generation can lead to overpressuring. Where saturation levels are reached, capillary pressures further reduce the effective permeability of the sediments and, thus, the prospects for drainage (Schowalter 1979). In muddy formations, mud volcanism and diatremes result from the effects of gas generation at depth (Hedberg 1974; Hovland and Judd 1988; K.M. Brown 1990). For example, a clear relationship has been demonstrated between abundant mud volcanoes and extensive methane hydrate accumulations in the Barbados accretionary prism (K.M. Brown and Westbrook 1988). Ebullition and expansion of the methane gas in the near-surface can be significant factors in driving diapirs to the surface particularly in subaerial environments and in shallow water depths (K.M. Brown 1990; J.C. Moore and Vrolijk 1992). This expansion can result in violent mud extrusion at the surface (e.g. Kugler 1939; Ridd 1970; Higgins and Saunders 1974). Other facets of this behaviour are discussed in sections 2.3.4 and 4.3.2.

In summary, although mechanical consolidation will initially dominate fluid production during burial in marine settings such as accretionary prisms (Figure 7.6) and deep sedimentary basins (Figure 7.13), at greater depths, chemically assisted porosity collapse, together with dehydration reactions and hydrocarbon generation (Spencer 1987), will take over. Fluid production is, consequently, predominantly controlled by stress state near the surface, with an increasing control exerted by temperature and fluid chemistry with depth (Figures 7.6 and 7.13). These chemical processes differ from mechanical fluid sources because they are also driven by temperature and fluid chemistry. For example, although consolidation-related fluid production will cease before fluid pressures rise to lithostatic levels, the primarily temperature-driven dehydration of clays or hydrocarbon generation need not. Indeed, chemical sources of fluid are probably primarily responsible for the formation of extensional shears and hydrofractures at deep levels in the earth's crust. Fault-valving, associated with gold deposition, was attributed by Sibson (1988) to the episodic cracking and sealing of reverse faults in response to the increase and subsequent release of fluid pressures generated by metamorphic chemical sources. Thus, with the fault-valving mechanism, it is the hydrogeological system that controls the tectonic response of the fault. Although Sibson suggested a deep source for the gold bearing fluids, clay dehydration and hydrocarbon generation may initiate a similar response in shallow-level faults in poorly lithified sediments.

In large-scale, natural systems, we thus have mechanical and chemical fluid sources and driving mechanisms for fluid flow that operate at different depths, on different time-scales and which are controlled by different parameters. Because both the mechanical and chemical sources of fluid result in changes in the pore geometry of the rocks, fluid production is nearly always coupled with significant permeability changes.

7.4 CONTROL OF LITHOLOGY AND BURIAL-RELATED CONSOLIDATION ON THE PERMEABILITY OF SEDIMENTARY UNITS

7.4.1 General

Lithology-related permeability variations can dominate regional flow patterns, leading to large-scale channelling of fluid flow through high-permeability aquifers or conduits. The intergranular permeability of a sediment is controlled by the number, size, distribution and interconnectivity of the voids it contains. Therefore, lithology, state of consolidation, cementation and extent of mineral transformations are all primary factors influencing permeability. At low effective stresses, consolidation is predominantly accommodated by grain-boundary sliding. A simple linear empirical relationship generally exists (Figure 7.14) between the log of hydraulic conductivity, log K, and void ratio, e. From Figure 7.14 it can be seen that a sand comprising large rounded grains of equal dimensions will generally have a considerably higher permeability than lithologies dominated by muds or clays.

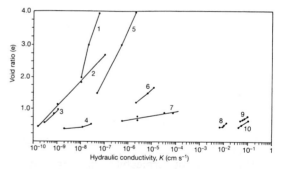

Figure 7.14 Plot of log of hydraulic conductivity, K, against e for a variety of sediment types. Note, the linear relationships for a wide variety of sediment types and permeabilities: 1, Silt – Boston; 2, sodium – Boston blue clay; 3, Vicksburg buckshot clay; 4, lean clay; 5, silt – Boston; 6, calcium kaolinite; 7, silt – North Carolina; 8–10, sand for dam filter. (Note that under near-surface and laboratory conditions $K\,\mathrm{m\,s^{-1}} \times 10^{-5} \approx k\,\mathrm{m^2}$.) Although e versus log K (or e versus log k) relationships commonly give the best straight-line fits, Lambe and Whitman (1979) have suggested that better fits occasionally can be made with K versus $(e^3 + e)$ or even a K versus $e^2/(1 + e)$ relationships for some lithologies.

The distribution of grain sizes in a sediment can have an important secondary effect on permeability. For example, Figure 7.15 is taken from work by Gal, Whittig and Faber (1990), who studied the hydraulic conductivity changes in coarse soils when different clay percentages were added. They determined that mixtures containing only sand- and silt-size fractions (0% clay) had a lower hydraulic conductivity than soils with a small clay-size fraction added. Hydraulic conductivities peak in this example for a clay fraction of approximately 8% and then subsequently fall to lower levels at very high clay percentages. The lower permeability of the pure sand and silt combination was attributed to the development of a denser packing arrangement in the absence of the clay component.

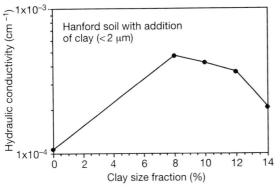

Figure 7.15 Hydraulic conductivity changes in a coarse soil as different percentages of clay are added. (Adapted from Gal, Whittig and Faber 1990.)

During burial, a given sediment will become progressively more consolidated with depth in response to increases in effective gravitational stress (section 2.2.3), and its permeability will likewise decrease (see Figure 7.14). In a typical sedimentary sequence, this progressive reduction leads to a form of permeability heterogeneity. Of considerable significance for the development of natural hydrogeological systems is that during burial, different lithologies experience varying mechanical processes that can strongly effect how their permeability develops.

In muds, any consolidation-related grain breakage generally does not appear to be a significant process and there is little change in the slope of the basic e versus $\log K$ relationship from low to moderate effective confining pressures. For example, a plot of $\log K$ versus $\log p'$ (Figure 7.16a) for the hydrostatically consolidated, clay-rich samples of Morrow, Shi and Byerlee (1984) shows almost straight-line relationships, even at relatively high hydrostatic confining pressures of 200 MPa. In contrast to mud-rich lithologies, once a sand has achieved its maximum packing density (for reasonably spherical grains, mechanical compaction can reduce intergranular volume to about 25%), further mechanical consolidation can initially be accommodated only by elastic deformation of grains,

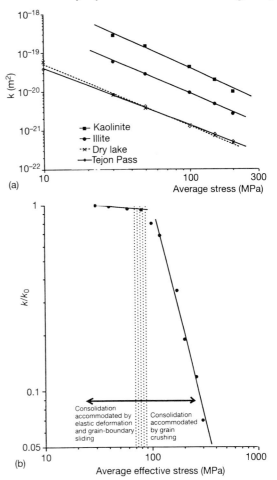

Figure 7.16 (a) Change in permeability of clay-rich samples with increasing effective confining stress. (Adapted from Morrow, Shi and Byerlee 1984.) (b) Change in permeability from an initial value k_0 of Ottawa Sand with increasing effective hydrostatic confining stress. (Adapted from Zoback and Byerlee 1976.) Note the significant change in slope of the otherwise linear relationship at about the point where grain crushing becomes significant.

and eventually, at high effective stresses, by the crushing of grains. The average effective-stress condition at the onset of grain breakage will vary with factors such as porosity, deviatoric stress, grain size, cementation and strain rate (Knipe 1986b; Zhang, Wong and Davis 1990). Grain crushing can cause an inflection in the generally linear relationship between e and log K, thereby increasing the rate of permeability reduction. This is illustrated in Figure 7.16b, taken from Zoback and Byerlee (1976). The inflection between the otherwise two relatively linear portions of the curves appears to be due to the onset of appreciable grain crushing and blockage of pores by grain fragments, which take place above confining pressures of 50–75 MPa.

Thus, in an untectonized, thick sedimentary section the onset of mechanical grain crushing can lead potentially to an increased rate of permeability reduction in deeply buried, hydrostatically pressured sandstones. Away from the local influence of faults, two factors may, however, reduce the individual significance of this particular permeability reduction mechanism. First, the common presence of muds and other materials with low permeabilities generally results in the development of significant overpressuring, particularly in the deeper regions of rapidly subsiding basins. Overpressuring substantially reduces the effective stress and potential for grain crushing. Second, at the elevated temperatures present at such depths, any grain breakage will also tend to be accompanied by intergranular pressure solution and cementation (e.g. Houseknecht 1987; Schmoker and Gautier 1988). Indeed, during burial these chemical processes compete with and can become significant even before the onset of large-scale grain breakage. Ultimately, however, a combination of these mechanical and chemical processes can result in the almost total destruction of pore volume and permeability in sandstones. Even in an untectonized basinal environment, important variations in the trends in hydrogeological behaviour can develop between lithologies as they are buried.

7.4.2 Permeability anisotropy resulting from consolidation

Where the permeability at a point in a porous medium varies with direction it is said to be anisotropic. As consolidation proceeds, anisotropy in intergranular permeability can become increasingly significant in lithologies containing grains with high aspect ratios (Arch and Maltman 1990). Anisotropic permeabilities may develop in muds, for example, because there tends to be a progressive alignment of the platy clay minerals perpendicular to the principal compressive stress during consolidation (Figure 7.17). Fluid-flow paths parallel to the plane of the aligned clay are then generally considerably less tortuous than the flow paths perpendicular to the general alignment in the clay minerals (Figure 7.17). England *et al.* (1987) used Pouisseille's law to relate permeability, k, to the **tortuosity**, T, of the fluid-flow path and the mean pore radius, ψ:

$$k = \frac{J\psi^2}{T^2} \tag{7.7}$$

where J is a constant, and T is the average total length of the tortuous fluid-flow route divided by

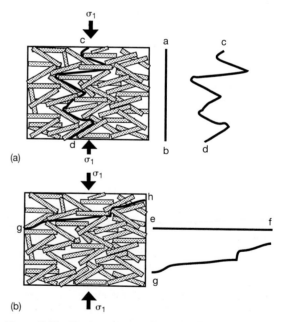

Figure 7.17 The importance of grain alignment for tortuosity of flow paths in muddy sediments. The direct paths ab and ef are shorter than the actual path the fluid moves along (paths cd and gh). Owing to the grain alignment perpendicular to the vertical principal compressive stress, the vertical tortuosity, T_v (where $T_v = cd/ab$), shown in (a) is greater than the horizontal tortuosity, T_h (where $T_h = gh/ef$), shown in (b).

the length of the direct route across the sample (Figure 7.17). Wilkinson and Shipley (1972) gave experimental examples of the vertical and horizontal permeability development during the consolidation of muds in oedometer cells with flow-path geometries similar to those shown schematically in Figure 7.17. From the results shown in Figure 7.18, it can be seen that both orientations show linear relationships between e and log K, but the vertical hydraulic conductivity, K_v, is lower than the horizontal hydraulic conductivity, K_h for a given void ratio. As the fluid properties will be constant, we can plot directly the permeability ratio $[\log(K_h/K_v)=\log(k_h/k_v)]$ for these data (see equation 7.4) to obtain a linear relationship for the magnitude of the permeability anisotropy for any given void ratio. Figure 7.19a shows that the anisotropy increases with increasing consolidation. Interestingly, the data also suggest that at a void ratio of approximately 1.73, the permeability of the unconsolidated soil would have been isotropic and that the maximum achievable value of K_h/K_v (at $e=0$) would have remained below 7. Indeed, core samples of natural clays and shales seldom have $K_h/K_v > 10$, with ratios of less than 3 being common (Freeze and Cherry 1979).

The ratio of the tortuosity difference between vertical and horizontal flow paths can also be obtained for the example shown in Figure 7.19a.

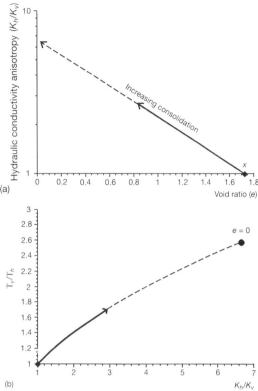

Figure 7.19 (a) The progressive development of anisotropy in hydraulic conductivity with one-dimensional consolidation (from data on the remoulded kaolin of Figure 7.18). (b) The progressive increase in the anisotropy of the average flow-path lengths (expressed as tortuosity, T) with consolidation (from the data shown a).

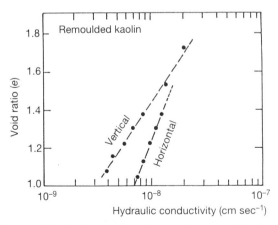

Figure 7.18 Void ratio (e) and hydraulic conductivity (K) relationships determined experimentally during one-dimensional consolidation of remoulded kaolin. (After Wilkinson and Shipley 1972.) Note that for a given void ratio the vertical hydraulic conductivity is less than the horizontal hydraulic conductivity.

For a given void ratio, J and ψ will be constant, so that from equation (7.7):

$$K_h/K_v = (l_v/l_h)^2, \qquad (7.8)$$

where l_v is the length of the average vertical tortuous flow-path length, and l_h is the average horizontal tortuous flow-path length. Figure 7.19b plots K_h/K_v against the tortuous flow path ratio, l_v/l_h, and we see that the maximum potential tortuosity ratio is just below 2.6. Thus, the maximum achievable anisotropy of permeability is not very substantial for one-dimensional consolidation of this particular mud. Such one-dimensional consolidation would be similar to the type of stress path a mud would experience in a subsiding basin away from any major structures. However, any deformation could considerably increase the anisotropy of permeability, and strongly affect how fluids interact with structures forming in the sediment.

7.4.3 Equivalent permeabilities and gross permeability anisotropy

In most sedimentary sequences, lithological variations will complicate the general burial-related trends in permeability. The interbedding of high- and low-permeability units, such as clays and sands, produces a layered heterogeneity. However, there is another type of gross anisotropy in hydrogeological properties when the system is considered on a larger scale (Freeze and Cherry 1979; Desbarats 1987). The gross vertical equivalent hydraulic conductivity, K_v, perpendicular to the layers in a sedimentary section is given in Figure 7.20 by equation (7.9). In contrast, the horizontal equivalent hydraulic conductivity, K_h, parallel to the layers, is given by equation (7.10) in Figure 7.20. In this case K_i represents the horizontal hydraulic conductivity of a given layer.

For all sets of possible varying $K_1, K_2, ..., K_n$ values it can be shown that $K_h > K_v$. Indeed it is not uncommon for such layered heterogeneity to lead to regional anisotropy values on the order of 100:1 or larger (Freeze and Cherry 1979). Thus, layered heterogeneity can produce a much larger gross anisotropy in gross formation permeability than the anisotropy in intergranular permeability resulting from consolidation-related effects in individual beds. Consequently, layered permeability heterogeneity has considerable importance for the general orientation of the pathways of fluid expulsion in both convergent and basinal environments.

7.5 PERMEABILITY VARIATIONS DUE TO DEFORMATION IN ACTIVE TECTONIC SYSTEMS: FRACTURES, FAULTS AND GOUGE

7.5.1 General

The following discussion is concerned mainly with sediments deforming in zones sufficiently localized that they can be regarded as brittle faults. Due to the potential complexities in the deformation processes affecting shallow fault zones there is a great difference between the permeability changes occurring during the faulting of sediments deforming by grain-boundary sliding, the opening and closing of a simple fracture (as for example in a hydrofracture or joint), and the gross permeability changes in a major fault zone, across which there can be considerable movement and in which, depending of the confining stress, thick gouges can form. Indeed, in some circumstances, fluid-pressure variations may cause faults to oscillate between high- and low-permeability phases as the deformation style changes from hydrofracture to cataclasis. The initial lithology, degree of lithification, stress state, total strain and the nature of the strain partitioning in and around a structure will all affect how its permeability evolves. For example, there are important and distinct differences in permeability evolution in tectonized muds and sands, with a transition between the processes dominating in clean sands and those in muds as the clay percentage rises. The following discussion almost exclusively refers to the two end-member systems, that is those dominated by clay-rich muds and clean sands that have been lithified to various degrees.

In cross-section, a major fault will consist of a highly strained shear zone, surrounded by a region of damaged, partially yielded sediment, with the amount of damage decreasing away from the central regions of the fault. Normal faults and thrusts tend, however, to have the damaged zones asymmetrically distributed about the zone of slip (Aydin 1978; Aydin and Johnson 1978; Webber et al. 1978), with more substantial damage being

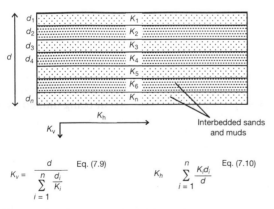

Figure 7.20 The gross vertical equivalent permeability, K_v, perpendicular to the layers in a sedimentary section is given by equation (7.9). Where d is the total vertical thickness of a sedimentary section with n layers, and where an individual layer i has a thickness d_i and a vertical permeability K_i. In contrast, the horizontal equivalent permeability, K_h, parallel to the layers is given by equation (7.10).

distributed across a wider zone in the hangingwall (Figure 7.21). At shallow depths, deformation of poorly lithified sediments in the damage zone and shear zones will be accommodated largely by grain-boundary sliding. In contrast, with increasing depth and higher effective confining stress, cataclasis and gouge development may occur in sandstones.

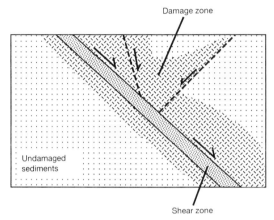

Figure 7.22 The hanging walls of normal faults tend to have a greater degree of fracturing and deformation concentrated in them than the footwalls.

7.5.2 Permeability changes in muddy fault zone materials

Although referring mostly to structures in convergent-margin settings, many of the mechanisms discussed here can be applied to other structures, such as listric normal faults (see also Section 6.3). There are considerable data which show that muds intrinsically tend to have low permeabilities (Figure 7.14), generally making them unsuitable as fluid conduits (Freeze and Cherry 1979). This, however, has been found not always to be the case. Major thrusts and, particularly, the basal décollement zones of accretionary prisms, are commonly confined to mud-rich units of one sort or another (Byrne 1984; Cowan 1985; Fisher and Byrne 1987; D.J. Prior and Behrmann 1990; Knipe, Agar and Prior 1991). Observations indicate that some of these low-angle, mud-rich faults can channel fluids laterally many tens of kilometres from deeply buried source regions in accretionary prisms (Gieskes, Vrolijk and Blanc 1990; Screaton, Wuthrich and Dreiss 1990; Vrolijk et al. 1990). Indeed, numerical studies (Screaton 1990) indicate that the permeabilities parallel to these fault zone conduits probably need to be at three to five orders of magnitude higher than the surrounding wall rocks. In some fashion, deformation of the muds can promote high permeabilities under certain circumstances (Moore, Orange and Kulm 1991) and the mechanisms by which this permeability enhancement is, at least episodically, achieved is an area of active research.

Progressive deformation of muds at a basal décollement beneath a thickening prism will tend to promote two different physical processes. First, deformation will affect the consolidation state of the mud, and second, deformation will greatly enhance grain alignment in the muds. The state of consolidation, in turn, affects the mean pore radius, φ, whereas grain alignment determines the differences in tortuosity, Γ, of the flow paths in the deforming media parallel to and across the fault zone. Pouisseille's law shows that $k \propto \varphi^2/\Gamma^2$ (equation 7.7; England et al. 1987). Hence, the three to five orders of magnitude of permeability enhancement parallel to the fault zone relative to the wall rocks required for long-distance lateral migration of fluids requires either net dilation of the fault zone (an increase in φ) and/or a greatly reduced tortuosity, Γ, parallel to the fault zone relative to flow paths perpendicular to the fault zone.

For a fault zone in a poorly lithified mud to be substantially dilated relative to the surrounding wall rocks it would need to have failed under average effective stresses substantially lower than those previously encountered by it and the surrounding wall rocks (Karig 1990; Brown and Bekins in preparation). Strain-softening effects, however, mean that dilated regions in muddy fault zones should be highly localized features, with permeability enhancement being limited as a result (Brown and Bekins in preparation). For example, Figure 7.9a shows the variation in water content across the dilated basal slip plane of a large gravitational slide affecting heavily overconsolidated London Clay. The total movement across this shear zone was about 0.5 m and it produced a softened zone less than 4 cm across. Thus, although the dilated region may have a high permeability, it will be extremely narrow and its effectiveness at conducting fluid will be consequently limited (Figure 7.9b). Rather than being a single simple zone of softening, the main failure zone may consist of a compact swarm of minor dilated shear zones only a few tens of centimetres in combined thickness, with intervening material that has not been plastically dilated (Figure 7.9b). Indeed, Karig (1990) recently postulated that some types of localized anastomozing scaly fabric in muddy accretionary faults form along this type of dilational stress path.

The development of a strong permeability anisotropy due to differences in flow-path tortuosity may be another important process. Arch and Maltman (1990) recently proposed that shear-related grain alignment in clays could promote sufficient differences in flow-path tortuosity, and the related permeability anisotropy, to account for the focusing of flow along muddy fault zones. As Figures 7.18 and 7.19 illustrate, normal burial-related one-dimensional consolidation on its own only offers a limited prospect of achieving $(K_v/K_h)^2$ ratios of 1000 or more. However, the combination of pure and simple shear in the regions surrounding the zone of failure may present such an opportunity. Whether the resulting clay fabric alignment is, of itself, sufficient to account for the several orders of magnitude permeability anisotropy in the fault zone is unclear, indeed it is likely that the effects of fabric anisotropy would have to be combined with dilation of the fault. As the average orientation of any permeable, anastomozing, dilated shear zones will predominantly have orientations similar to any general clay fabric alignment developed in the surrounding rocks (roughly in the plane of the fault), the tortuosity will be lower and equivalent permeability higher parallel to such a zone than perpendicular to such a zone. It is, thus, conceivable that a combination of fault-zone dilation and general fabric-related anisotropy may achieve the required fault-parallel permeability enhancement of several orders of magnitude.

7.5.3 Permeability changes in fault zones in sands and sandstones

Deformation in sands at very low confining stresses is dominated by grain-boundary sliding (with possible initial breakage of any pre-existing cements) and at moderate to high effective stresses by grain breakage (Handin *et al.* 1963; Zoback and Byerlee 1976; Logan and Rauenzahn 1987; Morrow and Byerlee 1989). Indeed, for all but the lowest effective confining stresses (commonly < 5 MPa for porous sands; K.M. Brown and Orange 1993), cataclastic deformation is the dominant control on the permeability evolution in sand-rich, fault-zone materials.

As already mentioned in connection with transient fluid sources, in low effective-stress environments the degree of overconsolidation controls whether the sands will dilate or consolidate during yielding and failure and, thus, whether the fault increases or decreases in permeability as a result of changing the mean pore diameter. Shear zones in heavily overconsolidated sands will tend to dilate and increase in permeability (Figure 7.21a), whereas slightly overconsolidated to normally consolidated equivalents will tend to be reduced in permeability during deformation (Figure 7.21b). Unless the sand grains have unusually flattened or elongated shapes, variations of tortuosity, Γ, with direction of fluid flow across a shear zone are unlikely to be significant. The equivalent permeability will, however, be slightly higher parallel to a dilated shear zone than perpendicular to it.

At moderate to high effective stresses, the initial yielding and deformation of a sand is associated with the development of intragranular extensional cracks (oriented parallel to the principal compressive stress) resulting from the high stresses at grain contacts (Logan 1979; Zhang, Wong and Davis 1990). During progressive shear these dilational cracks eventually link. At this stage the sand contains a zone of irregular, anastomozing, partially dilated fractures, some of which are throughgoing, and across most of which there is only minor displacement. Ultimately, as deformation is continued, strain across the fractures builds up, and the grains at the margins of the systems (especially in hanging-walls; Figure 7.22) become progressively abraded so that a gouge zone progressively develops (Figure 7.12).

Whether deformation results in a net increase in the permeability parallel to a fault zone in a sand depends on the initial porosity and permeability of the material. Chemical processes, such as diffusion mass transfer and mineral transformation, become increasingly significant with higher confining pressure and temperature.

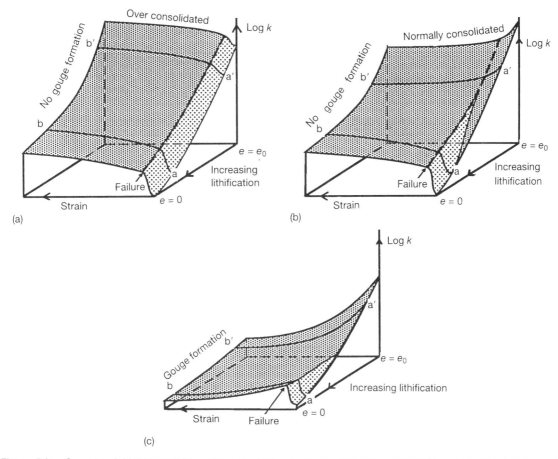

Figure 7.21 Conceptual, highly curviplanar three-dimensional permeability surfaces for fault zones cutting a sandstone lithology at different confining stresses and initial states of consolidation; (a) low confining stress and overconsolidated; (b) low confining stress and normally consolidated; (c) moderate to high confining stress with significant cataclasis and gouge formation at high strains. The permeability is plotted against initial degree of pre-strain lithification and increasing shear strain. A low state of lithification corresponds to a porous sandstone in which all the original pore space is preserved ($e = e_0$) and a high state of lithification corresponds to a sands in which nearly all the original porosity has been destroyed by cementation and/or diffusion mass transfer processes prior to deformation ($e = 0$). The lines ab and a′b′ correspond to the permeability profile across a major fault developed in a lithified sandstone (see for example Figure 7.21) in which there is an increasing degree of strain experienced by the rock from the relatively undeformed wall rock (corresponding to point a or a′) into the zone of damage, and ultimately the failure zone (corresponding to point b or b′).

A lithified sandstone may have almost all its porosity and permeability destroyed by cementation, and/or by mass diffusional transport mechanisms (pressure solution and resulting deposition). In well-lithified sandstones in which little porosity remains, the undeformed rock permeability will be very small and almost any deformation will tend to cause net dilation and a significant increase in permeability (Figure 7.21), particularly where open fractures exist.

Flow through fractures is often idealized as flow between two parallel plates (plane Pouisseille flow) and this leads to the so-called 'cubic law' (Snow 1968):

$$Q = \frac{-Lyd^3}{12\eta}\left(\frac{d\Phi}{dl}\right) \quad (7.11)$$

where the volume flow rate, Q, varies as the cube of the plate separation d (termed the aperture), Ly is the width of the fracture in the direction normal to fluid flow, and η is the dynamic viscosity of water. Thus, the permeability of a parallel plate fracture is $k = d^3/12$. In this simple form, however, the applicability of the parallel-plate model is limited to flow through fully open natural fractures, that is, where fracture walls are not in close proximity. There are, for example, reasonable correlations between experimental flow rates and theoretical parallel-plate flow rates at wide separations of rough-walled fractures (Figure 7.23a; Cook et al. 1990). At moderate to high effective confining stresses, however, the rough fracture walls will be in contact and only open to fluid flow at dilational jogs and between asperities that hold the walls apart. At this point, tortuosity (equation 7.8) becomes highly significant, with flow rates and permeabilities being reduced below parallel-plate predictions. A number of different modified Pouisseille flow equations exist, based around different types of statistical or fractal representations of the roughness of the fracture surface (K.M. Brown 1990; Cook et al. 1990; Liu and Sterling 1990). The theoretical investigations often modify the parallel-plate model to account for the variation in aperture in the natural fracture, commonly by assigning a mean aperture or hydraulic aperture parameter. The mean fracture aperture and permeability vary considerably with stress. Increasing effective normal confining stresses generally reduce fracture permeability in a highly non-linear way (Figure 7.23). The permeabilities of newly fractured rock are still higher than the original intact rock permeability, even with considerable effective normal stresses being applied across the fracture. The evolution in the permeability of such fractures with time, however, is by no means simple. For example, experiments indicate that repeated stress cycles tend to reduce the permeability of newly formed fractures, and that fractures also tend to heal with time at high pressures and temperatures (Handin et al. 1985; Brantley et al. 1990).

Thus, where the wall rocks originally have very low porosities and intergranular permeabilities, the fractured rock in the damage zone will generally be more permeable (Figure 7.24). At elevated temperatures and pressures, however, cementation and healing of fractures may mean that repeated movement is necessary to keep fractures open. In contrast to the damage zone, the permeability of gouge zones will be governed by grain-size reduction and the clay component. Experimental work by Morrow, Shi and Byerlee (1981) showed that at moderate to high effective stresses clay gouges have very low permeabilities and are likely to be less permeable than the

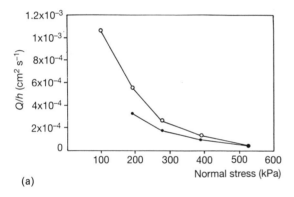

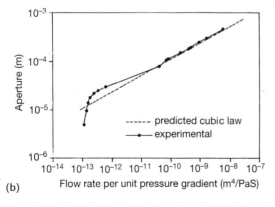

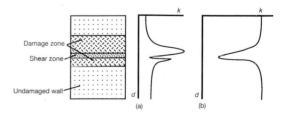

Figure 7.23 (a) Variation of fracture transmissivity (equal to KA, equation 7.3) with total normal stress. (From Wilbur and Amedei 1990.) (b) Comparison of experimental data on flow through a natural fracture with those predicted assuming smooth parallel plates. Residual aperture is assumed to be 5.0 mm. (From Cook et al. 1990.)

Figure 7.24 Permeability distribution across a fault in a lithified (a) and poorly lithified (b) sediment at moderate to high confining stresses, where cataclastic deformation leads to gouge formation.

surrounding fractured rock in the damage zone. Whether the gouge also has a lower permeability than the undamaged wall rock rather depends on how well lithified the wall rock is. Morrow, Shi and Byerlee (1981) found that the permeability of clay-rich San Andreas gouge was reduced to between 10^{-19} and 10^{-22} m^2 during deformation at confining stresses of 200 MPa. This is a very low permeability by any standards.

Porous, poorly lithified sands commonly have fairly high original permeabilities such that deformation involving grain-breakage cataclasis generally leads to permeability reduction (Figure 7.21c). This is ostensibly due to grain-size reduction, collapse of pore spaces and choking of the pore throats by small grain fragments. Pittman (1981) gives examples of permeabilities in a natural deformed sandstone decreasing by approximately four orders of magnitude as the percentage of granulated quartz in the sample increased from 0 to 35%, and the porosities dropped from 25–30% to 0–5% (see Figure 7.25). Experiments on Ottawa Sand undertaken by Zoback and Byerlee (1976) also show that a pore-volume reduction during cataclastic deformation is associated with a reduction in permeability by one to two orders of magnitude. In the experiments of Zoback and Byerlee, the sands were initially confined under different hydrostatic stresses and then failed by increasing the axial stress. The data of Zoback and Byerlee (1976) are recast in Figure 7.26 in terms of the average effective stress experienced by the various samples during the triaxial tests. From Figure 7.26 it can be demonstrated that the permeabilities at the failure point (peak stress) fell in an almost linear fashion as the initial hydrostatic confining stress was increased up to a confining stress of 75 MPa. Permeability then subsequently began to increase again as dilation during failure became significant at initial confining stresses higher than 75 MPa. Note that the 75-MPa increment also apparently corresponded to the onset of substantial grain crushing and permeability reduction in the isotopically consolidated sample: clearly the mode of deformation changes at this point.

Formations in which faults are developed will be associated with highly anisotropic equivalent permeabilities because of the permeability variations in different regions of the fault, and between the fault and the wall rock. In general, equivalent permeabilities in a direction parallel to a fault zone will be higher than those perpen-

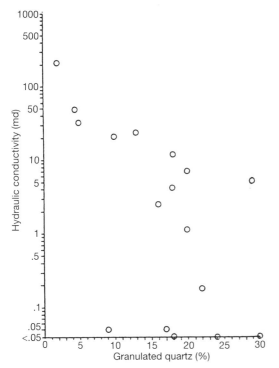

Figure 7.25 Reduction in permeability in a faulted sandstone as the percentage of granulated quartz associated with cataclasis increases. (From Pittman 1981.) Note that the hydraulic conductivity is given in units of millidarcies, where 1 mD = 10^{-6} cm s^{-1}.

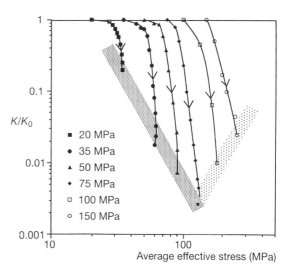

Figure 7.26 Reduction in permeability, k, from original value k_0, with progressive triaxial deformation of Ottawa Sand. (From Zoback and Byerlee 1976).

dicular to a fault zone. For example, where a fault zone cuts strata of low porosity and permeability, fluid flow will be enhanced through the fractures in a damaged zone in the hangingwall, but will be restricted across the gouge zone (Figure 7.24a). If, instead, a fault cross-cuts an originally porous, poorly-lithified sand, fluid will easily flow though the permeable wall rock in directions parallel to the fault zone, but will be significantly retarded for flow paths across cataclastic zones and any zone in which clay-rich gouges are developed (Figure 7.24b).

7.6 PERMEABILITY CHANGES AT LOW EFFECTIVE STRESSES

7.6.1 General

A number of processes can greatly increase the permeability during the unloading of sediments by either erosion or increased fluid pressures, particularly when more bonded materials dilate to form open fractures. They are discussed in section 6.2, with reference to stress paths. Regarding permeability, the processes can be divided into two broad groups depending on whether the extensional fractures form as load-parallel fractures developed under conditions where all the regional stresses are compressive (Lorenz, Teufel and Warpinski 1991) or, as in the case of extensional shears and hydrofractures, where the minimum regional stress is tensional (i.e. $\sigma_{min} < P_{H_2O}$). However, because the orientations of both load-parallel and tensional fractures are constrained to form perpendicular to the least principal stress, they produce a highly anisotropic permeability distribution in a sedimentary section, which greatly affects patterns of fluid flow.

7.6.2 Extensional failure where the least principal effective compressive stress is tensional

The generation and recognition of tensional stresses in a sediment mass was discussed in some detail in section 6.2. As explained there, in the context of stress paths, a common situation in which stresses are tensional is where the fluid pressure exceeds the least principal compressive effective stress. Hydrofracture occurs when $\sigma_3 - P_{H_2O} > S_T$, where P_{H_2O} is the pore fluid pressure and S_T is the tensile strength of the sediment. In fact, depending upon the magnitude of the differential stress, the effective least compressive stress and the tensile strength (S_T), two extensional failure modes are possible. Based on the geometry of the parabolic failure envelope predicted by the combined Griffith–Navier–Coulomb criteria (Sibson 1981a; Price and Cosgrove 1990):

1. if $\sigma'_3 > S_T$ and also $(\sigma'_1 - \sigma'_3) < 4S_T$, failure will occur in tension as true hydrofractures, with the fracture planes being oriented perpendicular to σ_3 (Figure 7.27); 2. if $0.8S_T < \sigma'_3 < S_T$ and also $4S_T < (\sigma'_1 - \sigma'_3) < 5.5S_T$, hybrid extension and shear may result (extensional shears; Figure 7.27).

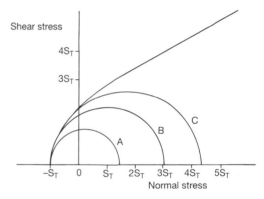

Figure 7.27 (a) Coulomb–Griffith failure envelope and stress circles for tensile failure (S_T). (b) Maximum differential stress for tensile failure. (c) Extensional shear. (From Etheridge 1983.)

Sibson (1981a) and Behrmann (1991) have proposed models that take into account two additional factors: (i) the significantly reduced tensional strengths that pre-existing hydrofractures can have relative to their host rock; and (ii) the differing orientation in the stress field of normal, wrench and thrust faults. Because of these factors, the minimum fluid pressure required at different depths for simultaneous shear and hydrofracture vary considerably between normal, strike-slip and thrust faults (Figure 7.28). In figure 7.28 the fluid pressure condition is given as $\lambda = P_{H_2O}/\sigma_v$, where σ_v is the total overburden load and P_{H_2O} the fluid pressure. For normal and strike-slip faults, simultaneous hydrofracture and displacement can initiate at fluid pressures below hydrostatic at shallow levels, increasing to near lithostatic pressure at depth. In contrast, on thrust

Permeability changes at low effective stress

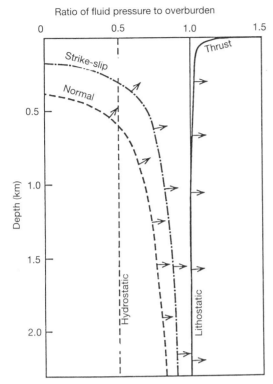

Figure 7.28 Minimum fluid pressures necessary for hydrofracture and displacement on normal, strike-slip and thrust faults. (Based on Sibson 1981a; Behrmann 1991; Moore and Vrolijk 1992.)

faults fluid pressures must be distinctly above lithostatic near the surface, decreasing to slightly above lithostatic with depth. Thus, vertically oriented hydrofractures can form relatively easily in shallow normal-fault and wrench-fault systems, but considerable overpressures are required to generate low-angle hydrofractures along thrust faults.

Recapitulating from section 6.2, extensional shears can be generated by any of the three main mechanisms shown in Figure 7.29: (a) an increase in the principal compressive stress (σ'_1) due to regional compression; (b) a decrease in the minimum compressive stress (σ'_3) caused by regional extension; and (c) an increase in fluid pressure (P_{H_2O}). The lateral displacement of the stress envelope caused by the fluid pressure increase, as depicted on Figure 7.29c is, however, a considerable oversimplification, because the stress envelope will not remain a constant diameter owing to the elastic properties of the sediments. Take, for example, sediments undergoing one-dimensional

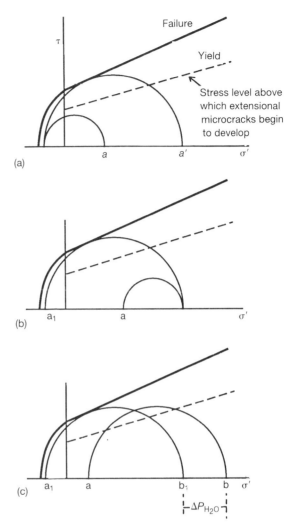

Figure 7.29 Mohr circles to show that extensional shears can form when: (a) the principal stress increases (a_0 to a); (b) the minimum stress is decreased by regional extension (b_0 to b); and (c) when the pore fluid pressures increase.

consolidation and burial in a laterally confined condition in a basinal environment where there is no tectonic component to the stress field. Under these conditions of plastic deformation the lateral stress, σ_h, is related to the vertical stress, σ_v, by the K_0 ratio ($K_0 = \sigma_h/\sigma_v$). For most clay-rich lithologies K_0 ratios lie between 0.5 and 0.7 (Lambe and Whitmann 1979; Price and Cosgrove 1990). Any increase in pore fluid pressure (or removal of overburden) will cause the elastic unloading of the sediments, with the vertical to lateral stress ratio now being controlled by the

Poisson's ratio, v, of the material (Magara 1981; Mandl and Harkness 1987).

For a sediment being unloaded by an increase in pore fluid pressure, P_{H_2O}, in a laterally confined manner, the change in effective horizontal stress is given by $\Delta\sigma'_h = -\Delta P_{H_2O}[P_{H_2O}/(1-P\sigma_{H_2O})]$ whereas the change in vertical stress is given by $\Delta\sigma'_v = -\Delta P_{H_2O}$. As values of v in the crust vary between 0.1 and 0.3 (Magara 1981) the rate of reduction in horizontal stress falls below the rate of vertical stress reduction. This leads to the stress envelope shrinking as the fluid pressures rise, and may lead to the inversion of stresses (Figure 7.30) so that eventually $\Delta\sigma'_h > \Delta\sigma'_v$ (Mandl and Harkness 1987; Lambe and Whitmann 1979). Indeed, both Mandl and Harkness (1987) and Lorenz, Teufel and Warpinski (1991) argued that whereas horizontal hydrofractures can form by increases in fluid pressures in tectonically relaxed basins, vertical hydrofractures cannot. For vertical hydrofractures to form, tectonic processes must reduce the minimum lateral stress in the basin at a rate that is roughly equal to, or more than, the rate of pore-pressure induced lateral stress increase (Figure 7.31).

In sedimentary basins, such as the USA Gulf Coast, the onset of significant overpressuring occurs at depths below 1–2 km, where smectite

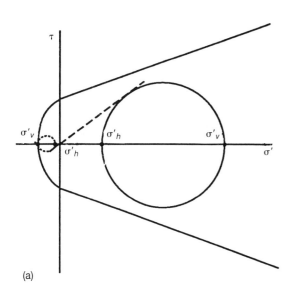

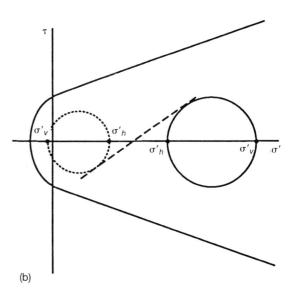

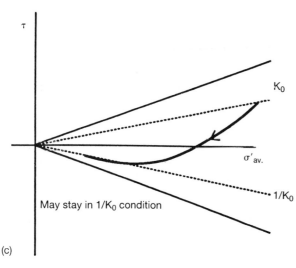

Figure 7.30 Idealized Mohr circles representing different basinal conditions (a) 'Tectonically relaxed' basin (Lorenz, Teufel and Warpinski 1991) (b) Basin previously having undergone burial related consolidation (Mandl and Harkness 1987) (c) Critical state behaviour. Any hydrofracturing will be at an isotropic stress state with orientation of open fractures being controlled by a pre-existing rock fabric. Owing to the lateral constraints of the surrounding sediments the horizontal stresses will decrease at a lower rate than the vertical stress as the pore fluid pressure increases in the deep regions of the sedimentary basin. This can lead to an inversion of the orientation of the principal stress axes, such that only horizontal hydrofractures may form. Various behaviours are possible depending on whether a typical Poisson's ratio (a) or the K_0 ratio for a yielding consolidating sediment (b and c) controls the original stress state in the basin.

Simultaneous pore fluid pressure increase and tectonic reduction in horizontal stress

$$\Delta\sigma'_h = \Delta\sigma'_v = -\Delta P_{H_2O}\left(\frac{\nu}{1-\nu}\right) - \Delta\sigma_t$$

where tectonic component of stress reduction, $-\Delta\sigma_t$,

$$\Delta\sigma_t = -\Delta P_{H_2O}\left(\frac{1-\nu}{\nu}\right)$$

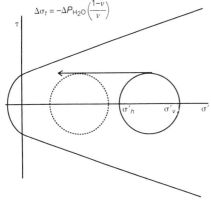

Figure 7.31 Idealized Mohr-circle representation of simultaneous increase in pore fluid pressure and decrease in horizontal tectonic stress. The stress circle can move only laterally without changing its size if the horizontal stress is reduced simultaneously with the pore-pressure increase.

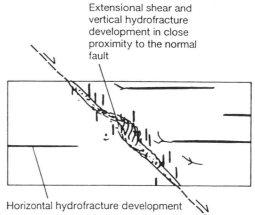

Figure 7.32 Schematic model of the variation in the nature and orientation of the hydrofractures with proximity to a major extensional fault.

dehydration and hydrocarbon generation begin to be significant sources of fluids (Magara 1975a; Fertl 1976). For the relatively weak sediments at these levels the fluid pressures will have to approach, but not exceed, lithostatic values for extensional shear and hydrofracture development (Figure 7.28). Because the elastic effects described above would tend to preclude vertical hydrofracture in the absence of tectonically related extension, the regions around normal faults are likely to be the localities of preferential extensional shear and vertical hydrofracture development (Figure 7.32). In contrast, fluids may be drained laterally away from regions distant to extensional faults along horizontally oriented hydrofractures (Figure 7.32). If individual extensional phases in basins are short-lived episodic events, pulses of hydrofracturing and fluid expulsion will be interspersed with periods of restricted vertical drainage, particularly if the vertical hydrofractures become healed by mineral cements.

Low-angle hydrofracture and extensional shear development in association with thrusts requires fluid pressures to rise to at least lithostatic levels, because any increase in fluid pressure would tend to further increase the horizontal stresses only (Figure 7.33). As Behrmann (1991) pointed out, however, strike-slip faults cutting thrust terranes could be prime sites of focused vertical fluid expulsion, with dilational jogs in these faults being prime sites for vertical open-fracture development and focused fluid advection. This would be particularly evident where strike-slip faults tap into the lithostatically pressured, low-angle hydrofracture systems associated with thrusts. Indeed, at shallower levels, the low fluid pressures required for hydrofracturing in strike-slip faults (Figure 7.28) may soon tend to cause the rapid deflation of thrust-related hydrofracture systems, so that expulsion events in these cases would be short lived.

7.6.3 Open-fracture development in the absence of regional tensional stresses: load-parallel extensional fractures

Lorenz, Teufel and Warpinski (1991) proposed that load-parallel extensional fractures (Figure 7.34a) can develop even when all the stresses are compressive and anisotropic. This load-parallel fracturing mechanism is, however, only significant at low confining stresses in lithified sediments and, therefore, still requires significant overpressuring in deeply buried formations. Indeed, the experimental examples that Lorenz, Teufel and Warpinski (1991) show for this mechanism are for uniaxial unconfined compression tests (zero lateral confining stress). As for yielding and failure at higher confining stresses, the extensional fractures of Lorenz, Teufel and Warpinski (1991) initiate with the coalescing of intragranular load-parallel micro-extensional fractures. A true tensional stress perpendicular to the fracture plane only exists at the crack tip, where it is

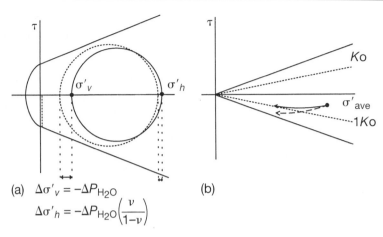

Figure 7.33 Mohr circle construction showing the significance of the state of lithification for the change in stress state due to increasing pore fluid pressures in an accretionary prism. The increase in pore fluid pressure will tend to increase horizontal stresses at a higher rate than vertical stresses in both lithified (a) and unlithified (b) sediments. This may cause failure of the prism prior to hydrofracture development. Unless failure significantly increases the permeability of the prism, fluid pressures will continue to rise. Considerably higher differential stresses can be supported at hydrofracture or extensional shear development in lithified (a) than unlithified (b) materials. Indeed, in a truly cohesionless sediment (b) the stress conditions at hydrofracture will be isotropic so that there will be little preferred orientation in any developing hydrofracture network. Compare with real data in Figure 6.17.

induced indirectly by the far-field compressive stress oriented parallel to the fracture plane (Granberg 1965, 1966, 1989; Costin 1987). Dilation and formation of open microfractures first begins when the load reaches between one-third and two-thirds of the compressive failure strength of the rock (Figure 7.29). The extensional fractures are best developed at low strains (commonly <1%), essentially as precursors to shear failure at higher strains. Lorenz, Teufel and Warpinski (1991) argue that this is the primary mechanism for forming open fractures in the subsurface. Load-parallel extensional fractures can be generated by any of three main mechanisms described for hydrofracture and extensional shear development, the only difference being that actual tensional stresses are not required for their formation. These extensional fractures will tend to form perpendicular to the minimum compressive stress and so will propagate horizontally if σ'_1 is horizontal and vertically if σ'_1 is vertical. Therefore, because their geometry and relationship to the regional stress pattern will be similar to those of true extensional fractures it would be difficult to separate them in the field.

Lorenz, Teufel and Warpinski (1991) put forward the (perhaps extreme) view that, as true tensional stresses have never been measured in the subsurface and overpressures never equal lithostatic values (even in highly overpressured formations), hydrofracturing does not exist as a significant natural mechanism in basins (the exception to this being almost instantaneous high-pressure fluid injection from a source outside of the formation of interest, as in the case of igneous and clastic sill and dyke formation and where a deeply buried highly pressured formation suddenly becomes linked to a shallow, low pressured formation). The absence of evidence for true tensional stresses in basins appears to support this view. Figure 7.28, however, demonstrates that lithostatic pressures are not required for hydrofacturing in basins. It could also be argued that true tensional stresses in the subsurface are likely only to exist for a short time near faults during episodic periods of active extensional faulting, so that hydrofracturing is similarly episodic and the affected regions are limited in spatial extent. A less extreme view would be to extend the field in which open fractures can develop to the low stress, positive stress side of the Griffith–Navier-Coulomb envelope to form a transitional spectrum of open-fracture response.

7.6.4 Stress amplification mechanisms

Stress amplification, as put forward by Eidelman and Rêches (1992) and Lorenz, Teufel and Warpinski (1991), is another complicating possibility for extensional fracture in the absence of a ten-

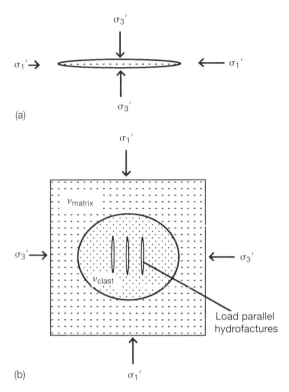

Figure 7.34 (a) Load-parallel extensional fractures can develop even though the minimum compressive stress is still slightly positive. (b) Stress amplification in rigid inclusions in a ductile matrix ($v_{clast} \gg v_{matrix}$) can cause load-parallel extensional fractures to develop in the inclusions even though the far-field stresses are all compressive. See also section 6.2

sional regional stress. Where lithified sediment bodies are enclosed inside a more ductile matrix, such as a mud (Figure 7.34), the resulting stress amplification can cause local tensional stresses inside the inclusion, which will be oriented perpendicular to the principal compressive stress. The stress amplification is greatly enhanced when the Poisson's ratio of the ductile matrix is large (i.e. in a highly incompetent matrix where v tends toward 0.5) and quite significant tensional stresses can be generated even though all the regional stresses are slightly compressive. At deep levels in the earth's crust, however, significant overpressuring would still be necessary for this mechanism to generate true extensional shears and hydrofractures. Note that the required differences in Poisson's ratios for this mechanism are quite extreme and it may be limited to certain special environments and formations. One such environment could be where partially lithified sand bodies are enclosed in highly overpressured muds. Such sand masses could be the product of either sedimentary or tectonic fault-zone processes. For example, due to decoupling along the main slip surface, vertical extensional fractures could form within stiff, lithified, sandstone inclusions in mélanges below the main detachment level of accretionary prisms where the principal stress is vertical. Above the detachment, horizontal hydrofractures may develop in response to the low-angled orientation of the principal compressive stress. Clearly, in mélanges, where there are numerous lithified sand inclusions in a poorly lithified muddy matrix, the discontinuous open fractures could significantly increase the overall permeability of the mélange, although the tortuosity of the flow paths would be quite significant.

7.7 EFFECT OF DEFORMATION ON THE TORTUOSITY OF FLOW PATHS AT DIFFERENT SCALES

The orientation, nature, interconnectivity and overall geometry of fracture systems can cause significant variation in the tortuosity of flow paths in the subsurface. Moreover, the tortuosity differences can occur on a variety of scales. K.M. Brown (1990), for example, made use of fractal models to define the roughness of individual fracture surfaces and the resulting tortuosity of flow paths. As mentioned previously, the significance of tortuosity effects at this microscopic scale greatly increases when the rough walls of individual fractures are in close contact (Figure 7.23). Thus, fluids follow less tortuous paths in open hydrofractures than in similar partially closed fractures subject to high normal stresses.

On an intermediate scale, the architecture of fracture systems depends very much on the nature of the processes forming them. Although commonly described as being fractal in nature (i.e. self-similar at a variety of scales), considerable differences in tortuosity should be expected between relatively well-ordered joint, hydrofracture and load-parallel extensional fracture systems and the highly anastomozing fluid conduits that would result from the complex deformation patterns around major faults. In the latter case, the flow paths are likely to be highly tortuous on a scale of metres to hundreds of metres, with low-permeability granulation seams and gouges being interspersed with permeable, partially open, fracture networks, or less deformed regions in which high intergranular permeabilities are preserved. The hydrogeological properties of individual fault zones also can be affected by whether they are thrust, strike-slip or normal in geometry, with the required stress conditions and potential for hydrofacturing being significantly different between the various fault types (Figure 7.28).

At the regional scale, the overall geometry of contractional, extensional or strike-slip tectonic environments will define how the various types of major faults, and intervening less deformed regions, are distributed and interconnected, defining a tortuosity in the fluid-flow paths on a scale of hundreds of metres to tens of kilometres. Tortuosity therefore exists on a variety of scales. Commonly, formations are found to be more permeable as the scale of the observations increases, particularly in fractured formations, as more widely spaced fractures come to be sampled. Lithological features, such as bedding and facies changes, also exert an important scale-dependent control. Because of tortuosity effects on a variety of scales, major features such as thrust belts will have a considerably anisotropic gross permeability, with flow paths generally having a lower tortuosity parallel to bedding and to the main through-going low-angle faults and hydrofracture networks. It is for this reason that significant, focused, lateral expulsion of fluids is observed in accretionary prisms, such as the Barbados prism (Arch and Maltman 1990; Screaton, Wathrich and Dreiss 1990; Brown and Bekins in preparation).

7.8 DISCUSSION: TRANSIENCE AND THE INTIMATE COUPLING OF HYDROGEOLOGICAL AND TECTONIC PROCESSES

By carrying chemical species and heat, and by determining how effective stresses are distributed within porous sediments, fluids exert a fundamental control on the development of many active tectonic systems. Obvious manifestations of fluid flow include such features as fluid vents along the surface traces of faults and mud volcanoes, while in the subsurface, mineral veins, together with thermal and exotic geochemical anomalies along faults, are good indicators of focused fluid flow.

The form of a hydrogeological system depends on both the nature and distribution of fluid sources driving the system, and the permeability distribution that controls the resulting flow patterns. By providing either stress discontinuities and barriers or conduits to fluid flow, structural and stratigraphical boundaries can compartmentalize fluid flow into 'hydrotectonic' domains. The processes operating within individual domains, and the intervening faults themselves, may be quite different. Examples of two such domains would include the offscraped and underthrust section in accretionary prisms, with the basal décollement forming their mutual boundary.

Meteoric water falling on upland regions is an important external source of fluids for many subaerial hydrogeological systems. In mountainous subaerial environments, topographic effects may drive meteoric water deep into the earth's crust along permeable conduits. Eventually this fluid may again surface at warm springs in surrounding lowland regions. Mechanical and temperature/chemical sources of fluid also exist within the porous sediments. The mechanical processes associated with consolidation initially dominate fluid production during burial of a porous, unlithified sediment mass. The rate of consolidation, however, falls off sharply with depth so that temperature and chemically assisted porosity collapse (mass diffusional transfer), together with dehydration reactions and hydrocarbon generation, become the significant fluid sources in deeply buried sediments. Fluid production is, consequently, controlled by stress state near the surface, with an increasing control exerted by temperature and chemistry with depth. Ultimately, as in the case of fault-valving, the chemical sources may create considerable overpressures in the deeper levels of tectonic systems (where low permeabilities inhibit fluid expulsion), with the build-up and release of these pressures being coupled to the episodic cracking and movement of faults that become destabilized by the reduced effective stresses. Because both the mechanical and chemical sources of fluid result in changes in the pore geometry of the rocks, fluid production is nearly always coupled with significant permeability changes and, as a consequence, a simultaneous change in fluid-flow patterns.

The significance of the coupling between tectonic and hydrogeological processes becomes particularly evident during periods of fault movement. Deformation generally involves changes in pore volume and permeability, causing the redistribution of fluids and the chemical species contained therein. A combination of the tectonic strain rate and diffusivity of the deforming sediments controls the magnitude of the resulting pore-pressure and effective-stress fluctuations. The diffusivity is, in turn, controlled by the permeability and stiffness of the porous framework, both of which will change as deformation proceeds. The magnitude of the fluctuations, and consequently the importance of the coupling, become most apparent at high strain rates in low-diffusivity materials. Depending on whether pore fluid pressures increase (associated with pore-volume reduction) or decrease (as a result of dilation), a fault may be, respectively, transitorily destabilized or stabilized during the initial critical period of yielding and failure. Marked permeability anisotropy can result in hydrogeological disturbances generated in one part of a tectonic system being channelled along faults, and destabilizing the structure as they migrate.

Even the shape of collisional belts is controlled by the intimate coupling between hydrogeological and tectonic processes. An accretionary prism, for example, must be strong enough to be pushed forward over the underthrust sediments against the frictional forces existing along the basal décollement. Both the strength of the prism interior and the basal décollement are largely controlled by fluid pressure. The degree of overpressuring in the prism is the product of the dynamic balance between the ability of the hydrogeological system to expel fluids to the surface and the rate at which fluids are produced in the prism by such processes as burial-related consolidation and mineral dehydration reactions. The hydrogeological and tectonic systems are coupled because the prism taper controls the rate of thickening and fluid production, whereas the rate of fluid production and permeability controls the fluid pressure and, thus, the prism taper. This type of circular pattern of interaction is common to many other aspects of 'hydrotectonic' coupling.

Transience in hydrogeological and tectonic systems can occur on a variety of time-scales, from the short time-scale fluctuations during an individual faulting event up to very long-period changes resulting from such factors as climatic change. Long-period changes in the nature of sediment input may, for example, exert a fundamental control on the development of a tectonic system. An accretionary prism's bulk diffusivity controls the rate at which overpressures generated in the prism can dissipate. The rate at which pore pressures can come to a new dynamic equilibrium within the deforming porous mass is strongly limited by the large volume and general low diffusivity of the prism, particularly if it is mud rich. Because fluid production and diffusivity are both strongly influenced by lithology, if there are changes in the thickness and types of sediment being subducted with time, for example, between glacial and interglacial periods, the prism will have to constantly readjust to different equilibrium configurations. If the rate of change in the nature of the incoming sedimentary section is faster than the readjustment rate of the pore-pressure distribution in the prism, the prism may actually never reach any truly 'steady-state' critical taper configuration. Similarly, significant periods of large-scale hydrotectonic transience can be generated in subsiding sedimentary basins in which rapid climate changes influence the rate of infilling and nature of the sediments, or where changes in the rates of tectonic extension and hence location of depocentres occur within different regions of the basin. Indeed, it is conceivable that for significant periods, tectonic systems, such as accretionary prisms, wallow in a state of chronic disequilibrium, with slowly but constantly changing pore-pressure distributions within them.

CHAPTER 8

Sediment deformation, dewatering and diagenesis: illustrations from selected mélange zones

TIM BYRNE

8.1 INTRODUCTION

The preceding chapters have examined various aspects of sediment deformation and dewatering in the context of different experimental, theoretical and geological conditions. The main purpose of this chapter is to consider some geological situations where the interaction of these processes is particularly well illustrated. The examples are taken from sediments that have accreted along plate boundaries in a submarine, and therefore water-rich, environment. A common feature in this setting is the occurrence of the materials known as shale-matrix mélanges and scaly clays (Hsü 1968; Biju-Duval *et al.* 1984; Raymond 1984; Taira *et al.* 1988; section 9.4.3). In recent years substantial progress has been made in the understanding of these materials, and it has increasingly emerged that they represent the result of the close interaction of deformation, dewatering and diagenesis (Labaume 1987; J.C. Moore and Byrne 1987; Byrne and Fisher 1990; Sample 1990; Knipe, Agar and Prior 1991; Labaume, Berty and Laurent 1991). These particular sediments form the subject of this chapter.

The recent work on accreting sediments has focused especially on relating fluid flow to deformation, and on documenting the geometry of any drainage conduits (Langseth and Moore 1990; Le Pichon, Henry and Lallemant 1990; Maltman *et al.* 1991). A significant effort has also been made in documenting the history of palaeofluid-flow conduits within ancient convergent margins (Cloos 1984; Engelder 1984; Fyfe and Kerrich 1985; Vrolijk *et al.* 1988; D. Fisher and Byrne 1990). These aspects are incorporated in the present review, as they help illustrate further the interdependence of the processes involved in producing mélanges.

The chapter also serves a second purpose, important in the understanding of deformation of sediments. It addresses a fundamental question: how can we distinguish tectonically induced deformation from gravitationally induced deformation in the rock record, if both processes can affect partly lithified sediments? The question has been well considered for mélanges (e.g. Greenly 1919; Hsü 1974; Silver and Beutner 1980; Raymond 1984), and is discussed in more general terms in Chapter 9. Shale-matrix mélanges illustrate the crux of the problem, because they commonly display a highly ductile appearance at the outcrop scale. There has consequently been a long-standing argument over whether to interpret mélanges as submarine gravity deposits or as submarine shear zones of tectonic origin (section 1.4).

In the following pages, two new approaches to this now classic conundrum of structural geology are proposed (Figure 8.1a). Put simply, the first suggested approach is to document the complete progressive deformation history and then tie different stages of the deformation history to different stages of dewatering and

The Geological Deformation of Sediments Edited by Alex Maltman Published in 1994 by Chapman & Hall ISBN 0 412 40590 3

tains ash layers; and a lower layer of hemipelagic sediments that does not contain ash. Based on drilling and seismic reflection data, the décollement below the Nankai prism appears to have propagated near the base of the ash-bearing hemipelagic sediments (Figure 8.2b).

Leg 131 of the Ocean Drilling Program drilled through the entire toe of the Nankai prism, including the frontal thrust and basal décollement, and penetrated the underthrust sediments to reach the oceanic basement (Taira, Hill and Firth, 1991). Moreover, relatively good core recovery allowed structural, physical property, geochemical and petrological data to be collected almost continuously. Over 70% of the 3000 measurements made by the structural geologists were oriented geographically with palaeomagnetic data. Information from Site 808, therefore, provides an exceptionally complete structural data set (Byrne et al. 1993a, b; Lallemant et al. 1993; Maltman et al. 1993b).

The results of Leg 131 provide four observations particularly relevant to the deformation and dewatering history of sediments. First, in the upper part of the prism, three generally time-transgressive structural features were documented (shown across the prism, or right to left in Figure 8.2b). These features are: (i) penetrative fabrics and kink-like structures; (ii) small, core-scale faults; and (iii) the frontal thrust. Second, the sediments above the décollement are overconsolidated relative to the sediments below the décollement, and probably relative to a normal marine sedimentary sequence (Taira, Hill and Firth, 1991). The sediments below the décollement appear to be underconsolidated and are considered to be overpressured. Third, the décollement is defined by a 20-m-thick zone of highly faulted sediments, with anomalously low porosities and textures that record cyclic periods of deformation and overpressuring. The brecciation and scaly clay fabric that are developed in this zone suggest similarities with shale-matrix mélanges, of which it may be an incipient example. Fourth, detailed computerized tomography (CT) scans of the core-scale faults, kink-like structures (Figure 8.3) and samples of the décollement show that these deformation zones are less porous than the host sediments (Byrne et al. 1993b).

Figure 8.2 summarizes the possible significance of these observations. The progressive development of the penetrative fabrics, small faults and the frontal thrust is interpreted using the stress-path diagram shown in Figure 8.2c (Schofield and Wroth 1968; Brandon 1984; Wong, Szeto and Zhang 1992). Stress path diagrams were explained in section 2.2.5 and developed in Chapter 6. In this interpretation, the sediments initially follow a uniaxial consolidation stress path (segment a, Figure 8.2c), resulting in a compactional bedding-parallel fabric. As the deformation front is approached, deviatoric stress increases, resulting in a steepening of the stress path (segment b, Figure 8.2c), and the principal stresses reorient (not shown in Figure 8.2c for simplicity). The new orientation of stresses results in a new compactive phyllosilicate fabric subperpendicular to bedding (Stage b, Figure 8.2b), as recorded by magnetic susceptibility and sonic velocity data (Byrne et al. 1993b; Owens 1993). As the mean and deviatoric stresses continue to increase, the stress path ultimately reaches the yield surface (marked by different porosities in Figure 8.2c) for the sediments, and the dominant deformation mechanism changes from intergranular dewatering and porosity loss to shear-enhanced compaction (e.g. Wong, Szeto and Zhang 1992). In the Nankai sediments, this transition is interpreted to be recorded by the change from penetrative fabrics to distributed faulting (stage c, Figure 8.2b). The increase in density within the fault zones argues for a strain-hardening stress path (e.g. b to c in Figure 8.2c). With higher mean and deviatoric stresses the state of stress of the sediments migrates along the yield surface (as the porosity decreases) and the sediments continue to fail, with strain hardening at individual faults, resulting in a pervasive fault fabric at the scale of the prism. Ultimately, the stress path crosses the critical state line and shear localization occurs (Stage d in Figure 8.2c). This stress state may correlate with the initiation of the frontal thrust (Stage d in Figure 8.2b).

The dewatering and deformation histories of the sediments within the décollement zone, in contrast, appear to be more protracted and less systematic (Byrne et al. 1993b; Maltman et al. 1993a). There are three important differences between the fabrics observed in the décollement and the structures from above the décollement. First, the shear surfaces in the décollement are typically much thinner, more irregular and discontinuous than fault surfaces 'from above the

CHAPTER 8

Sediment deformation, dewatering and diagenesis: illustrations from selected mélange zones

TIM BYRNE

8.1 INTRODUCTION

The preceding chapters have examined various aspects of sediment deformation and dewatering in the context of different experimental, theoretical and geological conditions. The main purpose of this chapter is to consider some geological situations where the interaction of these processes is particularly well illustrated. The examples are taken from sediments that have accreted along plate boundaries in a submarine, and therefore water-rich, environment. A common feature in this setting is the occurrence of the materials known as shale-matrix mélanges and scaly clays (Hsü 1968; Biju-Duval et al. 1984; Raymond 1984; Taira et al. 1988; section 9.4.3). In recent years substantial progress has been made in the understanding of these materials, and it has increasingly emerged that they represent the result of the close interaction of deformation, dewatering and diagenesis (Labaume 1987; J.C. Moore and Byrne 1987; Byrne and Fisher 1990; Sample 1990; Knipe, Agar and Prior 1991; Labaume, Berty and Laurent 1991). These particular sediments form the subject of this chapter.

The recent work on accreting sediments has focused especially on relating fluid flow to deformation, and on documenting the geometry of any drainage conduits (Langseth and Moore 1990; Le Pichon, Henry and Lallemant 1990; Maltman et al. 1991). A significant effort has also been made in documenting the history of palaeofluid-flow conduits within ancient convergent margins (Cloos 1984; Engelder 1984; Fyfe and Kerrich 1985; Vrolijk et al. 1988; D. Fisher and Byrne 1990). These aspects are incorporated in the present review, as they help illustrate further the interdependence of the processes involved in producing mélanges.

The chapter also serves a second purpose, important in the understanding of deformation of sediments. It addresses a fundamental question: how can we distinguish tectonically induced deformation from gravitationally induced deformation in the rock record, if both processes can affect partly lithified sediments? The question has been well considered for mélanges (e.g. Greenly 1919; Hsü 1974; Silver and Beutner 1980; Raymond 1984), and is discussed in more general terms in Chapter 9. Shale-matrix mélanges illustrate the crux of the problem, because they commonly display a highly ductile appearance at the outcrop scale. There has consequently been a long-standing argument over whether to interpret mélanges as submarine gravity deposits or as submarine shear zones of tectonic origin (section 1.4).

In the following pages, two new approaches to this now classic conundrum of structural geology are proposed (Figure 8.1a). Put simply, the first suggested approach is to document the complete progressive deformation history and then tie different stages of the deformation history to different stages of dewatering and

The Geological Deformation of Sediments Edited by Alex Maltman Published in 1994 by Chapman & Hall ISBN 0 412 40590 3

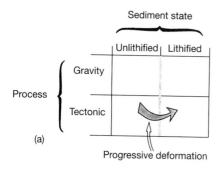

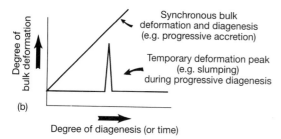

Figure 8.1 Schematic diagrams showing a conceptual difference between gravitationally induced deformation and tectonically induced deformation with respect to lithification. In (a) a single, progressive tectonic deformation persists through different stages of lithification. In (b) the generally different rates of tectonic and gravitational processes (e.g. submarine slumps) are keyed to the rate of diagenesis. Slumps and debris flows are considered generally to be short, episodic events that punctuate the diagenetic history. In contrast, tectonic events generally occur gradually and progressively, and at a rate comparable to diagenesis.

lithification of the sediments. In the cases presented below, the final stage of deformation is represented by a penetrative pressure-solution cleavage that is considered to be solely 'tectonic' in origin. In fact, because the 'tectonic' pressure-solution cleavage represents the culmination of a coaxial progressive deformation, the entire progressive deformation is interpreted to be a tectonic event. Gravitational forces are considered to have played a relatively minor role, even though the deformation was initiated in unlithified sediments.

A second, slightly different, approach is to consider the deformation history in the context of the progressive diagenetic history. For example, tectonic and diagenetic processes are inherently slow, relative to the time it takes to emplace a submarine slide or slump. Consequently, in a sedimentary sequence that is undergoing progressive diagenesis, tectonically induced deformational structures (e.g. veins, fault zones, kinks, folds, etc.) should form incrementally as diagenesis progresses. In contrast, a submarine slide will be associated with deformational structures that occur only once in the diagenetic history. That is, diagenetic reactions would occur before and after, but not during, the deformational events associated with sliding (Figure 8.1b).

In the following paragraphs this logic is applied to understanding the deformation of mélange-like sediments. To set the stage for this review, we begin with a summary of the deformation history of the toe region of the Nankai accretionary prism of southwest Japan, where recent drilling has yielded one of the more complete pictures of the processes occurring at modern convergent margins. In moving onland to ancient accretionary sequences, we first examine a sequence of rocks that are interpreted to have been initially offscraped. These rocks may represent an ancient analogue of the shallow levels of the Nankai accretionary prism. The materials also provide an 'end-member' for the discussion that follows on underthrust sediments, scaly fabrics and mélanges, because, being structurally coherent, they record a relatively straightforward deformation history.

8.2 PROGRESSIVE DEFORMATION AND DEWATERING IN THE NANKAI ACCRETIONARY PRISM

The Nankai accretionary prism, southwest Japan, represents one of the best studied, clastic-sediment-dominated convergence margins in the world ((Taira et al. 1991; Taira, Byrne and Ashi 1992; see also section 6.3). Active accretion of sediments is presently occurring at the Nankai Trough (Figure 8.2) as a result of convergence between the Philippine Sea plate and southwest Japan. The Philippine Sea plate is moving to the northwest at rate of 2–4 cm a^{-1} (Taira Hill and Firth 1991; Lallemant et al. 1993) and carries four gradational sedimentary sequences into the deformation zones (Figure 8.2b). These four sequences are: an upper layer of relatively coarse-grained (trough) turbidites; a layer of medium-grained and relatively fissile (outer-marginal trough-wedge) turbidite deposits; a layer of (Shikoku Basin) hemipelagic sediments that con-

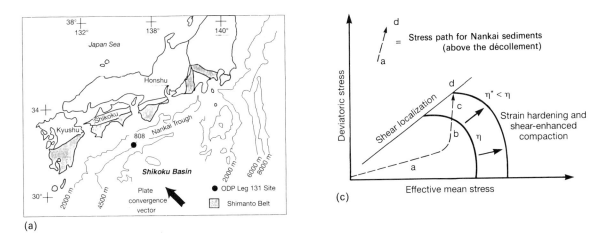

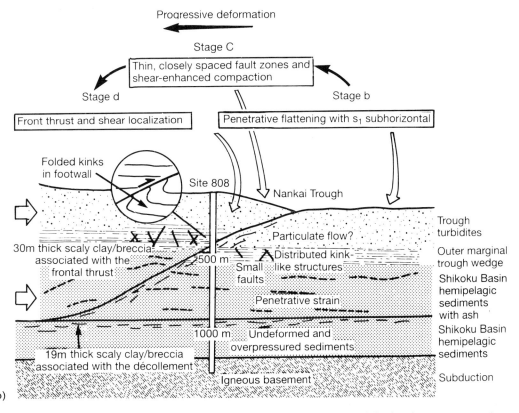

Figure 8.2 (a) Location map and (b) generalized cross-section of ODP Site 808. (c) Deviatoric versus mean stress diagram that summarizes the deformation history inferred for this site. Structural relations for the progression from stage b to stage d are given in Taira, Hill and Firth (1991), Maltman et al. (1993b) and Byrne et al. (1993b). In (b) the short, crossing heavy lines represent kink-like structures. The dashed lines schematically represent inferred ash layers and the short wavy lines show locations of scaly fabrics. Deviatoric stress versus mean stress in (c) is modified from Wong, Hiram and Zhang (1993).

tains ash layers; and a lower layer of hemipelagic sediments that does not contain ash. Based on drilling and seismic reflection data, the décollement below the Nankai prism appears to have propagated near the base of the ash-bearing hemipelagic sediments (Figure 8.2b).

Leg 131 of the Ocean Drilling Program drilled through the entire toe of the Nankai prism, including the frontal thrust and basal décollement, and penetrated the underthrust sediments to reach the oceanic basement (Taira, Hill and Firth, 1991). Moreover, relatively good core recovery allowed structural, physical property, geochemical and petrological data to be collected almost continuously. Over 70% of the 3000 measurements made by the structural geologists were oriented geographically with palaeomagnetic data. Information from Site 808, therefore, provides an exceptionally complete structural data set (Byrne et al. 1993a, b; Lallemant et al. 1993; Maltman et al. 1993b).

The results of Leg 131 provide four observations particularly relevant to the deformation and dewatering history of sediments. First, in the upper part of the prism, three generally time-transgressive structural features were documented (shown across the prism, or right to left in Figure 8.2b). These features are: (i) penetrative fabrics and kink-like structures; (ii) small, core-scale faults; and (iii) the frontal thrust. Second, the sediments above the décollement are overconsolidated relative to the sediments below the décollement, and probably relative to a normal marine sedimentary sequence (Taira, Hill and Firth, 1991). The sediments below the décollement appear to be underconsolidated and are considered to be overpressured. Third, the décollement is defined by a 20-m-thick zone of highly faulted sediments, with anomalously low porosities and textures that record cyclic periods of deformation and overpressuring. The brecciation and scaly clay fabric that are developed in this zone suggest similarities with shale-matrix mélanges, of which it may be an incipient example. Fourth, detailed computerized tomography (CT) scans of the core-scale faults, kink-like structures (Figure 8.3) and samples of the décollement show that these deformation zones are less porous than the host sediments (Byrne et al. 1993b).

Figure 8.2 summarizes the possible significance of these observations. The progressive development of the penetrative fabrics, small faults and the frontal thrust is interpreted using the stress-path diagram shown in Figure 8.2c (Schofield and Wroth 1968; Brandon 1984; Wong, Szeto and Zhang 1992). Stress path diagrams were explained in section 2.2.5 and developed in Chapter 6. In this interpretation, the sediments initially follow a uniaxial consolidation stress path (segment a, Figure 8.2c), resulting in a compactional bedding-parallel fabric. As the deformation front is approached, deviatoric stress increases, resulting in a steepening of the stress path (segment b, Figure 8.2c), and the principal stresses reorient (not shown in Figure 8.2c for simplicity). The new orientation of stresses results in a new compactive phyllosilicate fabric subperpendicular to bedding (Stage b, Figure 8.2b), as recorded by magnetic susceptibility and sonic velocity data (Byrne et al. 1993b; Owens 1993). As the mean and deviatoric stresses continue to increase, the stress path ultimately reaches the yield surface (marked by different porosities in Figure 8.2c) for the sediments, and the dominant deformation mechanism changes from intergranular dewatering and porosity loss to shear-enhanced compaction (e.g. Wong, Szeto and Zhang 1992). In the Nankai sediments, this transition is interpreted to be recorded by the change from penetrative fabrics to distributed faulting (stage c, Figure 8.2b). The increase in density within the fault zones argues for a strain-hardening stress path (e.g. b to c in Figure 8.2c). With higher mean and deviatoric stresses the state of stress of the sediments migrates along the yield surface (as the porosity decreases) and the sediments continue to fail, with strain hardening at individual faults, resulting in a pervasive fault fabric at the scale of the prism. Ultimately, the stress path crosses the critical state line and shear localization occurs (Stage d in Figure 8.2c). This stress state may correlate with the initiation of the frontal thrust (Stage d in Figure 8.2b).

The dewatering and deformation histories of the sediments within the décollement zone, in contrast, appear to be more protracted and less systematic (Byrne et al. 1993b; Maltman et al. 1993a). There are three important differences between the fabrics observed in the décollement and the structures from above the décollement. First, the shear surfaces in the décollement are typically much thinner, more irregular and discontinuous than fault surfaces from above the

Progressive deformation and dewatering **243**

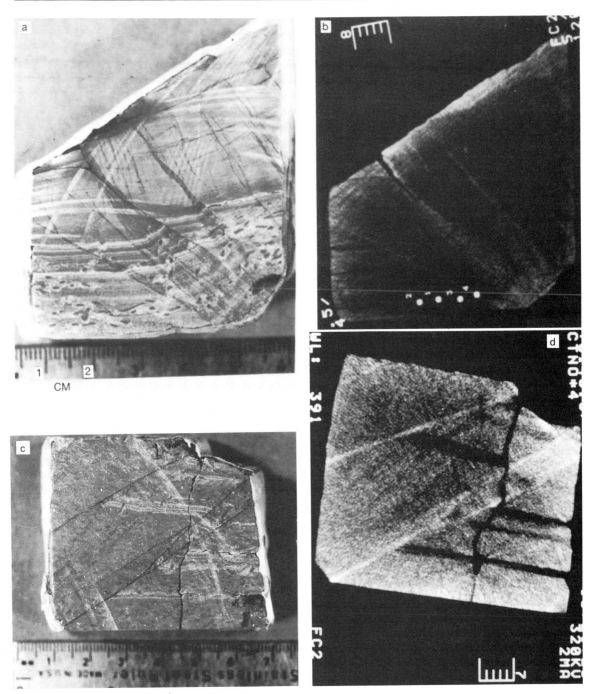

Figure 8.3 Photographs and computed tomography (CT) X-ray images of deformation structures from Site 808. (a) Core-scale photograph of kink-like bands and (b) a CT image of the same sample. (c) Core-scale photograph of several small thrust faults and (d) a CT image of the same sample. Note the close correlation between the bright areas in the CT image and the individual faults and bands. The bright areas show zones of relatively high density which is inferred to reflect a lower porosity. In (c) and (d) the fault zones appear to be more dense than the silty matrix that forms most of the sample. Similarly, the shale matrix appears to be more dense than the thin layers of sand (light areas in c and dark areas in d) that define bedding and are subhorizontal (i.e. parallel to the ruler). The amount of offset of the lower fault zone is about 3 cm. Photographs and CT images are parallel to the movement direction of the structures.

décollement. Second, the faults appear to form a three-dimensional web that results in a distinctive 'mottled' texture at both microscopic and submicroscopic scales. Finally, transmission electron microscopy observations reveal the presence of zones of brecciation and a collapsed phyllosilicate framework between individual shear surfaces. These observations suggest a cyclic history of shear-induced brecciation (and probably dilation) followed, at least locally, by pore collapse and phyllosilicate reorientation (Figure 8.4). All of these differences suggest a complex dewatering history for the décollement and probably account for the greater density of the décollement samples.

Byrne *et al.* (1993b) have proposed that these fundamental differences reflect the different structural positions and different fluid histories of the two suites of rocks. Specifically, they propose that the décollement zone is repeatedly deformed as overpressured fluids leak upward from the underlying Lower Shikoku basin sediments (Figure 8.2). Taira, Hill and Firth (1991) argue that the underthrust sediments are relatively water-rich and overpressured, based on the much higher porosities of these sediments. This interpretation is also supported by the reversed seismic polarity across the décollement (G.F. Moore *et al.* 1990). Sediments from the décollement zone, however, are more dense than any other sediments retrieved during Leg 131; thus, this zone was probably not fluid-rich during most of its deformation history. The décollement is therefore considered to be a zone of relatively low shear stress that fails episodically and cyclically as it dewaters. The complex, interconnected network, or web, of faults apparently facilitated dewatering in this zone. Local hydraulic gradients may have also driven these fluids up or down section.

In total, the Nankai accretionary prism sediments appear to be composed of three structural–hydrological regimes. The regime above the décollement is characterized by sediments that are progressively dewatered through the development of both penetrative fabrics and a closely spaced set of core-scale deformation structures (i.e. faults and kink-like structures). Individual structures probably strain harden (e.g. Guiraud and Seguret 1986; J.C. Moore and Byrne 1987), as shortening is accommodated by the formation of new structures elsewhere in the prism, resulting in a uniform distribution of structures in this regime (Taira *et al.* 1991). The décollement is characterized by a much higher density of structures per metre (probably an order of magnitude higher; Taira *et al.* 1991), and this regime is considered to be a zone of low stress and repeated failure. Individual deformation zones also probably strain harden in this regime but, because the materials are episodically overpressured, potential failure surfaces are as common as any fluid phase that is present. Hydrologically, the décollement retards the vertical flow of fluids and enhances the potential for overpressuring in the footwall because episodic failure progressively decreases the porosity, and thus probably also the hydraulic conductivity. Finally, the footwall contains very few tectonic structures and is essentially isolated from the subhorizontal stresses

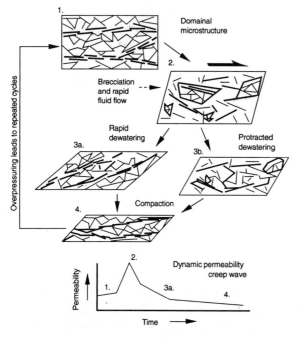

Figure 8.4 Interpretation of the cyclic and progressive development of scaly fabrics in the décollement zone at Nankai. (Based on Knipe, Agar and Prior 1991.) Shaded areas with thick lines are zones of locally reoriented phyllosilicates. Rapid and protracted dewatering paths are discussed in Knipe, Agar and Prior (1991). Only the rapid dewatering path is shown in the permeability–time curve in order to emphasize the progressive consolidation of the sediments. The recycling of the sediment through multiple deformation–compaction events is inferred to be driven by pulses of fluid overpressures. Each deformation–compaction event probably progressively consolidates the sediments, resulting in anomalously low porosities and permeabilities within the décollement zone.

related to plate convergence. This sediment regime is an important component of the tectonics of the Nankai prism, however, because it supplies the overpressured fluids that cause failure of the décollement at relatively low shear stresses.

8.3 PROGRESSIVE DEFORMATION OF COHERENT SEDIMENTS IN THE KODIAK ACCRETIONARY PRISM

A well-exposed suite of relatively coherent sedimentary rocks in the Kodiak accretionary prism, Alaska (Figure 8.5), records a deformation history similar to the offscraped sediments at Nankai. These uplifted sediments, however, provide a more complete picture of the consequences of deforming partly lithified sediments. The coherent sediments of interest here are part of the Upper Cretaceous to Lower Palaeocene Ghost Rocks Formation that crops out along the southeast coast of the Kodiak Islands (Figure 8.5b). In general, this formation consists of complexly deformed sandstone and shale, with interbedded volcanic rocks and hypabyssal intrusions. Although tectonic deformation has severely dismembered almost all the original stratigraphy, locally coherent sections are preserved and well exposed.

Detailed mapping of the coherent sections (Figure 8.5b; Byrne 1982, 1986) reveals sedimentary sequences consisting of thin (1–10 cm) sandstone and shale beds (e.g. Figure 8.6a and b) with local layers of pebbly mudstone and conglomerate. Based on the style and history of deformation, degree of metamorphism and structural setting, these relatively coherent sections are interpreted to be either offscraped sediments or slope basin deposits. The deformational history of the offscraped sequences is characterized by three coaxial deformations: conjugate folding, cleavage formation and conjugate strike-slip faulting.

The conjugate folds are unusual in their outcrop appearance in that they display a wide variety of styles and forms, suggesting deformation while the sediments were poorly lithified (Figure 8.6). Typically, the folds are disharmonically and/or polyclinally folded, and sandstone layers are locally folded, faulted, boudinaged, pinched-and-swollen, or all four types of structures may be present in the same outcrop or even

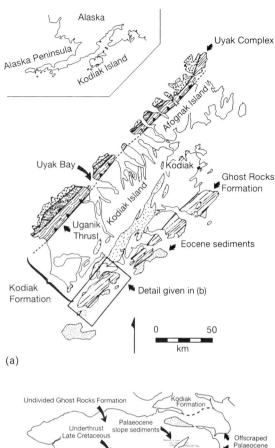

Figure 8.5 Geological maps of the Kodiak accretionary complex, Alaska. (a) The distribution of accreted sequences that are Cretaceous or younger. Units discussed in the text include the Uyak Complex, a relatively narrow section of the Kodiak Formation just below the Uganik Thrust and part of the Ghost Rocks Formation. (b) The distribution of Palaeocene offscraped sediments (including sediments with conjugate folds, see text) as well as slope basin sediments that overlie the underthrust sections of the Ghost Rocks Formation. Note the location of the Palaeocene décollement (barbs are on the upper plate).

in the same layer. In some cases, the pinch-and-swell structures appear to have formed as sediment literally flowed from the pinch area into the swells (e.g. Figure 8.6e). In other cases, classic 'box' folds with sharp angular hinges (Byrne 1982) give way to folds with rounded to circular

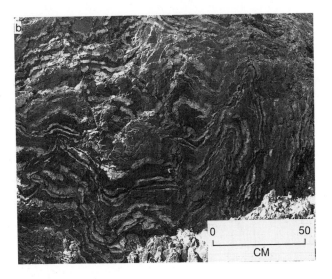

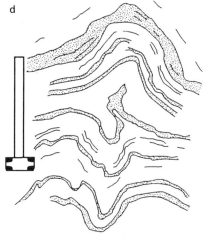

Figure 8.6 Examples of the style of deformation in the conjugate fold belt of the Ghost Rocks Formation. Note particularly the polyclinal and locally disharmonic nature of many of the folds in (a) and (b). (c and d). An outcrop-scale 'box' fold that suggests a very ductile style of deformation. In the middle of (e) note the abrupt thinning and thickening (or pinching and swelling) of a single folded layer. All of these folds are interpreted to have formed in partly lithified sediments.

hinges (e.g. Figure 8.6c and d). Overall, the mesoscopic-scale disruption of the folded layers is suggestive of deformation of weak, at best artly lithified, sediments. This interpretation is consistent with evidence of explosive brecciation and steam-mud injections along the contact of the sediments with a nearby andesite sill (Byrne 1984). Moreover, thin-section studies reveal shale intrusions in the hinge areas of some of the folds, and the shale appears to be dissolving synfolding calcite veins (Figure 8.7). Apparently, the shales (muds) were still undergoing diagenetic reactions that locally dissolved the adjacent $CaCO_3$ (probably through the production of CO_2 as discussed below). A partial state of lithification is also suggested by the 'ductile' swirl of shale just above the calcite vein in Figure 8.7c. A tectonic origin for this deformation is suggested by the progressive deformation history of the unit that contains the folds, as outlined below.

Although the conjugate folds are considered to have formed in partly lithified sediments, the folds are not chaotic, and 'S', 'Z' and 'M' fold profiles are present (Figure 8.6; Byrne 1986). The direction of maximum shortening during formation of the conjugate fold belt was determined from both individual conjugate folds and from the average orientation of the 'S' and 'Z' fold sets (Byrne 1982). These data show that when later deformations are removed the shortening axes trend subhorizontally about 319° (Figure 8.8).

A pressure-solution cleavage cross-cuts the conjugate folds, representing a subsequent stage of the deformation (Byrne 1982). The cleavage typically fans divergently around the axial surfaces of related mesoscopic and macroscopic folds, although areas of axial-planar fabric are also common. Incremental strain markers associated with the cleavage indicate that it approximately marks the plane of finite flattening and is therefore similar to many slaty cleavages. At a regional scale, the direction of maximum shortening indicated by the cleavage trends subhorizontally at 316°, which is similar to the axis of shortening determined from the conjugate folds (Figure 8.8).

A third deformation consists of a set of conjugate right-lateral and left-lateral strike-slip faults (Figure 8.8). Both types of faults have slickenlines that are nearly horizontal and the inferred shortening direction trends 312°, consistent with the direction of shortening determined for the folds and cleavage. This consistency in the shortening direction suggests that the three deformation phases are not distinct episodes, but different stages of one progressive event that was affecting materials with gradually changing mechanical properties, in this case, progressively lithifying sediments.

An important problem in interpreting the significance of such pre-lithification deformation is defining its possible origin – did the folds form from gravity induced slumping or from tectonic processes? Although there is no direct way of answering this question, three field observations suggest these folds are of tectonic origin. First, the development of the structures post-dates both the compaction of the sediments and the formation of calcareous concretions (Byrne 1982). If the folds formed during gravitational sliding, the slumped unit would have to be of sufficient size to cut deeply into a compacted and partly lithified sequence. Second, the northern boundary of the conjugate fold belt is gradational. There is no discontinuity or 'break-away' zone suggestive of a slump zone. Finally, the conjugate fold deformation is nearly coaxial with two later deformations that are clearly of tectonic origin (i.e. the pressure solution cleavage and the strike-slip faults). The simplest interpretation, therefore, is that the early conjugate folds that formed in partly lithified sediments record the early stages of a progressive tectonic deformation. This is an example of the approach shown diagrammatically in Figure 8.1a.

8.4 PROGRESSIVE DEFORMATION OF MÉLANGE TERRANES IN THE KODIAK ACCRETIONARY PRISM

The Kodiak accretionary prism also exposes at least three sedimentary sequences that were underthrust beneath the prism and experienced deformation while still unlithified. A more detailed description of the structural histories of these sequences is given elsewhere (Connelly 1978; Moore and Wheeler 1978; Byrne 1984; Vrolijk, Meyers and Moore, 1988; Fisher and Byrne 1987; Byrne and Fisher 1990; Sample 1990), but a few relevant aspects of these histories are summarized below. Fisher and Byrne (1987) proposed that underthrusting

248 *Sediment deformation, dewatering and diagenesis*

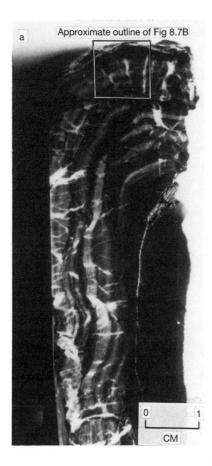
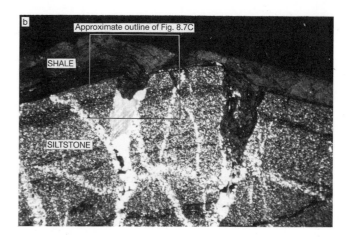
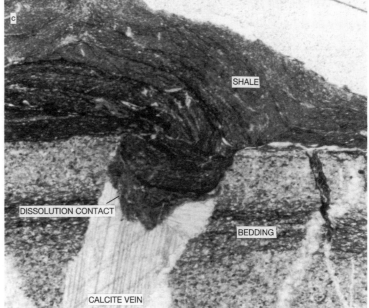

Figure 8.7 Photograph (a) and two photomicrographs (b and c) of the hinge area of a conjugate fold from the Ghost Rocks Formation: (a) is a photograph of a slabbed hand-sample; (b) and (c) show the veins and mud intrusions that developed in the hinge area. Note the well-developed swirls in the phyllosilicate fabric of the mud, and the irregular to sutured-like contact between the mud and the calcium carbonate vein in (c).

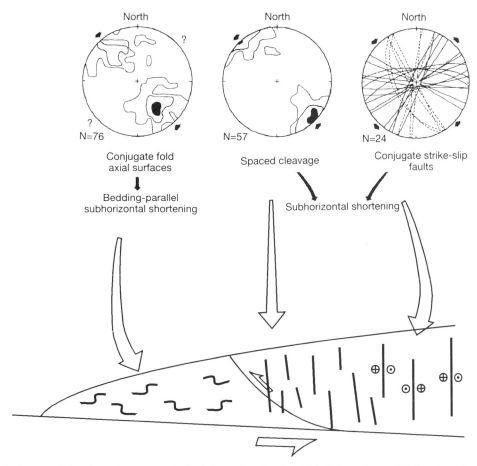

Figure 8.8 Structural data from the progressively deformed conjugate fold belt (from Byrne, 1986) and a schematic diagram showing the general structural setting and development of this belt. The conjugate folds are interpreted to have formed in partly lithified sediments, whereas the spaced (pressure-solution) cleavage and strike-slip faults formed in lithified sediments. All three deformations, however, are interpreted to be tectonic in origin because together they record a progressive coaxial deformation.

beneath an accretionary prism results in a unique tectono-stratigraphy (Figure 8.9). This consists of a major fault or décollement and an underlying tectonic mélange (Figures 8.9 and 8.10) that may grade downsection into structurally coherent sedimentary sequences or even oceanic crust. Examples of this tectonostratigraphy occur in the footwall of the Uganik Thrust, in sediments underlying the structurally coherent sediments of the Ghost Rocks Formation (discussed above) and in the Uyak Complex on the northwest side of the Kodiak Islands (Figure 8.5a and b). All of these sequences also record a progressive deformation indicating northwest-directed underthrusting.

The deformation of each of these underthrust sequences is characterized by four structural features that are relevant to understanding the deformation of partly lithified sediments. First, deformation is concentrated in the regional-scale mélange terranes that occur at the top of section, and the deformation is dominated by approximately layer-parallel shear of the partly lithified sediments. Second, this deformation is associated with an extensive development of carbonate-filled tensile fractures or veins that increases in abundance near the overlying décollement (Fisher and Byrne 1987, 1990). Third, the orientation of the veins, as well as a set of conjugate extensional shear zones in the Uyak Complex, indicate that σ_1 was steeply plunging during underthrusting (Byrne and Fisher 1990). Byrne and Fisher (1990) used these observations to propose that the décollement above the mélange was relatively weak and probably overpressured. This interpretation is also consistent with the results from the Nankai accretionary prism (section 8.2), which show an overpressured underthrust sequence that is capped by tectonic scaly fabrics (or incipient mélange) and a décollement (Figure 8.2b). Finally, the penetrative deformation history of these sedimentary sequences culminated with the development of a pressure-solution cleavage that is interpreted to be of tectonic origin (e.g. Figure 8.9b–g). In the following paragraphs I present in more detail the evidence for a partly lithified state of the sediments during deformation and the arguments for considering this deformation to be of tectonic origin.

The main evidence for a partly lithified state during deformation of the Kodiak mélanges comes from three sets of observations. First, like some of the conjugate folds described above, many of the structures of the mélange terranes have a 'ductile' or 'fluid' appearance at both outcrop and thin-section scales (Figures 8.9, 8.10 and 8.11). For example, several tuff horizons in the mélanges of the Ghost Rocks Formation display complex swirls and wisps that have a distinct (seahorse-like) shape (Figure 8.9b and c). Tuff horizons in the Uyak Complex also typically display ductile textures at an outcrop scale (Figure 8.11). Fisher and Byrne (1987) have described asymmetric sand inclusions that appear to have disaggregated through particulate flow (Figure 8.9; Fisher and Byrne 1987) and asymmetric swirls or spirals of phyllosilicates (e.g. Figure 8.9d) are common in the shale matrix.

Second, many of the extensional shear zones (or normal faults) in the Uyak Complex appear to have formed contemporaneously with the emplacement of shale diapirs, dykes and sills (Figure 8.12). The intrusions range in dimensions from relatively thin (tens of centimetres) up to hundreds of metres of coastal exposure (Byrne and Fisher 1990). Evidence that the shale intrusions were active at the time of deformation is provided by cross-cutting relations between the intrusions and the deformational fabrics. An excellent example of a mushroom-shaped shale diapir that cross-cuts the structural fabric defined by sheared bedding layers is shown in Figure 8.12a. Moreover, small protrusions of this diapir have intruded along two nearby extensional shear bands (Figure 8.12b), indicating that the shear bands formed before or during emplacement of the diapir (Byrne and Fisher 1990). In other outcrops, relatively large pieces of mélange with a well-developed fabric have been incorporated into the intrusions (Figure 8.12c). These pieces of included mélange are often asymmetric in cross-section and suggest northwest-directed underthrusting.

Finally, in all three of the underthrust sequences, microscopic-scale mud intrusions and three types of carbonate veins record increasing cohesion of the sand inclusions (Figure 8.13a). All of these structures cross-cut cataclastic shear zones and small faults that define the characteristic 'blocks' of the mélange (Fisher and Byrne 1987). In detail, the relations between these veins and mud intrusions illustrate a number of aspects of the deformation–fluid–diagenesis interplay mentioned in the other chapters, so they are explored below.

The earliest veins are filled by grains of mosaic calcite that contain large numbers of microscopic-scale clay inclusions. The inclusions appear to be pieces of the sediment matrix that mixed with the vein fluids as the vein formed. As a result, the vein appeas 'dirty' when viewed in plane-polarized light (Figure 8.13b). It is significant that the vein wall is very irregular and

Figure 8.9 (a) Schematic model for the formation of mélange terranes during underthrusting, and examples of the types of structures that form in this environment. (b and c) Examples of asymmetrical swirl tuff in the shale matrix of the Ghost Rocks mélange. The complex interdigitations of the shale and tuffs suggest a very ductile deformation for these rocks. (d) Photomicrograph showing a microscopic-scale fold in the mélange matrix, which also appears ductile at this scale. (e) Photomicrograph of the mélange matrix showing well-developed slip surfaces as well as an asymmetric sand inclusion. The inclusion is inferred to have formed as individual sand grains, or fragments of grains were disaggregated from the inclusion. (f) Outcrop-scale example of a sand inclusion that displays an asymmetry opposite to asymmetry shown in (e). In contrast to the inclusion shown in (e), this inclusion is inferred to have rotated as the matrix flowed in bulk shear. (g) Photomicrograph of a diagenetic pyrite framboid from the Uyak Complex with curved fibres of quartz and chlorite. This example records a non-coaxial strain history consistent with northwest-directed underthrusting (see enlargement for g in a; Byrne and Fisher 1990). Overall, the underthrust sequences record progressive deformation and lithification of sediments during a rotational strain regime. All photographs view to the northeast.

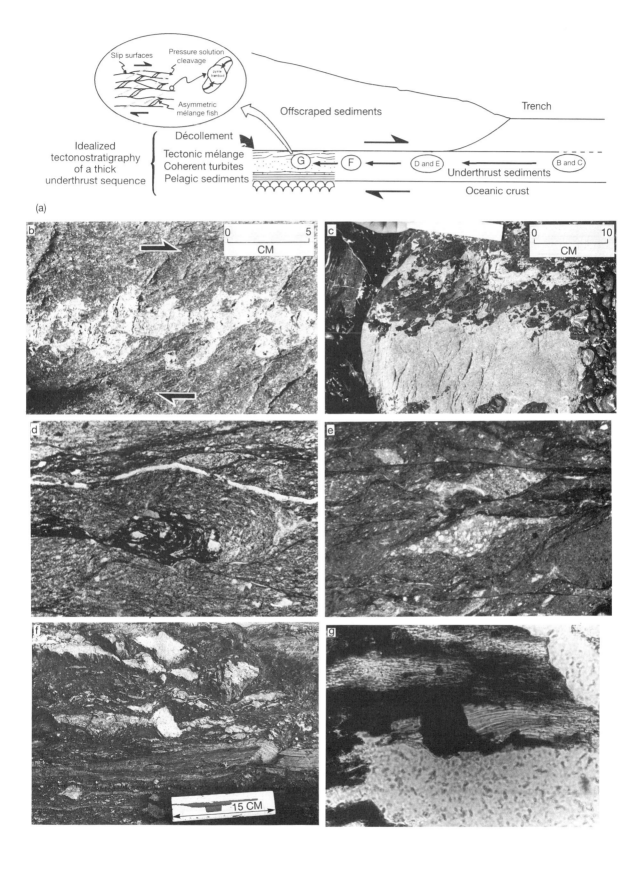

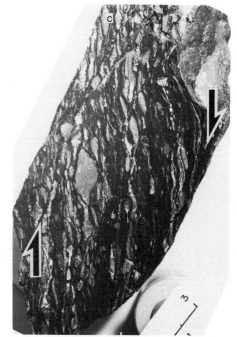

Figure 8.10 Photographs of shale-matrix mélange at outcrop scale and hand-sample scale. (a) A typical coastal exposure of the Ghost Rocks mélange. (b–d) Slabbed surfaces of the 'matrix' of this mélange. Note the abundance of sandstone inclusions at all scales and the angular to rounded shapes of the fragments. Note also the preservation of several carbonate vein fragments in (d).

Figure 8.11 Outcrop photograph of a folded tuff layer in the Uyak Complex. Note the extreme ductility suggested by the swirling and stretching of the tuff layer.

anastomoses around detrital grains of quartz and feldspar. Relatively large pieces of the sand wallrock (i.e. clusters of quartz and feldspar grains) are also commonly incorporated into the vein. All of these observations indicate that the veins formed at a time when the material bonding the sediment together (i.e. the matrix) was weaker than the framework grains, suggesting that the sand as a whole was incompletely lithified. At the same time, the presence of the veins themselves suggests that the sands were not completely unlithified. These earliest veins therefore appear to have formed during the initial stages of lithification and diagenesis.

This interpretation is also consistent with the local occurrence of microscopic shale intrusions that deform and dissolve the early carbonate veins. One of the best examples of these intrusions comes from an elongated sandstone inclusion that is coated with a thin veneer of shale (Figure 8.14). The carbonate veins in this example record extension parallel to the long axis of the sandstone inclusion and one of the veins has been expanded and dissolved by intruding shale. The expansion of the vein was accommodated in the surrounding sandstone by formation of a small right-lateral shear zone (left-central part of Figure 8.14c). The crack (vein + mud intrusion) is consequently wider in the area intruded by shale (Figure 8.14c). Also, as the shale was intruded, a rectangular piece of the calcite vein was plucked from the wall rock and incorporated into the shale. Serial thin-sections parallel to the slabbed surface in Figure 8.14c and d show that this plucked vein material is elongate perpendicular to the cut surfaces and that it originated in the 'hole' on the opposite vein wall (Figure 8.14d). The direction of opening of vein suggested by matching the plucked piece with the hole in the opposite vein wall is consistent with the small shear zone that formed as the vein expanded (compare Figure 8.14c and d).

An important aspect of the relation between the shale and the carbonate is that the contact is indented, or suture-like, and is interpreted to be a dissolution surface. This is especially clear along both the vein walls and around the plucked, roughly square piece of calcite (Figures 8.13b and 8.14d). Apparently, as the calcite-filled vein cracked and filled with shale, the shale partially dissolved both the plucked piece and the calcite remaining on the walls of the vein. About 40% of the original calcite vein has been dissolved by the shale, based on a comparison of the calcite vein thicknesses in areas affected and unaffected by the shale intrusion. At least some of this dissolved calcite appears to have been locally reprecipitated, because the tip of the shale intrusion is enriched in calcite and thin, post-intrusion calcite veins are localized in this region of the intrusion (see tip area of mud intrusion in Figure 8.14d).

The origin of the dissolution contact between the calcite and the intruding shale is difficult to interpret directly; however, a reasonable interpretation is that an acidic environment in the shale was produced by diagenetic processes within the shale. For example, studies of vertical sections of sands and shales in the Gulf Coast of Mexico (Boles and Franks 1979; Boles 1982) show that with increasing maturation (i.e. pressure and temperature), organic material degrades and produces CO_2, which, through bicarbonate reactions, results in the instability of calcium carbonate. This

254 *Sediment deformation, dewatering and diagenesis*

Figure 8.12 Outcrop photographs showing diapiric intrusions of shale and their relations to extensional shear bands in the Uyak Complex. (a) A relatively small intrusion with a mushroom-like cross-section. The largest documented intrusion occupies approximately 100 m of coastal exposure. (b) An extensional shear band that has been intruded by shale (centre of figure and along the bands). The intruding shale originates from the intrusion in (a) (located just to the right of b). (c) An isolated fragment of mélange (elongated from upper left to lower right) in a deformed diapir (darker, fine-grained material). The mélange fabric contains an inherited integral fabric that dips steeply to the right (northwest). All photographs view to the west or southwest.

process is important to the oil industry because it often results in secondary porosity in sandstones. In the suites studied in this paper, similar diagenetic reactions probably mobilized calcium relatively early. Calcium was presumably initially concentrated by the dissolution of fossil tests or the alteration of plagioclase feldspars. In any case, calcite precipitation probably occurred as the sand lenses failed and cracked, creating local areas of lower pressure. Presumably

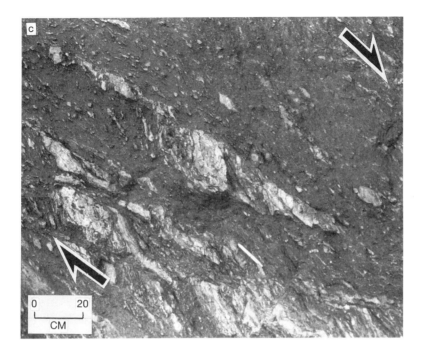

Figure 8.12. Contd.

the veins opened incrementally and at a rate comparable to the rate of growth of the calcite filling, otherwise the crack would have filled with shale. During or after the calcite was deposited, the shale that surrounded the sandstone inclusion, and initially probably contained more organic material, intruded and dissolved the ends of the recently precipitated calcite veins. In some cases, this dissolved calcite saturated the fluids in the shale and was reprecipitated as matrix or vein material. More importantly, these dissolution textures indicate that diagenetic processes were still active as the sand inclusions and calcite veins were forming.

These early veins and mud intrusions are in turn cross-cut by a second set of carbonate veins Figures 8.13b and 8.14d). These veins, like the earlier set, have irregular walls and the vein-forming crack appears to have anastomosed around the more rigid quartz and feldspar grains. These younger veins, however, are free of microscopic-scale inclusions and are clear when viewed in plain light. These textures are, therefore, interpreted to be evidence of progressive lithification in the sands. Apparently the sand(stone?) cement was still weaker than the quartz and feldspar framework grains, but not weak enough to mix with vein fluids.

The youngest set of veins is relatively rare, and consists of calcite veins that have sharp, straight interfaces with the sandstone wall rock. Individual quartz and feldspar grains are often cracked and separated by the vein. The calcite in the veins is typically clean and locally fibrous. Moreover, these veins often propagate into the shale that surrounds the sandstone inclusions. The veins also appear to grade into or are overprinted by quartz veins that are believed to be related to a later pressure-solution cleavage. The straight boundaries, fibrous textures, absence of inclusions and propagation into the shales suggest that the sandstone, and probably the shales, were relatively cohesive and perhaps fully lithified at the time these veins formed. Importantly, this youngest set of veins cross-cuts the two earlier vein sets. These textures provide additional evidence for a partially lithified state of the sandstones during the formation of the early veins.

Progressive lithification ⇓

Early – 'Dirty' calcite veins non-grain breaking
Middle – 'Clean' calcite veins, non-grain breaking
Late – 'Clean' calcite and quartz veins that break grains (pressure solution-related veins)

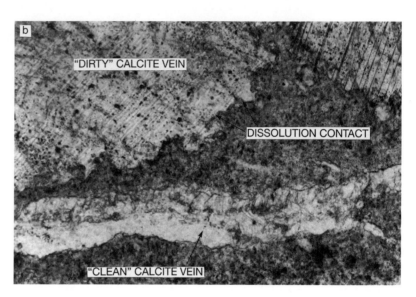

Figure 8.13 (a) Summary of the three types of veins recognized in the Kodiak Island mélange terranes (a) and a (b) Photomicrograph of dirty (top part of photograph) and clean (lower part of photograph) veins. Note also that the contact between the dirty vein and shale (lower half of the photograph) is irregular and suture-like whereas the contact between the clean vein and shale is much straighter and not suture-like. See text for discussion and Figure 8.14d for the location and scale of these veins.

8.5 DEFORMATION AND FLUID EVOLUTION IN AN ACCRETIONARY SEQUENCE IN WESTERN WASHINGTON

A well-exposed suite of deformed rocks on the Olympic Peninsula provides an example of another approach to understanding the relation between deformation and dewatering in sediments. In this area, Orange, Geddes and Moore (1993) have been able to compare the progressive vein chronology in a deformed sequence (the Hoh rock assemblage) with the progressive diagenetic history in a lithologically similar and probably correlative, but undeformed, sequence along strike (the Grays Harbor and Chehalis Basin sediments; Galloway 1974). The undeformed sequence therefore provides a reference frame for evaluating the role that deformation plays in modifying diagenesis and dewatering.

The Hoh rock assemblage is composed of Eocene to Miocene sediments that were substantially deformed and disrupted but only mildly metamorphosed (maximum conditions of laumontite grade) during accretion in the late Tertiary. Deformation structures include outcrop-scale and regional-scale faults and folds, as well as regional-scale belts of tectonic mélange. The style of deformation and relatively mild metamorphic conditions during deformation suggest that the Hoh sediments were offscraped or underplated at a shallow structural level (Orange, Geddes and Moore 1993).

Based on detailed field and petrographical studies combined with cathodoluminescence,

Orange, Geddes and Moore (1993) have documented a protracted history of cements and cross-cutting veins in the Hoh rocks. The history includes an early phase of distributed grain breakage, followed by carbonate cementation, several phases of carbonate veins with different petrographical or luminescence characteristics, and a final state of laumontite precipitation. Orange, Geddes and Moore also documented a progression from 'dirty' to 'clean' carbonate veins similar to the Kodiak examples (section 8.4), and considered the progression to reflect the progressive consolidation and lithification of the sediments.

One of the most important conclusions of Orange, Geddes and Moore (1993), however, is that despite the complexities suggested by the vein geometries and the subtle variations in composition, the vein history is essentially identical to the diagenetic history recorded in the undeformed sediments (Figure 8.15). Thus, at least in this case, deformation appears to have played a passive role in the fluid evolution. That is, the deformation processes only captured locally derived fluid phases; exotic sources are not needed to explain the fluid-flow history. On the other hand, based on the abundance of veins in the deformed Hoh assemblage, deformation probably played a more active role in dewatering the sediments. That is, in the deformed area deformation processes probably accentuated and localized the process associated with dewatering, relative to the undeformed region.

8.6 CONCLUSIONS

This chapter has reviewed a number of structural features from modern and ancient accretionary prisms, and particularly from shale-matrix mélanges, that appear to record the deformation of partly lithified sediments. These features include: penetrative fabrics, small faults and kink-like structures, unusually ductile mesoscopic-scale folds, 'wisps' and 'swirls' of tuff and shale, calcite veins and mud intrusions. Although many of these structures have been recognized and documented previously from unlithified sediments, they are here integrated into a progressive deformation. This integration suggests two new approaches to understanding the processes that cause the deformation of sediments.

In the first approach, different stages of sediment lithification are tied through cross-cutting textural relations to a single, progressive deformation. For example, the conjugate fold belt of the Ghost Rocks Formation in Alaska has deformation axes that are coaxial with two later deformations that were clearly formd by tectonic processes – a pressure-solution cleavage and strike-slip faulting. Consequently, it is proposed that the conjugate fold belt also formed due to tectonic processes even though the sediments were probably only partly lithified when the folds formed. Overall, this belt records the progressive lithification and deformation of sediments in a coaxial, irrotational strain regime.

A second example of this approach to understanding the geological deformation of sediments is provided by the mélange terranes in the Kodiak accretionary complex. In these sediments, the earliest deformation is recorded by asymmetric 'swirls' and 'wisps' of shale and by 'dirty' veins of calcite in sand inclusions. These early structures are overprinted by 'clean' crack-seal veins of calcite and quartz associated with penetrative pressure-solution cleavage. The cleavage also resulted in the growth of asymmetric pressure shadows. Because both the early and late deformations record non-coaxial strains consistent with underthrusting, the entire deformation is considered to record the progressive lithification and deformation of sediments in a rotational (shearing) strain regime. The example also illustrates the care that has to be used when using mineralized veins as criteria for the recognition of pre-lithification deformation (section 9.6.3).

Finally, the second proposed approach to understanding the deformation of sediments involves using progressive stages of diagenesis as a reference frame for the deformation history. This approach is similar to the way metamorphic petrologists use cross-cutting relations between different mineral assemblages and different deformational fabrics to document prograde and retrograde deformation. One of the best examples of this approach is the recent study of the complexly deformed sediments in the Hoh accretionary complex of the Olympic Mountains (Orange, Geddes and Moore 1993). These authors were able to tie different stages of deformation to specific types of veins and then correlate the

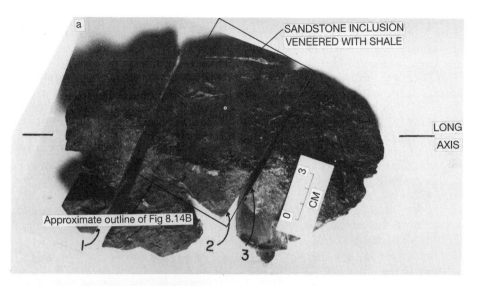
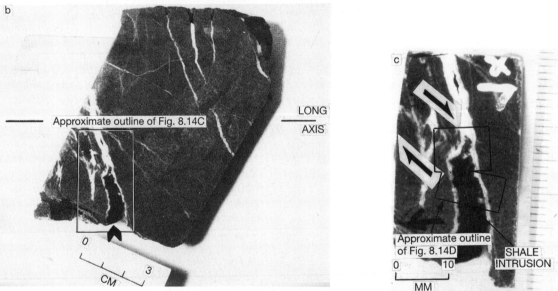

Figure 8.14 Photographs showing the textures and relations of calcite veins and mud intrusions in a sandstone inclusion in the Ghost Rocks mélange terrane. (a) The external texture of a mélange inclusion. The sandstone inclusion is black and polished because it is coated with a thin veneer of shale, as are all of the inclusions from the mélange. (b) A slabbed surface parallel to the long and intermediate axes of the inclusion (i.e parallel to the surface shown in (a) revealing multiple calcite veins that are nearly perpendicular to the long axis of the inclusion. (c) A more detailed photograph of one of the veins in (b). The calcite vein shown in (c) has been intruded by shale and, apparently, as the shale was emplaced the calcite vein cracked along its medial line and expanded, as shown in more detail in (d). Cracking of the vein and intrusion of the shale also caused part of the vein material to be plucked from one of the vein walls (see plucked calcite piece in d). The approximate plucking path (shown in d) is consistent with the direction of expansion inferred for the vein as the shale was intruded. That is, small sigmoidal veins to the left of the intrusion tip (see c) indicate that the calcite vein was expanded towards the upper left, parallel to the approximate plucking path in (d). Detailed studies of the contact between the shale and the calcite vein show a suture-like contact (e.g. Figure 8.13b), suggesting dissolution. Area calculations of the cracked and dissolved vein (lower part of c), and the undeformed vein (top part of c) suggest that about 40% of the vein material has been removed by dissolution. This dissolved carbonate is inferred to have mixed with the shale as it was intruded. This is evidenced by the increase in concentration of carbonate in the tip region of the shale intrusion (e.g. much lighter shale in the tip area shown in d).

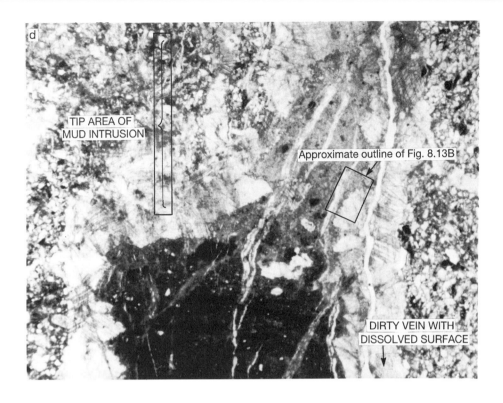

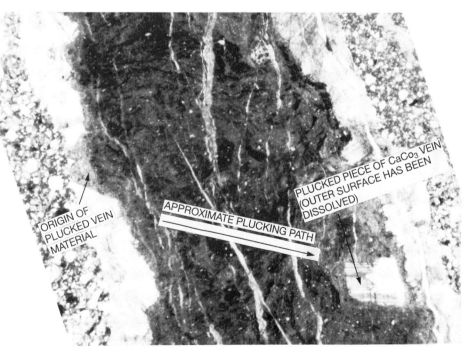

Figure 8.14. Contd.

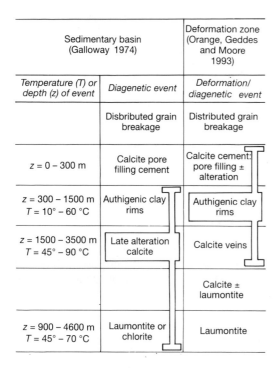

Figure 8.15 Diagram that compares the vein stratigraphy in the complexly deformed Hoh rock assemblage (Orange, Geddes and Moore 1993) with the diagenetic history of two, undeformed and probably equivalent sedimentary basins in the same area (Galloway 1974). (Modified from Orange, Geddes and Moore 1993.)

different veins directly to the progressive diagenetic history as preserved in undeformed sediments along strike. The veins also progress from 'dirty' to 'clean' with evolving diagenesis and deformation. Orange, Geddes and Moore (1993) therefore proposed that the tectonic deformation in the Hoh complex involved both unlithfied and lithified sediments.

In summary, the above review of selected examples of deformed sediments serves two purposes. First, the discussions are consistent with a number of recent studies that propose a complex interplay between the deformation, diagenesis and dewatering of sediments at submarine convergent margins (e.g. Langseth and Moore 1990). The summaries presented here also illustrate that the study of ancient well-exposed examples of such rocks may provide further insight into these processes (Guiraud and Seguret 1986; Maltman 1988; Sample 1990; Labaume, Berty and Laurent 1991). Second, the reviews suggest an approach for distinguishing the fundamental processes responsible for the formation of shale-matrix mélanges. Although such mélanges often appear to record complex geometries and kinematics, the structures are not chaotic. Rather than inferring submarine slumping and sliding, these units might be better understood in the context of interlinked progressive deformation, dewatering and diagenesis.

CHAPTER 9

Deformation structures preserved in rocks

ALEX MALTMAN

9.1 INTRODUCTION

The deformation processes outlined in the previous chapters give rise to a wide range of structures, most of which have the potential to become preserved during lithification. The problem facing the field geologist working with ancient rocks, therefore, is first to identify such structures correctly, and then to incorporate the mechanics of their origin into an understanding of the overall geology. A synoptic view of the possible structures is given in this chapter. It includes examples both from sediments and from materials that are now fully lithified, but it is not meant to be a catalogue, and it is certainly not comprehensive. A picture book approach has been avoided. Sims (1978) provided an extensive bibliography of reports of structures in very near-surface sediments, as did van Loon (1992), who set his numerous literature references in a historical context. In a general way, the following pages mention the structures in order of increasing scale. Emphasis is given to the *variety* of structures, by paying more attention to relatively newly recognized and less well-known features, and to those aspects that are potentially of use in wider interpretations. The maddening business of identifying pre-lithification structures in rocks is deferred until last.

9.2 TECHNIQUES OF EXAMINATION

The main differences between standard field geological techniques and those used for deformed sediments arise where the material is still poorly lithified and delicate. Although not directly concerned with deformation studies, methods of sampling, describing and analysing loose, on-land sediments are given by McGown et al. (1980) for civil engineering purposes, Gale and Hoare (1991) in the context of Quaternary studies, and Bullock and Murphy (1983) for agricultural soils. Submarine structures are usually seen in retrieved cores, techniques for which are dealt with in the *Initial Reports of the Ocean Drilling Program* (ODP).

A problem that continues to bedevil the analysis of structures seen in cores is the identification of coring disturbance. Improvements in corer design and quality control are reducing the problem (Parker 1991), but the geologist still has to make subjective assessments. Lundberg and Moore (1986) discussed criteria for recognizing coring artefacts: they were particularly wary of features that showed any geometrical relationship with the core axis. A standard problem of sampling on-land sediments is to somehow reinforce the material so that it can be transferred to the laboratory with minimal disturbance. The usual approach is to add some form of quick-drying liquid adhesive or cement, of specification appropriate to the sediment, climate and available time. Low-viscosity resins penetrate effectively but have longer hardening times; heating to accelerate hardening can induce disturbance. Holeywell and Tullis (1975) used a propane torch to harden epoxy in the field; to stabilize vertical surfaces in loose gravel, Hattingh, Rust and Reddening (1990) resorted to a combination of wire-netting, wallpaper paste and Portland cement! Once in the laboratory, more sophisticated impregnation techniques can be deployed, but even here, major

The Geological Deformation of Sediments Edited by Alex Maltman Published in 1994 by Chapman & Hall ISBN 0 412 40590 3

difficulties arise when detailed fabric studies are envisaged for wet or very poorly permeable sediments.

For the purposes of optical thin-section preparation, the technique of slowly replacing the pore water with molten Carbowax, a high molecular weight polyethylene glycol, has been used with success by many workers. The method was outlined by Maltman (1987a). However, the 40°C temperatures and slight volume change can produce artefacts; the sediment can behave capriciously and the ready volatilization precludes its use in the electron microscope. For the latter purpose, and for most other techniques of thin-section production, it remains necessary to dry the sediment, as the various impregnation agents fail to penetrate and displace the pore water. The trick is to achieve the drying with minimal fabric disturbance, and for this various combinations of air-, oven-, substitution-, freeze-, or critical-point-drying have been used, as described by Smart and Tovey (1982).

Even with intricate procedures like these, some sediments destined for optical thin-sections prove fragile during the initial immersion, especially if there are discontinuities in the specimen. Swartz and Lindsley-Griffin (1990) advocated some preliminary air-drying of such materials, and Palmer and Barton (1986) recommended some air-drying followed not by complete immersion, but by gradual dripping of the impregnation fluid on to the sample. For this, they recommended a two-component, cold-setting epoxy resin (Epo-tec 301), and found the addition of not more than 1% of dye (Waxoline or Keystone Oil) useful for defining pore space and microfractures. A number of workers have recommended the four-component epoxy resin known as SPURR (e.g. Jim 1985), with the impregnation being carried under vacuum. Awadallah (1991) advocated using a low-viscosity dimethacrylate called LR White.

Similar preparatory procedures are required for the electron microscopy of sediments, although preliminary attempts have been made to develop environmental cells that enable samples to remain wet while being examined (Fukami, Fukushima and Kohyama 1991). A well-established technique for electron microscopy of substitution-drying followed by freeze-drying was described in step-by-step detail by Baerwald, Burkett and Bennett (1991). A summary of the procedure is given here as Figure 9.1. Scanning or transmission electron microscopy is used as appropriate, and in recent years the advent of back-scattered electron imagery has proved particularly powerful (Figure 9.2). Besides improving resolution, this technique provides a visual contrast between grains of different composition and allows some quantification of the fabric elements.

A vital current trend in sediment microscopy is the improvement of the quantitative aspects, a task that is being greatly aided by computer-assisted image analysis. Techniques at the electron microscope scale are discussed by Smart et al. (1991). Provided that the particles or pores of interest have sufficient visual distinctiveness, the image can be scanned, digitized and processed to give information on distributions, sizes and shapes of pores and grains, and to quantify preferred orientations (e.g. Bhatia and Soliman 1991). Luo et al. (1992) described simple computer-assisted techniques for analysing the orientations of sediment particles.

Another trend is the development of techniques for detecting structures that are not readily visible. The anisotropy of magnetic susceptibility in titanomagnetite-bearing sediments is proving useful as a measure of finite strain and in detecting fabrics that are visually elusive (Hounslow 1990; Owens 1993). X-radiography has long been a standard technique for diagnosing sedimentary structures in cores and other packages where the features are obscure at the external surface. It is also useful for locating the presence of discrete deformational structures, as many of these have a sufficient porosity (density) difference to affect the X-ray transmission. A recent advance has been to link series of planar X-ray scans through a computer to give three-dimensional information on the structures; this computer-assisted tomography is variously known as CT or CAT scanning. Two-dimensional examples of shear zones in both experimentally deformed clays and similar structures from the Nankai accretionary prism are shown in Figure 8.3. In an interesting reversal of this application, Gilliland and Coles (1990) have suggested CT scanning to detect the presence in undeformed sediments of structures indicating core damage!

Step	Technique	Instrumentation or description	Step	Technique	Instrumentation or description
		Dehydration			
1	Subsampling of clay specimen and trimming	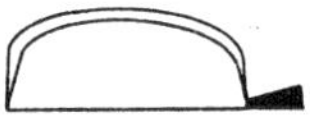 Wire knife	4	Replacement of amyl acetate with liquid CO_2	 Critical point apparatus
2	Replacement of interstitial water with ethyl alcohol	 Silver nitrate test; sealed glass containers	5	Critical point drying with CO_2	 Critical point apparatus
3	Replacement of ethyl alcohol with amyl acetate	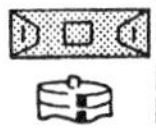 Protected specimen in lens paper or perforated sample holder	6	Placement and storage of individual specimen in desiccators	 Samples in vacuum desiccated or sealed jar with desiccant

	TEM only			SEM only	
7T	Placement of specimens into individual containers and into vacuum desiccator	 Plastic cup or Beem capsule; protective lens paper removed	7S	Fractured specimen	Razor blade
8T	Specimens impregnated under vacuum and removed	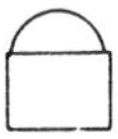 Vacuum desiccator; Spurr; epoxy resin	8S	Mounted specimen	Aluminium stub
9T	Specimens cured at 60° – 70 °C	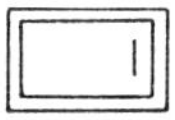 Oven	9S	Electrostatically clean fractured surface	CAB plastic tube; Sample
10T	Trimming of specimens prior to ultrathin sectioning	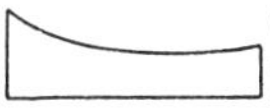 Glass knife or manually with a razor blade			
11T	Ultrathin sectioning with microtome	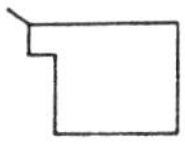 Diamond knife microtome			
12T	Placement of sections on copper grids	 Copper grid	10S	Sputter coating with palladium gold wire	Vacuum evaporator
13T	Light carbon sputtering of ultrathin sections	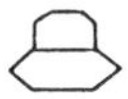 Vacuum evaporator			
14T	Ultrathin sections photomicroscopy	 TEM	11S	Surface feature photomicroscopy	 SEM

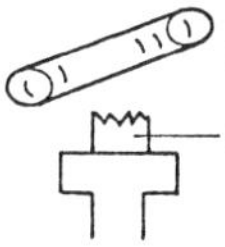

Figure 9.1 Summary of a typical sample preparation technique for the electron microscopy of sediments. The scheme involves critical-point dehydration of the sample (steps 1 to 6), followed by either transmission electron microscopy (TEM, steps 7T to 14T) or scanning electron microscopy (SEM, steps 7S to 11S). (From Baerwald, Burkett and Bennett (1991). Used with permission from Springer-Verlag.)

Figure 9.2 Example of a back-scattered electron image. Normal fault zones are filled with fine opal-A (pale colour), derived from the adjacent laminated diatomite (dark colour). Specimen from the Monterey Formation, Grefco Quarry, near Lompoc, California. (Photograph provided by Richard Brothers.)

9.3 MICROFABRICS

The term fabric is used informally here for patterns of grain orientation that are pervasive at the scale of observation. Such an arrangement normally comes about through deposition and consolidation, but tectonic fabrics in sediments are now being discovered (Maltman *et al.* 1993b). Microstructure is reserved for discrete modifications, at the microscopic scale, that result from deformation. Although the microfabric, especially in a fine-grained clay, may seem to be just too minute to be of general interest, its influence on bulk mechanical behaviour (e.g. Bennett, Bryant and Hulbert 1991) and later structures (e.g. Knipe, Agar and Prior 1991) is well established. Moreover, the microfabric commonly holds the explanation of fabrics visible macroscopically; in fact, it has been argued that sediment fabrics are particularly well suited to a fractal approach (C. A. Moore and Krepfl 1991).

Regarding clastic sediments, burial brings the grains closer together as pore water is expelled, and so far microfabric studies have focused on the interplay between dewatering and cementation, together with the controls on and the extent to which grain interpenetration occurs. Stephenson, Plumley and Palciauskas (1992) derived a 'burial constant' – the ratio of grain–cement contact diameter to the grain centre separation during normal burial – which predicts the maximum porosity possible at a given depth. However, determination of the actual fabric was found to be much less straightforward because of the influence of such factors as the timing of cementation with respect to burial, and the extent to which the cement helped support the burial load as opposed to merely being a passive infill. Rezak and Lavoie (1990) found that fabrics of carbonate sediments were found to depend significantly on the crystal habit of the grains, particularly in the more mixed near-shore sediments. For example, increasing consolidation of matrix-supported sediments brings the grains closer together – decreasing the pore size – but increasing the number of separate pores. In contrast, those materials with a matrix composed dominantly of aragonite needles showed the opposite trend, one result of which is a relatively high permeability for a given burial depth. Matrix-supported carbonates tend to be converted by increasing consolidation into a framework-supported fabric.

Most microfabric studies over the years have focused on clayey sediments, despite their comparative complexity and the fineness and delicacy of the materials. There has been particular debate on the configuration that clays adopt on settling and consolidation, and only the advent of electron microscopy has led to some consensus. Knowledge has by now advanced sufficiently for particular fabrics to be identified as 'signatures' of specific sedimentary environments and processes. Current thinking is reviewed by Bennett, O'Brien and Hulbert (1991), from which much of the following is derived.

An important recent advance is the realization that most clayey sediments are best viewed as collections of domains of clay particles rather than single platelets. A domain is defined as a multiple particle that behaves physically as an entity. It comprises subparallel platelets, typically stacked in one of three ways: vertically, in book-like form; in a staggered, offset pattern; or with a stair-step arrangement. Diagrammatic representations of these arrangements are given in Figure 9.3. On initial settling, the clay particles are more discrete, and are typically arranged in the edge-to-face or 'house-of-cards' pattern that is characteristic of the flocculated state (Figure 9.3d).

Bennett, O'Brien and Hulbert (1991) regarded the overall influences on sediment microfabric to be a combination of three fundamental groups of 'processes', namely physico-chemical, bio-organic and burial diagenesis. Each of these comprises various 'mechanisms' that are responsible for the specific particle-to-particle interactions. Their scheme is summarized in Table 9.1. The physico-chemical processes are significant during transport and sedimentation, and the interplay of electrochemical (arrangement of clay-surface bonds) and thermomechanical mechanisms (water temperatures) with interface dynamics (sediment and water motion) governs the details of the settled fabric. The electrochemical influence, for example, depends on the ionic arrangement at the clay surface and the electrolytic state of the water. While in fresh water the dominantly negative charges at the clay surfaces induce particle repulsion and inhibit the formation of domains, increasing salinity reduces the reach of the electrostatic repulsion at the same time as increasing the potential for van der Waal's bonding. Platelets in sea water are therefore more likely to be attracted to each other

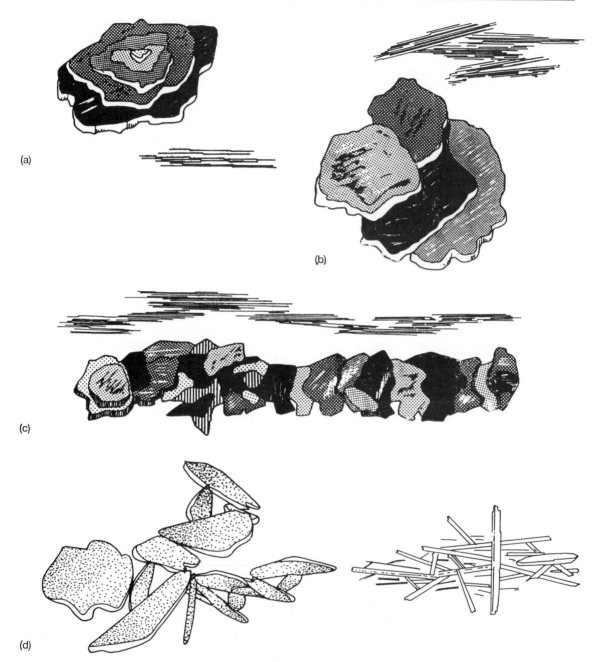

Figure 9.3 Diagrams to show details of clay fabric domains, in plan and cross-section views. (a) Domain consisting of a stacked, book-like arrangement of platelets. (b) Domain consisting of staggered, offset platelets. (c) Domain consisting of stepped, face-to-face platelets. (d) Random, edge-to-face arrangement of discrete platelets. (Based on Bennett, O'Brien and Hulbert (1991). Used with permission from Springer-Verlag.)

and to participate in domains. Although the initial aggregation tends to be edge-to-face, this weak, highly porous, flocculated state is vulnerable to collapse into the stronger face-to-face arrangement. Such reorientation will tend to give the offset pattern, and further energy input can convert this to the book form. In the presence of organic matter a flocculated state is promoted

Table 9.1 Summary of processes, sizes and time-scales associated with various clay fabrics. (From Bennett, O'Brien and Hulbert (1991). Used with permission from Springer-Verlag.) See text for further discussion.

Processes	Mechanisms	Fabric signatures (predominant)*	Scales		Remarks
			Size	Time	
Physico-chemical	Electrochemical	E–F	Atomic and molecular to ~4 μm	μs to ms	Two particles may rotate F–F
	Thermomechanical	F–F (some E–F)	Molecular to ⩽0.2 mm	ms to min	Initial contacts E–F then rotation to F–F; common in selective environments
	Interface dynamics	F–F and E–F	μm to ~⩽0.5 mm	s	Some large compound particles may be possible at high concentrations
Bio-organic	Biomechanical	E–F	~0.5 mm to >2.0 mm	s to min	Some F–F possible during bioturbation
	Biophysical	E–E and F–F	μm to mm	s to min	Some very large organic complexes possible
	Biochemical	Non-unique (unknown)	μm to mm	h to year	New chemicals formed, some altered
Burial diagenesis	Mass gravity	F–F localized swirl	cm to km	> year	Can operate over large physical scales
	Diagenesis–cementation	Non-unique (unknown)	Molecular	> year	New minerals formed, some altered, changes in morphology

*E–F, edge-to-face; E–E, edge-to-edge; F–F, face-to-face.

and likely to persist longer (Booth and Dahl 1986).

Bio-organic processes comprise a range of bio-physical (involving organic mucus or polymers), biochemical (organically controlled chemical changes) and biomechanical mechanisms, of which the most important, particularly in certain sedimentary environments, is that variety of the last group called bioturbation. Disturbance of the fabric by bottom-dwelling organisms, particularly during burrowing and feeding, can be intense. Some sediment packets pass several times through the guts of organisms. According to Bennet, O'Brien and Hulbert (1991), the disgorged fabric shows a randomness similar to the initial flocculated state, but is characterized by randomly oriented individual particles rather than domains (Figure 9.3d). It may contain an anomalously high proportion of silt grains.

Diagenetic processes modify the fabric during progressive burial, but the chemical aspects are poorly understood. Reynolds and Gorsline (1992), for example, noted the development of crenulate intergrowths during the diagenesis of muds from offshore California. They speculated that organic reactions may have been important to the diagenetic processes, but were unable to determine the details. The dominant physical effect of burial is to align increasingly the clay domains as pore water is expelled progressively. This intuitive idea is now well supported by electron microscope observations on natural clays (e.g. Faas and Crockett 1983). The growing preferred alignment with depth, normal to the burial stress, is responsible for much fissility in clayey sediments and rocks, but the rate of increase in intensity of the fabric can be highly irregular. The rate depends on the physico-chemical and biological factors mentioned above and, especially, the ease or otherwise with which the pore water is dissipated. If, for example, diagenetic reactions start to lock-up the framework before dewatering has allowed it to collapse, the sediment may achieve burial without developing much consolidation fabric. Bryant et al. (1990), for example, reporting on sediments retrieved from the Weddell Sea during ODP Leg 113, found that whereas clayey sediments showed a 74% porosity and a permeability of around 1×10^{-11} m^2 near the surface, at over 300 m of burial the porosity had only reduced to 50%, much greater than expected, yet the permeability, because of diagenetic clogging of pore throats, had fallen to a mere 1×10^{-17} m^2. This inability to expel water leads to overpressuring, underconsolidation and a fabric less intensely developed than normal. Nevertheless, many sediments do develop good particle alignment, which imparts a marked anisotropy to the material (Derbyshire et al. 1992). This can greatly influence subsequent processes. It has been suggested, for example, that mimetic overprint of consolidation fabrics during static, low-grade metamorphism may explain the commonness in orogenic belts of bedding-parallel fabrics without accompanying signs of synchronous tectonism (Maltman 1981; see also Biermann 1984). Excellent electron micrographs of the consolidation fabrics of clayey sediments, from various environments and states of consolidation, are contained in the papers collected by Bennett, Bryant and Hulbert (1991).

9.4 MICRO- TO MACROSCOPIC STRUCTURES

9.4.1 Shear zones

It is a characteristic feature of many unlithified sediments that deformation does not pervade the entire material equally, but tends to concentrate in discrete zones of shear displacement. This localization of strain is particularly noticeable in overconsolidated or dense sediments. Such behaviour is fundamental to much soil mechanics thinking, and the effects are well described in geology. Analyses of landslides, especially, have made much use of the fact that the bulk of the motion is commonly accomplished within isolated zones. In fact, it is tempting to use the presence of these discrete zones in sedimentary rocks as indicators of pre-lithification deformation, viewing their orientation as indicators of stress geometry, and using aspects of their appearance to infer deformation conditions. Unfortunately, things are not as simple as this. To start with, quite apart from the mechanical complexities, there is a formidable terminology problem. Numerous names have been applied to what is essentially the same kind of structure, and the same names have been used for quite unrelated features.

These localized surfaces have been called, for example, rupture zones, rupture layers, rupture bands, shear surfaces, shear bands, shear zones, shear discontinuities, slip surfaces, deformation bands, hydroplastic faults, microfaults and faults. The term deformation band has been used to cover these and other related structures, such as kinks (Maltman *et al.* 1993b) – in which manner shear band has also been used (Lundberg and Karig 1986); to some geologists shear band connotes a special (riedel shear) orientation (e.g. Chester and Logan 1987). There is no simple route through this miasma. *Caveat lector*! Although at the macroscopic scale, where the extent of grain breakage can be difficult to assess, these structures could reasonably be called faults, they are referred to here as shear zones, because they involve shear and on close examination are continuous over their finite width.

Identification of shear zones in a sedimentary rock should alert the observer to the likelihood of pre-lithification movement having taken place, but the zones must be seen to have formed without significant intragrain deformation. Similar looking structures do form in rocks, for example the deformation bands reported from Jurassic sandstones in Utah by Aydin and Johnson (1978). These, however, involve substantial fracturing and crushing of the quartz grains, betraying post-lithification movement (but see under web structure, section 9.5.3). The absence of shear zones does not mean that deformation took place after lithification. Arthur and Dunstan (1982) reasoned that distributed strains are promoted in sediments by cyclic deformation, where the stresses accumulate incrementally, cycle by cycle. Shear zones could form in the early stages of the deformation but later become obliterated. Osipov, Nikolaeva and Sokolov (1984) presented SEM micrographs of sheared thixotropic clays, frozen with liquid nitrogen at different stages of liquefaction and then freeze-dried, to argue that thixotropy is achieved not by microstructure destruction, but by the temporary loosening of selected bonds throughout the sediment. In their experiments, this homogeneous effect was able to obliterate any shear zones that had been produced before the liquefaction. High-porosity sediments, on the wet side of their critical state, should tend to deform more pervasively, although Maltman (1987b) argued that wet clays deform along shear zones at porosities up to 60% or more. High-porosity clayey sediments, which at the electron microscope scale showed zones of particle reorientation due to shearing and collapse, were described by Bennett, Bryant and Keller (1981).

Any initial elastic deformation of a sediment is unlikely to be recorded directly in the sediment. Early increments of plastic strain are accomplished homogeneously. Arch (1988) showed that 30% porosity ball-clays achieve between 8 and 12% bulk strain in this way, but if the material approaches a peak strength, shear zones begin to form (Vermeer 1982; Maltman 1987b). Briefly, it happens like this: strains begin to concentrate at some perturbation – perhaps a suitably oriented clay domain in the case of clays, or more highly spaced quartz grains in the case of sand; static friction is overcome as the clays begin to slide, or the quartz-grain framework dilates, leading to localized strain softening and propagation of the zone. The tip wanders through the nearby areas of greatest softening, but is constrained to a greater or lesser degree by the overall stress system and the boundary conditions. An anastomosing zone is produced, of overall planar form. The amount of additional strain depends on the ability of the key particles to rotate into energetically efficient orientations (Bardet and Proubet 1992), otherwise the zone will lock and force the initiation of new shear zones.

There has been much discussion on the overall orientation adopted by the shear zones with respect to the ambient stress geometry, with both soil mechanics (Vermeer and Lugen 1982) and mathematical treatments (Coleman and Hodgdon 1985; Mulhaus and Vardoulakis 1987). In a simple Mohr-Coulomb view of the structures, shear zones form when the stress magnitudes reach the failure envelope, where, assuming frictional behaviour, the zones form at angle of $45° - \phi/2$ to the greatest principal stress (σ_1). Thus, like viewing faults in rocks, it is possible to approximate from a fossilized example the general orientation of the originating stress. However, there are difficulties even greater than those in brittle fault analysis. The chief problems are as follows.

1. It has been argued that the original orientation depends not on the friction angle, ϕ, but on the dilatancy angle, a quantity derived from ratios of the principal plastic strains. The latter

is typically a smaller value, by as much as 30° or more. Some evidence suggests that it is fine sands that utilize the friction angle whereas coarse sands involve the dilatancy angle (Vermeer 1990).

2. If the deformation is rapid, water expulsion may be inefficient, leading to a situation analogous to an undrained laboratory test. Here, in theory, the shear zones will form at 45° to σ_1 (horizontal failure envelope). In practice, this may not be too much of a problem in that Atkinson and Richardson (1987) have shown that even in 'undrained' conditions, local drainage in association with the dilating shear zones prompts some friction and an angle less than 45°.

3. Where there is significant pre-existing anisotropy in the sediment, this can override the Mohr-Coulomb relations, even to the extent of inducing shear zones at angles greater than 45° to σ_1. Where the incipient shear zone is within +15° of the sedimentary fabric, the propagating tip appears to follow the anisotropy (Arch, Maltman and Knipe 1988).

4. The original orientation will be vulnerable to change during the ensuing volumetric strain, as Davison (1987) argued for early normal faults.

In short, provided that the possibility of certain complications are borne in mind, the principal compressive stress is likely to have been around 35° to shear zones in coarse sands, but closer to 25° in sands. Obliquities will be less in clays, reflecting the lower friction angles, unless there are reasons to think that the loading was rapid or that a pre-existing fabric attracted the zones.

The thickness of the shear zones is controlled to a large extent by grain size. Mulhaus and Vardoulakis (1987) reasoned mathematically that the thickness should be 'a small multiple' of particle size; Roscoe (1970) asserted that a typical thickness for sands is 10 times the grain diameter, and Vermeer (1990) put the figure at 15 grain diameters. The clays investigated by Maltman (1987b) had a particle size around 2 μm and developed shear zones between 20 and 50 μm in thickness. Scarpelli and Wood (1982) emphasized the variability of thickness along the zones and the narrowness towards the tip.

The morphology of shear zones is complex, particularly in clays. Arch, Maltman and Knipe (1989) used experimental evidence to suggest that in wetter clays more complex geometries (more splays and undulations) arise and with greater obliquity to any pre-existing fabric. Arthur and Dunstan (1982) devised triaxial experiments in which frictional plattens produced isolated shear zones whereas well-lubricated plattens promoted interfering slip along numerous zones. This importance of boundary conditions was also brought out by Scarpelli and Wood (1982), who produced more complex patterns in finer sands because of, as they put it, 'the degree of constraint perceived by the sand: the individual particle of fine sand is much less aware of the boundaries than is the individual particle of the [coarse sand]' (p. 473).

Due either to this same factor of grain size, or because clays are able to record preferred alignments better than clastic particles, the detailed form of shear zones and the substructure within them can be especially intricate in clayey sediments. A notable and often-cited contribution on this was the analysis by Morgenstern and Tchalenko (1967), based on the classic work of Riedel in 1927, of the structures that develop in kaolinite subjected to drained, simple shear. In summary, the first subshears form obliquely but synthetic with the overall shear direction, at an angle of $\phi/2$ – the so-called riedel shears (Figure 9.4). Although additional shear motion rotates σ_1

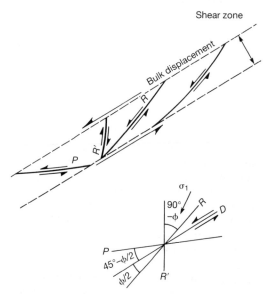

Figure 9.4 Schematic diagram to show the main substructures of shear zones, and a highly simplified Mohr-Coulomb analysis. R = riedel shears; R' = conjugate riedel shears; P = thrust shears; D = principal or displacement shears.

from its initial direction, and, theoretically, could induce a series of differently inclined shears, strain-softening effects commonly mean that within a modest angular range the major zone continues to mobilize. However, with progressive shear, successively formed riedel shears decrease in obliquity to the shear-zone margin, eventually making a small negative angle – the structures known as P or thrust shears. An array of sub-shear lenses develops, with attendant space–accommodation problems. The subshears are forced to link into a throughgoing principal shear, undulating in form but parallel overall to the main shear zone, along which further slip is now chiefly accommodated. The scenario was amplified by Mandl (1988, pp. 76–80).

Various antithetic and conjugate subshears also arise. The whole geometry has similarities with shear zones in ductile metamorphic rocks and in fault gouge, albeit with a slightly different terminology. Unfortunately, even the nomenclature for the substructure in sediments is not standard. Various letters and symbols are used; riedel shears are called shear bands, etc. (e.g. Will and Wilson 1989). Again, the reader has to be alert to the different usages.

The morphological features of shear zones outlined above have been widely reported, in soil mechanics (e.g. Scarpelli and Wood 1982) and clay mineralogy (Foster and De 1971), and in geological experiment (Maltman 1987b) and field-work (Maltman 1987c). Field examples have been used to interpret tectonic shear-senses in fault-controlled basins (Laville and Petit 1984) and superficial nappe tectonics (Labaume, Berty and Laurent 1991), and in an attempt to elucidate the emplacement of slump sheets (Eva 1992). Structures falling within this general family are forming today due to the tectonic shortening of accretionary prisms (Figure 9.5), and in the case of the Nankai example, they are documented at a mere 260 m below the sea floor (Maltman et al. 1993b). Knipe (1986a) discussed the microscopic appearance of these structures in extending slope-sediments. Even in tectonically quiet regions, shear zones can arise through differential compaction. Sediments at the Madeira abyssal plain, buried to only 12.5 m and still with a water content in excess of 50%, show shear zones, called microfaults by de Lange et al. (1988). Here they may be the propagating upper tips of deeper faults that have vertical displacements of tens of metres (Duin, Mesdag and Kok 1984).

A common association is for the shear zones to be accompanied by high-angle kink-bands. Although Foster and De (1971) asserted that in their experimentally sheared kaolinite the kinks formed after failure, Maltman (1977) noted kinks cut by shear zones, which formed at peak strength. In the examples from the Nankai prism, the kinks both deform and are cut by the shear zones, but most commonly the kinks simply abut the major zone. It seems that the two structures are approximately contemporary, with the kinks forming due to space constraints created during continued shear along the main zone (Maltman et al. 1993b).

Another noteworthy aspect of shear zones preserved in rocks is that they commonly stand proud of weathered surfaces. This is particularly true of sandstones, and, indeed, unlithified sand (Figure 9.6). A similar effect is commonly seen associated with post-lithification faults. Enhanced cementation is usually inferred, reflecting preferred fluid transfer along the faults. The explanation in unlithified sand, however, can be more elusive. For example, the shear zones (there is no grain breakage) visible in Figure 9.6b contain more cement, which is why the zones are tougher, but they actually have a lower porosity than the host sediment (counting cement as pore space). Why should fluids pass preferentially along zones which appear to be identical with the host in every way except that they are tighter? Of course, the zones may well have dilated during the shearing, but in these periglacial, lake-delta-front sediments the failure is likely to have been rapid, even instantaneous. Given the low solubility of the goethite cement and low near-surface reaction rates, it is hard to envisage how the cementation was accomplished within this fleeting period.

Experimental work suggests that in addition to the shear zones discussed above, a range of related microstructures can form in clayey sediments (e.g. Maltman 1977). For example, where the burial fabric is well developed, kink bands and crenulations develop, including very sharp, periodic cleavage-like varieties (Figure 9.7). Such structures are remarkably similar to those seen in rocks, and commonly but wrongly assumed to be restricted to post-lithification deformation.

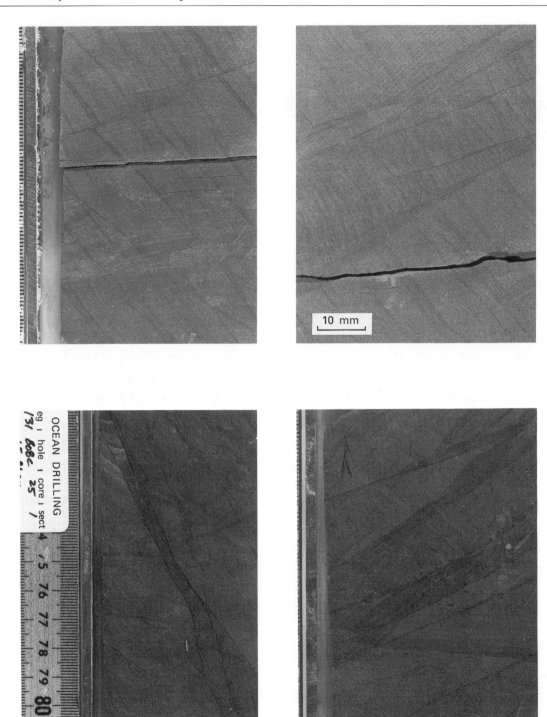

Figure 9.5 Examples of shear structures from a modern accretionary prism, Nankai, Japan. The structures are seen in the face of a split core, recovered by ODP drilling. Scale divisions are millimetres. Details of the structures are discussed in Maltman *et al.* (1993b).

Micro- to macroscopic structures

Figure 9.6 Shear zones in an unlithified Pleistocene sand deposit, Cardigan, west Wales: (a) shows the relative resistance of the zones to erosion; (b) shows the greater amount of cement (dark colour) within a shear zone.

9.4.2 Slickensides

This term is used here simply for a shear surface, so it includes the boundaries of the shear zones discussed in the previous section. The characteristic features of these slickensides are a shininess and the presence of hair-line grooves. The detailed morphology is varied. The term slickenside here has no implication of brittle fracture, as surfaces of this appearance are developed in laboratory shear tests on highly ductile clays, with water contents in excess of 50% (Maltman 1987b). Moreover, examples follow in which the mass shearing past the surface is not the neighbouring sediment but some other agent, such as ice.

The sheen of these surfaces appears to be due chiefly to an alignment of clay flakes parallel to the surface, rather analogous to the well-known lustre of mica-schists. The alignment may be

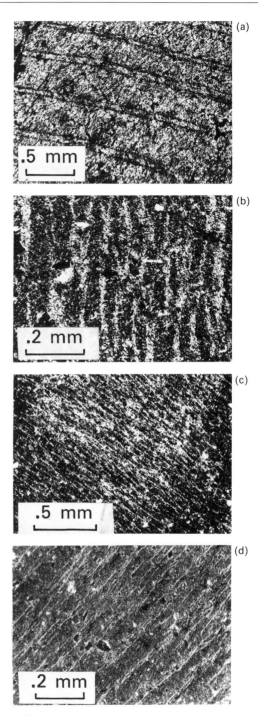

Figure 9.7 Some microstructures formed in the experimental deformation of clayey sediments: (a) kink-bands; (b) crenulations; (c and d) cleavage like structure. See Maltman (1977) for experimental details. (Reproduced with permission from Elsevier Science Publishers BV.)

partly the result of the clay framework collapsing during localized, enhanced dewatering adjacent to the surface, and partly due to shear-induced alignments. The steps that are commonly observed on the surfaces, as in rocks, are the intersections of various substructures with the main shear surface. Petit and Laville (1987) documented several geometries of intersection, as summarized in Figure 9.8. Note, however, that interpretation of shear directions from the steps might be precarious (Maltman 1987b). D.G. Moore (1980), for example, suggested an alternative explanation for certain morphologies that would imply an opposite sense of shear.

The striations have a typical appearance of hair-line grooves, tracking the direction of shear. Means (1987) has provided a preliminary classification of the detailed morphology (Figure 9.9). In silty clays, ploughing by some asperity, such as a protruding quartz grain, is clearly a likely process for producing grooves, but Means argued that the unnamed variety in Figure 9.9 could not involve simple abrasion. Will and Wilson (1989), referring to the structure as a ridge-in-groove type of lineation, showed that the lengths of some examples exceed the total shear displacement, clearly precluding a ploughing origin. Their explanation involved the progressive lateral interlinking of riedel and other substructures in a shear zone, of which the surface bears the striations. Moreover, Wilson and Will (1990) produced similar features in pure, ductile paraffin wax, further negating any scouring process and demonstrating that these features are by no means indicative of brittle fracture. Irrespective

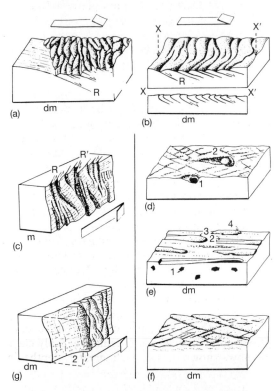

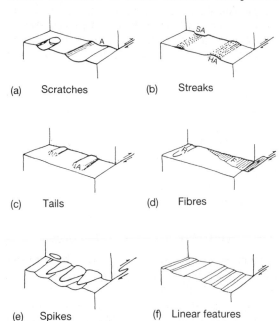

Figure 9.8 Types of shear surfaces in incompletely lithified sediment, according to Petit and Laville (1987). Arrows indicate the sense of movement of the missing block. (a–c) Many secondary riedel shears (R), commonly distorted and overturned, and occasional conjugate (R') shears. (d) Occasional grooves (2) closing up quickly behind a harder striating element (1). (e) A few grooves and trails left by clay pellets (1) which have been torn out (2) and spread along (3 and 4). (f) Bumpy surfaces due to the intersection of minor fracture patterns oblique to the slip direction. (g) Pads (1) accumulated at the front of a now removed block. dm (decimetres) and m (metres) indicate the approximate scale. (Reproduced with permission from the Geological Society.)

Figure 9.9 Classification of slickenlines according to Means (1987). (a) Grooves or scratches resulting from asperity (A) ploughing. Upper groove with sheared-off asperity at its down-slip end. (b) Debris streaks trailing soft asperity (SA) or accumulated in lee of hard asperity (HA). (c) Tails of erosion-sheltered material on the down-slip side of hard asperities. (d) Fibres (F) and rods (R) of vein material filling potential voids behind steps or asperities, respectively. (e) Spikes on a slickolite surface. (f) Linear feature in materials with no asperities. (Reproduced with permission from Pergamon Press.)

of how exactly the grooves are produced, it does seem that they record the movement direction along the surface in the same way as rocks. Inverse analysis of the stress tensor has been used successfully on the striated surfaces of both on-land (Guiraud and Seguret 1987) and off-shore unlithified sediments (Lallemant *et al.* 1993).

A particular variety of surface was termed pedogenic slickenside by Gray and Nickelsen (1989). Subaerial clayey sediments with more than 30% expandable clays that are exposed to seasonal wetting and drying undergo repeated differential expansion and shrinkage. This induces localized shearing and the production of undulating, grooved surfaces bearing a dull patina. The examples described by Gray and Nickelsen (1989) are from Palaeozoic redbed sequences in the Appalachian foreland, where they exercised some influence on the progress of subsequent tectonic deformation.

On a larger scale, striations a few millimetres across and many metres in length are known to form on modern, exposed mud surfaces, due to the ploughing action of passing tools. According to Dionne (1974), these implements can range from ice bodies to boats! Because shearing beneath an ice-sheet would also deform the underlying material to some extent, rather than just the exposed surface, Dionne argued that glacial grooving alone must represent the passing of drift ice. Some polishing may accompany these striations, due to framework collapse under the load of the tool. Dionne (1974) reviewed occurrences of such striated surfaces, in both recent sediments and ancient rocks.

9.4.3 Scaly clay and related features

A macroscopic feature that has long been recognized in highly deformed clayey sediments and their lithified equivalents has, in recent years, returned to prominence. It is termed scaly fabric or scaly foliation, and comprises the material called scaly clay. The fabric typically appears as subparallel undulations of shiny surfaces anastomosing around narrow lenticles of less fissile material (Figure 9.10). The overall orientation may or may not be related to bedding. The fabric is also observed at the microscopic scale. It seems likely that in many cases scaly clay comes about through the coalescing of numerous shear zones during progressive shear, although Agar, Prior

and Behrmann (1989) have warned against inferring deformation mechanisms from the macroscopic appearance alone. They advocated using the term solely as a loose field term, partly because it grades into fissility and what has been called microscaliness (Lundberg and Moore 1986). In addition, Agar, Prior and Behrmann (1989) documented, at the electron microscope scale, a variety of microstructures from scaly clays that were macroscopically identical.

The fabric seems to be characteristic of clayey sediments that have undergone high shear-strain, whatever the geological setting. Lash (1989) documented the progressive development of scaly clay in stratal disruption zones preserved in Ordovician mudstones in the central Appalachians. The porous clay fabric of undeformed clays passed through a transition zone that recorded fabric collapse and the initiation of shear zones, into highly deformed sediments with complex arrays of very compact scaly foliae showing intense alignment of the clays. The inferred origin involved initial deformation in wet, weak horizons, with the propagating shear zones extending the amount of fabric collapse and strain hardening. Thus a zone of scaly fabric extends until it meets material sufficiently dewatered to resist the shear; its width is inversely proportional to the water-content gradient between deforming and more consolidated sediment.

J.C. Moore *et al.* (1986) concluded from their study of scaly clays from several fore-arcs sampled by the DSDP that the fabric develops preferentially in underconsolidated smectite muds, at temperatures less than 25°C, burial pressures less than 4 MPa and at strain rates of approximately $1 \times 10^{-13} \, s^{-1}$. The extent to which these conditions apply generally is speculative. Scaly clays are now known from much greater depths in accretionary prisms – such as the example in Figure 9.10b – and the feature is widely distributed in olistostromes (Agar, Prior and Behrmann 1989). Scaly clays are even reported from drumlins (Menzies and Maltman 1991)! Because these intrinsically fragile lithologies have been further weakened during the shearing, probably dropping to their residual strength, where they outcrop on modern hillslopes they are particularly vulnerable to slope instability (van den Berg 1987).

Scaly clays highly analogous to those formed in other geological environments (e.g. section 8.2),

276 Deformation structures preserved in rocks

Figure 9.10 Scaly clays. (a) In outcrop. Cenozoic argille scagliose, Passo Della Cisa, north Italy. (b) In an ODP core, 360 m below the sea floor, Nankai prism, Japan.

are common in the landslide at Portuguese Bend, southern California, described by Larue and Hudleston (1987). The material continues to move, largely on a number of bedding-subparallel discontinuities. The surfaces develop preferentially in the relatively weak volcanogenic clays, within which a variety of fabrics has developed. Shales near the slide surface are experiencing brittle fracture, and the more ductile clays are undergoing boudinage. Sandy material displays

web structure (section 9.5.3). Larue and Hudleston recognized six different types of slide-breccia, each representing a different location and strain regime within the sliding massa (Figure 9.11). A host of other structures is also present, reflecting different strain conditions (Larue and Hudleston, in preparation).

There is, therefore, assuming that such structures can be preserved during lithification, the potential for gaining some understanding of the dynamics of ancient examples. Kinematic indicators, such as striations on scaly surfaces, fold

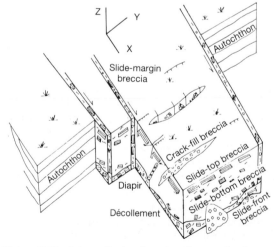

Figure 9.11 Schematic drawing of environments of brecciation in an idealized landslide complex. The drawing is weighted towards breccia types and thus overemphasizes their distribution. (From Larue and Hudleston (1987). Reproduced from the Journal of Geology, with permission from the University of Chicago Press.)

orientations and vergences, clast alignments and asymmetric tails of blocks, were used by K.M. Brown and Orange (1993) to reconstruct the geometry of mélange – diapir emplacement. Similar features are likely in submarine slides, and Maltman (1988) suggested that in clayey palaeoslides, shear zones of different styles and orientations might record indications of the geometry and kinematics of the movement. However, although in principle this remains possible, in practice, in the Silurian slides of North Wales, at least, complex mutual overprinting of the structures and the frequency of local aberrations have so far made the shear zones impossible to disentangle (Eva 1992).

In addition to the information preserved in an allochthonous mass by the scaly clay and shear zones, there is the possibility of extracting movement data from clasts. The use of data on clast nature, distribution and orientation is well established in till studies (van der Meer 1987), but less so in other deformed sediments. Rockslides, for example, are generally perceived to lack fabric, although Gates (1987) gave examples of boulder orientations recording the known movement of recent slides. One example is summarized in Figure 9.12, namely the Gros Ventre slide in western Wyoming. Until the recent Mount St Helens volcanic eruption, this was the largest rockslide ever witnessed in the USA. Interpretation of the orientations seen in the rose diagrams suggests that most of the boulders floated in a saturated, high-density matrix, and rotated into the flow lines in order to offer least shear resistance during the sliding. The degree of alignment reflects the distance of travel. However, in the toe region, the transverse alignment was explained not by contraction, as is commonly invoked for slumps, but by increased basal shear stresses at the front of the slide producing a rolling motion – which had been noticed by eye-witnesses. There is evidence that the rolling effect decreases upwards with continued sliding, so that the intensity of transverse clast orientation should be inversely proportional to the distance of movement. In this way, Gates (1987) argued that clast fabrics in rockslides can preserve useful indications of the direction and dynamics of transport, especially as the fabric will remain after other movement features have been erased. Kohlbeck et al. (1994) used similar arguments for clast orientations in lahars, and Harris (1981) sug-

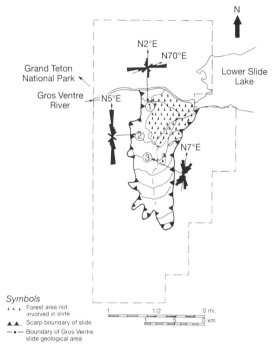

Figure 9.12 Clast orientations, represented in rose diagrams, in the rockslide at Gros Ventre, Teton County, Wyoming. (From Gates 1987.) Note the dominantly bimodal distribution in the central part of the slide, and different unimodal patterns at the head and margin. (Used with permission from the Association of Engineering Geologists.)

gested that certain microstructures, in particular sand-grain alignments, may be diagnostic of solifluction deposits, formed through periglacial downslope creep.

9.5 MACRO- TO MESOSCOPIC STRUCTURES

9.5.1 Faults, folds and related structures

Faults are very commonly reported from deformed sediments, in most cases being the larger scale manifestations of the shear zones discussed in an earlier section. The sediments appear brittle at this scale of observation: it is unlikely that the structures involve much grain breakage, unless the material was approaching lithification at the time of movement. If the fault is thought to have formed in highly porous sediments, its present dip may be significantly less than originally, due to the intervening consolidation (Davison 1987).

Figure 9.13 Structures in subglacial sediments due to ice-induced shear. Thrust and related displacements are accentuated by ink lines. Boginia sand pit, near Łódź, Poland.

Pickering (1983) described normal faults from the Jurassic Boulder Beds of northeast Scotland. Closely spaced faults were interpreted as representing the basal region of gravity slides, whereas isolated examples reflected slope oversteepening, draping over protuberances, or hydraulic fracture. Reverse faults were also reported, possibly reflecting contraction in the toe region of a slide. Normal faults, commonly of listric form, from the Namurian of western Ireland, were illustrated by Crans, Mandl and Harembourne (1980) as part of their mechanical analysis of growth faulting. Van Loon and Wiggers (1976) reported paired extensional faults that created graben-like effects in Holocene lagoonal sediments of The Netherlands. In contrast, faults in surficial sediments of the Ganges River described by Chakrabati (1981) showed reverse geometry, thought to be the result of hydrodynamic current drag. Contractional faults are known from submarine slope deposits (Roberts 1989) and subglacial sediments (Figure 9.13); the example shown in Figure 9.14 brings pegmatitic gneiss over loose sands and gravels!

imbrication, they inferred contraction within sheets that were gliding down opposite sides of submarine channels. An imbricate unit of similar appearance occurs in the Silurian rocks of North Wales (Eva 1992), and may have developed during downslope sliding.

Where slope failure is involved, it may be possible, if the relevant geological parameters are sufficiently well preserved, to interpret some of the likely physical conditions. Booth, Sangrey and Fugate (1985) extended the Coulomb approach outlined in section 1.3.3 to derive a safety factor which included excess pore pressure, $P_{H_2O_e}$ the bouyant (submerged) unit weight of sediment, UW, sediment thickness, z, and the friction angle with respect to effective stress, ϕ':

$$F = \left(1 - \frac{P_{H_2O_e}}{UW}\cos 2\theta\right)\frac{\tan \phi'}{\tan \theta}$$

They combined this with a consolidation factor and derived the nomogram, linking sedimentation rate, time, slope angle (θ), and likely slope and sediment conditions, that is reproduced here

Figure 9.14 The Lichar thrust, Rakhiol, northern Pakistan. This major neotectonic thrust, still in motion, brings high-grade Precambrian (Nanga Parbat) gneiss over unlithified glaciofluvial sediments. (Further details are given in Owen (1989a). Photograph provided by Lewis Owen.)

Sigmoidal contractional faults forming imbricate slices within discrete sedimentary horizons were described by Shanmugan, Moiola and Sales (1988), from Upper Palaeozoic rocks in the Ouachita Mountains, Arkansas, and likened to the duplexes of thrust belts. However, tectonism was not thought to be involved. Because some adjacent horizons show opposing senses of

as Figure 9.15. The device illustrates the link between theory and practice in that not only can it be used to predict modern stability conditions, but it furthers the possibility of interpreting dynamic conditions from ancient structures.

Fitches et al. (1986) reasoned that some of the largely bedding-parallel, slickensided, and in places mineralized, discontinuities in Lower

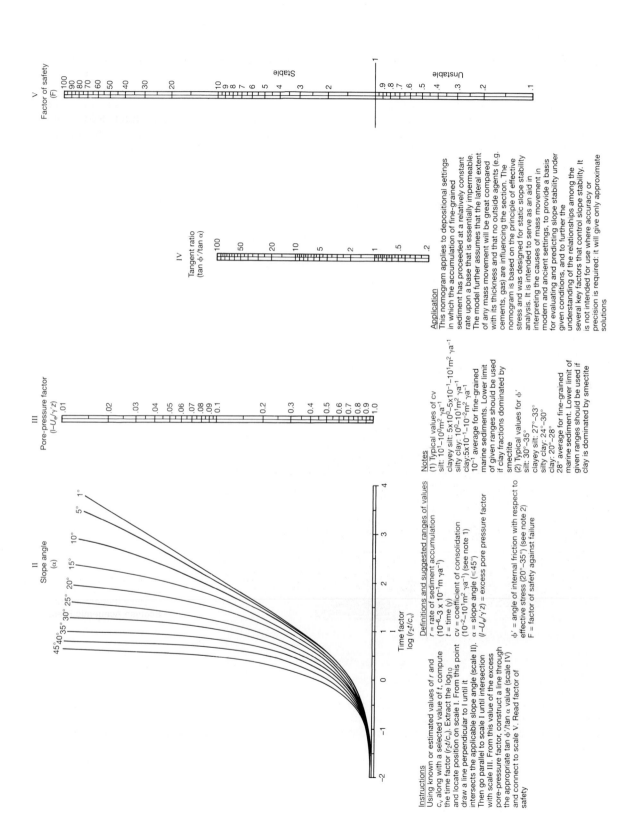

Palaeozoic Welsh rocks were recording gravity driven detachment before complete lithification. Milici et al. (1982) reported similar, subtle bedding-parallel faults from Upper Palaeozoic shales and siltstones associated with coal deposits, where they were thought to represent an early stage of décollement related to Valley and Ridge tectonism. Henderson, Wright and Henderson (1989) argued that gold-bearing bedding-parallel veins in Nova Scotia pre-date the onset of tectonism. The extent to which tensile tectonic stresses directly affect near-surface sediments has yet to be determined, although section 6.2 discusses various implications for sediment accumulation in extensional settings. Maltman (1988) suggested that the small-scale extensional faults shown in Figure 9.16 might be tectonic, in that the delicate laminations of the sediment record neither differential compaction nor slope instability, whereas the faults do parallel structures caused by regional extension.

Folds are widely reported from deformed sediment, of extremely different scales and styles. Recumbent folds in Ordovician rocks of the English Lake District, interpreted as slump folds by Webb and Cooper (1988), have typical wavelengths of around 500 m, and these are thought to be parasitic to a slump fold with a wavelength of several kilometres. The upright folds forming in the Gulf of Oman, associated with the convergence of the Eurasian and Arabian plates, have amplitudes of up to 2.5 km and wavelengths exceeding 10 km (White 1977; White and Klitgord 1976). Regarding fold style, in sediment as in rock deformation, this is largely influenced by the viscosities of the folding layers, and the viscosity ratio in particular. Although sediment viscosities are clearly much lower than for most rocks, the ratios are of similar range, and consequently a full range of fold styles can result. A common perception is of pre-lithification folds being highly rounded and 'ductile-looking', but this is just one segment of a spectrum that extends to markedly angular kink-folds. Figure 9.17 shows two examples of folds thought to be due to subaqueous slumping, one

Figure 9.16 Pre-lithification extensional structures in the Carboniferous Granton 'shrimp-bed', near Edinburgh. The thin laminations and fine preservation of delicate organisms suggest stagnant lagoonal conditions with negligible palaeoslope. The structures parallel extensional faults in the basement, and therefore may be due to regional tensile tectonic stresses reaching high in the sediment sequence. (From Hesselbo and Trewin (1984). Photograph provided by Stephen Hesselbo; used with permission from the Scottish Academic Press.)

Figure 9.15 Nomograph developed by Booth, Sangrey and Fugate (1985) for interpreting slope stability in modern and ancient fine-grained deposits. The nomogram:

1. relates variables that influence slope stability;
2. helps interpret the cause of palaeo-movements;
3. helps evaluate the stability of present-day slopes.

One of the worked examples provided involves a 25-m section of Lower Jurassic rocks containing an apparently allochthonous block of shale. The homogeneity and clay grain-size of the host material suggests quiet, deep-water sedimentation, at a rate of about 50 mm ka^{-1} over a period of 500 ka, perhaps on a slope of 5°. Such values lead to a log time factor of -1.00 and tangent ratios, even with low friction angles, that give factor of safety values in excess of 4. This inherent stability of the slope sediments implies either that the block is actually in place or that a significant external agent has to be invoked. See Booth, Sangrey and Fugate (1985) for further details. (Reproduced with permission from the Society of Economic Paleontologists and Mineralogists.)

Figure 9.17 Variations in the style of slump folds. (a) Angular, chevron-like style, Pliocene, near Chikura, Bozo Peninsula, Japan. (b) Rounded, bulbous style, Silurian, near Colwyn Bay, Wales.

markedly angular and the other rounded and bulbous in style.

Although, just as in rocks, kink bands are most common in sediments that posses anisotropy, van Loon, Brodzikowski and Gotowala (1984) documented their occurrence in a range of sediments and circumstances, including almost pure sands and poorly lithified limestones. Using indirect sedimentological arguments, they derived likely strain rates (between 2.5×10^{-6} and $1 \times 10^{-10}\,\mathrm{s}^{-1}$) that are surprisingly similar to values typically inferred for rock deformation. In a further article (van Loon, Brodzikowski and Gotowala 1985) they emphasized the geometrical similarities between sediment and rock kinking, although pointing out that fluid pressure would substantially influence strain conditions and hence kink angles in sediments. In general, kink folds can be analysed in the same way as rocks, although the stress systems are more likely to be localized, and less representative of regional effects.

Rounded, tight to isoclinal folds were discussed by Decker (1990), from Eocene volcaniclastic rocks in northwest Wyoming (Figure 9.18, type 1). The folds, of amplitudes from 1 to 100 m and wavelengths from 10 to 100 m, are disharmonic, variably oriented and separated from non-folded beds by local discontinuities rather than a major detachment. All this led Decker to propose an origin involving slow, gravity driven buckling within a 'rigidity' zone overlying a narrow zone of 'depth creep', rather than slope failure. The concept is illustrated in Figure 9.19. In an attempt to determine the time-scales that might be involved in this kind of folding, Decker combined likely values of material properties, such as cohesion and friction angle, with measured values, such as slope and thickness, into an infinite-slope analysis (section 1.3.3). He presented a series of graphs relating the various parameters, which could form the basis for analysing similar folds in other regions. He found

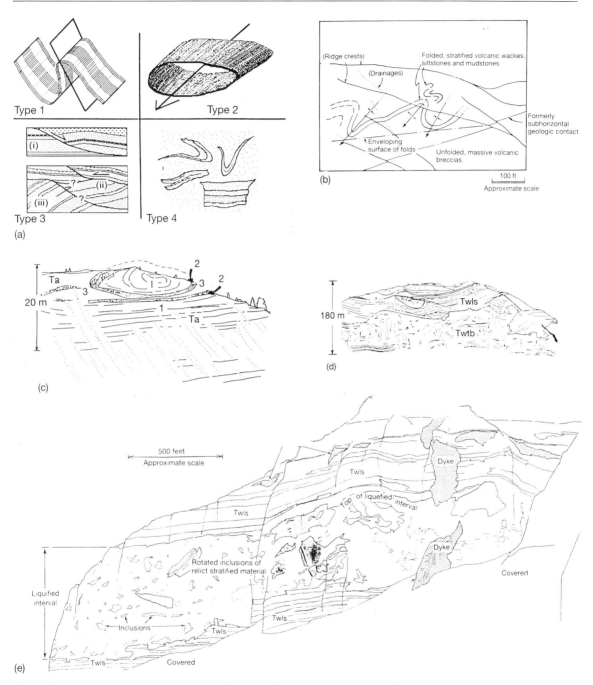

Figure 9.18 Structural styles recognized by Decker (1990) to be associated with liquefaction of Eocene volcanic rocks in northwest Wyoming. (a) A summary of the four styles. Type 1 is characterized by tight to isoclinal overturned folds; type 2 by layer-parallel shearing structures, including sheath folds; type 3 by gently dipping truncation surfaces (i, low-angle normal faults; ii and iii, thrust or undetermined slip, respectively); and type 4 by liquefaction and chaotic diomembormont. (b), (c), (d) and (e), respectively, are field representations of structural styles 1, 2, 3 and 4. The numbers 1–3 in (c) indicate the oldest to youngest beds, to illustrate the anticlinal nature of the sheath fold. The letters Ta in (c), Twlb in (d) and Twls in (d) and (e) refer to local formation names. See also Figure 9.19. (Used with permission of P. L. Decker.)

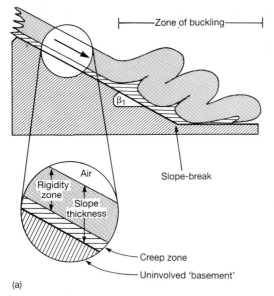

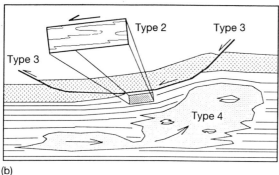

Figure 9.19 The conception of Decker (1990) of the deformational setting of structures illustrated in Figure 9.20. (a) Schematic representation of fold development (type 1 structures) in a rigidity zone overlying a zone of depth creep. Buckling is induced by the difference in creep rates above and below the slope break. (b) Schematic diagram showing hypothetical relation of type 4 liquefaction and dismemberment to secondary deformation in structurally higher domains of other types. Gravitational instability is generated by equilibration of upper strata to deeper liquefaction and flow. For simplicity, no type 1 folding is depicted, but could occur in place of the type 3 faulting in a buckled layer overlying a zone of depth creep. The layer-parallel shearing recorded by type 2 structures could occur either in response to drag forces in the footwall of gently dipping faults, as shown here, or within a zone of depth creep. (Used with permission of P. L. Decker.)

that the creep rates were more sensitive to slope angle than the mechanical properties, and derived folding durations of between 1 and 100 years. These rates indicate, as Decker put it, 'fairly rapid creep rather than catastrophic or million-year orogenic processes'.

Also reported by Decker (1990) were recumbent intrafolial folds, including some of markedly sheath-like form (Figure 9.18, type 2). These were thought likely to have originated within the zone of depth creep, perhaps around local perturbations. Sheath-like folds formed during the thrusting of accreting plate-margin sediments were reported by Hibbard and Karig (1987), and in association with subglacial thrusting by Owen (1988b; Figure 9.20). The highly distended, isoclinal folds described from the Gondwana sediments of eastern India by Ghosh and Mukhopadhyay (1986) not only represent similar, low-viscosity materials, but also, they suggested, an exaggeration of style by gravitational flattening during subsequent consolidation. The authors argued that although laminar shear during gravity gliding can provide sufficient strain, the lack of marked asymmetry and any consistent sense of overturning in the folds, suggested that the amount of downslope translation was not great. Although, in general, slump folds seem to be more common in clastic rocks, they are known in evaporites (Wardlaw 1972) and in carbonates – even shallow-water varieties for which there are no indications of any submarine slope (Buggisch and Heinitz 1984)!

The idea of using structures such as slump folds to estimate palaeoslopes has been around since at least the time of O.T. Jones (section 9.6.1). Most of the attempts that have been made are based on the assumption that the folds will develop: (i) with their axes parallel to the strike of the slope and at right-angles to the slump transportation; and (ii) with an asymmetry that shows downslope vergence. Woodcock (1979b) assessed different techniques of dealing with such structures. Although the above assumptions may be a reasonable first approximation in some cases, many natural situations will present complications. For example, the likely inhomogeneity of continuing slump strain together with irregularities in the bathymetric slope will conspire to scatter the axial distributions. At the same time, continuing inhomogeneous movement may passively rotate early-formed folds into upright or even upslope orientations. Consequently, careful statistical handling of field orientation data is vital. Eva and Maltman (in press) provide a recent example of the difficulties of deducing palaeoslopes from ancient rocks.

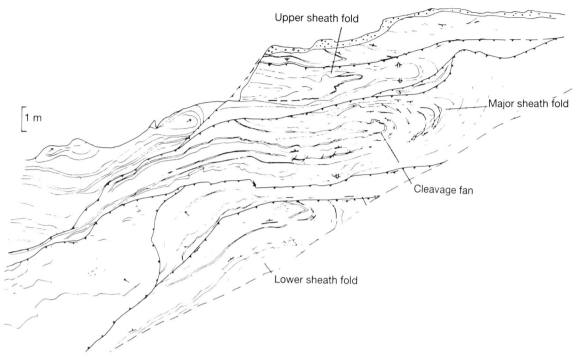

Figure 9.20 Sheath folds and thrusts in glacial deposits, Skardu Basin, Baltistan. (From Owen (1988b). Used with permission from A. A. Balkema.)

Pre-lithification folds can develop spaced, axial-planar foliations, probably through processes of dewatering and liquefaction, and near-surface folds can produce pervasive fabrics (Elliot and Williams 1988). The topic is developed further in section 9.6. Lineations on bedding planes involved in slump folds have long been recognized, and are probably the surface intersections of shear zones along which the sediment failed during slumping (Maltman 1988). None of the examples so far reported appear to be pervasive, nevertheless, Gibling and Stuart (1988), reporting lineations in slumps in Jurassic carbonate rocks in Portugal, remarked that such features in the past may have been interpreted as tectonic lineations.

The viscosity contrasts that influence the folding of unlithified sediments also give rise to boudinage. Examples are reported, for example, by Visser, Colliston and Tereblanche (1984). McCrossan (1958) noted the attendant development of tension fissures in the folded beds. An interesting example of boudinage that involves diagenesis was described by Brueckner, Snyder and Boudreau (1987) from the Miocene Monterey Formation of western California. The scenario they envisaged for the originating processes, summarized in Figure 9.21, has much in common with that mentioned in section 1.2.5. Clayey sediments containing silica (opal-A), derived from radiolaria, behaved in a ductile manner because of their high porosity, and the high solubility of this disordered silica promoted pressure solution. The low porosity and solubility of clay-poor, opal-CT cherts, on the other hand, gave a more brittle style of deformation and inhibited pressure solution. The different ductilities enabled the layers to deform into boudins, the shape of which was modified by the differential solution.

The various structures mentioned above typically formed in association with one another. For example, Stel (1976) invoked the diapirism of overpressured clay to induce the range of associated faults, folds and pull-apart structures summarized in Figure 9.22. A wide variety of structures due to glacial and related processes was described and illustrated by Dionne (1971), who also provided extensive references to earlier descriptions, and by Banham (1975). Longva and Bakkejord (1990) described faults, folds and related structures resulting from scouring by

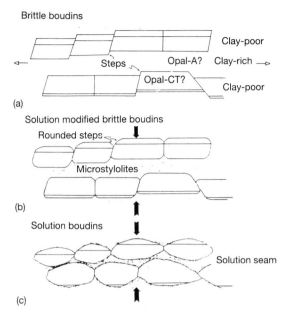

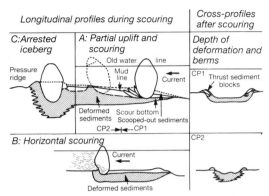

Figure 9.21 Summary of the diagenetic–mechanical explanation proposed by Brueckner, Snyder and Boudreau (1987) for the shapes of boudins in the upper Palaeozoic siliceous rocks of north Nevada. Clay helps retard conversion of opal-A to opal-CT, thus preserving in the clay-rich layers a lower diagenetic state and higher porosity than the adjacent clay-poor layers. The latter are therefore stronger and less ductile, and respond to layer-parallel extension by forming boudins. Differential solution, partly utilizing the extension fracture, then modified the boudin shape. (Used with permission from Pergamon Press.)

Figure 9.23 Deformation structures forming due to various types of iceberg scour, simplified from Longva and Bakkejord (1990). The cross-profiles CP1 and CP2 show how a 'fast' moving iceberg in the initial stages of ploughing forms berms of thrust sediment blocks (CP1) whereas a retarded iceberg (dashed line in a) forms berms of remoulded sediment. In (b), after first grounding contact with the slope, all eroded material is carried into suspension. In (c), the ridge in front of the iceberg is higher than that behind it because it is a combination of plough-up and pressure ridge. The zig-zag line bordering the deformed zone indicates liquefied layers that have penetrated into the surrounding sediments. (Reproduced with permission from Elsevier Science Publishers BV.)

the deformation was induced at depths up to four times greater than the channel itself. The source of this deformation in fossilized examples may therefore be far from obvious. From the contractional tectonic environment, Lash (1985) de-

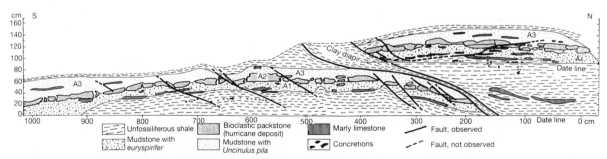

Figure 9.22 Cross-section of Lower Devonian rocks northwest of Colle, Cantabrian Mountains, Spain. Stel's explanation of the structures involved no palaeoslope or external stresses but overpressuring due to differential in-place consolidation of the sediments. For example, unit A is thought to have been poorly permeable and responsible for overpressuring and localized diapirism. The main diapir (indicated) lifted and thrust the overlying section over a footwall sequence which, as a consequence of the increased loading, underwent extension, including boudinage and localized failure. (From Stel (1976). Used with permission from Kluwer Academic Publishers.)

icebergs! Their preliminary analysis of how the structures relate to different ploughing mechanisms is summarized in Figure 9.23. Note that although the scour itself is vulnerable to erosion, scribed the interaction of boudinage, stratal disruption, scaly clay, etc., in Ordovician sequences from eastern Pennsylvania, inferring the scenario of fluctuating effective stresses indicated in

Macro- to mesoscopic structures 287

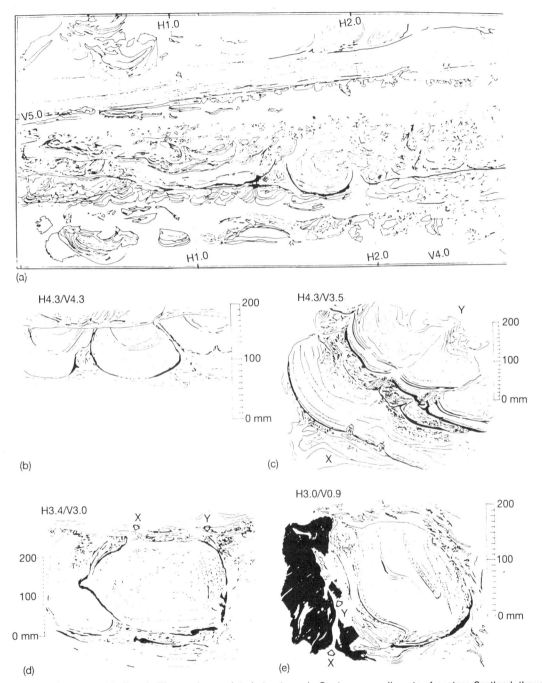

Figure 9.24 Sketches of ball-and-pillow and associated structures in Quaternary sediments of eastern Scotland, thought to have deformed in response to earthquake shock. The grid lines in (a) are 1 m apart. (From Davenport and Ringrose (1987). Reproduced with permission from the Geological Society.)

section 1.2.5. A superb photographic compendium of the range of structures encountered at a convergent plate margin, in this case the Shimanto Belt of southwest Japan, was provided by Taira, Byrne and Ashi (1992). This atlas illustrates many of the features discussed in the present chapter, from materials at various stages of lithification.

9.5.2 Liquefaction structures

The principles of liquefaction were outlined in section 1.3.2. A wide range of structures can result from the phenomenon, some representatives of which are included in section 4.3. The Eocene volcaniclastic rocks of northwest Wyoming mentioned earlier with respect to folds, also contain abundant liquefaction structures (Decker 1990; Figure 9.18, type 4). Their characteristics are typical of liquefaction features in other lithologies. The structures comprise a chaotic arrangement of irregularly shaped, commonly delicate inclusions, scattered in a massive matrix showing very variable grain size and content. The margins of the liquefied domains are very irregular, commonly gradational but locally sharp, and show complex interleaving with neighbouring, relatively undisturbed units. Among the reasons why Decker inferred that liquefaction produced the structures, rather than other processes that had been proposed previously, were: the disruption affected already stratified sequences rather than occurring during deposition; the intensity of the dismemberment; and the lack of evidence of lateral transport.

Other structures commonly associated with near-surface liquefaction include ball-and-pillow structure (Figure 9.24) and sedimentary dykes (Figure 9.25). Note that the mere presence of sedimentary dykes does not in itself have palaeoenvironmental significance, as overpressured sediments are so very widespread. They are even known from below glaciers (Amårk 1986). They can form after substantial burial, and in association with regional tectonism. The only restriction is that the sediment is able to respond to overpressures by liquefying. Winslow (1983) recorded a particularly intensive development of sedimentary dykes in the Cenozoic fold-and-thrust belt of southern Chile. The dykes occur in swarms, contain clasts as much as 25 cm across, and reach hundreds of metres in length. Restriction of the dykes to the toe areas of thrust hangingwalls led Winslow to relate generation of the dykes to the emplacement above of successive thrust sheets. Structural relationships, such as cross-cutting of joints by the dykes, and dykes by the cleavage, allowed the deformational history summarized in Figure 9.26 to be deduced. Huang (1988) paid particular attention to the morphology of sedimentary dykes preserved in Cretaceous deposits in southeast France, relating bifurcation and branching patterns to the sense of shear that was operating across the fractures at the time of dyke injection. Some of the arrangements he found useful are illustrated in Figure 9.27.

Ronnlund (1989) described a quantitative approach to analysing liquefaction structures. He presented a series of curves that related wavelength and thickness ratios of observed structures to viscosity ratios. For a series of Late Precambrian troughs and ridges from Finnmark, northern Norway, interpreted as liquefaction structures, he derived a viscosity ratio between the layers of 10:1. He went on to argue that this implies (if the rising suspension had a viscosity close to that of water (10^{-3} Pa S)) a viscosity for the water-saturated sand of the upper layer of around 10^{-3} to 10^{-2} Pa S and a period of less than 5 min to generate the structures. The assumptions involved in constructing the curves (e.g. two layers only, Newtonian behaviour) restrict the accuracy of these figures, but the approach represents an admirable advance towards achieving a fuller interpretation of existing liquefaction structures.

Because much liquefaction arises from seismic activity, the possibility of determining palaeoseismicity from the structures has received much attention (see section 1.3.2). Sims (1975) listed criteria for identifying structures due to earthquake-induced liquefaction. These included: restriction to single layers separated by undisturbed beds and correlatable over a wide area; location in beds too level at the time for slope

Figure 9.25 Sedimentary dykes, Eocene Shimanto Belt, Point Gyoto, Shikoku, Japan. (Photograph courtesy of Asahiko Taira.)

Macro- to mesoscopic structures   **289**

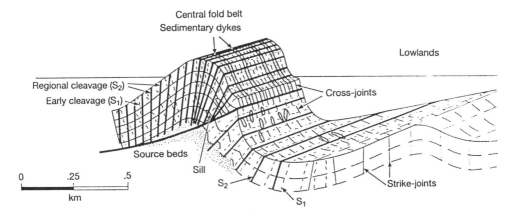

Figure 9.26 Schematic diagram to show the conception of Winslow (1983) for the relations between sedimentary dykes and regional thrusting, folding and cleavages. (Used with permission of M. A. Winslow.)

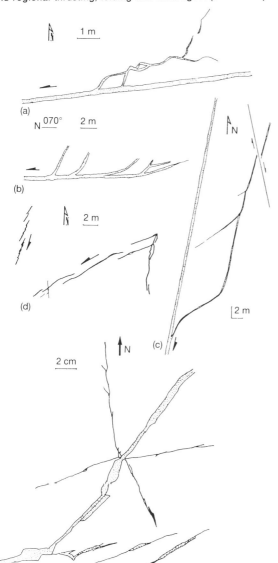

failure to have occurred; and occurrence in an area known from some other evidence to have been seismically active. Using these kinds of principles, attempts have been made to identify palaeoseismicity in both recent sediments (e.g. Visser and Joubert 1990; Amick and Gelinas 1991) and in more ancient equivalents (e.g. Scott and Price 1988; Amand and Jain 1987), although clearly the reliability of the guides falls away rapidly with older materials. Tuttle and Seeber (1991) recorded cross-cutting liquefaction structures, indicative of multiple seismic events, and Allen (1986) derived an approximate relationship between the frequency of observed liquefaction structures and the palaeoseismicity of sedimentary basins. The types of structures used by Ringrose (1988) to infer a major earthquake at the end of the glacial period in Scotland are illustrated in Figure 9.28. A selection of the liquefaction structures reported by Audermard and Santis (1991) from northern Venezuela, the result of a swarm of seismic events that included two $>5\,M_b$ earthquakes about 25 km distant, is shown in Figure 9.29.

9.5.3 Dewatering structures

Many of the structures mentioned above grade into those that are due mainly to dewatering

Figure 9.27 Sketches of sedimentary dykes showing their use as shear sense indicators. In (a), the bifurcation is taken to indicate sinistral shear; in (b) and (c) the branches imply sinistral and dextral motion, respectively, and the en échelon patterns in (d) suggest a principal compression approximately from the north east. (From Huang (1988). Reproduced with permission from Pergamon Press.)

290 *Deformation structures preserved in rocks*

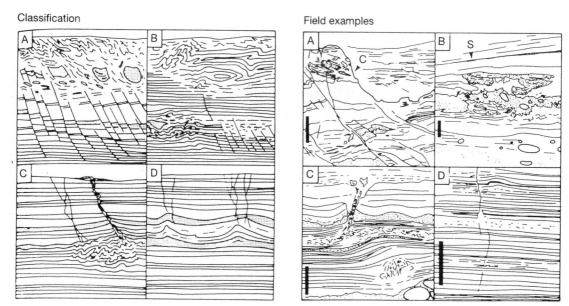

Figure 9.28 Structures due to earthquake shock, from Ringrose (1988). The sediment sequences consist of laminated silts and sands (stippled) with occasional gravel casts (heavy outline in the field examples). Classification scheme: A, fault-grading stratigraphy (closely spaced faults with throws increasing upwards, and passing upwards into broken-up and then completely liquefied sediment) – ball-and-pillow structures developed in the upper liquefied zone; B, confined-layer deformation with incipient ball-and-pillow structures or incomplete elements of fault-grading stratigraphy – surface-layer deformation involving only contortions of layering; C, incipient (discontinuous) confined-layer deformation with injection structures involving upward movement of liquefied material; D, flame and fisssure structures only – the term 'fissure' is used for small, near-vertical, irregular fractures that do not involve injection of sediment or significant displacement of layers. Field examples from Quaternary sediments, Glen Roy, Scotland; A, fault-grading stratigraphy passing upwards into a liquefied zone with well-developed ball-and-pillow structures (top left) – the deformed lake sediment has been eroded by channel 'C' and then infilled by undeformed sediment; B, incipient ball-and-pillow structures in a confined layer – the surface at the time of deformation is thought to have been at 'S', above which lies undeformed sediment; C, injection structure rising from a confined-layer deformation zone (centre) – a very localized confined-layer deformation is seen at the lower right; D, fissuring and slight deformation (small ruck-fold at top). Scale: classification box approximately 2 m across; field examples – bar = 0.1 m. (Reproduced with permission from Blackwell Scientific Publications.)

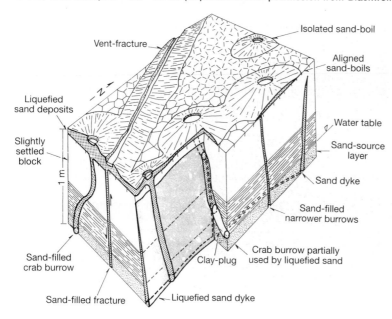

Figure 9.29 Block diagram, from Audemard and Santis (1991), of liquefaction features in the Tocuyo Delta, showing the tendency for the liquefied sand to follow pre-existing fissures and conduits.

of the sediment. Examples of the latter from near-surface sediments are given in section 4.3.4. Some recently recognized dewatering structures may have a greater depth distribution, such as that known as vein structure or, more recently, mud-filled veins (Taira, Byrne and Ashi 1992). This feature is now widely reported from modern oceanic sediments and on-land sedimentary rocks (Lundberg and Moore 1986; Lindsley-Griffin, Kemp and Swartz 1990). There is no question that vein structure can form in highly unlithified sediments – Kimura, Koga and Fujoka (1989) reported it from within a few metres of the sea floor – but the catalogue of Lindsley-Griffin, Kemp and Swartz (1990) includes examples from numerous burial depths between 8 and 634 m.

Mud-filled veins consist of periodically spaced, subparallel curviplanar bands, typically with a length:width ratio of 10:1 or more, and arrayed along a particular zone. The detailed morphology is variable (Ogawa and Miyata 1985; Ogawa, Ashi and Fujioka 1992; Figures 9.30 and 9.31). The zones containing the vein structures are typically parallel to bedding, but can show a low obliquity, commonly in symmetrically opposed, conjugate sets. A characteristic field appearance is for each vein to be similar to the host sediment but darker. There is none of the pronounced

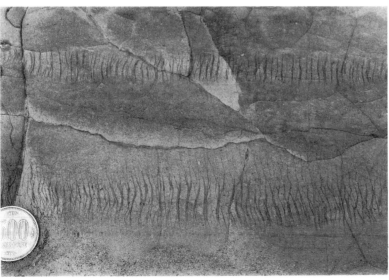

Figure 9.30 Vein structure or mud-filled veins. Miocene Ishido Group, Bozo Peninsula, Japan.

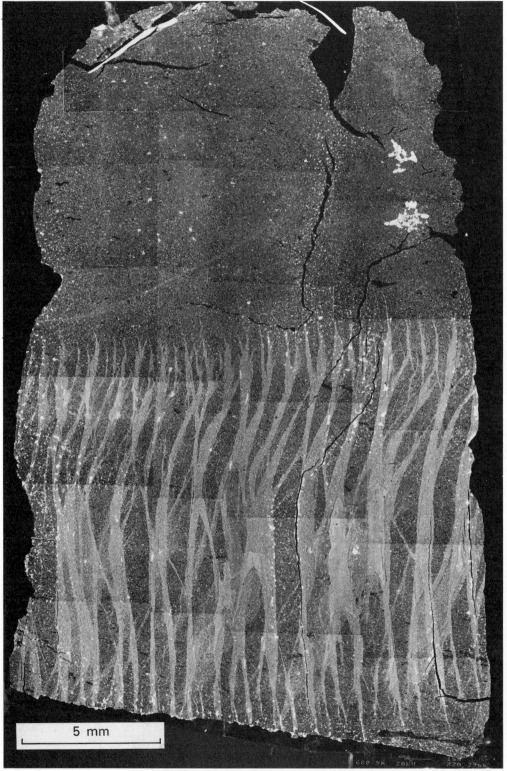

Figure 9.31 Scanning electron micrograph of mud-filled veins, in an array within diatomaceous muds. The veins show 'fish-tail' bifurcated terminations, some interlinking, and a slight, normal-sense, displacement. Sample from offshore Peru, sampled by Leg 112 of the Ocean Drilling Program. (Photograph provided by Richard Brothers.)

quartz or calcite mineralization that is typical of the geometrically analogous tension gashes found in rocks.

Electron microscopy of Miocene examples from southeast Japan (Pickering, Agar and Prior 1990) suggests that concentration of very finely disseminated pyrite is the chief reason for the darker colour, and that the vein-fill is derived from the local host material. Although the modern examples are all from accretionary prisms, the consensus is that the structures arise through extension during downslope motion coupled with fluid activity. Thus it seems that converging plates cause the production of sediments with properties appropriate for vein structure generation rather than any particular kind of tectonism.

The common occurrence of the Miocene examples in southeast Japan immediately below slumped horizons, and with associated sedimentary dykes, suggests that rapid loading and pore-pressure increase facilitates development of the structure. However, Knipe (1986c) has argued that the structure can arise without overpressuring. He presented two genetic models: one in which overpressuring promoted hydraulic fracture of the sediment followed by fabric collapse or rapid fluid flow within the vein; the other involved shear or tensional failure associated with slope instability to generate the veins, without rapid dewatering. Knipe, Agar and Prior (1991) suggested that the veins could affect sediment dewatering patterns in different ways, depending on whether the fluid was being driven out of or into the vein, as summarized in Figure 9.32. A situation where veins lack connection to high permeability pathways, and merely allow slow seepage was offered as an explanation of the 'ghost' veins described by Kemp (1990). These subtle, irregular networks of sediment finer than the host are illustrated in Figure 9.33. The intensity of their development in cores retrieved from the Peru fore-arc during ODP Leg 112 suggests that this kind of slow, semi-pervasive drainage plays an important role there, at least, in sediment consolidation. Vein structure seems to result chiefly from dewatering, but this process grades through water moving with entrained sediment, such as along the normal faults illustrated in Figure 9.34, to sediment liquefaction and the structures described earlier.

Increasingly reported from sandstones is a feature termed web structure (Lundberg and Moore 1986). It provides an illustration of the difficulty of defining lithification (see section 1.1.1), because examples have been described from sediments buried less than 300 m, but which at the microscopic scale show evidence of quartz cataclasis (Lucas and Moore 1986). The typical field appearance on bedding surfaces is shown in Figure 9.35. Lucas and Moore (1986) suggested that repeated strain hardening linked to fluctuating pore pressures encourages the grain breakage, and Knipe, Agar and Prior (1991) presented electron micrographs of multiple cementation textures within webs that supported the idea. Their conception of the transience of fluid pressures and related properties during deformation, and their implications for fabric development, is illustrated in Figure

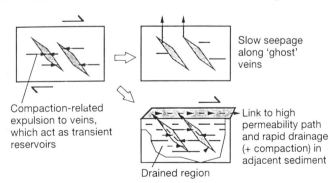

Figure 9.32 Diagrams to show possible interrelationships between fluid expulsion and vein evolution. The fluid ingress associated with the vein dilation may be linked to the collapse and compaction of the fabric adjacent to the vein. Fluid expulsion from the vein may either be by slow fluid flow along 'ghost veins' or by propagation and linking of the veins to a more extensive migration pathway. Such a linking process may induce a more extensive collapse and compaction in the adjacent sediment. (From Knipe, Agar and Prior (1991). Reproduced with permission from the Royal Society.)

Figure 9.33 Back-scattered electron image of ghost veins in diatomaceous sediment recovered from offshore Peru during ODP Leg 112. The pale colour of the veins is due to the relatively fine sediment there, probably very fine opal. Further details in Kemp (1990). (Photograph provided by Alan Kemp.)

Recognition of sediment deformation structures

Figure 9.34 Sediment entrained along normal faults, which presumably acted as rapid dewatering channels. Cretaceous Pigeon Point Formation, west-central California.

Figure 9.35 Field appearance of web structure, preserved in Miocene sandstones, Bozo Peninsula, Japan.

8.4. Lucas and Moore's (1986) view of the progressive development of cataclastic shear zones in quartz-rich sediments is shown in Figure 9.36.

Cylindrical structures preserved in rocks are commonly interpreted as biological in origin, but in some cases the absence of supporting evidence coupled with extraordinary dimensions – for example, 0.5 cm in diameter and 9 m in length – has led to doubt (Bromley *et al.* 1975). At the same time, it was mentioned in section 1.3.4 that pore water may in some circumstances escape along cylindrical conduits, and that many examples are known from the modern sea floor. It seems likely, therefore, that some of the enigmatic cylinders in rocks may be fossilized dewatering structures.

9.6 RECOGNITION OF SEDIMENT DEFORMATION

9.6.1 General

Interpretations of structures in sediments that are still unlithified will tend to be directed at details of the processes rather than identifying the state of the material at the time of deformation, as this is closely bracketed. However, examination of structures seen in ancient sedimentary or metasedimentary rocks prompts more basic questions. To what extent were the materials lithified at the time of deformation? In what general burial and stress environments did the deformation take place? The ideas developed in the previous chapters can be applied only if structures are recognized for what they are. The

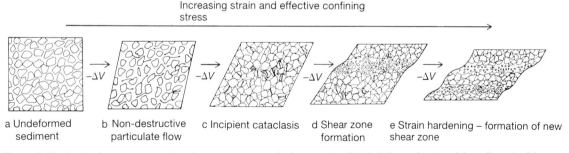

Figure 9.36 Model for shear zone development and cataclasis in sands. (a) Undeformed water-rich sediments. (b) Internal deformation with particulate flow. (c) Incipient cataclasis. Increasing confining stress inhibits grain-boundary sliding and causes localized cataclasis where stress concentrations exceed grain strength. (d) Propagation of discrete shear zones. (e) Strain hardening of initial shear zone and development of new shear zone in weaker adjacent material. (From Lucas and Moore (1986). Used with permission of J. Casey Moore.)

state of lithification may sometimes be clear – for example, if late-stage diagenetic or metamorphic minerals have demonstrably been involved in the deformation – but if these features are not visible, or they are overprinting earlier structures, then the situation can be difficult. As Chapter 1 emphasized, tectonic deformation of rocks at depth cannot be assumed, or just guessed, as being more likely. Judging by modern earth processes, gravity driven processes and near-surface sediment deformation, for example, are too widespread for that.

But how can pre-lithification deformation be identified? The question has been asked many times over the decades (e.g. P.F. Williams, Collins and Wiltshire 1969; Helwig 1970; Cowan 1982; Elliott and Williams 1988). In fact most papers on pre-lithification structures include some discussion about how they were recognized. Unfortunately, many earlier discussions of the problem have been aimed at the meaningless distinction between 'tectonic' and 'soft-sediment' deformation (see section 1.1.1). Moreover, the latter term has been used frequently merely as a synonym for gravitational slumping. Some features that have been used as hallmarks of a pre-lithification origin do not hold up to scrutiny. Rather than merely catalogue here the numerous past efforts, one historical episode is recounted in some detail in Historical vignette 3. Apart from encapsulating the conceptual difficulties of recognizing pre-lithification deformation, this vitriolic dispute, between two leading British academics of the first half of this century, rehearsed the arguments on identification criteria.

Historical Vignette 3: Possible slump deformation in North Wales

The early interests of P.G.H. Boswell were diverse. They included the link between geology and the new science of soil mechanics, including properties such as the thixotropy and dilatancy of clays. Boswell conducted research into other aspects of applied geology, and introduced into English the term 'isopachytes', now widely used in oil geology and elsewhere as isopachs. The considerable irony in all this will become apparent shortly. In 1917, Boswell became the first Professor of Geology at the University of Liverpool, and felt it prudent to undertake field studies close to his new University. For this he selected the nearby but little known Denbigh Moors region of North Wales. Boswell's investigations there soon revealed puzzling things about the structures, which he began to explain by thrust movements. It seems that once Boswell had embarked upon this direction of geological thinking, it became increasingly easy to invoke tectonically induced dislocations to explain other problematic structures. Features such as 'confused masses of highly cleaved and crushed flags, mélanges of shattered rock, often 'balled-up into spheroidal masses' and sharply separated from normal beds' were diagnosed as being due to thrusting (Boswell 1928). Other geologists were shown the structures and seem to have accepted this interpretation.

These must have been happy times for Boswell. He built up a respected department at Liverpool and then became head of the department at Imperial College, London. He received honours such as election to the Royal Society in 1932, and his work in North Wales was much respected. He was publishing numerous papers on the area, and knew the geology better than anyone else. It must have seemed like 'his ground'. Enter, however, in 1935, O.T. Jones. In some ways, Jones was a long-standing rival of Boswell, in that he also had a long and successful history of geological field work in Lower Palaeozoic rocks (in Central Wales) and he was head of a rival department at Cambridge. Both were leading figures in the Geological Society and both were Fellows of the Royal Society. But Jones had seen rocks elsewhere with features that had registered with him the importance of subaqueous sliding as a process. The concept seems to have stayed dormant in Jones's mind until his 'active interest was awakened on reading Boswell's papers and studying the photographs' from the Denbigh region. It appeared to Jones 'as though the sediments had been violently stirred while still in a pasty condition'. He eventually gave a report on the Denbigh Moors phenomena – the benchmark paper of 1937 – to the Geological Society, saying that 'there is no doubt that the disturbance of the contorted rocks had been completed before the deposition of the overlying sediments' (Jones 1937). 'A long-standing engagement' prevented Boswell from attending the lecture.

Jones's interpretations were to a large extent intuitive; in any case there was little direct information available at the time on how sediments would behave during sliding. Much of his evidence, as is often still the case, depended on describing the variability and intricacy of the structures, and viewing them as incompatible with the relatively ordered results expected from tectonism. His mapping had shown that there were many disturbed horizons, some less than a metre in thickness and others hundreds of metres thick, covering a hundred square kilometres or more. Jones had no problem envisaging submarine sliding taking place at all these sorts of scales, but to Boswell this was a major difficulty. Other objections raised by Boswell in his written discussion were readily viewed by Jones as a natural part of the sliding process: a discordant base, for example, noted from one of the presumed slides, was simply due to the moving material 'digging in' to the floor during rapid movement. Also prescient of many discussions today, Jones remarked that 'some confusion was necessarily introduced by the word tectonics. Small stresses acting on weak, unconsolidated materials give rise to structures difficult to distinguish from the effects of large 'tectonic' stresses operating on consolidated rocks'.

The next shot was fired by Boswell, in a paper which was published back-to-back with Jones's! He concluded that 'if features characteristic of slumping were once present, they have been almost, if not entirely, obliterated' (Boswell 1937). He evidently was happy to accept submarine sliding as a viable process, but not on a large scale. 'If one small area is accepted as due to slumping, then the similarity of the features elsewhere means that the entire region must have been affected', which Boswell viewed as unrealistic. Boswell's next tack was to emphasize stratigraphical arguments. He felt that on the basis of the observed structures alone 'it may be difficult or almost impossible to distinguish between the effects of contemporaneous sliding and those of later earth-movement'. However, he asserted, 'the interpretation...of large-scale slumping is not...in harmony with the stratigraphic evidence' (Boswell and Double 1938). He began to grasp at stratigraphy as his major weapon in the battle, and went on to spend much energy in refining it.

Jones gave a second paper devoted to submarine sliding, an influential and much cited publication (Jones 1939a). Now, just as Boswell's descriptions tended to have thrusting as a premise, so Jones's writings assumed the importance of slumping. He began, 'the distinguishing feature of the district is the occurrence of large numbers of slumped beds. These are so clearly displayed that this district may be regarded as the most favourable area for the detailed study of such occurrences'. Jones went on to detail what he saw as the supporting evidence, and to discuss how he envisaged the processes happening. He provided detailed maps and numerous field sketches, a few of which are reproduced here as Figure 9.37. He argued that the upper boundaries of the disturbed sheets were critical. The sharpness of the junctions, with no signs of movement along them, the slight grading upwards of the overlying sediment, indicating *in situ* deposition, and the lithological similarity between the disturbed and the overlying materials were, to Jones, wholly incompatible with a tectonic origin. He specified 39 localities where such crucial upper contacts could be clearly examined.

To most of the listeners at the Geological Society the evidence was overwhelming, but Boswell was less than happy. Again he was absent from the meeting, and bitterness and sarcasm tainted his written response. Boswell was becoming entrenched in a line of defence based on detailed graptolite stratigraphy. He announced that he had succeeded in collecting numerous graptolites from horizons inferred by Jones to be slumped, and that the sequence was complete and intact overall. To Jones, himself a graptolite worker of renown, the resolution of the graptolite zonation was insufficiently fine for the stratigraphical arguments to carry weight in the present argument. He believed that the observed structures and lithologies held the key. Boswell, on the other hand, appeared convinced that the graptolite succession was no problem for his own structural interpretations but undermined the idea of slumping. Indeed, he persevered for much of the rest of his life with his studies on the graptolite zones of the region. According to Alan Wood (personal communication 1987), Boswell felt he had been driven into a corner, and he was taking the dispute very personally. He fervently hoped that something would eventually emerge from his toils with the graptolites that would prove Jones wrong.

The strain on Boswell began to show. In 1938 he resigned his position at Imperial College, because of poor health. A few years after that he retired to a somewhat reclusive life in North Wales. His scientific interests remained, however, and these were to see a series of ironic twists. He had had a long-standing interest in sediment mechanics; now

Jones was publishing a paper on how muddy sediments consolidated (Jones 1939b). Perhaps this once again seemed to Boswell like Jones entering his own time-honoured territory. Jones, with a series of simplifying assumptions and a complete neglect of diagenesis, argued that consolidation would typically take about 5000 years. With this in mind for the Denbigh Moors deposits, 'it is not surprising that these rocks could take on and retain the remarkable and beautiful structures which they acquired in the process of sliding on the sea-floor' (Jones 1939b). Boswell would soon attempt to refute this line of approach in an expanded effort on sediment mechanics, but for the moment the battle remained centred on the field evidence. It was at Jones's critical upper contacts that Boswell had to aim, and he began by listing 22 localities where contacts between disturbed and undisturbed material could be seen (Boswell and Double 1940). In these, the passing of jointing across the junctions without modification was felt to 'furnish the best evidence in favour of contemporaneous movement (slumping) of the deposits, but this "most favourable evidence" is only visible at four of the localities'. Hence, it followed that slumping was only of local significance. Boswell was not prepared to accept Jones' upper contact argument. 'The evidence for contemporaneous sliding of sediments should be unequivocal if deductions relating to a large area or great thickness of deposits are made from it. The gradual passage from disturbed deposits up into normal beds is not unequivocal evidence. Only if the contact is a plane of erosion and the graptolite faunas immediately above and below the disturbed material are identical, is the evidence irrefragable. Such requirements are rarely fulfilled'.

By 1943, Boswell believed that most of the structures 'exemplify the effects of earth movements' (Boswell 1943) but did allow small-scale sliding at a few localities, affecting horizons up to about 5 m in thickness. Any hint in this of a growing compromise, or of the disagreement settling into a stable equilibrium, were dashed by Boswell's charge that the 'disturbed beds cannot be mapped as a unit' and that 'Professor Jones has included many outcrops of "normal" cleaved rocks in his disturbed beds'. Jones replied briefly but bitingly. 'I am grateful to find that Professor Boswell adduces some evidence that the disturbances were contemporaneous'. However, 'his conversion to the view is still only partial' and he 'betrays some confusion' about the evidence, particularly that 'crucial test of the date of origin of the "disturbed beds", the upper contacts'. The reasons why the upper boundaries had to be depositional were recounted by Jones, before he launched into comments of a more personal nature regarding the alleged omission of disturbed beds from his map. 'It is not without an element of humour that Professor Boswell charges me with having missed some slumped beds' (Jones 1943). 'After such a promising start it is disappointing to find that Professor Boswell has recorded a beautiful series of slumped beds' that occurred nearby, 'as if they were normal beds'.

Boswell had by now, after 20 years of study, reported on virtually the entire Denbigh Moors area. It remained for him to synthesize the information (Boswell 1949). He produced a mainly detailed descriptive account, but reviewed and discussed the evidence regarding the disturbed beds. Although he accepted slumping in certain situations, such as where normal bedding can be seen passing into gentle folds, and through overfolds and recumbent folds into 'a tumbled unstratified mass', in other cases 'the evidence as to the origin is by no means always so clear'. Many folds show a consistent appearance throughout a supposed slide sheet, the self-same characteristics as folds in known tectonic belts. The lack of fractures at fold crests, Boswell argued, implies low deformation rates incompatible with sliding. Some fractures in disturbed beds were filled with quartz, which seemed to imply advanced lithification. Concretions must have formed and hardened before disturbance, because they now lay in a variety of orientations. Disoriented, jointed and allegedly cleaved fragments within disturbed material implied post-cleavage disturbance, Boswell argued. He appreciated that particular difficulties of recognition arise where 'both contemporaneous and subsequent movements have operated' in the same place, and also because it is possible that 'the earth-movements took place at no distant date after the deposition of the sediments, that is, at a time when the shaley or clayey beds were still only partly consolidated and in a plastic state'. He envisaged a situation where sediments at depth were undergoing tectonic deformation, with their varying degrees of lithification influencing the nature of the junctions between disturbed and undisturbed units, and the same movements triggering some local sliding at the sea floor. To Boswell, 'the crux of the question is, not whether the disturbed deposits are all due to contemporaneous or to subsequent movement, but which of them are the result of one or other, or both, causes'. Although Boswell was failing to envision the vast scale and frequency now documented for the submarine sliding process, this general scenario is close to that visualized today for many sedimentary piles.

Meanwhile, the skirmishing between the two adversaries had switched back to the related field of sediment mechanics. Jones (1944) expanded his earlier account of sediment consolidation; Boswell responded with a series of papers on sediment thixotropy, eventually to be synthesized in his book *'Muddy Sediments'* (Boswell 1961a). Discussing the extent to which clays might produce, for example, landslides, Boswell taunted that 'even a slight acquaintance with the physical properties and behaviour of sediments should give us pause before we theorize about, for example, their rate of compaction and their ability to slide under their own weight'.

By 1953, Boswell felt it appropriate to use his continuing graptolite work to launch yet another salvo on the origin of the Denbigh Moors structures. In his article entitled 'The alleged subaqueous sliding of large sheets of sediment in the Silurian rocks of North Wales', he claimed that although disturbances in many regions were now being attributed to subaqueous sliding, invariably 'no indisputable evidence is provided' (Boswell 1953). None of the intricate schemes of graptolite zonation that he had assembled, and none of his detailed maps of graptolite distributions even hinted at such vast disturbances. These data demanded, according to Boswell, either abandoning belief in graptolites as time indicators or in the reality of Jones's major slump sheets. Jones fired up his sarcasm in reply. 'Professor Boswell alleges "no indisputable evidence" that ... the disturbances are due to subaqueous sliding. It is not quite clear whether he disbelieves that they are subaqueous or are due to sliding, or both' (Jones 1953). 'The evidence is as conclusive as any geological evidence can be. It has been stated before but apparently it needs to be repeated'. In summary, Jones's evidence was that there were numerous clear exposures of disturbed horizons sharply overlying lithologically similar, normal sediments, but with pieces of the underlying material detached, disoriented, and 'sometimes much crumpled'. The overlying normal sediments commonly had graded bases, which adhered closely to the irregularities in the upper surface of the disturbed bed. 'It is certain that there has been no sliding on that surface since it formed – it is an original undisturbed contact'. Pieces containing the contact could be hammered off. The graptolite zonation was irrelevant.

Boswell's next published contribution concerned itself mainly with a theoretical analysis of compaction and porosity loss in clayey sediments. It is a remarkably similar treatment, given the circumstances, to Jones's 1944 paper on the compaction of muddy sediments. Jones's article was not, however, referred to by Boswell. Perhaps it is worth reminding ourselves at this point that this protracted and spiteful quarrel was not between two ambitious geological upstarts, but two worldly gentlemen who had already reached the highest academic pinnacles their country had to offer! However, by now, the Denbigh Moors structures seem to have been just about wrung dry of argument. After all, it had been going on for 15 years. Despite its bitterness, it had raised numerous useful points, many of equal relevance today.

Unfortunately, the antagonism did not end here; it expanded into the use of graptolites, and to field-mapping techniques. At one point, Jones (1955) reviewed once more the overall Lower Palaeozoic evolution of Wales, citing, incidentally, 10 of his own papers and none of Boswell's, but making much use of isopachytes – the device which, as mentioned earlier, Boswell had first introduced to English usage! Jones's synthesis was couched in terms of the accepted model of the time, the 'Welsh Geosyncline'. Boswell's next article was entitled 'The case against a Lower Palaeozoic geosyncline in Wales' (Boswell 1961b)! In this he argued that the area was really a complex system of connecting depressions rather than one enormous trench, and would be better referred to as a 'Welsh Basin'. It was to be his last publication. After gradually failing health, Boswell died in 1960 in North Wales, not far from the area that had proved his geological nemesis. By now, the idea of widespread slumping accounting for the structures in the Denbigh Moors was widely accepted, and Boswell had become virtually isolated. Jones's interpretation was later further vindicated by the systematic British Geological Survey investigations of the region (Warren *et al.* 1984). However, as a final irony, when Jones died in 1967, plate tectonics ideas were just being applied to Wales. This new thinking heralded the demise of the notion of a Welsh Geosyncline, which Jones had long championed and the swift acceptance of the idea of a 'Welsh Basin'.

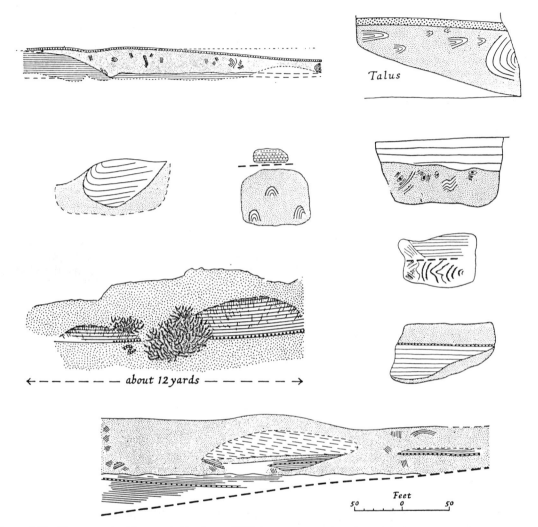

Figure 9.37 Diagrams of evidence used by Jones (1939a) to support a slumping origin for structures seen in Upper Silurian rocks in North Wales. The diagrams (scale shown where provided in the original) illustrate relations of the inferred slumped masses with adjacent beds and features within them. (Reproduced with permission from the Geological Society.)

9.6.2 Importance of recognizing pre-lithification deformation

It is axiomatic that the geologist is aiming for as full an understanding as possible of what he observes. Failure to identify pre-lithification deformation could lead to his assigned age for the structures being awry by hundreds of millions of years (historical vignette 1, section 1.1.1). The environment of deformation will be wrongly construed. Cataclastically deformed quartz, for example, does not necessarily signify tectonic deformation after substantial burial: it is known from landslides (Larve and Hudleston 1987) and from 230 m below the modern sea floor. The vast scale of modern mass-movement raises the possibility of allochthonous sequences being unrecognized and regional sedimentary settings wrongly interpreted. In an often cited article on the 'mimicry' of sediment deformation post-lithification structures, by Hendry and Stauffer (1977) pointed out the possibility of reading erroneous younging directions from these structures, if they are not correctly identified, leading to an incorrect structural geometry. Attempts to quantify the deformation processes will be based on inappropriate relationships.

If structures are recognized to be the result of pre-lithification deformation, it then becomes necessary to identify the setting and the processes responsible. The range of possible deformational environments and the similarity of the resulting structures is evident from the preceding chapters. To illustrate the point further, Schultz (1986) remarked that the geometry and scale of the ancient rockslides he described from the southern Appalachians, involving blocks up to 5 km long spread over an area of 25 km^2, could lead to the structures mistakenly being interpreted as a thrust complex. Eyles and Clark (1985) have discussed the potential confusion, even if an overall glacial setting is correctly diagnosed, between structures interpreted as resulting from subaerial permafrost and those due to glaciomarine processes, which leads to differing views of the glacial palaeoenvironment. Delineating the extent of ice-cover depends on the correct distinction between structures due to glacial shearing and those resulting from proglacial slumping (Visser, Colliston and Tereblanche 1984). Deciding whether a structure forming on a slope-apron through gravitational disturbance or as a result of tectonic shearing within an accretionary prism will affect reconstructions of the plate tectonic setting (Cowan 1982). Take central Newfoundland for example. Various conflicting scenarios have been put forward for the tectonic evolution of the region, depending on whether structures are interpreted as gravity driven or tectonic, and whether they formed in material that was poorly, partially or fully lithified (Blewett 1991).

If economic deposits are involved, then wrong diagnosis of the structures could lead to a flawed exploration strategy. For example, Green (1982) noted the association of uranium mineralization, in Jurassic limestones of New Mexico, with intraformational folds and associated structures that he interpreted as the result of differential sediment loading. The detrital organic content throughout the limestone had provided the necessary reducing environment, but only the deformed horizons were able to transmit the fluids sufficiently for precipitation to take place. Because Green felt it unlikely that uranium deposits would be found where sediment deformation had not occurred, future exploration would depend on the correct identification of the early structures. Garlick (1988) described an analogous situation in Proterozoic carbonate-bearing quartzites of Montana and Idaho, where the precipitation of various metallic sulphides was induced by algal debris in areas where differential gravitational loading was occurring.

9.6.3 Outline of possible criteria

Unfortunately, no new solution to the long-standing problem of recognizing sediment deformation is revealed here! Possible approaches and discriminatory criteria are discussed below, but these merely expose the sheer difficulty of correct diagnoses. Practicable criteria are scarce, and in consequence it is often impossible to prove a pre-lithification origin. In general terms, the kinds of approaches that can be used fall into four overlapping groups, which are discussed in turn below. Sections 8.2–4 gave illustrations of their use in practice.

Association with a diagnostic feature

Many structures due to liquefaction, of the kind illustrated in Figures 9.28 and 9.29, can be recognized with reasonable confidence. The cause of the liquefaction may be far from clear (section 9.5.2) and later overprint may confuse the appearance, but the ductility required to generate this style of structure in rocks would be recorded, in the case of post-lithification movements, by intragrain deformation unless it had been obscured by later events. To this extent, injection of liquefied material into neighbouring structures can be indicative of their early origin, and such features, especially sedimentary dykes, have been widely used in the past. However, there are pitfalls. The discordant mass must be due to injection, and not just a passive infill of a pre-existing fissure (section 1.3.4). The mass must have been liquefied, with minimal grain breakage. It is possible for cemented rocks to disaggregate under elevated fluid pressures (section 2.3.3), but usually with significant cataclasis. Therefore, the likelihood of intragrain deformation having been subsequently masked should be minimal.

In the field, cataclastic shear zones can mimic dykes, and the difference is not always clear. Williams (1983), for example, interpreted dyke-like structures in an Ordovician mélange of

Newfoundland, which had been viewed as liquefaction features by previous workers, to be the result of faulting combined with metamorphic differentiation. To be diagnostic, a sedimentary dyke must follow rather than truncate the participating grains. Also, to identify the pre-lithification origin of a structure through its association with an undoubted injection feature, the cogenetic relationship between the two structures must be unambiguous. Although less diagnostic and dangerous to use alone, the vein structure described in section 9.5.3 and the shear zones in clayey materials discussed in section 9.4.1 seem to be more characteristic of sediment rather than rock deformation. In both cases, the sedimentary particles merely rotate and are not themselves bent.

The converse approach to assessing a doubtful structure is to seek features diagnostic of post-lithification deformation. If the deformation has affected metamorphic minerals, or has plastically deformed igneous or mineralized bodies, then a prelithification origin can be eliminated. Note, however, that although sedimentary deformation usually takes place by intergranular movements, cataclasis is not necessarily diagnostic of rock deformation (sections 6.2.6 and 9.5.3).

Reference to a known time-plane

This approach uses simple cross-cutting relations to date the structure in question with respect to some feature whose depth or stage of lithification is known. Ghent and Henderson (1965), for example, argued for the synsedimentary origin of convolute lamination because of instances where it is cross-cut by undisturbed burrows. Of course, biological burrows are only relevant to assessing possible near-surface structures, and even here there are potential pitfalls. Burrows have to be separated from tubes bored into rock. As Elliott and Williams (1988) pointed out, the two features are normally distinguishable by the margins of a burrow not truncating grains. Also, the burrow must be of definite organic origin, and not a possible dewatering conduit (section 9.5.3), as this could have formed at considerable depth.

The main shortcoming of this approach is that there are few features known to have sufficient restriction in range. A case in point is detrital remnant magnetization (Lund and Karlin 1990). It has been argued that this is acquired only after substantial consolidation, during which dewatering has allowed the magnetic particles to collapse into good alignment (Yagishita, Westgate and Pearce 1981). On this basis, the magnetization pattern of deformed sediments will be either dispersed or clustered depending on whether or not the deformation took place before consolidation. However, as those authors concede, the burial stage at which magnetization is acquired is controversial. As it could even be synsedimentary, this potentially useful approach is undermined!

Tremlett (1982) advocated the time-plane approach for dealing with Lower Palaeozoic structures in west-central Wales, relating them to the chlorite–mica stacks thought, following Craig, Fitches and Maltman (1982), to originate during diagenesis. Certain concretions appeared to be cross-cut by the stacks and were therefore inferred to be of high-level origin, whereas structures that disturbed the preferred orientation of the stacks were taken to represent post-diagenetic deformation. However, there are major difficulties with this criterion because: (i) it has been argued that chlorite–mica stacks have a very protracted evolution (Milodowski and Zalasiewicz 1991); and (ii) their orientation during diagenetic growth may be strongly influenced by the alignment of pre-existing nucleii. Hence it is conceivable that where, for example, the long axes of chlorite–mica stacks can be traced around a fold, they were folded to give that pattern but grew mimetically over an earlier, folded, fabric. In an apparently clearer example involving authigenic Fechlorite, Hurst and Buller (1984) substantiated the early formation of some Palaeogene dish structures from the North Sea by showing their overprint by the chlorite in a delicate honeycombed network.

Elliott and Williams (1988) closely argued the case for the preferred dimensional orientation of consolidating minerals being a useful marker plane for recognizing surficial slumping. They marshalled evidence that well-developed consolidation fabrics are unlikely to develop at burial depths less than about 100 m. On this basis, prelithification folds of bedding-parallel fabrics must have developed at depths greater than this figure, and are unlikely to be slump folds unless the moving sheets are of at least this thickness. However, although the onset of fabric development could be delayed until well below such depths, through overpressuring or early cementation, it is by no means demonstrated that bedding-parallel fabrics cannot form at shallower levels. Some of the Silurian slump folds in North Wales that appear from a number of lines of evidence to have been open-cast and no more than a few metres thick (Eva 1992) clearly deform a fabric. Even if there has been later mimetic replacement, there must have been some preferred orientation in existence before the folding.

The possibility of mimetic overprint is also relevant to analysis involving the orientation of concretions (section 4.3.9). Some of these growths are particularly likely to follow already existing fabrics, so that, for instance, concretions tracing around a fold closure do not necessarily imply a pre-folding age. Things are clearer if the concretion has demonstrably participated in the deformation – if it is bent or fractured – but a further shortcoming is that the range of burial conditions may be insufficiently restricted for concretion growth to be regarded as a timeplane. Coniglio (1987), for example, described chert formation during the lithification of Cambro-Ordovician marine deposits in western Newfoundland, that extended from precipitation close to the sediment surface to deep concretions that post-dated limestone lithification. Calcite concretions from the Devonian black shales of southwest Ontario were shown (Coniglio and Cameron 1990) to have a more restricted range – from within a few centimetres of the sediment–water interface to a depth of a few hundred metres.

Even though most concretions do form at shallow levels of burial, the depth range may be large enough to allow considerable overlap with the formation of other structures. Davies and Cave (1976) reported concretions that appeared to have ploughed into the adjacent bedding surfaces, the inference being that the concretion growth was accomplished before the host material had acquired sufficient strength to resist the scouring. However, Cave (personal communication) added that other concretions, identical to those that had participated in the slumping, embody a slumped bedding-fabric, suggesting that concretion growth and slumping were diachronous. At one place, concretion growth might pre-date slumping whereas at another it is the converse. Concretions, therefore, have use as time-markers, but each situation has to be assessed independently. Other potential markers include deformed oolites (Sarkar, Chanda and Bhattacharya 1982) and lapilli. Boulter (1983), for example, argued that because the lapilli in folded early Proterozoic rocks of the Pilbara region, Western Australia, still show $X=Y>Z$ compactional strains, the folding must have preceded the main consolidation. Section 8.3 illustrates the value of cross-cutting relations in deciphering the structural evolution of complexly deformed sediments.

Characteristic aspects

Asserting that structures arising from sediment deformation have observable characteristics is perhaps the most common approach to identifying them, but, with the possible exception of liquefaction phenomena, it is possibly also the most precarious. Although some features may be useful when combined with other criteria, the utmost caution is urged. As Williams (1986) emphasized, and as should be clear from the foregoing chapters, aspects such as the shape, style and scale of the structures are in no way diagnostic by themselves. The presence of slickensided limbs has been used as a recognition criterion, but grooving and shininess can be acquired in both very wet sediments (section 9.4.2) and rocks. The presence of fold interference is not in itself helpful – all three textbook patterns were documented from one locality of sediment folding by Tobisch (1984). Elliott and Williams (1988) have discussed the dangers of indiscriminately relying on such things as boudinage, disruption and balling-up of beds, and curvilinear fold hinges to identify or to eliminate pre-lithification deformation. All of these have been used in the past. The latter authors did suggest that opposing vergence of neighbouring folds may be indicative of sediment deformation, but only if the overlying and underlying beds are undeformed.

Three aspects of folds and kindred structures have been used as criteria sufficiently widely to warrant more detailed discussion. The first of these is the restriction of the structures to isolated horizons – the kinds of intraformational structures that have long intrigued geologists, as recounted in Historical vignette 2, section 1.3.2, and illustrated in the frontispiece. The intraformational nature alone is not relevant; post-lithification tectonism can be extremely localized. Even the intricate and highly localized folds shown in Figure 9.38 may be tectonic (Kirkland and Anderson 1970). It is the nature of the top and bottom contacts that is critical, but their assessment is not straightforward. To be diagnostic of sediment deformation, the junctions should both truncate the structures and be demonstrably of pre-lithification origin. As Jones (1939a) appreciated (Historical vignette 3, section 9.6.1), a depositional upper contact is persuasive, provided there is no significant time gap. But much sediment deformation is not open-cast, and even mass movement can generate high-strain zones at depth (Coniglio 1986). Slickensiding, and even mineralization (see below), of the bounding surfaces does not imply rock deformation, and neither, of course, does their absence indicate a pre-lithification origin. To be truly diagnostic, the truncating surfaces should be cross-cut by a feature such as a burrow or definite liquefaction feature. In other words, intraformational structures are suggestive of pre-lithification movement if they are truncated by the bounding surfaces, and the certainty increases with the amount of evidence against the junctions having formed in rock.

The second aspect is the association of the structure with brittle, mineralized fractures. This has been touted frequently as a good criterion, but often fallaciously. Needless to say, the cogenetic association between the fractures and the structure in question must be unambiguous. The brittle behaviour must be demonstrated through significant grain breakage – failure surfaces in very weak, water-rich sediment can be surprisingly narrow, and appear 'brittle'. Excellent illustrations of differences between vein-wall textures produced in partially, lithified and lithified sediments are given in section 8.4.

The value of the mineralization depends to some extent on the nature of the mineral and, in particular, its solubility. Carbonates and some sulphides are common fracture-filling materials, but both can precipitate in surficial sediments and are therefore unhelpful as criteria. Daley (1972), for example, described fractures filled with calcite and pyrite in surficial, subaqueous sediments, in the Oligocene deposits of the Isle of Wight. K.M. Brown and Orange (1993) described an intricate sequence of calcite–quartz–laumontite deposition during the emplacement of a mélange diapir, and Genna (1988) reported calcite-filled shear-zones in unlithified Eocene sediments in Minervois (Figure 9.39). At the Peru convergent margin, Suess *et al.* (1988) reported calcite and barite veins forming at burial depths around 30 m, in sediments with porosities in excess of 77% and shear strengths less than 100 kPa. Section 8.4 describes calcite-filled veins that are participating in the tectonic deformation of partly lithified sediments. The calcite-filled fracture shown in Figure 9.40 formed in still friable, subaerial sands that have never been buried more than a few metres (K. Brodzikowski, personal communication).

The other common fracture infill is quartz, and here, in the absence of definitive isotope or fluid inclusion data (e.g. Vrolijk 1987), the situation is complex. The low solubility of quartz makes it generally less likely to fill near-surface transient fractures, and is therefore something more of a guide to somewhat deeper deformation, although not necessarily of lithified rocks. Fisher and Byrne (1990), for example, described quartz-filled

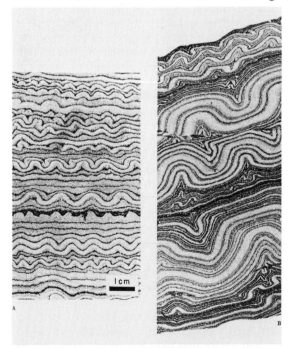

Figure 9.38 Microfolds of anhydrite and calcite (dark) laminae, in the Upper Permian Castile Formation of West Texas. Although volume changes associated with gypsum–anhydrite transitions have been widely invoked to explain structures such as these, Kirkland and Anderson (1970) argued for an origin due to tectonically induced, differential layer-parallel shortening. (From Kirkland and Anderson (1970). Reproduced with permission from D. W. Kirkland.)

Recognition of sediment deformation structures 305

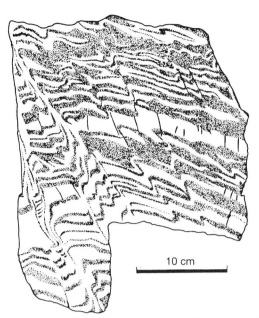

Figure 9.39 Example of calcite veins (black strokes) following shear zones, in argillaceous Eocene sands, Minervois, France. Clayey laminae are shown in white; sandy laminae in grey. (From Genna (1988). Reproduced with permission from the Académie des Sciences, France.)

fractures from Cretaceous turbidites of Kodiak, Alaska, that closely follow but do not truncate grain boundaries, from which they inferred a genesis before complete lithification. Quartz can be transported and precipitated subaerially – over periods of tens of years – so the judgement becomes more a matter of gauging the length of time a fracture remained available for infill, and whether sufficient volumes of fluids could have been transmitted. Based on the equation given by Fournier and Potter (1982), at near-surface temperatures and pressures, and assuming no unusual chemical or microbiological activity, a given volume of quartz requires the passage of around 108 times that volume of fluid.

The main difficulty with using quartz as a criterion is that it readily replaces pre-existing minerals. Although the quartz observed in a structure may have formed at some depth, it may have replaced an earlier mineral, perhaps calcite or aragonite, which formed together with the structure at near-surface conditions. In short, the observation of quartz-veins is not a criterion of rock deformation, but if the quartz is primary and is filling a fracture unlikely to have been long-lived, it normally suggests processes at some depth.

(a)

(b)

Figure 9.40 Calcite-filled zones in glacial Pleistocene sands, near Szchukwin, south of Łódź, central Poland. (a) A mineralized shear zone. (b) The kink-band is also mineralized, but in a more diffuse style.

The third matter concerns folds with axial-planar foliations. This also has been cited frequently as a reliable criterion, but it, too, is complex. It does not apply to flat-lying folds with non-fanning foliations, as post-folding cosolidation can generate a fabric which is fortuitously parallel with the axial surface (Maltman 1981). Many flat-lying slump folds with axial-planar fabrics can be interpreted in this way (e.g. J.C. Moore and Geigle 1974). The mechanism clearly does not apply to folds that are not horizontal, but most slump folds do not have this attitude. Farrell and Eaton (1988) interpreted a fabric axial-planar to flat-lying slump folds in Miocene limestones in Cyprus as being due to a combination of liquefaction and consolidation, but, significantly, they failed to detect a fabric in the associated inclined or upright folds.

306 *Deformation structures preserved in rocks*

Almost all the foliations that have been reported from inclined or upright sediment folds are of a spaced nature. C.M. Bell (1981) described pre-lithification folds with an axial-planar cleavage that appears to be an array of microfaults. Corbett (1973) reported axial-planar foliations in parts of slump folds, but of irregular development and consisting mainly of fine sandstone stringers. The flat-lying, axial-planar foliations described by Tobisch (1984) involve aligned phyllosilicates, but only in the silt–sand domains that chiefly define the structure. The stringers form from the disruption of the thin bedding laminae. Tobisch surmised this to result from liquefaction and expulsion of the sediment parallel to the axial surfaces, which prompted some particle reorientation. Similar features are apparent in Figure 9.41. O'Connor (1980) reported a fabric of similar appearance, but apparently with a spacing of only a millimetre. Clearly, foliations can arise axial-planar to inclined folds, but with a distinctive, spaced aspect. The glacially induced fold included in Figure 9.20, from Quaternary sediments of the Karakoram (Owen 1988b), has a fanning axial-planar cleavage, but it, too, has a spaced aspect and relationships with flame structures.

The question is whether or not pre-lithification folds can develop axial-planar **pervasive** fabrics. It would appear possible in surficial wet sediments. Yagishita and Morris (1979) have shown that flow during folding of cross-bedding can orient the quartz grains, and Williams, Collins and Wiltshire (1969) documented a pervasive fabric axial-planar to steeply inclined folds in Devonian rocks on the coast of New South Wales. However, these too are thought to have formed in water-rich, high-level sediments, and essentially are a form of convolute lamination. The extent to which an axial-planar fabric can be acquired in partially consolidated sediment remains to be demonstrated. There is, as yet, no documentation of such a feature having formed before lithification but at depths more than a few metres. However, unless it is shown to be impossible, it would be presumptuous to use the presence of an axial-planar fabric to eliminate a wide range of lithification states.

Overall association

Assessing the sedimentary and deformational setting of structures under consideration may seem

Figure 9.41 Folds with spaced axial-planar foliations, and approximately parallel sedimentary dykes. Core sample, Magnus Sandstone Formation, Upper Jurassic, northernmost Viking Graben, North Sea. (Specimen provided by Dennis Wood.)

a cumbersomely indirect approach, but, especially in view of the uncertainties of the more direct methods outlined above, it can be useful. Some degree of sediment deformation is possible in virtually every depositional environment, but an appraisal of the sedimentary setting will provide some indication of whether the deformation is likely to be dominated by, say, mass movement processes or tectonic contraction. As Mills (1983) noted for synsedimentary deformation, the analysis should involve all available information, whether lithological, structural or palaeontological. In the glacial setting, for example, a careful assessment of the facies associations, as described by Brodzikowski and van Loon (1991), will help identify whether structures are the result of subglacial shear or some form of proglacial process.

Similarly, bearing in mind the correlation problems outlined by Williams (1985), the analysis of a structure will be more reliable if it is seen as one of a family of cogenetic structures, which together are recording the overall deformational setting. The similarity between structures of very different origins has been illustrated a number of times in the foregoing pages. The structures induced by a moving ice sheet are virtually identical, at least in geometry, with those formed in fold-and-thrust belts. The similarity between these and structures formed in accretionary prisms has been widely mentioned, and these duplicate the contractional parts of a gravitational sliding mass, where even the scale may be similar. Interpretation of these features in ancient rocks, especially where exposure is inadequate, cannot be carried out satisfactorily from individual structures: it has to involve whatever can be deduced about the overall setting. The information forthcoming from the structures in turn provides further input to the geological understanding of the region.

Some of the guidelines mentioned above can be used with more confidence than others, but all require care. None can be used alone. The important thing is that any use of criteria is informed – that the worker is alert to the range of possible processes and has an understanding of how they operate. In many ways, this thinking encapsulates the *raison d'etre* of this book. To recognize sediment deformation 'the geologist has to assess the cumulative weight of several not infallible criteria' (Fitches and Maltman 1978). This remark was made over 15 years ago, but still holds true. What has changed since then is the new realization of the breadth and importance of sediment deformation, and the consequent exploration of the mechanisms by which it takes place. These developments have led, among other things, to an improved understanding of the recognition guidelines and hence to an awareness of their fallibilities. Today, the criteria for recognizing sediment deformation are being applied more cautiously, but their value is far better appreciated.

Appendix: List of symbols

This list includes the main symbols used in the text, but is not intended to be comprehensive. References to examples of usage are indicated in parentheses. See also Table 6.1 for additional symbols used in that chapter. A prime (') after a symbol denotes its effective value, i.e. with allowance for pore fluid pressure.

A	area (equations 1.15, 2.1)	K_0	stress ratio for no lateral strain (equation 2.18)
α	thermal compressibility (equation 6.5)	k	permeability (equation 7.4)
B	constant relating porosity and cohesion (equation 3.1)	K	hydraulic conductivity (equation 7.4)
γ	shear strain (equation 1.6)	l	length (equations 2.2, 7.8)
$\dot{\gamma}$	shear strain rate (Figure 3.4)	L_b	rheological constant (equation 3.3)
C	compressibility (equation 2.3)	λ	pore fluid/total pressure ratio (equation 1.14)
C_m	compressibility of mineral particles (equation 2.12)	μ	coefficient of viscosity (equations 1.6; 3.2)
C_s	compressibility of particle framework (equation 2.12)	ν	Poisson's ratio (equation 6.3)
C_f	compressibility of pore fluid (equation 7.6)	P	pressure
C_v	coefficient of consolidation (equation 2.13)	P_{H_2O}	pore fluid pressure (equations 1.12, 2.8)
c	cohesion (equation 1.5)	$P_{H_2O_e}$	overpressure (excess fluid pressure) (equation 2.8)
c_0	cohesion at zero porosity (equation 3.1)	$P_{H_2O_h}$	hydrostatic fluid pressure (equation 2.8)
D	diffusivity (equation 7.5)	p'	mean effective stress (equation 2.17)
E	Young's modulus (equation 6.3)	$\tan \phi$	coefficient of sliding friction (equation 1.5)
e	void ratio (equations 1.2, 6.1)	Q	fluid flow rate (equations 7.3, 7.9)
ε	strain (equation 2.2)	q	differential (deviatoric) stress (equation 2.16)
$\dot{\varepsilon}$	strain rate (equation 3.2)	r	resistance to deformation (section 2.11)
η	porosity (equations 1.2, 6.2)	r_c	strong-bonding resistance (equation 2.11)
θ	slope inclination (equation 1.15), or normal to slope (equation 2.5)	r_i	intergranular friction resistance (equation 2.11)
F	force (equation 2.1)	r_m	weak-bonding resistance (equation 2.11)
Φ	pressure head (equation 7.1)	ρ	density (equation 2.4)
ϕ	angle of internal friction (equation 1.5)	S_s	specific storage (equation 7.5)
g	gravitational acceleration (equation 2.4)	S_T	tensile strength (equation 6.14)
H	hydraulic head (equation 7.1)	σ_1, σ_3	maximum and minimum principle stresses (equations 2.5, 2.6)
h	height (equation 1.19)	σ_c	consolidation stress (section 6.2.2)
J	constant relating permeability to pore radius and tortuosity (equation 7.7)	σ_n	stress normal to plane of shear stress (equation 1.5)
K	horizontal/vertical stress ratio (equation 1.1)	σ_v	vertical stress (equation 6.1)

σ_h	horizontal stress (equation 6.4)	V	volume (Figure 1.3)
T	temperature (equation 6.5)	W	weight (equation 1.15)
t	time equations 1.6, 2.10)	w	water content (equation 1.3)
Γ	tortuosity (equation 7.7)	w_{wet}	wet water content (equation 1.4)
τ	shear stress (equations 1.5, 5.1)	x	rheological constant (equation 1.11)
τ_y	yield stress (equations 3.2, 3.3)	ψ	mean pore radius (equation 7.7)
U	velocity (Figure 5.20)	Z	elevation head (equation 7.1)
UW	unit weight (equation 2.14)	z	depth (equation 2.4)

References

Aam, K. (1988) Ekofisk subsidence: the problem, the solution and the future. *Proceedings of the International Conference on the Behaviour of Offshore Structures*, Tapir, Trondheim, vol. 2, pp. 19-40.

Abbott, D.H., Embley, R.W. and Hobart, M.A. (1985) Correlation of shear strength, hydraulic conductivity, and thermal gradients with sediment disturbance: South Pass region, Mississippi Delta. *Geo-Marine Letters*, 5, 113-119.

Abdulraheen, A., Roegiers, J-C. and Zaman, M. (1992) Mechanics of pore collapse and compaction in weak porous rocks, in *Rock Mechanics, Proceedings of the 33rd US Symposium* (eds J.R. Tillerson and W.R.W. Wawersik), A.A. Balkema, Rotterdam, pp. 233-242.

Abele, G. (1972) Kinematik und Morphologie spät und postglazialer Bergsturze in den Alpen. *Zeitschrift fur Geomorphologie*, 14, 138-149.

Aber, J.S. (1988) Ice-starved hills of Saskatchewan compared with the Mississippi Delta mudlumps – implications for glaciotectonic models, in *Glaciotectonics: Forms and Processes* (ed. D.G. Croot) A.A. Balkema, Rotterdam, 212 pp.

Aber, J.S., Croot, D.G. and Fenton, M.M. (1989) *Glaciotectonic Landforms and Structures*. Kluwer Academic Publishers, Dordrecht, 200 pp.

Addis, M.A. (1987) Mechanisms of sediment compaction responsible for oil field subsidence. Unpublished PhD thesis, University of London, 561 pp.

Addis, M.A. (1989) The behaviour and modelling of weak rocks, in *Rock at Great Depth* (eds V. Maury and D. Fourmaintraux), A.A. Balkema, Rotterdam, 891-898.

Agar, S.M. (1988) Shearing of partially consolidated sediments in a lower trench slope setting, Shimanto Belt, SW Japan. *Journal of Structural Geology*, 10, 21-32.

Agar, S.M., Prior, D.J. and Behrmann, J.H. (1988) Back-scattered electron imagery of the tectonic fabrics of some fine-grained sediments: implications for fabric nomenclature and deformation processes. *Geology*, 17, 901-904.

Ahuja, L.R., Cassel, D.K., Bruce, R.R. and Barnes, B.B. (1989) Evaluation of spatial distribution of hydraulic conductivity using effective porosity data. *Soil Science* 148, 404-411.

Aiban, S.A. and Znidaric, D. (1989) Evaluation of the flow pump and constant head techniques for permeability measurements. *Geotechnique*, 39, 655-666.

Aksu, A.E. and Hiscott, R.N. (1989) Slides and debris flows on the high-latitude continental slopes of Baffin Bay. *Geology*, 17, 885-888.

Al-Tabbaa, A. and Wood, D. Muir (1991) Horizontal drainage during consolidation: insights gained from analyses of a simple problem. *Geotechnique*, 41, 571-585.

Allen, J.R.L. (1970) The sequence of sedimentary structures in turbidites, with special reference to dunes. *Scottish Journal of Geology*, 6, 146-161.

Allen, J.R.L. (1977) The possible mechanics of convolute lamination in graded sand beds. *Journal of the Geological Society*, 134, 19-31.

Allen, J.R.L. (1984) *Sedimentary Structures, their Character and Physical Basis*. Developments in Sedimentology 30, Elsevier, Amsterdam, 663 pp.

Allen, J.R.L. (1985) *Principles of Physical Sedimentology*. Allen & Unwin, London, 272 pp.

Allen, J.R.L. (1986a) Earthquake magnitude frequency, epicentral distance, and soft sediment deformation in sedimentary basins. *Sedimentary Geology* 46, 67-75.

Allen, J.R.L. (1986b) Pedogenic calcrete in the Old Red Sandstone facies (Late Silurian-Early Carboniferous) of the Anglo-Welsh area, southern Britain, in *Palaeosols: their Recognition and Interpretation:* (ed. V.P. Wright), Blackwells Scientific Publications, Oxford, 58-86.

Allen, J.R.L. and Banks, N.L. (1972) An interpretation and analysis of recumbent-folded deformed cross-bedding. *Sedimentology*, 19, 257-283.

Alley, R.B. (1989a) Water-pressure coupling of sliding and bed deformation: II. Velocity-depth profiles. *Journal of Glaciology*, 119, 119-29.

Alley, R.B. (1989b) Water-pressure coupling of sliding and bed deformation: I. Water system. *Journal of Glaciology*, 119, 108-119.

Alley, R.B. (1991) Deforming-bed origin for southern Laurentide till sheets? *Journal of Glaciology*, 37, 67-76.

Alley, R.B., Blankenship, D.D., Bentley, C.R. and Rooney, S.T. (1986) Deformation of till beneath ice-stream B, West Antarctica. *Nature*, 322, 57-59.

Alley, R.B., Blankenship, D.D., Bentley, C.R. and Rooney, S.T. (1987a) Till beneath Ice Stream B. Till deformation: evidence and implications. *Journal of Geophysical Research*, **92**, 8921–8929.

Alley, R.B., Blankenship, D.D., Rooney, S.T. and Bentley, C.R. (1987b) Till beneath ice stream B. 4. A coupled ice-till flow model. *Journal of Geophysical Research*, **92**, 8931–8940.

Alley, R.B., Blankenship, D.D., Rooney, S.T. and Bentley, C.R. (1989a) Sedimentation beneath ice-shelves – the view from Ice Stream B. *Marine Geology*, **85**, 101–120.

Alley, R.B., Blankenship, D.D., Rooney, S.T. and Bentley, C.R. (1989b) Water-pressure coupling of sliding and bed deformation. III. Application to Ice Stream B. *Journal of Glaciology*, **119**, 131–139.

Allison, P.A. (1988) The role of anoxia in the decay and mineralization of proteinaceous macro-fossils. *Paleobiology*, **14**, 139–154.

Allison, R.J. and Brunsden, D. (1990) Some mudslide movement patterns. *Earth Surface Processes and Landforms*, **15**, 297–311.

Almagor, G. (1984) Salt-controlled slumping on the Mediterranean slope of central Israel. *Marine Geophysical Researches*, **6**, 227–243.

Amand, A., and Jain, A. (1987) Earthquake and deformational structures (seismites) in Holocene sediments from the Himalayan–Sandaman Arc, India. *Tectonophysics*, **133**, 105–120.

Amårk, M. (1986) Clastic dykes formed beneath an active glacier. *Geologiska Förengen i Stockholm Förenhandlingar*, **108**, 13–20.

Amick, D., and Gelinas, R. (1991) The search for evidence of large prehistoric earthquakes along the Atlantic seaboard. *Science*, **251**, 655–658.

Andersen, M.A., Foged N. and Pedersen, H.F. (1992) The rate-type compaction of a weak North Sea chalk, *Rock Mechanics, Proceedings of the 33rd U.S. Symposium* (eds J.R. Tillerson and W.R.W. Wawersik), A.A. Balkema, Rotterdam, 253–261.

Anketell, J.M., Cegla, J. and Dzulinski, S. (1970) On the deformational structures in systems with reversed density gradients. *Annales de la Société Géologique de Pologne*, **40**, 3–29.

Aramaki, S. (1963) Geology of Asama volcano. *Journal of the Faculty of Science, Tokyo University*, **14**, 229–443.

Arch, J. (1988) An experimental study of deformation microstructures in soft sediments. Unpublished PhD thesis, University College of Wales, Aberystwyth, 387 pp.

Arch, J. and Maltman, A.J. (1990) Anisotropic permeability and tortuosity in deformed wet sediments. *Journal of Geophysical Research*, **95**, 9035–9046.

Arch, J., Maltman, A.J. and Knipe, R.J. (1988) Shear-zone geometries in experimentally deformed clays: the influence of water content, strain rate and primary fabric. *Journal of Structural Geology*, **10**, 91–99.

Archanguelsky, A.D. (1930) Slides of sediments on the Black Sea bottom and the importance of this phenomenon for geology. *Bulletin of the Society for Nature, Moscow*, **38**, 38–80.

Archer, J.B. (1984) Clastic intrusions in deep-sea fan deposits of the Rosroe Formation, lower Ordovician, Western Ireland. *Journal of Sedimentary Petrology*, **54**, 1197–1205.

Archer, J.S. and Wall, C.G. (1986) *Petroleum Engineering Principles and Practice*. Graham & Trotman, London, 362 pp.

Arnott, R.W.C. and Hand, B.M. (1989) Bedforms, primary structures and grain fabric in the presence of suspended sediment rain. *Journal of Sedimentary Petrology*, **59**, 1062–1069.

Arthur, J.R.F. and Dunstan, T. (1982) Rupture layers in granular media, in *Deformation and Failure of Granular Materials* (Eds P.A. Vermeer and H.J. Luger), A.A. Balkema, Rotterdam, 453–459.

Aspler, L.B. and Donaldson, J.A. (1986) Penecontemporaneous sandstone dikes, Nonacho Basin (early Proterozoic, Northwest Territories): horizontal injection in vertical, tabular fissures. *Canadian Journal of Earth Sciences*, **23**, 827–838.

Astin, T.R. and Rogers, D.A. (1991) 'Subaqueous shrinkage cracks' in the Devonian of Scotland reinterpreted. *Journal of Sedimentary Petrology*, **61**, 850–859.

Astin, T.R. and Rogers, D.A. (1993) 'Subaqueous shrinkage cracks' in the Devonian of Scotland reinterpreted – reply. *Journal of Sedimentary Petrology*, **63**, 566–567.

Astin, T.R. and Scotchman, I.C. (1988) The diagenetic history of some septarian concretions from the Kimmeridge Clay, England. *Sedimentology*, **35**, 349–368.

ASTM (American Society for Testing and Materials) (1991) Standard terminology, relation to soil, rocks, and contained fluids. *Annual Book of ASTM Standards*, **04.08**, 129–156.

Athy, L.F. (1930) Density, porosity, and compaction of sedimentary rocks. *Bulletin of the American Association of Petroleum Geologists*, **14**, 25–35.

Atkinson, J.H. (1987) Basic mechanics of soils, in *Ground Engineer's Reference Handbook* (ed. F.G. Bell), Butterworth, London, pp. 3/1–3/35.

Atkinson, J.H. and Bransby, P.L. (1978) *The Mechanics of Soils: an Introduction to Critical State Soil Mechanics*. McGraw-Hill, London, 375 pp.

Atkinson, J.H. and Richardson, D. (1987) The effect of local drainage in shear zones on the undrained strength of overconsolidated clay, *Geotechnique*, **37**, 393–403.

Atkinson, J.H., Lau, W.H.W. and Powell, J.J.M. (1991) Measurement of soil strength in simple shear tests. *Canadian Geotechnical Journal*, **28**, 255–262.

Atkinson, J.H., Richardson, D. and Robinson, P.J. (1987) Compression and extension of K_0 normally consolidated kaolin clay. *Journal of Geotechnical Engineering*, **113**, 1468–1482.

Audemard, F.A. and Santis, F. de, (1991) Survey of liquefaction structures induced by recent moderate earthquakes. *Bulletin of the International Association of Engineering Geologists*, **44**, 5–16.

Audet, D.M. and McConnell, J.D.C. (1992) Forward modelling of porosity and pore pressure evolution in sedimentary basins. *Basin Research*, **4**, 147–162.

Awadallah, S.A. (1991) A simple technique for vacuum impregnation of unconsolidated, fine-grained sediments. *Journal of Sedimentary Petrology*, **61**, 632–633.

Axen, G.J. (1992) Pore pressure, stress increase, and fault weakening in low-angle normal faulting. *Journal of Geophysical Research*, **97**, 8979–8991.

Aydin, A. and Johnson, A.M. (1978) Small faults formed as deformation bands in sandstones. *Pure and Applied Geophysics*, **116**, 913–930.

Aydin, A. and Johnson, A.M. (1983) Analysis of faulting in porous sandstones. *Journal of Structural Geology*, **5**, 19–31.

Bacon, C.R. and Duffield, W.A. (1978) Soft-sediment deformation near the margin of a basalt sill in the Pliocene Coso Formation, Inyo County, California. *Geological Society of America Abstracts with Programs*, **10**, 94.

Baerwald, R.J., Burkett, P.J. and Bennett, R.H. (1991) Submarine sediment preparation for electron microscopy, in *Microstructure of Fine-grained Sediments* (eds R.H. Bennett, W.R. Bryant and M.H. Hulbert), Springer-Verlag, New York, 307–320.

Bagnold, R.A. (1954) Experiments on a gravity-free dispersion of large solid spheres in a Newtonian fluid under shear. *Proceedings of the Royal Society of London, Series A*, **225**, 49–63.

Bagnold, R.A. (1956) The flow of cohesionless grains in fluids. *Proceedings of the Royal Society of London, Series A*, **249**, 235–297.

Bagnold, R.A. (1962) Auto-suspension of transported sediment: turbidity currents. *Proceedings of the Royal Society of London, Series A*, **265**, 315–319.

Bailey, E.B., Collett, L.W. and Field, R.M. (1928) Paleozoic submarine landslips near Quebec City. *Journal of Geology*, **36**, 577–614.

Bakken, B. (1987) Sedimentology and syndepositional deformation of the Ross Slide, Western Irish Namurian Basin, Ireland. Unpublished Candidatus Scientiarum thesis, Geological Institute, University of Bergen.

Baldwin, B. and Butler, C.O. (1985) Compaction curves. *Bulletin of the American Association of Petroleum Geologists*, **69**, 622–626.

Ballance, P.F. and Hayward, B.F. (1983) Piha Formation volcanic conglomerates (Waitakere Group, Lower Miocene, Auckland) – a sedimentological conundrum. *Geological Society of New Zealand Miscellaneous Publications*, **30A**.

Bangs, N.L.B., Westbrook, G.K., Ladd, J.W. and Buhl, P. (1990) Seismic velocities from the Barbados Ridge Complex: indicators of high pore fluid pressures in an accretionary complex. *Journal of Geophysical Research*, **95**, 8767–8782.

Banham, P.H. (1975) Glaciotectonic structures; a general discussion with particular reference to the contorted drift of Norfolk, in *Ice ages: Ancient and Modern* (eds A. Wright and F. Moseley), *Geological Journal Special Issue*, **6**, 69–94.

Baraza, J., Lee, H.J., Kayen, R.E. and Hampton, M.R. (1990) Geotechnical characteristics and slope stability on the Ebro margin, western Mediterranean. *Marine Geology*, **95**, 379–393.

Barber, A., Tjokrosapoetro, S. and Charlton, T. (1986) Mud volcanoes, shale diapirs and mélanges in accretionary complexes, Eastern Indonesia. *Bulletin of the American Association of Petroleum Geologists*, **70**, 1729–1741.

Barclay, W.J., Glover, B.W. and Mendum, J.R. (1993) 'Subaqueous shrinkage cracks' in the Devonian of Scotland reinterpreted – discussion. *Journal of Sedimentary Petrology*, **63**, 564–565.

Bardet, J.P. and Proubet, J. (1992) Shear band analysis in idealized granular material. *Journal of Engineering Mechanics*, **118**, 397–415.

Barker, C. (1972) Aquathermal pressuring – the role of temperature in the development of abnormal pressure zones. *Bulletin of the American Association of Petroleum Geologists*, **56**, 2068–2071.

Barker, C. and Horsfield, B. (1982) Mechanical versus thermal cause of abnormally high pore pressures in shales. *Bulletin of the American Association of Petroleum Geologists*, **66**, 99–100.

Barker, J.A. (1981) *Dictionary of Soil Mechanics and Foundation Engineering*. Construction Press, London, 210 pp.

Barnes, H.A. and Walters, K. (1985) The yield stress myth. *Rheologica Acta*, **24**, 323–326.

Barnes, H.A., Hutton, J.F. and Walters, K. (1991) *Introduction to Rheology*. Elsevier, Amsterdam, 199 pp.

Barnes, P.M. and Lewis, K.B. (1991) Sheet slides and rotational failures on a convergent margin: the Kidnappers Slide, New Zealand. *Sedimentology*, **38**, 205–221.

Bauer, E.R. and Hill, J.L. (1987) *Analysis of Cutter Roof Failure Induced by Clastic Dykes in an Underground Coal Mine*. United States Bureau of Mines Reports of Investigations, Washington, DC, 23 pp.

Baulig, H. (1956) Peneplaines et pediplaines. *Société Belge Etudes Géographie*, **25**, 25–58.

Bayer, U. and Wetzel, A. (1985) Compactional behavior of fine-grained sediments – examples from Deep

Sea Drilling Project cores. *Geologische Rundschau*, **78**, 807–819.
Bayly, M.B. (1992) *Mechanics in Structural Geology*. Springer-Verlag, New York, 253 pp.
Bayly, M.B. (1993) Chemical change in deforming materials, *Oxford Monographs on Geology and Geophysics* **21**, 256.
Bear, J. (1972) *Dynamics of Fluids in Porous Media*. Elsevier, Amsterdam, 764 pp.
Beard, D.C. and Weyl, P.K. (1973) Influence of texture on porosity and permeability of unconsolidated sand. *Bulletin of the American Association of Petroleum Geologists*, **57**, 349–369.
Beazley, K.M. (1980) Industrial aqueous suspensions, in *Rheometry: Industrial Applications* (ed. K. Walters), Research Studies Press, Chichester, 339–413.
Bebout, G.E. (1991a) Field-based evidence for devolatization in subduction zones: implications for arc magmatism. *Science*, **251**, 413–416.
Bebout, G.E. (1991b) Geometry and mechanisms of fluid flow at 15 to 45 km depths in an Early Cretaceous accretionary complex. *Geophysical Research Letters*, **18**, 923–926.
Bebout, G.E. and Barton, M. (1989) Fluid flow and metasomatism in a subduction zone hydrothermal system: Catalina Schist terrane, California. *Geology*, **17**, 976–980.
Becker, A. (1993) An attempt to define a 'neotectonic period' for central and northern Europe. *Geologische Rundschau* **82**, 67–83.
Becker, K., Sakai, H. and 23 others (1990) Drilling deep into young oceanic crust, Hole 504B, Costa Rica Rift. *Reviews of Geophysics*, **27**, 79–102.
Been, K. and Sills, G. (1981) Self-weight consolidation of soft soils: an experimental and theoretical study. *Geotechnique*, **31**, 519–534.
Been, K., Jefferies, M.G. and Hackey, J. (1991) The critical state of sands. *Geotechnique*, **41**, 365–381.
Beget, J.E. (1986) Modelling the influence of till rheology on the flow and profile of the Michigan lobe, southern Laurentide ice sheet, U.S.A. *Journal of Glaciology*, **32**, 235–241.
Beget, J.E. and Kienle, J. (1992) Cyclic formation of debris avalanches at Mount St. Augustine volcano. *Nature*, **356**, 701–704.
Behrman, J.H. (1991) Conditions for hydrofracture and the fluid permeability of accretionary wedges. *Earth and Planetary Science Letters*, **107**, 550–558.
Behrman, J.H., Brown, K., Moore, J.C., Mascle, A., Taylor, E. and the shipboard scientifc party (1988) Evolution of structures and fabrics in the Barbados accretionary prism: insights from Leg 110 of the Ocean Drilling Program. *Journal of Structural Geology*, **10**, 577–591.
Bekins, B. and Dreiss, S.J. (1992) A simplified analysis of parameters controlling dewatering in accretionary prisms. *Earth and Planetary Science Letters*, **109**, 275–287.
Belitz, K. and Bredehoeft, J.D. (1988) Hydrodynamics of Denver Basin: explanations of subnormal fluid pressures, *Bulletin of the American Association of Petroleum Geologists*, **72**, 1334–1359.
Bell, C.M. (1981) Soft-sediment deformation of sandstone related to the Dwyka glaciation in South Africa. *Sedimentology*, **28**, 321–329.
Bell, F.G. (1987) Engineering behaviour of soils and rocks, in *Ground Engineer's Reference Handbook* (ed. F.G. Bell), Butterworths, London, pp. 9/1–9/34.
Bell, J.S. (1990) The stress regime of the Scotian shelf offshore eastern Canada to 6 kilometres depth and implications for rock mechanics and hydrocarbon migration, in *Rock at Great Depth* (eds V. Maury and D. Fourmaintraux), A.A. Balkema, Rotterdam, 1243–1265.
Belloni, L. and Morris, D. (1991) Earthquake-induced shallow slides in volcanic debris soils. *Geotechnique*, **41**, 539–552.
Benest, H. (1899) Submarine gullies, river outlets and fresh water escapes beneath the sea level. *Geographical Journal*, **14**, 394–413.
Bennett, R.H., Bryant, W.R. and Hulbert, M.H. (eds) (1991) *Microstructure of Fine-grained Sediments*. Springer-Verlag, New York, 582 pp.
Bennett, R.H., Bryant, W.R. and Keller, G.H. (1981) Clay fabric of selected submarine sediments: fundamental properties and models. *Journal of Sedimentary Petrology*, **51**, 217–232.
Bennett, R.H., O'Brien, N.R. and Hulbert, M.H. (1991) Determinants of clay and shale microfabric signatures: processes and mechanisms, in *Microstructure of Fine-grained Sediments* (eds R.H. Bennet, W.R. Bryant and M.H. Hulbert), Springer-Verlag, New York, 5–32.
Bennett, R.H., Lehman, L., Hulbert, M.H. et al. (1985) Interrelationships of organic carbon and submarine sediment geotechnical properties. *Marine Geotechnology*, **6**, 61–98.
Berner, R.A. (1968) Calcium carbonate concretions formed by the decomposition of organic matter. *Science*, **159**, 195–7.
Besly, B.M. and Fielding, C.R. (1989) Palaeosols in Westphalian coal-bearing and red-bed sequences, Central and Northern England. *Palaeogeography, Palaeoclimatology, Palaeoecology*, **70**, 303–330.
Bethke, C.M. (1989) Modelling subsurface flow in sedimentary basins. *Geologische Rundschau*, **78**, 129–154.
Bhatia, S.K. and Soliman, A. (1991) The application of image analysis techniques to microstructure studies in geotechnical engineering, in *Microstructure of Fine-grained Sediments* (eds R. H. Bennett, W. R. Bryant and M.H. Hulbert), Springer-Verlag, New York, 367–378.

Biermann, C. (1984) On the parallelism of bedding and cleavage in deformed rocks from the internal zone of the Beltic Cordillera, SE Spain. *Geologie en Mijnbouw*, **63**, 75–83.

Biju-Duval, B. and Moore, J.C. and the shipboard scientific party (1984) *Initial Reports of the Deep Sea Drilling Project*, **78A**. United States Government Printing Office, Washington, DC, 621 pp.

Biot, M.A. (1956) General solutions of the equations of elasticity and consolidation for a porous material. *Journal of Applied Mechanics*, **23**, 91–96.

Birch, F. (1966) Compressibility: elastic constants, in *Handbook of Physical Constants* (ed. S.P. Clark). *Geological Society of America Memoir*, **97** (section 7), 97–173.

Bird, P. (1984) Hydration-phase diagrams and friction of montmorillonite under laboratory and geologic conditions, with implications for shale compaction, slope stability and strength, of fault gouge. *Tectonophysics*, **107**, 235–260.

Bishop, A.W. (1966) The strength of soils as engineering materials. *Geotechnique*, **16**, 91–130.

Bishop, A.W. (1976) The influence of system response on the observed pore pressure response to an undrained change in stress in saturated rock. *Geotechnique*, **26**, 372–375.

Bishop, A.W. and Hight, D.W. (1977) The value of Poisson's ratio in saturated soils and rocks stressed under undrained conditions. *Geotechnique*, **27**, 369–384.

Bishop, A.W. and Lovebury, H.T. (1969) Creep characteristics of two undisturbed clays. *Proceedings 7th International Conference Soil Mechanics and Foundation Engineering, Mexico, 1969*, pp. 24–28.

Bishop, A.W., Kumapley, N.K. and El-Ruwayih, A. (1975) The influence of pore-water tension on the strength of clay. *Philosophical Transactions of the Royal Society of London, Series A*, **319**, 511–554.

Bjørlykke, K., Ramm, M. and Saigal, G.C. (1989) Sandstone diagenesis and porosity modification during basin evolution. *Geologische Rundschau*, **78**, 243–268.

Blake, E.W. and Clarke, G.K.C. (1989) *In situ* bed strain measurements beneath a surge-type glacier. *Eos (Transactions of the American Geophysical Union)*, **70**, 1084.

Blake, E.W., Clarke, G.K.C. and Gérin, M.C. (1992) Tools for examining subglacial bed deformation. *Journal of Glaciology*, **38**, 388–396.

Blake, S. (1987) Sediment entrainment in viscous fluids: can crystals be erupted from magma chamber flows? *Journal of Geology*, **95**, 397–406.

Blankenship, D.D., Bentley, C.R., Rooney, S.T. and Alley, R.B. (1986) Seismic measurements reveal a saturated porous layer beneath an active Antarctic ice stream. *Nature*, **322**, 54–57.

Blanton, T.L. (1983) The relation between recovery deformation and *in situ* stress magnitudes. *Proceedings Society of Petroleum Engineers–Department of Energy Symposium on Low Permeability Gas Reservoirs, Denver, CO, 14–16 March 1983*, pp. 213–218.

Blewett, R.S. (1991) Slump folds and early structures, northeastern Newfoundland Appalachians, re-examined. *Journal of Geology*, **99**, 547–557.

Blikra, L.H. and Nesje, A. (1991) Quaternary glaciofluvial and glaciomarine deposits and post-glacial sedimentation related to steep mountain slopes and fjord margins, western Norway. *Excursion Guidebook, 12th Regional Meeting: International Association of Sedimentologists, Bergen*.

Bloch, S. (1991) Empirical prediction of porosity and permeability in sandstones. *Bulletin of the American Association of Petroleum Geologists*, **75**, 1145–1160.

Boles, J.R. (1982) Active albitization of plagioclase, Gulf Coast Tertiary. *American Journal of Science*, **282**, 165–180.

Boles, J.R. and Franks, S.G. (1979) Clay diagenesis in Wilcox sandstones of southwest Texas: implications of smectite diagenesis on sandstone cementation. *Journal of Sedimentary Petrology*, **49**, 55–70.

Bolt, B.A., Horn, W.L., Macdonald, G.A. and Scott, R.F. (1975) *Geological Hazards*. Springer-Verlag, Berlin, 328 pp.

Booth, J.S. and Dahl, A.G. (1986) A note on the relationships between organic matter and some geotechnical properties of a marine sediment. *Marine Geotechnology*, **6**, 281–297.

Booth, J.S., Sangrey, D.A. and Fugate, J.K. (1985) A nomogram for interpreting slope stability of fine-grained deposits in modern and ancient marine environments. *Journal of Sedimentary Petrology*, **55**, 29–36.

Bornhold, B.D. and Prior, D.B. (1989) Sediment blocks on the sea floor in British Columbia fjords. *Geo-Marine Letters*, **9**, 135–144.

Bornhold, B.D. and Prior, D.B. (1990) Morphology and sedimentary processes on the subaqueous Noeick River delta, British Columbia, Canada, in *Coarse-grained deltas* (eds A. Colella and D.B. Prior) *Special Publications of the International Association of Sedimentologists*, **10**, Blackwell Scientific Publications, Oxford, 169–181.

Borradaile, G.J. (1977) Compaction estimate limits from sand dike orientations. *Journal of Sedimentary Petrology*, **47**, 1598–1601.

Borradaile, G.J. (1981) Particulate flow of rock and the formation of cleavage. *Tectonophysics*, **71**, 305–321.

Bosellini, A. (1984) Progradation geometries of carbonate platforms: examples from the Triassic of the Dolomites, northern Italy. *Sedimentology*, **31**, 1–24.

Bosscher, P.J., Bruxvoort, G.P. and Kelley, T. (1987) Influence of discontinuous joints on permeability. *Journal of Geotechnical Engineering*, **114**, 1318–1331.

Boswell, P.G.H. (1928) The Salopian rocks and tectonics of the district south-west of Ruthin

(Denbighshire). *Quarterly Journal of the Geological Society of London*, **83**, 689–710.

Boswell, P.G.H. (1937) The tectonic problems of an area of Salopian rocks in north-western Denbighshire. *Quarterly Journal of the Geological Society of London*, **93**, 284–321.

Boswell, P.G.H. (1943) The Salopian rocks and geological structure of the country around Eglwys-fach and Glan Conway, N.W. Denbighshire. *Proceedings of the Geologists Assiciation*, **54**, 93–112.

Boswell, P.G.H. (1949) *The Middle Silurian Rocks of North Wales*. Edward Arnold & Co., London, 448 pp.

Boswell, P.G.H. (1953) The alleged subaqueous sliding of large sheets of sediment in the Silurian rocks of North Wales. *Liverpool and Manchester Geological Journal* **1**, 148–152.

Boswell, P.G.H. (1961a) *Muddy Sediments: some Geotechnical Studies for Engineers, Geologists, and Scientists*. Heffer, Cambridge, 140 pp.

Boswell, P.G.H. (1961b) The case against a Lower Paleozoic geosyncline in Wales. *Liverpool and Manchester Geological Journal*, **2**, 612–625.

Boswell, P.G.H. and Double, I.S. (1938) The Ludlow rocks and structure of the neighbourhood of Llanfair Talhairn and Llansannan, Denbighshire. *Proceedings of the Liverpool Geological Society*, **17**, 277–311.

Boswell, P.G.H. and Double, I.S. (1940) The geology of an area of Salopian rocks west of the Conway Valley, in the neighbourhood of Llanrwst, Denbighshire. *Proceedings of the Geologists Association*, **51**, 151–187.

Boudreau, M.E., Brueckner, H.K. and Snyder, W.S. (1984) The effects of silica diagenesis on the evolution of deformational structures in the Havallah Sequence, Nevada. *Geological Society of America, 97th Annual General Meeting, Abstracts with Programs*, **16**, 451.

Boulter, C.A. (1983) Compaction-sensitive accretionary lapilli: a means for recognizing soft-sediment deformation. *Journal of the Geological Society*, **140**, 789–94.

Boulton, G.S. (1976) The origin of glacially fluted surfaces – observations and theory. *Journal of Glaciology*, **76**, 287–309.

Boulton, G.S. (1978) Boulder shapes and grain-size distributions of debris as indicators of transport paths through a glacier and till genesis. *Sedimentology*, **25**, 773–799.

Boulton, G.S. (1979) Processes of glacier erosion of different substrates. *Journal of Glaciology*, **89**, 15–38.

Boulton, G.S. (1986) A paradigm shift in glaciology. *Nature*, **322**, 18.

Boulton, G.S. (1987) A theory of drumlin formation by subglacial sediment deformation, in *Drumlin Symposium* (eds J. Menzies and J.J. Rose), A.A. Balkema, Rotterdam, 25–80.

Boulton, G.S. and Dent, D.L. (1974) The nature and rates of post-depositional changes in recently deposited till from south-east Iceland. *Geografiska Annaler*, **56A**, 121–134.

Boulton, G.S. and Dobbie, K.E. (1993) Consolidation of sediments by glaciers: relations between sediment geotechnics, soft-bed glacier dynamics and subglacial ground-water flow. *Journal of Glaciology*, **39**, 26–44.

Boulton, G.S. and Hindmarsh, R.C.A. (1987) Sediment deformation beneath glaciers: rheology and geological consequences. *Journal of Geophysical Research*, **92**, 9059–9082.

Boulton, G.S. and Jones, A.S. (1979) Stability of temperate ice caps and ice sheets resting on beds of deformable sediment. *Journal of Glaciology*, **90**, 29–43.

Boulton, G.S., Dent, D.L. and Morris, E.M. (1974) Subglacial shearing and crushing, and the role of water pressures in tills from south-east Iceland. *Geografiska Annaler*, **56A**, 135–145.

Boulton, G.S., Morris, E.M., Armstrong, A.A. and Thomas, A. (1979) Direct measurement of stress at the base of a glacier. *Journal of Glaciology*, **86**, 3–24.

Bouma, A.H. (1962) *Sedimentology of some Flysch Deposits: a Graphic Approach to Facies Interpretation*. Elsevier, Amsterdam, 168 pp.

Bouma, A.H., Sangrey, D.A., Coleman, J. et al. (eds) (1982) Offshore geologic hazards; a short course presented at Rice University, May 1981. *American Association of Petroleum Geologists Continuing Education Course Note Series*, **18**.

Bowles, J.E. (1978) *Engineering Properties of Soils and their Measurement*. McGraw-Hill, New York, 113 pp., plus data sheets.

Boyce, R.E. (1976) Definitions and laboratory techniques of compressional sound velocity parameters and wet-water content, wet-bulk density, and porosity parameters by gravimetric and gamma-ray attenuation techniques. *Initial Reports of the Deep Sea Drilling Project*, **33**, United States Government Printing Office, Washington, DC, 931–958.

Brabb, E.E. and Harrod, B.L. (eds) (1989) Landslides: extent and economic significance. *Proceedings of the 28th International Geological Congress symposium on landslides, Washington, DC*. A.A. Balkema, Rotterdam, 399 pp.

Brace, W.F., Paulding, B.W. and Scholz, C. (1966) Dilatancy in the fracture of crystalline rocks. *Journal of Geophysical Research*, **71**, 3939–3953.

Brandon, M. (1984) Deformational processes affecting unlithified sediments at active margins: a field study and structural model. Unpublished PhD thesis, University of Washington, Seattle, 179 pp.

Brandon, M. (1989) Deformational styles in a sequence of olistostromal mélanges, Pacific Rim complex, western Vancouver Island, Canada. *Geological Society of America Bulletin*, **101**, 1520–1542.

Brantley, S.L., Evans, B., Hickman, S.H. and Crerar, D.A. (1990) Healing of microcracks in quartz: implications for fluid flow. *Geology*, **18**, 136–139.

Bray, C.J. and Karig, D.E. (1985) Porosity of sediments in accretionary prisms and some implications for dewatering processes. *Journal of Geophysical Research*, **90**, 768–778.

Bray, C.J. and Karig, D.E. (1986) Physical properties of sediments from the Nankai Trough, DSDP Leg 87A, Sites 582 and 583. *Initial Reports of the Deep Sea Drilling Project*, **87**, United States Government Printing Office, Washington, 817–842.

Breckles, I.M. and van Eekelen, H.A.M. (1982) Relationship between horizontal stress and depth in sedimentary basins. *Journal of Petroleum Technology*, **34**, 2191–2199.

Bredehoeft, J. and Hanshaw, B. (1968) On the maintenance of anomalous fluid pressures: 1. Thick sedimentary sequences. *Geological Society of America Bulletin*, **79**, 1097–1106.

Bredehoeft, J.D., Djevanshir, R.D. and Belitz, H.R. (1988) Lateral fluid flow in a compacting sand-shale sequence: South Caspian Basin. *American Association of Petroleum Geologists Bulletin*, **72**, 416–24.

Bredehoeft, J.D., Wolff, R.G., Keys, W.S. and Shuter, E. (1976) Hydraulic fracturing to determine the regional *in situ* stress field, Piceance Basin, Colorado. *Geological Society of America Bulletin*, **87**, 250–258.

Brodzikowski, K. (1981) The role of dilatancy in the deformational process of consolidated sediments. *Annales de la Société Géologique de Pologne*, **51**, 83–98.

Brodzikowski, K. and Haluszczak, A. (1987) Flame structures and associated deformations in Quaternary glaciolacustrine and glaciodeltaic deposits: examples from central Poland, in *Deformation of Sediments and Sedimentary Rocks* (eds M.E. Jones and R.M.F. Preston), *Geological Society of London Special Publication*, **29**, 279–286.

Brodzikowski, K. and van Loon, A.J. (1991) *Glacigenic Sediments*. Elsevier, Amsterdam, 674 pp.

Brodzikowski, K., Krzyszkowski, D. and van Loon, A. (1987) Endogenic processes as a cause of penecontemporaneous soft-sediment deformations in the fluviolacustrine Czyżów Series (Kleszczów Graben, central Poland), in *Deformation of Sediments and Sedimentary Rocks* (eds M.E. Jones and R.M.F. Preston), *Geological Society of London Special Publication*, **29**, 269–278.

Bromhead, E.N. (1992) *The Stability of Slopes*, 2nd edn. Chapman & Hall, London, 411 pp.

Bromley, R.G., Curran, H.A., Frey, R.W., Gutschick, R.C. and Suttner, L.J. (1975) Problems in interpreting unusually large burrows, in *The study of trace fossils: A synthesis of principles, problems, and procedures in ichnology* (ed. R.W. Frey), Springer-Verlag, New York, 351–376.

Brooker, E.W. and Ireland, H.O. (1965) Earth pressures at rest related to stress history. *Canadian Geotechnical Journal*, **11**, 1–15.

Brown, E.T. and Hoek, E. (1978) Trends in relationships between measured *in situ* stresses and depth. *International Journal of Rock Mechanics, Mineral Science and Geomechanical Abstracts*, **15**, 211–215.

Brown, E.T. and Yu, H.S. (1988) A model for the ductile yield of porous rock. *International Journal for Numerical and Analytical Methods in Geomechanics*, **12**, 679–688.

Brown, K.M. (1990) The nature and hydrogeologic significance of mud diapirs and diatremes for accretionary systems. *Journal of Geophysical Research*, **95**, 8969–8982.

Brown, K.M. and Behrmann, J. (1990) Genesis and evolution of small-scale structures in the toe of the Barbados Ridge accretionary wedge. *Proceedings of the Ocean Drilling Program, Scientific Results*, **11**. Ocean Drilling Program, College Station, TX, 229–243.

Brown, K.M. and Orange, D. (1993) Structural aspects of diapiric mélange emplacement: the Duck Creek Diapir. *Journal of Structural Geology*, **15**, 831–847.

Brown, K.M. and Westbrook, G.K. (1988) Mud diapirism and subcretion in the Barbados Ridge complex. *Tectonics*, **7**, 613–640.

Brown, N.E., Hallet, B. and Booth, D.B. (1987) Rapid soft-bed sliding of the Puget Sound glacial ice lobe. *Journal of Geophysical Research*, **92**, 8985–8997.

Brown, P.E., Chambers, A.D. and Becker, S.M. (1987) A large soft-sediment fold in the Lilloise Intrusion, East Greenland, in *Origins of Igneous Layering* (ed. I. Parsons), *NATO Advanced Study Institutes Series, Series C, Mathematical and Physical Sciences*, **196**, 125–143.

Brown, S.R. (1990) Surface roughness and physical properties of fractures, in *Rock Mechanics and Challenges* (eds W.A. Hustrulid and G.A. Johnson), A. A. Balkema, Rotterdam, 269–276.

Brown, T.C. (1913) Notes on the origin of certain Palaeozoic sediments, illustrated by the Cambrian and Ordovician rocks of Center County, Pennsylvania. *Journal of Geology*, **21**, 232–250.

Brownhead, E.N. (1986) *The Stability of Slopes*. Chapman & Hall, London, pp. 373.

Bruce, C.H. (1973) Pressurized shale and related sediment deformation mechanisms for development of regional contemporaneous faults. *Bulletin of the American Association of Petroleum Geologists*, **57**, 878–886.

Bruce, C.H. (1984) Smectite dehydration – its relation to structural development and hydrocarbon accumulation in northern gulf of Mexico basin. *Bulletin of the American Association of Petroleum Geologists*, **68**, 673–683.

Brueckner, H.K, Synder, W.S. and Boudreau, M. (1987) Diagenetic controls on the structural evolution of siliceous sediments in the Golconda allochthon, Nevada, U.S.A. *Journal of Structural Geology*, **9**, 403–417.

Brugman, M.M. (1983) Properties of debris-laden ice: application to the flow response of the glaciers of Mount St Helens. *Annals of Glaciology*, **4**, 297 pp.

Brunsden, D. (ed.) (1971) Slopes: form and process. *Institute of British Geographers Special Publication*, **3**, 178 pp.

Brunsden, D. (1979) Mass movements, in *Process in Geomorphology* (eds C. Embleton and J. Thornes), Edward Arnold, London, pp. 130–186.

Brunsden, D. (1984) Mudslides, in *Slope Instability* (eds D. Brunsden and D.B. Prior), John Wiley & Sons, Chichester, 363–418.

Brunsden, D. and Lin J-C. (1991) The concept of topographic equilibrium in neotectonic terrains, in *Neotectonics and Resources* (eds J. Cosgrove and M.E. Jones), Belhaven, London, 120–143.

Brunsden, D. and Prior, D.B. (eds) (1984) *Slope Instability*. John Wiley & Sons, Chichester, 620 pp.

Bryant, S., Code, C. and Meller, D. (1993) Permeability prediction from geological models. *American Association of Petroleum Geologists Bulletin*, **77**, 1338–50.

Bryant, W.R., Bennett, R.H., Burkett, P.J. and Rack, F.R. (1990) Microfabric and physical property characteristics of a consolidated clay section: ODP Site 697, Weddell Sea, in *Microstructure of Fine-grained Sediments* (eds R.H. Bennett, W.R. Bryant and M.H. Hulbert) Springer-Verlag, New York, 73–92.

Bryant, W.R., Bennett, R.H. and Katherman, C.E. (1981) Shear strength, consolidation, porosity and permeability of oceanic sediments, in *The Oceanic Lithosphere* (ed. C. Emiliani) John Wiley, New York 555–616.

BSI (British Standards Institute) (1991) *Methods of Test for Soils for Civil Engineering Purposes*. BS 1377, HMSO, London.

Bucher, W.H. (1956) The role of gravity in orogenesis. *Geological Society of America Bulletin*, **67**, 1295–1318.

Buckley, D.E. (1989) Small fractures in deep sea sediments: indications of pore fluid migration along compactional faults, in *Advances in Underwater Technology* (ed. T.J. Freeman), *Ocean Science and Offshore Engineering*, **18**, Graham and Trotman, London, 115–135.

Buckley, D.E. and Grant, A.C. (1985) Fault-like features in abyssal plain sediments; possible dewatering structures. *Journal of Geophysical Research*, **90**, 9173–9180.

Bugge, T., Belderson, R.H. and Kenyon, N.H. (1988) The Storegga slide. *Philosophical Transactions of the Royal Society of London*, **325**, 357–388.

Buggisch, W. and Heinitz, W. (1984) Slump folds and other early deformations in the early Cambrian of the Western and Central Antiatlas (Morocco). *Geologische Rundschau*, **73**, 809–818.

Bullard, T.F. and Lettis, W.R. (1993) Quaternary fold deformation associated with blind thrust faulting, Los Angeles basin, California. *Journal of Geophysical Research*, **98**, 8349–8369.

Bullock, P. and Murphy, C.P. (eds) (1983) *Soil Micromorphology*, vol. 1. *Techniques and Applications*. A.B. Academic Publishers, Berkhamstead, 376 pp.

Burland, J.B. (1989) The ninth Bjerrum Memorial Lecture: 'Small is beautiful' – the stiffness of soils at small strains. *Canadian Geotechnical Journal*, **26**, 499–516.

Burland, J.B. (1990) On the compressibility and shear strength of natural clays. *Geotechnique*, **40**, 329–378.

Burst, J.F. (1976) Argillaceous sediment dewatering. *Annual Review of Earth and Planetary Science*, **4**, 293–318.

Byerlee, J. (1978) Friction of Rocks. *Pure and Applied Geophysics*, **116**, 615–26.

Byrne, T. (1982) Structural evolution of coherent terranes in the Ghost Rocks Formation, Kodiak Island, Alaska, in *Trench and Forearc Geology: Sedimentation and Tectonics in Modern and Ancient Subduction Zones* (ed. J.K. Leggett), *Geological Society of London Special Publication*, **10**, 21–51.

Byrne, T. (1984) Early deformation in mélange terranes of the Ghost Rocks Formation, Kodiak Island, Alaska, in *Mélanges: their Nature, Origin and Significance* (ed. L. Raymond), *Geological Society of America Special Paper*, **198**, 21–52.

Byrne, T. (1986) Eocene underplating along the Kodiak shelf, Alaska: implications and regional correlations. *Tectonics*, **5**, 403–421.

Byrne, T. and Fisher, D. (1990) Evidence for a weak and overpressured décollement beneath sediment-dominated accretionary prisms. *Journal of Geophysical Research*, **95**, 9081–9097.

Byrne, T., Bruckmann, W., Owens, W., Lallemant, S. and Maltman, A.J. (1993a) Correlation of structural fabrics, velocity anisotropy and magnetic susceptibility data. *Proceedings of the Ocean Drilling Program, Scientific Results*, **131**, Ocean Drilling Program, College Station, TX, 365–78.

Byrne, T., Maltman, A., Stephenson, E., Soh, W. and Knipe, R. (1993b). Deformation structures and fluid flow in the toe region of the Nankai accretionary prism. *Proceedings of the Ocean Drilling Program, Scientific Results*, **131**, Ocean Drilling Program, College Station, TX, 83–192.

Cakmak, A.S. and Heron, I. (eds) (1989) *Solid Dynamics*

and Liquefaction. Computational Mechanics Publications, Southampton, 362 pp.

Campbell, C.S. (1989) The stress tensor for simple shear flow of a granular material. *Journal of Fluid Mechanics*, **203**, 449–473.

Carpenter, G. (1981) Coincident sediment/slump clathrate complexes on the U.S. continental slope. *Geo-Marine Letters*, **1**, 29–32.

Carson, B. (1983) Convergent plate-margin sedimentation: deposition, deformation, and lithification in subduction zones, in *Revolution in the Earth Sciences, Advances in the Past Half-century* (ed. S.J. Boardman). Kendall-Hunt, Dubuque, IA, 136–166.

Carson, B. and Berglund, P.L. (1986) Sediment deformation and dewatering under horizontal compression: experimental results. *Geological Society of America Memoir*, **166**, 135–150.

Carson, B., von Huene, R. and Arthur, M. (1982) Small-scale deformation structures and physical properties related to convergence in Japan slope sediments. *Tectonics*, **1**, 277–302.

Carson, B., Westbrook, G.K. and the shipboard scientific party, in press. *Proceedings of the Ocean Drilling Program, Scientific Results*, **146**, Ocean Drilling Program, College Station, TX.

Carson, M.A. and Kirby, M.J. (1972) *Hillslope Form and Process.* Cambridge University Press, London, 475 pp.

Carstens, H. and Dypvik, H. (1981) Abnormal formation pressures and shale porosity. *Bulletin of the American Association of Petroleum Geologists*, **65**, 344–350.

Carter, R.M. (1975) Mass-emplaced sand-fingers at Mararoa construction site, southern New Zealand. *Sedimentology*, **22**, 275–288.

Castle, R.O., Yerkes, R.F. and Young, T.L. (1973) Ground rupture in the Baldwin Hills: an alternative explanation. *Bulletin of the American Association of Engineering Geologists*, **10**, 21–46.

Castro, G. (1975) Liquefaction and cyclic mobility of saturated sands. *American Association of Civil Engineers, Journal of Geotechnical Engineering Division*, **101**, 551–569.

Castro, G. (1987) On the behavior of soils during earthquakes: liquefaction, in *Soil Dynamics of Liquefaction* (ed. A.S. Cakmak), Elsevier, Amsterdam, 169–204.

Cello, G. and Nur, A. (1988) Emplacement of foreland thrust systems. *Tectonics*, **7**, 261–271.

Chakrabati, A. (1981) Kink-like structures and penecontemporaneous thrusting in the foreset laminae of mega-ripples. *Indian Journal of Earth Sciences*, **8**, 29–34.

Chan, Y. (1964) Preliminary study on the geothermal gradients and formation of reservoir pressures of oil and gas fields in Northern Taiwan. *Petroleum Geology of Taiwan*, **3**, 127–139.

Chaney, R.C. and Demars, K.R. (eds) (1985) Strength testing of marine sediments and *in situ* measurements. *ASTM (American Society for Testing and Materials) Special Technical Paper*, **883**, 558 pp.

Chapman, R.E. (1972) Clays with abnormal interstitial pore fluid pressures. *Bulletin of the American Association of Petroleum Geologists*, **56**, 790–795.

Chapman, R.E. (1974) Clay diapirism and overthrust faulting. *Geological Society of America Bulletin*, **85**, 1597–1602.

Chapman, R.E. (1981) *Geology and water. An Introduction to Fluid Mechanics for Geologists.* Martinus Nijhoff/Dr W. Junk Publishers, The Hague, 228 pp.

Chapman, R.E. (1983) *Petroleum Geology.* Elsevier, Amsterdam, 415 pp.

Charles, J.A. (1982) An appraisal of the influence of a curved failure envelope on slope stability. *Geotechnique*, **32**, 389–392.

Charlez, P. and Heugas, O. (1991) Evaluation of optimal mudweight in soft shale levels, in *Rock Mechanics as a Multidisciplinary Science* (ed. J-C. Roegiers), A.A. Balkema, Rotterdam, 1005–1014.

Cheel, R.J. and Rust, B.R. (1986) A sequence of soft-sediment deformation (dewatering) structures in Late Quaternary subaqueous outwash near Ottawa, Canada. *Sedimentary Geology*, **47**, 77–93.

Cheng, D.C.-. and Richmond, R.A. (1978) Some observations on the rheological behaviour of dense suspensions. *Rheologica Acta*, **17**, 446–453.

Chester, F.M. and Logan, J.M. (1987) Composite planar fabric of gouge from the Punchbowl Fault, California. *Journal of Structural Geology*, **9**, 621–634.

Chilingarian, G.V. and Wolf, K.H. (1975) *Compaction of Coarse-grained Sediments*, 1. Developments in Sedimentology. Elsevier, Amsterdam, 562 pp.

Chown, E.H. and Gobeil, A. (1990) Clastic dykes of the Chibougamau Formation; distribution and origin. *Canadian Journal of Earth Sciences*, **27**, 1111–1114.

Chowns, T.M. and Elkins, J.E. (1974) The origin of quartz geodes and cauliflower cherts through the silification of anhydrite nodules. *Journal of Sedimentary Petrology*, **44**, 885–903.

Clark, J.I. and Gilliott, J.E. (1985) The role of composition and fabric of soils in selected geotechnical engineering case histories. *Applied Clay Science*, **1**, 173–191.

Clarke, G.K.C. (1987a) A short history of scientific investigations on glaciers. *Journal of Glaciology*, Special Issue, 1–21.

Clarke, G.K.C. (1987b) Subglacial till: a physical framework for its properties and processes. *Journal of Geophysical Research*, **92**, 9023–9036.

Clarke, G.K.C. (1987c) Fast glacier flow: ice streams, surging and tidewater glaciers. *Journal of Geophysical Research*, **92**, 8835–8841.

Clarke, G.K.C., Collins, S.G. and Thompson, D.E. (1984) Flow, thermal structure, and subglacial conditions of a surge-type glacier. *Canadian Journal of Earth Sciences*, **21**, 232–240.

Clarke, G.K.C., Meldrum, R.D. and Collins, S.G. (1986) Measuring glacier motion fluctuations using a computer-controlled survey system. *Canadian Journal of Earth Sciences*, **23**, 727–733.

Clayton, L., Mickelson, D.M. and Attig, J.W. (1989) Evidence against pervasively deformed bed material beneath rapidly moving lobes of the southern Laurentide ice sheet. *Sedimentary Geology*, **62**, 203–208.

Clendenin, C.W. and Duane, M.J. (1990) Focused fluid flow and Ozark Mississippi Valley-type deposits. *Geology*, **18**, 116–119.

Clenell, B. (1992) The melanges of Sabah, Malaysia. Unpublished PhD thesis, University of London, 450 pp.

Clifton, H.E. (1984) Sedimentation units in stratified resedimented conglomerates, Paleocene submarine canyon fill, Point Lobos, California, in *Sedimentology of Gravels and Conglomerates* (eds E.H. Koster and R.J. Steel), *Canadian Society of Petroleum Geologists Memoir*, **10**, 429–441.

Cloos, M. (1984) Landward dipping reflectors in accretionary wedges: active dewatering conduits? *Geology*, **9**, 519–523.

Coleman, B.D. and Hodgdon, M.L. (1985) On shear bands in ductile materials. *Archive for Rational Mechanics and Analysis*, **90**, 219–247.

Coleman, J.M. and Garrison, L.E. (1977) Geological aspects of marine slope stability, northwestern Gulf of Mexico. *Marine Geotechnology*, **2**, 9–44.

Coleman, J.M., Prior, D.B. and Lindsay, J.F. (1983) Deltaic influences on shelf edge instability processes, in *The shelfbreak: Critical Interface on Continental Margins*, (eds D.J. Stanley and G.T. Moore), *Society of Economic Paleontologists and Mineralogists Special Publication*, **33**, 121–137.

Collinson, J.D. and Thompson, D.B. (1989) *Sedimentary Structures*, 2nd edn. Chapman & Hall, London, 207 pp.

Collinson, J.D., Bevins, R.E. and Clemmensen, L.B. (1989) Post-glacial mass flow and associated deposits preserved in palaeovalleys: the Late Precambrian Moraenesó Formation, North Greenland. *Meddelelser om Grønland Geoscience*, **21**, 3–26.

Collinson, J.D., Martinsen, O.J., Bakken, B. and Kloster, A. (1991) Early infill of the Western Irish Namurian Basin: a complex relationship between turbidites and deltas. *Basin Research*, **3**, 223–242.

Colten-Bradley, V.A. (1987) Role of pressure in smectite dehydration – effects on geopressure and smectite–illite transformation. *Bulletin of the American Association of Petroleum Geologists*, **71**, 1414–1427.

Conaghan, P.J., Mountjoy, E.W. and Edgecomb, D.R. (1976) Nubrigyn algal reef (Devonian), eastern Australia: allochthonous blocks and megabreccias. *Geological Society of America Bulletin*, **87**, 515–530.

Coniglio, M. (1986) Synsedimentary submarine slope failure and tectonic deformation in deep-water carbonates, Cow Head Group, western Newfoundland. *Canadian Journal of Earth Sciences*, **23**, 476–490.

Coniglio, M. (1987) Biogenic chert in the Cow Head Group (Cambro-Ordovician), western Newfoundland. *Sedimentology*, **34**, 813–823.

Coniglio, M. and Cameron, J.S. (1990) Early diagenesis in a potential oil shale: evidence from calcite concretions in the Upper Devonian Kettle Point Formation, southwestern Ontario. *Bulletin of Canadian Petroleum Geology*, **38**, 64–77.

Connelly, W. (1978) Uyak complex, Kodiak Island, Alaska: A Cretaceous subduction complex. *Geological Society of America Bulletin*, **89**, 755–769.

Cook, A.M., Myer, L.R., Cook, N.G.W. and Doyle, F.M. (1990) The effects of tortuosity on flow though a natural fracture, in *Rock Mechanics and Challenges* (eds W.A. Hustrulid and G.A. Johnson), A.A. Balkema, Rotterdam, 371–378.

Coop, M.R. (1990) The mechanics of uncemented carbonate sands. *Geotechnique*, **40**, 607–626.

Corbett, K.D. (1973) Open-cast slump sheets and their relationship to sandstone beds in an Upper Cambrian flysch sequence, Tasmania. *Journal of Sedimentary Petrology*, **43**, 147–159.

Costa-Filho, L. (1984) A note on the influence of fissures on the deformation characteristics of London Clay. *Geotechnique*, **34**, 268–272.

Costin, L.S. (1987) Time-dependent deformation and failure, in *Fracture Mechanics of Rock* (ed. B.K. Atkinson) Academic Press, London, 167–215.

Cotton, C.A, and Te Punga, M.T. (1955) Solifluction and periglacially modified landforms at Wellington, New Zealand. *Transactions of the Royal Society of New Zealand*, **82**, 1001–1031.

Coutard, J.P. and Mucher, H.J. (1985) Deformations of laminated silt loam due to repeated freezing and thawing cycles. *Earth Surface Processes and Landforms*, **10**, 309–310.

Cowan, D.S. (1982) Deformation of partly-dewatered and consolidated Franciscan sediments near Piedras Blancas Point, California, in *Trench–Forearc Geology: Sedimentation and Tectonics in Modern and Ancient Subduction Zones* (ed. J.K. Leggett), *Geological Society of London Special Publication*, **10**, 439–457.

Cowan, D.S. (1985) Structural styles in Mesozoic and Cenozoic mélanges in the western Cordillera of North America. *Geological Society of America Bulletin*, **96**, 451–462.

Cox, S.F., Etheridge, M.A. and Wall V.J. (1986) The role of fluids in syntectonic mass transport and the localization of metamorphic vein-type ore deposits. *Ore Geology Reviews*, **2**, 65–86.

Craig, J. (1985) Tectonic evolution of the area between Borth and Cardigan, Dyfed, west Wales. Unpublished PhD thesis, University College of Wales, Aberystwyth, 3 vols.

Craig, J., Fitches, W.R. and Maltman, A.J. (1982) Chlorite–mica stacks in low-strain rocks from central Wales. *Geological Magazine*, **119**, 243–256.

Crandell, D.R. (1971) Postglacial lahars from Mount Rainier volcano, Washington. *United States Geological Survey Professional Paper*, **667**, 75 pp.

Crans, W., Mandl, G., and Harembourne, J. (1980) On the theory of growth faulting: a geomechanical model based on gravity sliding. *Journal of Petroleum Geology*, **2**, 265–307.

Croot, D.G. (1987) Glacio-tectonic structures: a mesoscale model of thin-skinned thrust sheets? *Journal of Structural Geology*, **9**, 787–808.

Croot, D.G. (ed.) (1988) *Glaciotectonics: Proceedings of the Glaciotectonics Working Group.* A.A. Balkema, Rotterdam, 220 pp.

Crozier, M.J. (1973) Techniques for the morphometric analysis of landslips. *Zeitschrift für Geomorphologie Dynamique*, **17**, 78–101.

Cruden, D.M. (1991) A simple definition of a landslide. *Bulletin of the International Association of Engineering Geology*, **43**, 27–29.

Dahlen, F.A. (1990) Critical taper model of fold-and-thrust belts and accretionary wedges. *Annual Reviews in Earth and Planetary Sciences*, **18**, 55–90.

Dailly, G. (1976) A possible mechanism relating progradation, growth faulting, clay diapirism and overthrusting in a regressive sequence of sediments. *Bulletin of Canadian Petroleum Geology*, **24**, 92–116.

Daines, S. (1982) Aquathermal pressuring and geopressure evaluation. *Bulletin of the American Association of Petroleum Geologists*, **66**, 931–939.

Daley, B. (1972) Slumping and microjointing in an Oligocene lagoonal limestone. *Sedimentary Geology*, **7**, 35–46.

Dalrymple, R.W. (1979) Wave-induced liquefaction: a modern example from the Bay of Fundy. *Sedimentology*, **26**, 835–844.

Dana, J.D. (1849) *Geology, Volume 10 of U.S. Exploring Expedition 1838–1842, under the command of Charles Wilkes.* C. Sherman, Philadelphia, 18 vols.

Daramola, O. (1980) On estimating K_0 for overconsolidated granular soils. *Geotechnique*, **30**, 310–313.

Darcy, H. (1856) *Les fontaines publiques de la Ville de Dijon*, Victor Damant, Paris.

Darwin, C. (1846) *The Geology of the Voyage of the Beagle, part 3: Geological Observations on South America.* Smith, Elder and Company, London, 279 pp.

Das, B.M. (1985) *Advanced Soil Mechanics*, McGraw-Hill, New York, 511pp.

Davenport, C.A. and Ringrose, P.S. (1987) Deformation of Scottish Quaternary sediment sequences by strong earthquake motions, in *Deformation of sediments and sedimentary rocks* (eds. M.E. Jones and R.M.F. Preston). *Geological Society of London Special Publication*, **29**, 299–314.

Davies, W. and Cave, R. (1976) Folding and cleavage determined during sedimentation. *Sedimentary Geology*, **15**, 89–133.

Davis, F.A., Suppe, J. and Dahlen, F.A. (1983) Mechanics of fold and thrust belts and accretionary wedges: cohesive Coulomb theory. *Journal of Geophysical Research*, **88**, 11153–11172.

Davis, R.O. and Berrill, J.B. (1983) Comparison of a liquefaction theory with field observations. *Geotechnique*, **33**, 455–460.

Davison, I. (1987) Normal fault geometry related to sediment compaction and burial. *Journal of Structural Geology*, **9**, 391–402.

De Boer, P.L. (1979) Convolute lamination in modern sands of the estuary of the Oosterschelde, The Netherlands, formed as the result of entrapped air. *Sedimentology*, **26**, 283–294.

De Freitas, M.H. and Watters, R.J. (1973) Some field examples of toppling failure. *Geotechnique*, **23**, 495–514.

De Lange, G.J., Kok, P.T.J., Ebbing, J. and van der Klugt, P. (1988) Microfault structures in unconsolidated Upper Quaternary Sediment from the Madiera abyssal plain (eastern North Atlantic) *Marine Geology*, **80**, 155–159.

De Terra, H. (1931) Structural features in gliding strata. *American Journal of Science*, **21**, 204–213.

De Waal, J.A. (1986) On the rate type compaction behaviour of sandstone reservoir rock. PhD thesis, Delft University.

Decker, P.L. (1990) Structural style and mechanics of liquefaction-related deformation in the Lower Absaroka Volcanic Supergroup (Eocene), east-central Absaroka Range, Wyoming. *Geological Society of America Special Paper*, **240**, 80 pp.

Demars, K.R. and Chaney, R.C. (eds) (1990) Geotechnical engineering of ocean waste disposal. *Special Technical Paper*, **1087**, 310 pp.

Demars, K.R. and Nacci, V.A. (1978) Significance of Deep Sea Drilling Project sediment physical property data. *Marine Geotechnology*, **3**, 151–170.

Denness, B. (1984) *Seabed Mechanics.* Graham and Trotman, London, 281 pp.

Derbyshire, E. (1978) A pilot study of till microfabrics using the scanning electron microscope, in *Scanning Electron Microscopy* (ed. W.B. Whalley), Geo-Abstracts, Norwich, 41–59.

Derbyshire, E., Edge, M.J. and Love, M. (1985) Soil fabric variability in some glacial diamicts, in *Glacial Tills '85* (ed. M.C. Forde), Engineering Technics Press, Edinburgh, 169–175.

Derbyshire, E., Unwin, D.J., Fang, X.M. and Langford, N. (1992) The Fourier frequency domain representation of sediment fabric anisotropy. *Computers and Geosciences*, **18** p. 63–73.

Desbarats, A.J. (1987) Numerical estimation of effective permeability in sand–shale formations. *Water Resources Research*, **23**, 273–286.

Desrochers, A. and Al-Aasm, I.S. (1993) The formation of septarian concretions in Queen Charlotte Islands, British Columbia: evidence for microbially and hydrothermally mediated reactions at shallow burial depth. *Journal of Sedimentary Petrology*, **63**, 282–294.

Devgun, J.S. (1989) Suitability of unconsolidated sediments for hosting low-level radioactive waste disposal facilities. *Canadian Journal of Civil Engineering*, **16**, 560–567.

Diller, G.S. (1890) Sandstone dykes. *Geological Society of America Bulletin*, **1**, 411–442.

Dingle, R.V. (1977) The anatomy of a large submarine slump on a sheared continental margin (SE Africa). *Journal of the Geological Society*, **134**, 293–310.

Dingle, R.V. (1980) Large allochthonous sediment masses and their role in the construction of the continental slope and rise off SW Africa. *Marine Geology*, **37**, 333–354.

Dionne, J.C. (1971) Contorted structures in unconsolidated Quaternary deposits, Lake Saint-Jean and Saguenay regions, Quebec. *Revue de Geographie Montreal*, **25**, 5–34.

Dionne, J.C. (1973) Miniature mud volcanoes and other injection features in tidal flats, James Bay, Quebec. *Canadian Journal of Earth Sciences*, **13**, 422–428.

Dionne, J. (1974) Polished and striated mud surfaces in the St Lawrence Tidal flats, Quebec. *Canadian Journal of Earth Sciences*, **11**, 860–866.

Dionne, J.C. and Shilts, W.W. (1974) A Pleistocene clastic dyke, Upper Chaudiere Valley, Quebec. *Canadian Journal of Earth Sciences*, **11**, 1594–1605.

Dixon, R.J. (1990) The Moel-y-Golfa Andesite: an Ordovician (Caradoc) intrusion into unconsolidated conglomeratic sediments, Breidden Hills Inlier, Welsh Borderland. *Geological Journal*, **25**, 35–46.

Doe, T.W. and Dott, R.H. Jr. (1980) Genetic significance of deformed cross bedding – with examples from the Navajo and Weber Sandstones of Utah. *Journal of Sedimentary Petrology*, **50**, 793–812.

Donaghe, R.T., Chaney, R.C. and Silver, M.L. (1988) Advanced triaxial testing of soil and rock. *American Society for Testing and Materials Special Technical Publication*, **977**, 896 pp.

Donovan, R.M. and Archer, R. (1975) Some sedimentological consequences of a fall in the level of Haweswater, Cumbria. *Proceedings of the Yorkshire Geological Society*, **40**, 547–562.

Donovan, R.M. and Foster, R.J. (1972) Subaqueous shrinkage cracks from the Caithness Flagstone Series (Middle Devonian) of northeast Scotland. *Journal of Sedimentary Petrology*, **42**, 885–903.

Dott, R.H. Jr. (1963) Dynamics of subaqueous gravity depositional processes. *Bulletin of the American Association of Petroleum Geologists*, **47**, 104–128.

Dowdeswell, J.A, and Sharpe, M.J. (1986) Characteristics of pebble fabrics in modern terrestrial glacigenic sediments. *Sedimentology*, **33**, 699–710.

Dreimanis, A. and Vagners, U.J. (1971) Bimodal distribution of rock and mineral fragments in basal tills, in *Till: a Symposium* (ed. R.P. Goldthwait), Ohio State University Press, Columbus, pp. 237–250.

Drewry, D.J. (1986) *Glacial Geologic Processes*. Edward Arnold, London, 276pp.

Drozdowski, E. (1983), Load deformations in melt-out till and underlying laminated till. An example from northern Poland, in *Tills and Related Deposits, Genesis/Petrology/Application/Stratigraphy*. Proceedings of the INQUA symposium on the genesis and lithology and Quaternary deposits, USA 1981 and Argentina 1982 (eds E.B. Evenson, C. Schluchter and J. Rabassa), A.A. Balkema, Rotterdam, 119–124.

Duck, R.W. (1990) S.E.M. study of clastic fabrics preserved in calcareous concretions from the late-Devensian Errol Beds, Tayside. *Scottish Journal of Geology*, **26**, 33–39.

Dudgeon, C.R. and Yong, K.C. (1969) Extraction of water from unconsolidated sediments: a literature survey. *New South Wales University Water Resources Laboratory Report*, **111**, 30 pp.

Duin, E.J.T., Mesdag, C.S. and Kok, P.T.J. (1984) Faulting in Madeira abyssal plain sediments. *Marine Geology*, **56**, 299–308.

Dunn, D.E., LaFountain, L.J. and Jackson, R.E. (1973) Porosity dependence and mechanism of brittle fracture in sandstone. *Journal of Geophysical Research*, **78**, 2403–2417.

Dunn, R.J. and Mitchell, J.K. (1884) Fluid conductivity testing of fine-grained soils. *American Society of Civil Engineers, Journal of the Geotechnical Engineering Division*, **110**, 1648–1665.

Dunne, T. (1990) Hydrology, mechanics, and geomorphic implications of erosion by subsurface flow, in *Groundwater Geomorphology: the Role of Subsurface Water in Earth-Surface Processes and Landforms* (eds C.G. Higgins and D.R. Coates), *Geological Society of America Special Paper*, **252**, 1–28.

Dzevanshir, R.D., Buryakovskiy, L.A. and Chilingarian, G.V. (1986) Simple quantitative evaluation of porosity of argillaceous sediments at various depths of burial. *Sedimentary Geology*, **46**, 169–175.

Echelmeyer, K. and Wang Zhongxiang (1987) Direct observation of basal sliding and deformation of basal drift at sub-freezing temperatures. *Journal of Glaciology*, **33**, 83–98.

Edge, M.J. and Sills, G.C. (1989) The development of layered sediments in the laboratory as an illustration of possible field processes. *Quarterly Journal of Engineering Geology*, **22**, 271–279.

Edwards, M.B. (1976) Growth faults in Upper Triassic deltaic sediments, Svalbard. *Bulletin of the American Association of Petroleum Geologists*, **60**, 341–355.

Ehrenberg, S.N. (1989) Assessing the relative importance of compaction processes and cementation to reduction of porosity in sandstones: discussion. *Bulletin of the American Association of Petroleum Geologists*, **73**, 1274–1276.

Eidelman, A. and Reches, Z. (1992) Fractured pebbles – a new stress indicator. *Geology*, **20**, 307–310.

Eigenbrod, K.D. and Burak, J.B. (1991) Effective stress paths and pore pressure responses during undrained shear along the bedding planes of varved Fort William clay. *Canadian Geotechnical Journal*, **28**, 804–811.

Einsele, G. (1982) Mechanism of sill intrusion into soft sediment and expulsion of pore water. *Initial Reports of the Deep Sea Drilling Project*, **64**, 1169–1176.

Einsele, G. (1989) *In situ* water contents, liquid limits, and submarine mass flows due to a high liquefaction potential of slope sediments (results from DSDP and subaerial counterparts). *Geologische Rundschau*, **78**, 821–840.

Elliott, C.G. and Williams, P.F. (1988) Sediment slump structures: a review of diagnostic criteria and application to an example from Newfoundland. *Journal of Structural Geology*, **10**, 171–182.

Elliott, W.C., Aronson, J.L., Matisoff, G. and Gautier, J.L. (1991) Kinematics of the smectite to illite transformation in the Denver Basin: clay mineral, K–Ar data, and mathematical model results. *Bulletin of the American Association of Petroleum Geologists*, **75**, 436–462.

Elson, J.A. (1975) Origin of a clastic dyke at St Ludger, Quebec: an alternative hypothesis. *Canadian Journal of Earth Sciences*, **12**, 1048–1053.

Embly, R.W. and Hayes, D.E. (1974) Giant submarine slump south of the Canaries. *Geological Society of America, Abstracts with Programs*, **6**, 721.

Engelder, T. (1979) The nature of deformation within the outer limits of the central Appalachian foreland fold and thrust belt in New York State. *Tectonophysics*, **55**, 289–310.

Engelder, T. (1984) The role of pore water circulation during the deformation of foreland fold and thrust belts. *Journal of Geophysical Research*, **89**, 4319–4326.

Engelder, T. (1985) Loading paths to joint propagation during a tectonic cycle: an example from the Appalachian Plateau, U.S.A. *Journal of Structural Geology*, **7**, 459–476.

Engelder, T. (1987) Joints and shear fractures in rock, in *Fracture Mechanisms of Rock* (ed. B.K. Atkinson). Academic Press, London, 27–69.

Engelder, T. and Geiser, P. (1980) On the use of regional joint sets as trajectories of paleostress fields during the development of the Appalachian Plateau, New York. *Journal of Geophysical Research*, **85**, 6319–6341.

Engelder, T. and Lacazette, A. (1990) Natural hydraulic fracturing, in *Rock Joints* (eds N. Barton and O. Stephansson). A.A. Balkema, Rotterdam, 35–43.

Engelder, T. and Marshak, S. (1985) Disjunctive cleavage formed at shallow depths in sedimentary rocks. *Journal of Structural Geology*, **7**, 327–343.

Engelder, T. and Oertel, G. (1985) Correlation between abnormal pore pressure and tectonic jointing in the Devonian Catskill Delta. *Geology*, **13**, 863–866.

Engelhardt, H., Harrison, W.D. and Kamb, B. (1978) Basal sliding and conditions at the glacier bed as revealed by borehole photography. *Journal of Glaciology*, **20**, 469–508.

Engelhardt, H., Humphrey, M., Kamb, B. and Fahnestock, M. (1990) Physical conditions at the base of a fast moving Antarctic ice stream. *Science*, **248**, 57–59.

England, W.A., Mackenzie, A.S., Mann, D.M. and Quigley, T.M. (1987) The movement and entrapment of fluids in the subsurface. *Journal of the Geological Society*, **144**, 327–347.

Enriquez-Reyes, M.D.P. and Jones, M.E. (1991) On the nature of the scaly texture developed in mélange deposits, in *Rock Mechanics as a Multidisciplinary Science* (ed. J-C. Roegiers), A.A. Balkema, Rotterdam, 713–722.

Ervine, W.B. and Bell, J.S. (1987) Subsurface *in-situ* stress magnitudes from oil-well drilling records: an example from the Venture area, offshore eastern Canada. *Canadian Journal of Earth Sciences*, **24**, 1748–1759.

Eschman, D.F. (1982) An unusual ball and pillow structure locality in Michigan. *Journal of Geology*, **90**, 742–744.

Espinosa, R.D., Repetto, P.C. and Muhunthan, B. (1992) General framework for stability analysis of slopes. *Geotechnique*, **42**, 427–441.

Esrig, M.I. and Kirby, R.C. (1978) Implications of gas content for predicting the stability of submarine slopes. *Marine Geology*, **2**, 81–100.

Eugster, H.P. (1969) Inorganic bedded cherts from the Magadi area, Kenya. *Contributions to Mineralogy and Petrology*, **22**, 1–31.

Eugster, H.P. and Hardie, L.A. (1978) Saline lakes, in *Lakes; Chemistry, Geology, Physics* (ed. A. Lerman), Springer-Verlag, New York, 237–293.

Eva, S.J. (1992) Sediment deformation in the Silurian rocks of North Wales. Unpublished PhD thesis, University College of Wales, Aberystwyth, 561 pp.

Eva, S.J. and Maltman, A.J., in press, Slump fold and palaeoslope orientations in Upper Silurian rocks south of Colwyn Bay, North Wales. *Geological Magazine*.

Evans, K.F. and Engelder, T. (1989) Some problems in estimating horizontal stress magnitudes in 'thrust' regimes. *International Journal Rock Mechanics and Mining Science*, **26**, 647–660.

Evans, K.F., Engelder, T. and Plumb, R.A. (1989) Appalachian stress study 1. A detailed description of *in situ* stress variations in Devonian shales of the Appalachian Plateau. *Journal of Geophysical Research*, **94**, 7129–7154.

Evans, K.F., Oertel, G. and Engelder, T. (1989) Appalachian stress study 2. Analysis of Devonian shale core: some implications for the nature of contemporary stress variations and Alleghanian deformation in Devonian rocks. *Journal of Geophysical Research*, **94**, 7154–7170.

Evans, S.G. and Brooks, G.R. (1991) Prehistoric debris avalanches from Mount Cayley volcano, British Columbia. *Canadian Journal of Earth Sciences*, **28**, 1365–1374.

Evenson, E.B. (1971) The relationship of macro- and microfabrics of till and the genesis of glacial landforms in Jefferson County, Wisconsin, in *Till: a Symposium* (ed. R.P. Goldthwait) Ohio State University Press, Columbus, pp. 345–364.

Eyles, N. and Clark, B.M. (1985) Gravity-induced soft-sediment deformation in glaciomarine sequences of the Upper Proterozoic Port Askaig Formation, Scotland. *Sedimentology* **32**, 789–814.

Faas, R.W. (1982) Gravitational compaction patterns determined from sediment cores recovered during the DSDP Leg 67, Guatemalan transect: continental slope. *Initial Reports of the Deep Sea Drilling Project*, **67**, United States Government Printing Office, Washington, DC, 639–644.

Faas, R.W, and Crockett, D.S. (1983) Clay fabric development in a deep-sea core: Site 515, Deep Sea Drilling Project Leg 72. *Initial Reports of the Deep Sea Drilling Project*, **72**, United States Government Printing Office, Washington, DC, 519–525.

Fahnestock, M. and Humphrey, N. (1988) Borehole water level measurements, Columbia Glacier, Alaska. *Ice*, **86**, 25–26.

Fang, W.W., Langseth, M.G. and Schultheiss, P.J. (1993) Analysis and application of *in situ* pore pressure measurements in marine sediments. *Journal of Geophysical Research*, **98**, 7921–7938.

Farmer, I. (1983) *Engineering Behaviour of Rocks*, 2nd edn, Chapman and Hall, London, 208 pp.

Farrell, S.G. (1984) A dislocation model applied to slump structures. *Journal of Structural Geology*, **6**, 727–736.

Farrell, S.G. and Eaton, S. (1987) Slump strain in the Tertiary of Cyprus and the Spanish Pyrenees. Definition of palaeoslopes and models of soft-sediment deformation. In: *Deformation of Sediments and Sedimentary rocks*. (eds. M.E. Jones and R.M.F. Preston), *Geological Society of London Special Publication*, **29** 181–196.

Farrell, S.G. and Eaton, S. (1988) Foliations developed during slump deformation of Miocene marine sediments, Cyprus. *Journal of Structural Geology*, **10**, 561–576.

Feda, J. (1982) *Mechanics of particulate materials*, Elsevier, Amsterdam, 447 pp.

Feda, J. (1992) *Creep of Soils and related Phenomena. Developments in Geotechnical Engineering*, **14**, Elsevier, Amsterdam, 422 pp.

Feeser, V. (1988) On the mechanics of glaciotectonic contortions of clays, in *Glaciotectonics: Forms and Processes* (ed. D.G. Croot), A.A. Balkema, Rotterdam, 63–76.

Feeser, V., Moran, K. and Bruckmann, W. (1993) Stress-regime-controlled yield and strength behavior of sediment from the frontal part of the Nankai accretionary prism. *Proceedings of the Ocean Drilling Program, Scientific Results*, **131**, Ocean Drilling Program, College Station, TX, 261–274.

Ferguson, H.F. (1967) Valley stress release in the Allegheny Plateau. *Engineering Geology*, **4**, 63–71.

Fernwik, N. and Haug, M. (1990) Evaluation of *in situ* permeability testing methods. *Journal of Geotechnical Engineering*, **116**, 297–311.

Fertl, W.H. (1973) Significance of shale gas as an indicator of abnormal pressures. *Society of Petroleum Engineers Paper*, **4230**.

Fertl, W.H. (1976) *Abnormal Pressures: Implications to Exploration Drilling and Production of Oil and Gas Reserves*. Elsevier, New York, 382 pp.

Fischbein, S.A, (1987) Analysis and interpretation of ice-deformed sediments from Harrison Bay, Alaska. *United States Geological Survey Open File Report OF-0262*, 73 pp.

Fisher, A.T. and Hounslow, M.W. (1990a) Transient fluid flow through the toe of the Barbados accretionary complex: constraints from Ocean Drilling Program Leg 110 heat flow studies and simple models. *Journal of Geophysical Research*, **95**, 8845–8858.

Fisher, A.T. and Hounslow, M.W. (1990b) Heat flow through the toe of the Barbados accretionary complex. *Proceedings of the Ocean Drilling Program, Scientific Results*, **110**, Ocean Drilling Program, College Station, TX, pp. 345–354.

Fisher, D. and Byrne, T. (1987) Structural evolution of underthrust sediments, Kodiak Islands, Alaska. *Tectonics*, **6**, 775–793.

Fisher, D. and Byrne, T. (1990) The character and distribution of mineralized fractures in the Kodiak Formation, Alaska: implications for fluid flow in an underthrust sequence. *Journal of Geophysical Research*, **95**, 9069–9080.

Fisher, R.V. (1971) Features of coarse-grained, high-concentration fluids and their deposits. *Journal of Sedimentary Petrology*, **41**, 916–927.

Fisher, R.V. and Schminke, H.U. (1984), *Pyroclastic Rocks*. Springer-Verlag, Amsterdam, 472 pp.

Fisher, R.V. and Smith, G.A. (1991) *Sedimentation in Volcanic Settings. Society for Sedimentary Geology Special Publication* **45**, Tulsa, OK, 275 pp.

Fitches, W.R. and Maltman, A.J. (1978) Deformation of soft sediments, Conference report. *Journal of the Geological Society*, **135**, 245–251.

Fitches, W.R., Cave, R., Craig, J. and Maltman, A.J. (1986) Early veins as evidence of detachment in the Lower Palaeozoic rocks of the Welsh Basin. *Journal of Structural Geology*, **8**, 607–620.

Fleming, R.W. and Johnson A.M. (1975) Rates of seasonal creep of silty clay soil. *Quarterly Journal of Engineering Geology*, **8**, 1–29.

Fleming, R.W. and Varnes, D.J. (1991) Slope movements, in *The Heritage of Engineering Geology: the First Hundred Years* (ed. G.A. Kiersch), Geological Society of America, Centennial Special vol. 3, Boulder, CO, 201–218.

Flint, R.F. (1971) *Glacial and Quaternary Geology*. John Wiley & Sons, New York, 892 pp.

Foster, R.H. and De, P.K. (1971) Optical and electron microscopic investigation of shear-induced structures in lightly consolidated (soft) and heavily consolidated (hard) kaolinite. *Clays and Clay Minerals*, **19**, 31–47.

Fournier, R.O. and Potter, R.W., II (1982) An equation correlating the solubility of quartz in water from 25°C to 900°C at pressures up to 10,000 bars. *Geochimica et Cosmochimica Acta*, **46**, 1969–1973.

Fraser, H.J. (1935) Experimental study of the porosity and permeability of clastic sediments. *Journal of Geology*, **43**, 910–1010.

Freeze, R.A. and Cherry, J.A. (1979) *Groundwater*. Prentice-Hall Inc., Englewood Cliffs, NJ, 604 pp.

Frossard, E. (1979) Effect of sand grain shape on interparticle friction; indirect measurements of Rowe's stress dilatancy theory. *Geotechnique* **29**, 341–50.

Fryer, P., Ambos, E.L. and Hussong, G.M. (1985) Origin and emplacement of Mariana forearc seamounts. *Geology*, **13**, 774–777.

Fryer, P., Pearce, J.A., Stokking, L.B. et al. (1990) Summary of results from Leg 125. *Proceedings of the Ocean Drilling Program Scientific Results*, **125**, Ocean Drilling Program, College Station, TX, pp. 367–380.

Fujioka, K. and Taira, A. (1989) Tectono-sedimentary settings of seep biological communities – a synthesis from the Japanese subduction zones, in *Sedimentary Facies in the Active Plate Margin* (eds. A. Taira and F. Masuda), Terra Publishing, Tokyo, 577–602.

Fukami, A., Fukushima, K. and Kohyama, N. (1991) Observation technique for wet clay minerals using film-sealed environmental cell equipment attached to high-resolution electron microscope, in *Microstructure of fine-grained sediments* (eds R.H. Bennett, W. R. Bryant and M.H. Hulbert), Springer-Verlag, New York, 321–331.

Fyfe, W.S. and Kerrich, R. (1985) Fluids and thrusting. *Chemical Geology*, **49**, 353–362.

Fyfe, W.S., Price, N.J. and Thompson, A.B. (1978) *Fluids in the Earth's Crust*. Elsevier, Amsterdam, 383 pp.

Gal, M.L., Whittig, L.D. and Faber, B.A. (1990) Influence of clay on water movement in coarse-textured soils. *Clays and Clay Minerals*, **38**, 144–150.

Gale, S.J. and Hoare, P.G. (1991) *Quaternary Sediments: Petrographic Methods for the Study of Unlithified Rocks*. Bellhaven Press, London, 323 pp.

Galloway, W.E. (1974) Deposition and diagenetic alteration of sandstone in Northeast Pacific arc-related basins: Implications for graywacke genesis. *Geological Society of America Bulletin*, **85**, 379–390.

Galloway, W.E. (1986) Growth faults and fault-related structures of prograding terrigenous clastic continental margins. *Gulf Coast Association of Geological Societies Transactions*, **36**, 121–128.

Gardner, J. (1970) Rockfall: a geomorphic process in high mountain terrain. *Albertan Geography*, **6**, 15–20.

Garfunkel, Z. (1984) Large-scale submarine rotational slumps and growth faults in the eastern Mediterranean. *Marine Geology*, **55**, 305–324.

Garga, V.K. and Khan, M.A. (1991) Laboratory evaluation of K_0 for over-consolidated clays. *Canadian Geotechnical Journal*, **28**, 650–659.

Garlick, W.G. (1988) Algal mats, load structures, and synsedimentary sulfides in Revett quartzites of Montana and Idaho. *Economic Geology and the Bulletin of the Society of Economic Geologists*, **83**, 1259–1278.

Gates, W.C.B. (1987) The fabric of rockslide avalanche deposits. *Bulletin of the Association of Engineering Geologists*, **24**, 389–402.

Gawthorpe, R.L. and Clemmey, H. (1985) Geometry of submarine slides in the Bowland Basin (Dinantian) and their relation to debris flows. *Journal of the Geological Society*, **142** 555–565.

Genç, S. (1993) Structural and geomorphological aspects of the Çatak landslide, NE Turkey. *Quarterly Journal of Engineering Geology*, **26**, 99–108.

Genna, A. (1988) Déformations synsédimentaires, hydroplastiques liées à la tectonique pyrenéene compressive dans la molasse Eocene du Minervois. *Comptes Rendus de l'Academie des Sciences, Sciences de la Terre*, **306**, 1109–1114.

Georgiannou, V.N., Burland, J.B. and Hight, D.W. (1990) The undrained behaviour of clayey sands in triaxial compression and extension. *Geotechnique*, **40**, 431–449 pp.

Ghent, E.D. and Henderson, R.A. (1965) Significance of burrowing structures in the origin of convolute lamination. *Nature*, **207**, 1286–1287.

Ghosh, S.K. and Mukhopadhyay, A. (1986) Soft-sediment recumbent folding in a slump-generated bed in Jharia Basin, eastern India. *Journal of the Geological Society of India*, **27**, 194–201.

Gibling, M.R. and Stuart, C.J. (1988) Carbonate slide deposits in the Middle Jurassic of Portugal. *Sedimentary Geology*, **57**, 59–73.

Gibson, R.E. (1953) Experimental determination of the true cohesion and true angle of internal friction in clays. *Proceedings of the Third International Conference on Soil Mechanics and Foundation Engineering*, **1**, 126–130.

Gibson, R.E. (1958) The progress in consolidation in a clay layer increasing in thickness with time. *Geotechnique*, **8**, 171–182.

Gibson, R.E., England, G.L. and Husey, M.J.L. (1967) The theory of one-dimensional consolidation of saturated clays. 1. Finite non-linear consolidation of thin homogeneous layers. *Geotechnique*, **17**, 261–273.

Gibson, R.E. Schiffman, R.L. and Whitman, R.V. (1989) On two definitions of excess pore water pressure. *Geotechnique*, **39**, 169–171.

Gieskes, J.M., Blanc, G., Vrolijk, P., Elderfield, H. and Barnes, R. (1990) Interstitial water chemistry – major constituents. *Proceedings of the Ocean Drilling Program, Scientific Results*, **110**, Ocean Drilling Program, College Station, TX, 115–178.

Gieskes, J.M., Vrolijk, P. and Blanc, G. (1990) Hydrogeochemisty of the northern Barbados accretionary complex transect, Ocean Drilling Program Leg 110. *Journal of Geophysical Research*, **95**, 8809–8818.

Gilbert, P.A. (1991) Rapid water content loss by computer-controlled microwave drying. *Journal of Geotechnical Engineering*, **117**, 118–138.

Gill, W.D. and Kuenen, Ph.H. (1958) Sand volcanoes on slumps in the Carboniferous of County Clare, Ireland. *Quarterly Journal of the Geological Society of London* **75**, 251–291.

Gilliland, R.E. and Coles, M.E. (1990) Use of CT scanning in the investigation of damage to unconsolidated cores. *Proceedings, Ninth SPE Symposium on Formation Damage Control, Lafayette, LA* (ed. A. Ghalambour), 83–90.

Glennie, K.W. and Buller, A. (1983) The Permian Weissliegend of N.W. Europe: the partial deformation of aeolian sand caused by the Zechstein transgression. *Sedimentary Geology*, **35**, 43–81.

Glicken, H. (1982) Criteria for identification of large volcanic debris avalanches. *Eos (Transactions of the American Geophysical Union)*, **63**, 1141.

Gloppen, T.G. and Steel, R.J. (1981) The deposits, internal structure and geometry in six alluvial fan – fan delta bodies (Devonian–Norway): a study of bedding sequences in conglomerates, in *Recent and Ancient Non-marine Depositional Environments: Models for Exploration* (eds F.G. Ethridge and R. Flores), *Society of Economic Paleontologists and Mineralogists Special Publication*, **31**, 49–69.

Goldsmith, A.S. (1989) Permeability decline and compressibility in sandstone reservoir rocks, in *Rock at Great Depth* (eds V. Maury and D. Fourmaintraux), A.A. Balkema, Rotterdam, 923–928.

Goldthwait, R.P. and Matsch, C.L. (eds) (1989) *Genetic Classification of Glacigenic Deposits. Final report of the INQA Commission on Genesis and Lithology of Quaternary Deposits.* A.A. Balkema, Rotterdam, 304 pp.

Goodfellow, G.E. (1887) The Sonora earthquake. *Science*, **10**, 81–82.

Govier, G.W. and Aziz, K. (1972) *The Flow of Complex Mixtures in Pipes*. Van Nostrand Reinhold, New York, 792pp.

Grabau, A.W. (1913) *Principles of Stratigraphy*. A.G. Seiler and Co., New York, 1185 pp.

Graham, J. (1984) Methods of stability analysis, in *Slope Instability* (eds D. Brunsden and D.B. Prior) John Wiley & Sons, Chichester, 171–215.

Graham, J., Crooks, J.H.A. and Bell, A.L. (1983) Time effects on the stress–strain behaviour of natural soft clays. *Geotechnique*, **33**, 327–340.

Gramberg, J. (1965) Axial cleavage fracturing, a significant process in mining and in geology. *Engineering Geology* **1**, 31–72.

Gramberg, J. (1966) A theory on the occurrence of various types of vertical and sub-vertical joints in the earth's crust. *Proceedings of the 1st Congress of International Society for Rock Mechanics*, Lisbon **1**, 443–50.

Gramberg, J. (1989) *A non-conventional view on rock mechanics and fracture mechanics*. A.A. Balkema, Rotterdam, pp. 256.

Gratier, J.P. (1987) Pressure solution-deposition creep and associated tectonic differentiation in sedimentary rocks, in *Deformation of Sediments and Sedimentary Rocks* (eds M.E. Jones and R.M.F. Preston) *Geological Society of London Special Publication*, **29**, 25–38.

Gray, M.B. and Nickelsen, R.P. (1989) Pedogenic slickensides, indicators of strain and deformation processes in redbed sequences of the Appalachian foreland. *Geology*, **17**, 72–75.

Green, M.W. (1982) Origin of intraformational folds in the Jurassic Todilto Limestone, Ambrosia Lake uranium mining district, McKinley and Valencia counties, New Mexico. *United States Geological Survey Open File Report*, 29 pp.

Greenly, E. (1919) *The Geology of Anglesey*, 2 vols. Memoirs of the Geological Survey of England and Wales, HMSO, London.

Gresley, W.S. (1898) Clay-veins vertically intersecting coal measures. *Bulletin of the Geological Society of America*, **9**, 35–58.

Gretener, P.E. (1981) Pore pressure: fundamentals, general ramifications, and implications for structural geology (revised). *American Association of Petroleum Geologists Education Course Notes Series*, **4**, 131 pp.

Griffiths, F.J. and Joshi, R.C. (1991) Change in pore size distribution owing to secondary consolidation of clays. *Canadian Geotechnical Journal*, **28**, 20–24.

Guiraud, M. and Seguret, M. (1986) Soft sediment microfaulting related to compaction with the fluvio-deltaic infill of the Soria basin (North Spain). *Comptes Rendues, Académie de Science, Paris*, **302**, 793.

Guiraud, M. and Seguret, M. (1987) Soft-sediment microfaulting related to compaction within the fluvio-deltaic infill of the Soria strike-slip basin, northern Spain, in, *Deformation of Sediments and Sedimentary Rocks* (eds M.E. Jones and R.M.F. Preston), *Geological Society of London Special Publication*, **29**, 123–136.

Hadding, A. (1931) On subaqueous slides. *Geologiska Föreningen i Stockholm Förhandlingar*, **53** 377–393.

Hahn, F.F. (1913) Untermeerische Gleitung bei Trenton Falls (Nord Amerika) ünd ihr Verhaltnis zu ähnlichen Störungsbildern. *Neues Jahrbuch für Mineralogie, Geologie, und Paläontologie*, **36**, p. 1–41.

Haldorsen, S. (1981) Grain-size distribution of subglacial till and its relation to glacial crushing and abrasion. *Boreas*, **10**, 91–105.

Haldorsen, S. (1983) The characteristics and genesis of Norwegian tills, in *Glacial Deposits in North-west Europe* (ed. J. Ehlers), A. A. Balkema, Rotterdam, 11–17.

Hall, J. (1812) On the vertical position and convolutions of certain strata and their relation with granite. *Transactions of the Royal Society of Edinburgh*, **7**, 79–108.

Hamilton, D.H. and Mechan, R.L. (1971) Ground rupture in the Baldwin Hills. *Science*, **172**, 333–344.

Hampton, M.A. (1979) Buoyancy in debris flows. *Journal of Sedimentary Petrology*, **49**, 753–758.

Handin, J., Hager, R.V., Friedman, M. and Feather, J.M. (1963) Experimental deformation of sedimentary rocks under confining pressure: Pore pressure tests. *Bulletin of the American Association of Petroleum Geologists*, **47**, 717–755.

Hanor, J.S. (1980) Dissolved methane in sedimentary brines: potential effect on the PVT properties of fluid inclusions. *Economic Geology*, **75**, 603–617.

Hansen, B.L. and Langway, C.G. (1966) Deep-core drilling and ice-core analysis, Camp Century, Greenland, 1961–1966. *Antarctic Journal of the United States*, **1**, 207–208.

Hansen, M.J. (1984) Strategies for classification of landslides, in *Slope Instability* (eds D. Brunsden and D.B. Prior), John Wiley and Sons, Chichester, 1–25.

Hanshaw, B. and Bredehoeft, J. (1968) On the maintenance of abnormal fluid pressures, II; source layer at depth *Geological Society of America Bulletin*, **79**, 1107–1122.

Hanshaw, B. and Zen, E. (1965) Osmotic equilibrium and overthrust faulting. *Geological Society of America Bulletin*, **76**, 1379–1387.

Hardin, B.O. (1989) Low-stress dilation test. *Journal of Geotechnical Engineering*, **115**, 769–87.

Hardin, B.O. and Blandford, G.E. (1989) Elasticity of particulate materials. *Journal of Geotechnical Engineering*, **115**, 788–787.

Harp, E.L., Wells, W.G. and Sarmiento, J.G. (1990) Pore pressure response during failure in soils. *Geological Society of America Bulletin*, **102**, 428–438.

Harrington, P.K. (1985) Formation of pockmarks by pore-water escape. *Geo-Marine Letters*, **5**, 193–197.

Harris, C. (1981) Microstructures in solifluction sediments from South Wales and North Norway. *Bulletyn Peryglacial*, **28**, 221–226.

Harrison, W. (1958) Marginal zones of vanished glaciers reconstructed from the pre-consolidation pressure values of overridden silts. *Journal of Geology*, **66**, 72–95.

Harrison, W.J. and Summa, L.L. (1991). Paleohydrology of the Gulf of Mexico Basin. *American Journal of Science*, **291**, 109–176.

Hart, J.K. (1990) Proglacial glaciotectonic deformation and the origin of the Cromer Ridge push moraine complex, north Norfolk, England. *Boreas*, **19**, 165–180.

Hart, J.K. and Boulton, G.S. (1991) The interrelation of glaciotectonic and glaciodepositional processes within the glacial environment. *Quaternary Science Reviews*, **10**, 335–350.

Hart, J.K. Hindmarsh, R.C.A. and Boulton, G.S. (1990) Styles of subglacial glaciotectonic deformation within the context of the Anglian ice-sheet. *Earth Surface Processes and Landforms*, **15**, 227–241.

Hattingh, J., Rust, I.C. and Reddering, J.S.V. (1990) A technique for preserving structures in unconsolidated gravels in relief peels. *Journal of Sedimentary Petrology*, **60**, 626–627.

Haxby, W.F. and Turcotte, D.L. (1976) Stress induced by the addition or removal of overburden and associated thermal effects. *Geology*, **14**, 181–185.

Hedberg, H.D. (1974) Relation of methane generation to undercompacted shales, shale diapirs and mud volcanoes. *Bulletin of the American Association of Petroleum Geologists*, **58**, 661–673.

Heezen, B.C. and Drake, C.L. (1964) Grand Banks slump. *Bulletin of the American Association of Petroleum Geologists*, **48**, 221–225.

Heim, A. (1908) Uber rezente ünd fossile subaquatische Rurschungen ünd deren lithologische Bedeutung. *Neues Jahrbuch für Mineralogie, Geologie, und Paläontologie*, **2**, 137–157.

Heim, A. (1932) *Bergsturz und Menschenleben*. Fretz und Wasmuth, Zurich, 218 pp.

Hein, F.J. (1982) Depositional mechanisms of deep-sea coarse clastic sediments. *Canadian Journal of Earth Sciences*, **19**, 267–287.

Hein, F.J. (1985) Fine-grained slope and basin deposits, California continental borderland: facies, depositional mechanisms and geotechnical properties. *Sedimentary Geology*, **67**, 237–262.

Helwig, J. (1970) Slump folds and early structures, northeastern Newfoundland Appalachians. Journal of Geology, **78**, 172–187.

Henderson, J.R., Wright T.O. and Henderson, M.N. (1989) Mechanics of formation of gold-bearing quartz veins, Nova Scotia, Canada – comment. *Tectonophysics*, **166**, 351–352.

Hendry, H.E. and Stauffer, M.R. (1977) Penecontemporaneous folds in cross-bedding. Inversion of facing criteria and mimicry of tectonic folds. *Geological Society of America Bulletin*, **88**, 809–812.

Henkel, D.J. (1960) The shear strength of saturated remolded clays. *Proceedings of the Research Conference on the Shear Strength of Cohesive Soils, American Society of Civil Engineers*, 533–554.

Henkel, D.J. (1970) The role of waves in causing submarine landslides. *Geotechnique*, **20**, 75–80.

Hesse, R. and Chough, S.K. (1981) The Northwest Atlantic mid-ocean channel of the Labrador Sea: II. Deposition of parallel-laminated levee-muds from the viscous sub-layer of low-density currents. *Sedimentology*, **27**, 697–711.

Hesse, R. and Reading, H.G. (1978) Subaqueous clastic fissure eruptions and other examples of sedimentary transposition in the lacustrine Horton Bluff Formation (Mississipian), Nova Scotia, Canada, in *Modern and Ancient Lake Sediments* (eds A. Matter and M.E. Tucker), *Special Publications of the International Association of Sedimentologists*, **2**, Blackwell Scientific Publications, Oxford, 241–257.

Hesselbo, S.P. and Palmer, T.J. (1992) Reworked early diagenetic concretions and the bioerosional origin of a regional discontinuity within British Jurassic marine mudstones. *Sedimentology*, **39**, 1045–1065.

Hesselbo, S.P. and Trewin, N.H. (1984) Deposition, diagenesis and structures of the Cheese Bay shrimp bed, Lower Carboniferous, East Lothian. *Scottish Journal of Geology*, **20**, 281–296.

Hibbard, J. and Karig, D.E. (1987) Sheath-like folds and progressive deformation in Tertiary sedimentary rocks of the Shimanto accretionary complex, Japan. *Journal of Structural Geology*, **9**, 845–857.

Hickman, S.H., Healy, J.H. and Zoback, M.D. (1985) *In situ* stress, natural fracture distribution, and bore hole elongation in the Auburn geothermal well, Auburn, New York. *Journal of Geophysical Research*, **90**, 5497–5512.

Higgins, G.E. and Saunders, J.B. (1974) Mud volcanoes – their nature and origin: contribution to the geology and paleobiology of the Caribbean and adjacent areas. *Verhandlungen der Naturförschenden Gesellschaft in Basel*, **84**, 101–152.

Higgs, B. (1988) Syn-sedimentary structural controls on basin deformation in the Gulf of Corinth, Greece. *Basin Research*, **1**, 155–165.

Hill, A.R. (1973) The distribution of drumlins in County Down, Ireland. *Annals of the Association of Geographers*, **63**, 226–240.

Hill, P.R. (1984) Sedimentary facies of the Nova Scotian upper and middle continental slope, offshore eastern Canada. *Sedimentology*, **31**, 293–309.

Hill, P.R., Moran, K.M. and Blasco, S.M. (1982) Creep deformation of slope sediments in the Canadian Beaufort Sea. *Geo-Marine Letters*, **2**, 163–170.

Hird, C.C. and Hassona, F.A.K. (1990) Some factors affecting the liquefaction and flow of saturated sands in laboratory tests. *Engineering Geology*, **28**, 149–170.

Hiscott, R.N. (1979) Clastic sills and dikes associated with deep-water sandstones, Tourelle Formation, Ordovician, Quebec. *Journal of Sedimentary Petrology*, **49**, 1–9.

Hiscott, R.N. and Middleton, G.V. (1979) Depositional mechanisms of thick-bedded sandstones at the base of a submarine slope, Tourelle Formation (Lower Ordovician), Quebec, Canada, in *Geology of Continental Slopes* (eds L.J. Doyle and O.H. Pilkey), *Society of Economic Paleontologists and Mineralogists Special Publication*, **27**, 307–326.

Hitchon, B. (1984) Geothermal gradients, hydrodynamics and hydrocarbon occurrences, Alberta, Canada. *Bulletin of the American Association of Petroleum Geologists*, **68**, 713–743.

Holdsworth, G. and Bull, C. (1970) The flow law of cold ice: investigations on Meserve Glacier, Antarctica. *International Association of Scientific Hydrology*, **86**, 204–216.

Holeywell, R.C. and Tullis, T.E. (1975) Mineral reorientation and slaty cleavage in the Martinsburg Formation, Lehigh Gap, Pennsylvania. *Geological Society of America Bulletin*, **86**, 1296–1304.

Holler, P. (1986) Fracture activity: a possible triggering mechanism for slope instabilities in the Eastern Atlantic? *Geo-Marine Letters*, **5**, 211–216.

Holmes, A (1965) *Principles of Physical Geology*, 2nd edn., Thomas Nelson, London, 1288 pp.

Hooper, E.C.D. (1991) Fluid migration along growth faults in compacting sediments. *Journal of Petroleum Geology*, **14**, 161–180.

Horowitz, D.H. (1982) Geometry and origin of large-scale deformation structures in some ancient wind-blown sand deposits. *Sedimentology*, **29**, 155–180.

Hoshino, K. (1981) Consolidation and strength of the soft sedimentary rocks, in *Proceedings of an International Symposium on Weak Rock, 21–24 September, Tokyo*, vol. 1, A.A. Balkema, Rotterdam, 155–160.

Hou, G. (1988) Physical properties and mechanical state of an artificial silty clay as a simulation of deformation of deep sea sediments at convergent plate boundaries. Unpublished MSc dissertation, Cornell University, Ithaca, NY.

Hounslow, M.W. (1990) Grain fabric measured using magnetic susceptibility anisotropy in deformed sediments of the Barbados accretionary prism: Leg 110. *Proceedings of the Ocean Drilling Program, Scientific Results*, **110**, Ocean Drilling Program, College Station, TX, 257–275.

Houseknecht, D.W. (1987) Assessing the relative importance of compaction processes and cementation to reduction of porosity in sandstones. *Bulletin of the American Association of Petroleum Geologists*, **71**, 633–642.

Hovland, M. (1990) Do carbonate reefs form due to fluid seepage? *Terra Nova*, **2**, 8–18.

Hovland, M. and Judd, A.G. (1988) *Seabed Pockmarks and Seepages: Impact on Geology, Biology and the Marine Environment*. Graham & Trotman, London, 293 pp.

Hsieh, P. and Bredehoeft, J.D. (1981) A reservoir analysis of the Denver Earthquakes: a case of induced seismicity. *Journal of Geophysical Research*, **86**, 903–920.

Hsü, K. (1968) Principles of mélanges and their bearing on the Franciscan–Knoxville paradox. *Geological Society of America Bulletin*, **79**, 1063–1074.

Hsü, K. (1974) Mélanges and their distinction from olistostromes, in *Modern and Ancient Geosynclinal Sedimentation* (eds R. Dott and R. Shaver), *Society of Economic Paleontologists and Mineralogists Special Publication*, **19**, 321–333.

Huang, Q. (1988) Geometry and tectonic significance of Albian sedimentary dykes in the Sisteron area, SE France. *Journal of Structural Geology*, **10**, 453–462.

Hubbert, M.K. (1940) Theory of groundwater motion. *Journal of Geology*, **48**, 785–944.

Hubbert, M.K. (1956) Darcy's law and the field equations of flow of underground fluids. *Transactions of the American Institute of Mining and Metallurgical Engineering*, **207**, 222–239.

Hubbert, M.K. and Ruby, W.W. (1959) Role of fluid pressure in the mechanics of overthrust faulting I: mechanics of fluid filled porous solids and its applications to overthrust faulting. *Geological Society of America Bulletin*, **70**, 115–166.

Huffman, M.E., Scott, R.G. and Lorens, P.J. (1969) Geologic investigation of landslides along the Middle Fork Eel River, California. *Geological Society of America Abstracts with Programs*, **7**, 111–112.

Humphrey, N., Kamb, B., Fahnestock, M. and Engelhardt, H. (1993) Characteristics of the bed of the Lower Columbia Glacier, Alaska. *Journal of Geophysical Research*, **98**, 837–846.

Hunt, C.B., Robinson, T.W., Bowles, W.A. and Washburn, A.L. (1966) Hydrological basin, Death Valley, California. *United States Geological Survey Professional Paper*, **494B**, 138 pp.

Hunt, J.M. (1960) Generation and migration of petroleum from abnormally pressured fluid compartments. *Bulletin of the American Association of Petroleum Geologists*, **74**, 1–12.

Hurley, M.T. and Schultheiss, P.J. (1990) Sea-bed shear moduli from measurements of tidally-induced pore-pressures, in *Shear Waves in Marine Sediments* (eds J. Hovem, M. Richardson and R. Stel), Kluwer Academic Publishers, Hingham, MA, 411–418.

Hurst, A. and Buller, A.T. (1984) Dish structures in some Paleocene deep-sea sandstones (Norwegian sector, North Sea): origin of the dish-forming clays and their effect on reservoir quality. *Journal of Sedimentary Petrology*, **54**, 1206–1211.

Hutchinson, J.N. (1986) A sliding-consolidation model for flowslides. *Canadian Geotechnical Journal*, **23**, 115–126.

Hutchinson, J.N. (1988) General report: morphological and geotechnical parameters of landslides in relation to geology and hydrogeology, in *Proceedings of the Fifth International Symposium on Landslides* (ed. C. Bonnard), A.A. Balkema, Rotterdam, 3–55.

Hutchinson, J.N. (ed.) (1989) Debris flows. *Bulletin of the International Association of Engineering Geology*, **40**, 5–126.

Hutton, J. (1788) Theory of the earth; or an investigation of the laws observable in the composition, dissolution, and restoration of land upon the globe. *Transactions of the Royal Society of Edinburgh*, **1**, 209–304.

IAEG (International Association of Engineering Geologists), Commission on Land slides (1990) Suggested nomenclature for landslides. *Bulletin of the International Association of Engineering Geologists*, **41**, 13–16.

Ineson, J.R. (1985) Submarine glide blocks from the Lower Cretaceous of the Antarctic Peninsula. *Sedimentology*, **32**, 659–70.

Ingersoll, R.V. (1988) Tectonics of sedimentary basins. *Geological Society of America Bulletin*, **100**, 1704–1719.

Inman, D.L. (1963) Physical properties and mechanics of sedimentation, in *Submarine Geology* (ed. Shepard, F.P.), Harper & Row, New York, 101–151.

Isaacs, C.M., Pisciotto, K.A. and Garrison, R.E. (1983) Facies and diagenesis of the Monetery Formation, California: a summary, in *Siliceous Deposits in the Pacific Region* (eds A. Iijima, J.R. Hein and R. Siever), *Developments in Sedimentology*, **36**, Elsevier, Amsterdam, 247–282.

Iverson, N.R. (1991) Potential effects of subglacial water pressure fluctuations on quarrying. *Journal of Glaciology*, **125**, 27–36.

Iverson, N.R. (1985) A constitutive equation for mass-movement behavior. *Journal of Geology*, **93**, 143–160.

Jackson, M.P.A. and Talbot, C.J. (1994) Advances in salt tectonics, in *Continental Deformation* (ed. P.L. Hancock), Pergamon, Oxford **17**, 159–179.

Jacobi, R.D. (1976) Sediment slides on the north-western continental margin of Africa. *Marine Geology*, **22**, 157–173.

Jaeger, J.C. (1969) *Elasticity, Fracture and Flow.* Methuen, London, 152 pp.

Jaeger, J.C. and Cook, N. (1979) *Fundamentals of Rock Mechanics.* Chapman & Hall, London, 585 pp.

James, N.P. and Ginsburg, R.N. (eds) (1979) The seaward margin of the Belize barrier and atoll reefs. *Special Publications of the International Association of Sedimentologists*, **3**, Blackwell Scientific Publications, Oxford, 193 pp.

Jamiolkowski, M., Ladd C.C., Germaine, J.T. and Lancellotta, R. (1985) New developments of field and laboratory testing of soils. *Proceedings of the 11th International Conference on Soil Mechanics and Foundation Engineering, San Francisco, CA*, 57–153.

Jamison, W.R. (1992) Stress spaces and stress paths. *Journal of Structural Geology*, **14**, 1111–1120.

Janbu, N. (1985) Soil models for offshore engineering. *Geotechnique*, **35**, 241–281.

Jansen, E., Befring, S., Bugge, T., et al. (1987) Large submarine slides on the Norwegian continental margin: sediments, transport and timing. *Marine Geology*, **78**, 77–107.

Jenkins, J.P. (1984) *Thesaurus of Rock and Soil Mechanics.* Pergamon, Oxford, 62 pp.

Jewell, R.A. (1989) Direct shear tests on sand. *Geotechnique*, **39**, 309–322.

Jibson, R.W. and Keefer, D.K. (1993) Analysis of the seismic origin of landslides: example from the New Madrid seismic zone. *Geological Society of America Bulletin*, **105**, 521–536.

Jim, C.Y. (1985) Impregnation of moist and dry unconsolidated clay sample(s) using SPURR resin for microstructural studies. *Journal of Sedimentary Petrology*, **55**, 597–599.

Johnson, A.M. and Rodine, J.R. (1984) Debris flow, in *Slope Instability* (eds D. Brunsden and D.B. Prior), John Wiley & Sons, Chichester, 257–361.

Johnson, J.P., Rhett, D.W. and Seimers, W.T. (1989) Rock mechanics of the Ekofisk reservoir in the evaluation of subsidence. *Offshore Technology Conference, Houston, Texas, OTC5621*, 39–50.

Johnson, M.S. and Heron, D., Jr. (1965) Slump features in the McBean Formation and younger beds, Riley Cut, Calhoun County, South Carolina. *Geologic Notes (Geological Survey of South Carolina)*, **9**, 37–44.

Johnson, S.Y. (1986) Water-escape structures in coarse-grained, volcaniclastic, fluvial deposits of the Ellensburg Formation, south-central Washington. *Journal of Sedimentary Petrology*, **56**, 905–910.

Johnston, W.A. (1915) Rainy River District, Ontario; surficial geology and soils. *Geological Survey of Canada Memoir*, **82**, 1–123.

Jones, A.S. (1979) The flow of ice over a till bed. *Journal of Glaciology*, **87**, 393–395.

Jones, B.G. (1972) Deformation structures in siltstone resulting from the migration of an Upper Devonian aeolian dune. *Journal of Sedimentary Petrology*, **42**, 935–940.

Jones, J.A.A. (1990) Piping effects in humid lands, in *Groundwater Geomorphology: the role of Subsurface Water in Eath-surface Processes and Landform* (eds C.G. Higgins and D.R. Coates), *Geological Society of America Special Paper*, **252**, 111–138.

Jones, M.E. (1991) Topographic elevation, erosion rate and strain rate: a possible relationship, in *Neotectonics and Resources* (eds J. Cosgrove and M.E. Jones), Belhaven, London, 144–147.

Jones, M.E. and Addis, M.J. (1984) Volume change during sediment diagenesis and the development of growth faults. *Marine and Petroleum Geology*, **1**, 118–122.

Jones, M.E. and Addis, M.J. (1986) The application of stress path and critical state analysis to sediment deformation. *Journal of Structural Geology*, **8**, 575–580.

Jones, M.E. and Leddra, M.A. (1989) Compaction and flow of porous rocks at depth, in *Rock at Great Depth* (eds V. Maury and D. Fourmaintraux), A.A. Balkema, Rotterdam, 891–898.

Jones, M.E., Leddra, M.J. and Addis, M.A. (1987) *Reservoir Compaction and Sea Floor Subsidence due to Hydrocarbon Extraction.* Offshore Technology Report OTH 87 276, HMSO, London, 175 pp.

Jones, M.E., Leddra, M.J., Berget, O.P., Goldsmith, A.S. and Tappel, I. (1990) The geotechnical characteristics of weak North Sea reservoir rocks. *North Sea Oil and Gas Reservoirs*, II. Graham and Trotman, London, 201–211.

Jones, M.E., Leddra, M.J., Goldsmith, A.S. and Edwards, D. (1992) *The Geomechanical Characteristics of Reservoirs and Reservoir Rocks.* Offshore Technology Report OTH 90 333, HMSO, London, 202 pp.

Jones, M.E., Leddra, M.J., Goldsmith, A.S. and Yassir N. (1991) Mechanisms of compaction and flow in porous sedimentary rocks, in *Neotectonics and Resources* (eds J. Cosgrove and M.E. Jones), Belhaven, London, 16–42.

Jones, M.E., Leddra, M.J. and Potts, D. (1990) Ground motions due to hydrocarbon production from the chalk. *Proceedings of the International Chalk Symposium*, Thomas Telford, London, 341–347.

Jones, O.T. (1937) On the sliding or slumping of submarine sediments in Denbighshire, N Wales, during the Ludlow period. *Quarterly Journal of the Geological Society of London*, **93**, 241–282.

Jones, O.T. (1939a) The geology of the Colwyn Bay district: a study of submarine slumping during the Salopian period. *Quarterly Journal of the Geological Society of London*, **95**, 335–382.

Jones, O.T. (1939b) The consolidation of muddy sediments. *Geological Magazine*, **76**, 170–171.

Jones, O.T. (1943) A comment on a new area of slumped beds in Denbighshire. *Geological Magazine*, **80**, 66–68.

Jones, O.T. (1944) The compaction of muddy sediments. *Quarterly Journal of the Geological Society of London*, **100**, 137–160.

Jones, O.T. (1953) On submarine slumping in the Lower Ludlow rocks of North Wales. *Geological Magazine*, **90**, 220–221.

Jones, O.T. (1955) The geological evolution of Wales and adjacent regions. William Smith Lecture. *Quarterly Journal of the Geological Society of London*, **111**, 323–351.

Jørstad, F.A. (1968) Waves generated by landslides in Norwegian fjords and lakes. *Norwegian Geotechnical Institute Publication*, **79**, 1–20.

Kagami, H., Karig, D.E., and the shipboard scientific party (1986) *Initial Reports of the Deep Sea Drilling Project*, **87**, United States Government Printing Office, Washington, DC, 985pp.

Kamb, W.V. (1987) Glacier surge mechanism based on linked cavity configuration of basal water conduit system. *Journal of Geophysical Research*, **92**, 9083–9100.

Kamb, W.B. (1991) Rheological non-linearity and flow instability in the deforming-bed mechanism of ice-stream motion. *Journal of Geophysical Research*, **96**, 16585–16595.

Kamb, W.B., Raymond, C.F., Harrison, W.D. *et al.* (1985) Glacier surge mechanism: 1982–1983 surge of Variegated Glacier, Alaska. *Science*, **227**, 469–479.

Karig, D.E. (1986) Physical properties and mechanical state of accreted sediments in the Nankai Trough, Southwest Japan Arc. *Geological Society of America Memoir*, **166**, 117–33.

Karig, D.E. (1990) Experimental and observational constraints on the mechanical behavior in the toes of accretionary prisms, in *Deformation Mechanisms, Rheology and Tectonics* (eds R.J, Knipe and E.H. Rutter), Geological Society of London Special Publication, **54**, 383–398.

Karig, D.E. (1993) Reconsolidation tests and sonic velocity measurements of clay-rich sediments from the Nankai Trough. *Proceedings of the Ocean Drilling Program, Scientific Results*, **131**, Ocean Drilling Program, College Station, TX, 247–260.

Karig, D.E. and Hou, G. (1987) Anelastic consolidation of sediments and some implications for joint formation. *Eos Transactions of the American Geophysical Union* (abstracts), **68**, 1463.

Karig, D.E. and Hou, G. (1992) High-stress consolidation experiments and their geologic implications. *Journal of Geophysical Research*, **97**, 289–300.

Karig, D.E. and Lundberg, N. (1990) Deformation bands from the toe of the Nankai accretionary prism. *Journal of Geophysical Research*, **95**, 9099–9109.

Kastner, M. (1981) Authigenic silicates in deep-sea sediments: formation and diagenesis, in *The Sea* (ed. C. Emiliani), vol. 7, Wiley Interscience, New York, 915–980.

Kastner, M., Elderfield, H. and Martin, J.B. (1991) Fluids in convergent margins: what do we know about their composition, origin, role in diagenesis and importance for oceanic chemical fluxes? *Philosophical Transactions of the Royal Society of London, Series. A*, **335**, 275–288.

Kastner, M., Elderfield, H., Jenkins, W.J., Gieskes, J.M. and Gamo, T. (1993) Geochemical and isotopic evidence for fluid flow in the western Nankai subduction zone. *Proceedings of the Ocean Drilling Program, Scientific Results*, **131**, Ocean Drilling Program, College Station, TX, pp. 397–413.

Kayen, R.E. and Lee, H.J. (1990) Sea-floor landslides in regions of gas hydrate, a global change perspective (Abstract.) *Bulletin of the American Association of Petroleum Geologists*, **74**, 982.

Keedwell, M.J. (1984), *Rheology and Soil Mechanics*. Elsevier Applied Science, London and New York, 323 pp.

Keedwell, M.J. (1988) *Rheology and Soil Mechanics: International Conference Proceedings*. Elsevier, Amsterdam, 215 pp.

Keene, J.B. (1976) The distribution, mineralogy, and petrography of biogenic and authigenic silica from the Pacific Basin. Unpublished PhD thesis, Scripps Institution of Oceanography, University of California at San Diego, 264 pp.

Kemmis, T.J. (1981) Importance of the regelation process to certain properties of basal tills deposited by the Laurentide ice sheet in Iowa and Illinois, U.S.A., *Annals of Glaciology*, **2**, 147–152.

Kemp, A.E.S. (1990) Fluid flow in 'vein structures' in Peru forearc basins: evidence from back-scattered electron microscope studies. *Proceedings of the Ocean Drilling Program, Scientific Results*, **112**, Ocean Drilling Program, College Station, TX, 33–41.

Kennedy, W.J. (1986) Sedimentology of Late Cretaceous–Palaeocene chalk reservoirs, North Sea central graben, in *Petroleum Geology of North-west Europe*, **1** (eds J. Brooks and K.W. Glennie), Graham & Trotman, London 469–481.

Kenney, C. (1984) Properties and behaviours of soils relevant to slope instability, in *Slope Instability* (eds D. Brunsden and D.B. Prior), John Wiley & Sons, Chichester, 27–65.

Kenney, T.C. (1967) The influence of mineral composition on the residual strength of natural soils. *Proceedings, International Geotechnical Conference, Oslo*, **1**, 123–129.

Keren, R. and Shainberg, I. (1975) Water vapor isotherms and heat of immersion Na/Ca-montmorillonite systems I: homoionic clays. *Clays and Clay Minerals*, **23**, 193–200.

Kimura, G., Koga, K. and Fujioka, K. (1989) Deformed soft sediments at the junction between the Mariana and Yap trenches. *Journal of Structural Geology*, **11**, 463–472.

Kindle, E.M. (1917) Deformation of unconsolidated beds in Nova Scotia and Southern Idaho. *Geological Society of America Bulletin*, **28**, 323–334.

King, F.H. (1899) Principles and conditions of the movements of ground water. *Nineteenth Annual Report of the United States Geological Survey, for 1897–1898*, part 2, 59–294.

Kinkaldie, L. (1992) The effect of soil joints on soil mass properties. *Bulletin of the Association of Engineering Geologists*, **29**, 415–420.

Kirkland, D.W. and Anderson, R.Y. (1970) Microfolding in the Castille and Todilto evaporites, Texas and New Mexico. *Geological Society of America Bulletin*, **81**, 3259–3282.

Kitamura, R. and Shinchi, M. (1988), Application of Markov model to the K_0 compression and consolidation of particulate material, in *Micromechanics of Granular Materials* (eds M. Satake, and J.T. Jenkins), Studies in Applied Mechanics, **20**, Tohaku University, Department of Civil Engineering, Sendai, Japan, 225–234.

Knipe, R.J. (1986a) Faulting mechanisms in slope sediments: examples from Deep Sea Drilling Project cores. *Geological Society of America Memoir*, **166**, 45–54.

Knipe, R.J. (1986b) Deformation mechanism path diagrams for sediments undergoing lithification. *Geological Society of America Memoir*, **166**, 151–160.

Knipe, R.J. (1986c) Microstructural evolution of vein arrays preserved in Deep Sea Drilling Project cores from the Japan Trench, Leg 57. *Geological Society of America Memoir*, **166**, 75–87.

Knipe, R.J. and Needham, D.T. (1986) Deformation processes in accretionary wedges – examples from the SW margin of the Southern Uplands, Scotland, in *Collision Tectonics* (eds M.P. Coward & A.C. Ries), *Geological Society of London Special Publication*, **19**, 51–65.

Knipe, R.J., Agar, S.M. and Prior, D.J. (1991) The microstructural evolution of fluid flow paths in semi-lithified sediments from subduction complexes. *Philosophical Transactions of the Royal Society of London Series A*, **335**, 261–273.

Koerner, R.M. (1970) Effect of particle characteristics on soil strength. *American Society of Civil Engineers, Journal of Soil Mechanics and Foundations Division, Proceedings*, **SM4**, 1121–1233.

Kohlbeck, F., Mojica, J., and Scheidegger, A.E. (1994) clast orientations of the 1985 lahars of the Nevado del Ruiz, Colombia, and implications for depositional processes. *Sedimentary Geology*, **88**, 175–183.

Kohler, J. and Proksch, R. (1991) *In situ* measurement of subglacial till deformation beneath Storglaciaren, N Sweden, *Eos (Transactions of the American Geophysical Union)* **72**, 158.

Kokelaar, B.P. (1982) Fluidization of wet sediments during the emplacement and cooling of various igneous bodies. *Journal of the Geological Society*, **139**, 21–33.

Kokelaar, B.P. (1986) Magma–water interactions in subaqueous and emergent basaltic volcanism. *Bulletin of Volcanology*, **48**, 275–289.

Kokelaar, B.P., Bevins, R.E. and Roach, R.A. (1985) Submarine silicic volcanism and associated sedimentary and tectonic processes, Ramsey Island, SW Wales, *Journal of the Geological Society*, **142**, 591–613.

Korosec, M.A., Rigby, J.G. and Stoffel, K.L. (1980) The 1980 eruptions of Mount St. Helens, Washington. *Washington State Department of Natural Resources, Division of Geology and Earth Sciences, Information Circular*, **71**, 1–27.

Kozarski, S. and Szuprycynski, J. (1973) Glacial forms and deposits in the Sidujokull deglaciation area. *Geographia Polonica*, **26**, 255–311.

Kraft, L.M., Jr., Helfrich, S.C., Suhayada, J.N., and Marin, J.E. (1985) Soil responses to ocean waves. *Marine Geotechnology*, **6**, 173–203.

Kruit, C., Brouwer, J., Knox, G., Schollnberger, W. and van Vliet, A. (1975) Une excursion aux cones d'alluvions – un eau profonde d'age Tertiaire prés de San Sebastian (province de Guipuzcoa, Espagne). *Excursion Guidebook, part 23, 9th International Sedimentological Congress (IAS)*, Nice.

Kugler, H.G. (1939) Visit to Russian oil districts. *Journal of the Institute of Petroleum*, **25**, 68–88.

Kulm, L.D., Suess, E., Moore, J.C. et al. (1986) Oregon subduction zone: venting, fauna, and carbonates. *Science*, **231**, 561–566.

Kutter, B.L. and Sathialingam, N. (1992) Elastic-viscoplastic modelling of the rate-dependent behaviour of clays. *Geotechnique*, **42**, 427–441.

Kvenvolden, K.A. (1985) Comparison of marine gas hydrates in sediments of an active and passive continental margin. *Marine and Petroleum Geology*, **2**, 65–70.

Kvenvolden, K.A. (1988) Methane hydrates and global climate. *Global Biogeochemical Cycles*, **2**, 221–229.

Kvenvolden, K.A. and Kastner, M. (1990) Gas hydrates of the Peruvian outer continental margin. *Proceedings of the Ocean Drilling Program, Scientific Results*, **112**, Ocean Drilling Program, College Station, TX, 517–526.

Labaume, P., Bonsquet, J.C. and Lanzafame, G. (1990) Early deformation at a submarine compressive front, the Quaternary Catalania foredeep south of Mt. Etna, Sicily. *Tectonophysics*, **117**, 349–66.

Labaume, P. (1987) Syndiagenetic deformation of a turbiditic series related to submarine gravity nappe emplacement (Autapie nappe, French Alps), in *Deformation of Sediments and Sedimentary Rocks* (eds M.E. Jones and R.M.F. Preston), *Geological Society of London Special Publication*, **29**, 147–163.

Labaume, P., Berty, C. and Laurent, Ph. (1991) Syndiagenetic evolution of shear structures in superficial nappes: an example from the Northern Apennines (NW Italy). *Journal of Structural Geology*, **13**, 385–398.

Ladd, G.E. (1935) Landslides, subsidences and rockfalls. *American Railway Engineering Association Bulletin*, **37**, 1091–1162.

Ladd, J.W., Westbrook, G.K., Buhl, P. and Bangs, N. (1990) Wide-aperture seismic reflection profiles across the Barbados Ridge Complex. *Proceedings of the Ocean Drilling Program, Scientific Results*, **110**, Ocean Drilling Program, College Station, TX, 3–6.

Lade, P.V. and Overton, D.D. (1991) Cementation effects in frictional materials. *Journal of Geotechnical Engineering*, **115**, 1373–1387.

Laird, M.G. (1970) Vertical sheet structure – a new indicator of sedimentary fabric. *Journal of Sedimentary Petrology*, **40**, 428–438.

Lallemant, S.J., Byrne, T., Maltman, A., Karig, D. and Henry, P. (1993) Stress tensors at the toe of the Nankai accretionary prism: an application of inverse methods to slickenlined faults. *Proceedings of the Ocean Drilling Program, Scientific Results*, **131**, Ocean Drilling Program, College Station, TX, 103–122.

Lambe, W.T. and Whitman, R.V. (1979) *Soil Mechanics, SI Version*. John Wiley & Sons, New York, 553 pp.

Lang, S.C. and Fielding, C.R. (1991) Facies architecture of a Devonian soft-sediment-deformed alluvial sequence, Broken River Province, Northeastern Australia, in *The Three-dimensional Architecture of Terrigenous Clastic Sediments and its Implications for Hydrocarbon Discovery and Recovery* (eds A.D. Miall and N. Tyler). *Society of Economic Paleontologists and Mineralogists, Concepts in Sedimentology and Paleontology*, **3**, 122–132.

Langseth, M.G. and Moore, J.C. (1990) Introduction to special section on the role of fluids in sediment accretion, deformation and diagenesis, and metamorphism in subduction zones. *Journal of Geophysical Research*, **95**, 8737–8741.

Larsen, S.C., Agnew, D.C. and Hager, B.M. (1993) Strain accumulation in the Santa Barbara Channel: 1970–1988. *Journal of Geophysical Research*, **98**, 2119–2133.

Larue, D.K. and Hudleston, P.J. (1987) Foliated breccias in the active Portugese Bend landslide complex, California: bearing on mélange genesis. *Journal of Geology*, **95**, 407–422.

Lasemi, Z., Sandberg, P.A. and Boardman, M.R. (1990) New microstructural criteria for differentiation of compaction and early cementation of fine-grained limestones. *Geology*, **18**, 370–373.

Lash, G. (1985) Accretion-related deformation of an ancient (early Paleozoic) trench-fill deposit, central Appalachian orogen. *Geological Society of America Bulletin*, **96**, 1167–1178.

Lash, G. (1989) Documentation and significance of progressive microfabric changes in middle Ordovician trench mudstones. *Geological Society of America Bulletin*, **101**, 1268–1279.

Laville, E. and Petit, J.P. (1984) Role of syn-sedimentary strike-slip faults in the formation of the Moroccan Triassic basins. *Geology*, **12**, 424–427.

Lawson, D.E. (1979) *Sedimentological Analysis of the Western Terminus Region of the Matanuska Glacier, Alaska*. (CRREL) Cold Regions Research and Engineering Laboratory Report 79-9.

Le Pichon, X., Henry, P. and Lallemant, S. (1990) Water flow in the Barbados accretionary complex. *Journal of Geophysical Research*, **95**, 8945–8969.

Le Pichon, X., Foucher, J.P., Boulegue, J., et al. (1990) Mud volcano field seaward of the Barbados accretionary complex: a submersible survey. *Journal of Geophysical Research*, **95**, 8931–8944.

Lea, P.D. (1990) Pleistocene glacial tectonism and sedimentation on a macrotidal piedmont coast, Ekuk Bluffs, southwestern Alaska. *Geological Society of America Bulletin*, **102**, 1230–1245.

Leach, D.L. and Rowan, E.L. (1986) Genetic link between Ouachita foldbelt tectonism and Mississippi Valley-type lead–zinc deposits of the Ozarks. *Geology*, **14**, 931–935.

Leat, P.T. and Thompson, R.N. (1988) Miocene hydrovolcanism in NW Colorado, USA, fuelled by explosive mixing of basic magma and wet unconsolidated sediment. *Bulletin of Volcanology*, **50**, 229–243.

Leddra, M.J. (1990) Deformation of chalk through compaction and flow. Unpublished PhD thesis, University of London, 370 pp.

Leddra, M.J. and Jones, M.E. (1990) Steady-state flow during undrained loading of the Chalk. *Proceedings of the International Chalk Symposium*, Thomas Telford, London, 117–124.

Leddra, M.J., Goldsmith, A.S. and Jones, M.E. (1990) Compaction and shear deformation of weakly-cemented high-porosity rock, in *The engineering geology of weak rock* (eds J.C. Cripps and C. Moon), *Conference Proceedings of the Geological Society of London*, 45–59.

Leddra, M.J., Petley, D.N. and Jones, M.E. (1992) Fabric changes induced in a cemented shale through consolidation and shear, in *Rock mechanics, Proceedings of the 33rd U.S. Symposium* (eds J.R. Tillerson and W.R.W. Wawersik) A.A. Balkema, Rotterdam, 917–926.

Lee, H.J., Olson, H.W. and von Huene, R. (1973) Physical properties of deformed sediments from site 181. *Initial Reports of the Deep Sea Drilling Project*, **18**, United States Government Printing Office, Washington, DC, p. 897–901.

Leeder, M. (1987) Sedimentary deformation structures and palaeotectonic analysis of sedimentary basins, with a case-study from the Carboniferous of northern England, in *Deformation of Sediments and Sedimentary Rocks* (eds M.E. Jones and R.MF. Preston), *Geological Society of London Special Publication*, **29**, 137–146.

Lerouiel, S. and Vaughan, P.R. (1990) The general and congruent effects of structure in natural soils and rocks. *Geotechnique*, **40**, 467–488.

Lewis, D.W. and Titheridge, D.G. (1978) Small scale sedimentary structures resulting from foot impressions in dune sands. *Journal of Sedimentary Petrology*, **48**, 835–837.

Lewis, K.B. (1971) Slumping on continental slope inclined at 1–4°. *Sedimentology*, **16**, 97–110.

Lin, J-C. (1991) The structural landforms of the coast range of eastern Taiwan, in *Neotectonics and Resources* (eds J. Cosgrove and M.E. Jones), Belhaven, London, 65–74.

Lindsley-Griffin, N., Kemp, A.E.S. and Swartz, J.F. (1990) Vein structures of the Peru margin, Leg 112, *Proceedings of the Ocean Drilling Program, Scientific Results*, **112**, Ocean Drilling Program, College Station, TX, 3–16.

Liu, H. and Sterling, R.L. (1990) Statistical description of the surface roughness of rock joints, in *Rock Mechanics: Contributions and Challenges* (eds. W.A. Mostrulid and G.A. Johnson) Proceedings of the 31st Symposium on Rock Mechanics, Golden, CO, 277–84.

Lingle, C.S. and Brown, T.J. (1987) A subglacial aquifer model and water pressure dependent basal sliding relationship for a West Antarctic ice stream, in *Dynamics of the West Antarctic Ice Sheet* (eds C.J. van der Veen and J. Oerlemans), Riedel Publishers, Dordrecht, 249–285.

Loe, N.K., Leddra. N.M. and Jones, M.E. (1992) Strain states during stress path testing of the Chalk, in *Rock Mechanics, Proceedings of the 33rd U.S. Symposium* (eds J.R. Tillerson and W.R.W. Wawersik), A.A. Balkema, Rotterdam, 927–936.

Logan, J.M. (1979) Brittle phenomena. *Reviews in Geophysics and Space Physics*, **17**, 1121–1132.

Logan, J.M. and Rauenzahn, K.A. (1987) Frictional resistance of gouge mixtures of quartz and montmorillonite on velocity, composition and fabric. *Tectonophysics*, **144**, 87–108.

Logan, W.E. (1863), *Report on the Geology of Canada*. John Lovell (Printer), Montreal, 464 pp.

Longva, O. and Bakkejord, K.J. (1990) Iceberg deformation and erosion in soft sediments, southeastern Norway. *Marine Geology*, **92**, 87–104.

Lorenz, B.E. (1984) Mud–magma interactions in the Dunnage Mélange, Newfoundland, in *Marginal Basin Geology* (eds B.P. Kokelaar and M.F. Howells), *Geological Society of London Special Publication*, **16**, 271–277.

Lorenz, J.C., Teufel, L.W. and Warpinski, N.R. (1991) Regional fractures I: a mechanism for the formation of regional fractures at depth in flat-lying reservoirs. *Bulletin of the American Association of Petroleum Geologists*, **75**, 1714–1737.

Lorenz, J.C., Warpinski, N.R., Teufel, L.W. et al. (1988) Results of the Multiwell experiment – in situ stresses, natural fractures, and other geological controls on reservoirs. *Eos (Transactions of the American Geophysical Union)* **69**, 823–826.

Love, M. and Derbyshire, E. (1985) Microfabric of glacial soils and its quantitative measurement, in *Glacial Tills '85* (ed. M.C. Forde), Engineering Technics Press, Edinburgh, 129–134.

Lowe, D.R. (1975) Water escape structures in coarse grained sediments. *Sedimentology*, **22**, 157–204.

Lowe, D.R. (1976a) Subaqueous liquefied and fluidized sediment flows and their deposits. *Sedimentology*, **23**, 285–308.

Lowe, D.R. (1976b) Grain flow and grain flow deposits. *Journal of Sedimentary Petrology*, **23**, 188–199.

Lowe, D.R. (1979) Sediment gravity flows: their classification and some problems of application to natural flows and deposits, in *Geology of Continental Slopes* (eds L.J. Doyle and O.H. Pilkey) *Society of Economic Paleontologists and Mineralogists Special Publication*, **27**, 75–82.

Lowe, D.R. (1982) Sediment gravity flows II: depositional models with special reference to the deposits of high-density turbidity currents. *Journal of Sedimentary Petrology*, **52**, 279–297.

Lowe, D.R. and Lopiccolo, R.D. (1974) The characteristics and origins of dish and pillar structures. *Journal of Sedimentary Petrology*, **44**, 484–501.

Lu, N.Z., Suhayada, J.N., Prior, D.B., et al. (1991) Sediment thixotropy and submarine mass movement, Huanghe Delta, China. *Geo-Marine Letters*, **11**, 9–15.

Lucas, S.E. and Moore, J.C. (1986) Cataclastic deformation in accretionary wedges: Deep Sea Drilling Project Leg 66, southern Mexico, and on-land examples from Barbados and Kodiak Islands. *Geological Society of America Memoir*, **166**, 89–103.

Luckman, B.H. (1971) The role of snow avalanches in the evolution of alpine talus slopes, in *Slopes: Form and Process* (ed. D. Brunsden), *Institute of British Geographers Special Publication,* **3**, 93–110.

Lund, S.P. and Karlin, R. (1990) Introduction to the special section on physical and biogeochemical processes responsible for the magnetization of sediments. *Journal of Geophysical Research,* **95**, 4353–4354.

Lundberg, N. and Karig, D.E. (1986) Structural features in cores from the Nankai Trough, Deep Sea Drilling Project Leg 87A. *Initial Reports of the Deep Sea Drilling Project,* **87**, US Government Printing Office, Washington, 797–808.

Lundberg, N. and Moore, J.C. (1986) Macroscopic structural features in Deep Sea Drilling Project cores from forearc regions. *Geological Society of America Memoir,* **166**, 13–44.

Luo, D., Macleod, J.E.S., Leng, X. and Smart, P. (1992) Automated orientation analysis of particles of soil microstructures. *Geotechnique,* **42**, 97–107.

Lupini, J.F., Skinner, A.E. and Vaughan, P.R. (1981) The drained residual strength of cohesive soils. *Geotechnique,* **31**, 181–213.

Lyell, C., (1838) *Elements of Geology,* John Murray, London.

Maclintock, P. and Dreimanis, A. (1964) Reorientation of till fabric by overriding glacier in the St. Lawrence valley. *American Journal of Science,* **262**, 133–142.

Magara, K. (1971) Permeability considerations in the generation of abnormal pressures. *Journal of the Society of Petroleum Engineers Paper,* **11**, 236–242.

Magara, K. (1975a) Reevaluation of montmorillonite dehydration as the cause of abnormal pressures and hydrocarbon migration. *Bulletin of the American Association of Petroleum Geologists,* **59**, 292–303.

Magara, K. (1975b) Importance of the aquathermal pressuring effect in the Gulf Coast. *Bulletin of the American Association of Petroleum Geologists,* **59**, 2037–2045.

Magara, K. (1976) Water expulsion from clastic sediments during compaction – directions and volumes. *Bulletin of the American Association or Petroleum Geologists,* **60**, 543–553.

Magara, K. (1980) Comparison of porosity–depth relationships of shale and sandstone. *Journal of Petroleum Geology,* **3**, 175–185.

Magara, K. (1981) Mechanisms of natural fracturing in a sedimentary basin. *Bulletin of the American Association of Petroleum Geologists* **65**, 123–132.

Mahaney, W.C. (1990) Macrofabrics and quartz microstructures confirm glacial origin of Sunnybrook drift in the Lake Ontario basin. *Geology,* **18**, 145–148.

Major, J.J. (1993) Rheometry of natural sediment slurries. *Proceedings of the ASCE National Conference on Hydraulic Engineering, San Francisco, CA* (unpag.)

Major, J.J. and Pierson, T.C. (1990) Rheological analysis of fine-grained natural debris-flow material, in *Hydraulics/Hydrology of Arid Lands. Proceedings of International Symposium, San Diego,* CA, 225–231.

Maliva, R.G. and Siever, R. (1988) Diagenetic replacement controlled by force of crystallization. *Geology,* **16**, 688–691.

Maltman, A.J. (1977) Some microstructures of experimentally deformed argillaceous sediments. *Tectonophysics,* **39**, 417–436.

Maltman, A.J. (1981) Primary bedding-parallel fabrics in structural geology. *Journal of the Geological Society,* **138**, 475–483.

Maltman, A.J. (1984) On the term 'soft-sediment deformation'. *Journal of Structural Geology,* **6**, 589–592.

Maltman, A.J. (1987a) A laboratory technique for investigating the deformation microstructures of water-rich sediments, in *Deformation of Sediments and Sedimentary Rocks* (eds M.E. Jones and R.M.F. Preston), *Geological Society of London Special Publication,* **29**, 71–76.

Maltman, A.J. (1987b) Shear zones in argillaceous sediments – an experimental study, in *Deformation of Sediments and Sedimentary rocks* (eds. M.E. Jones and R.M.F. Preston), *Geological Society of London Special Publication,* **29**, 77–87.

Maltman, A.J. (1987c) Microstructures in deformed sediments, Denbigh Moors, North Wales. *Geological Journal,* **22**, 87–94.

Maltman, A.J. (1988) The importance of shear zones in naturally deformed wet sediments. *Tectonophysics,* **145**, 163–175.

Maltman, A.J. (1994) Prelithification deformation, in *Continental Deformation* (ed. P.L. Hancock), Pergamon Press, Oxford, 143–158.

Maltman, A., Byrne, T., Karig, D.E., Lallemant, S. and the shipboard scientific party (1991) Structural geological evidence from ODP Leg 131 regarding fluid flow in the Nankai prism, Japan. *Earth and Planetary Science Letters,* **109**, 463–8.

Maltman, A.J., Byrne, T., Karig, D.E. and Lallemant, S. (1993a) Deformation at the toe of an active accretionary prism: synopsis of results from ODP Leg 131, Nankai, SW Japan. *Journal of Structural Geology,* **15**, 949–964.

Maltman, A.J., Byrne, T., Karig, D.E., *et al.* (1993b) Deformation structures at Site 808, Nankai accretionary prism, Japan. *Proceedings of the Ocean Drilling Program, Scientific Results,* **131**, Ocean Drilling Program, College Station, TX, 123–134.

Mandl, G. (1988) *Mechanics of Tectonic Faulting: Models and Basic Concepts.* Elsevier, Amsterdam, 407 pp.

Mandl, G., and Crans, W. (1981) Gravitational gliding in deltas, in *Thrust and Nappe Tectonics* (eds K.R. McClay and N.J. Price), *Geological Society of London Special Publication,* **9**, 41–54.

Mandl, G. and Harkness, R.M. (1987) Hydrocarbon migration by hydraulic fracturing, in *Deformation of*

Sediments and Sedimentary Rocks (eds M.E. Jones and R.M.F. Preston), *Geological Society of London Special Publication*, **29**, 39–53.

Manheim, F.T. (1971) The diffusion of ions in unconsolidated sediments. *Earth and Planetary Science Letters*, **9**, 307–309.

March, A. (1932) Mathematische Theorie der Regelungen nach der Korngestalt bei affiner Deformation. *Zeitschrift Kristallographie, Mineralogie und Petrographie*, **1**, 285–297.

Marshall, J.D. (1982) Isotopic composition of displacive fibrous calcite veins: reversals in pore-water composition trends during burial diagenesis. *Journal of Sedimentary Petrology*, **52**, 615–630.

Martill, D.M. and Hudson, J.D. (1989) Injection clastic dykes in the Lower Oxford Clay (Jurassic) of central England: relationship to compaction and concretion formation. *Sedimentology*, **36**, 1127–1133.

Martin, R.G. and Bouma, A.H. (1982) Active diapirism and slope steepening, northern Gulf of Mexico continental slope. *Marine Geotechnology*, **5**, 63–91.

Martinez, R., McVay, M., Bloomquist, D. and Townsend, F.C. (1987) Consolidation of slurried soils. *Proceedings of the 1987 National Conference on Hydraulic Engineering* (ed. R.M. Ragun), 285–290.

Martini, A. (1967) Preliminary experimental studies on frost weathering of certain rock types from the West Sudetes. *Biuletyn Peryglacjalny*, **16**, 147–194.

Martinsen, O.J. (1987) Sedimentology and syndepositional deformation of the Gull Island Formation (Namurian R1), western Irish Namurian Basin, Ireland – with notes on the basin evolution. Unpublished Candidatus Scientarium thesis, Geological Institute, University of Bergen, 327pp.

Martinsen, O.J. (1989) Styles of soft-deformation on a Namurian (Carboniferous) delta slope, Western Irish Namurian Basin, Ireland, in *Deltas: Sites and Traps for Fossil Fuels* (eds M.K.G. Whateley and K.T. Pickering), *Geological Society of London Special Publication*, **41**, 167–177.

Martinsen, O.J. and Bakken, B. (1990) Extensional and compressional zones in slumps and slides in the Namurian of County Clare, Ireland. *Journal of the Geological Society*, **147**, 153–164.

Mascle, A., Bonhold, B. and Renard, V. (1973) Diapiric structures of the Niger Delta. *Bulletin of the American Association of Petroleum Geologists*, **57**, 1672–1678.

Mascle, A., Moore, J.C. and the shipboard scientific party, (1988) *Proceedings of the Ocean Drilling Program, Initial Reports*, **110**, Ocean Drilling Program, College Station, TX, 601 pp.

Mase, C.W. and Smith, L. (1987) Effects of frictional heating on the thermal, hydrogeologic, and mechanical response of a fault. *Journal of Geophysical Research*, **92**, 6249–6272.

Masson, D.G. (1991) Fault patterns at outer trench walls. *Marine Geophysical Researches*, **13**, 209–225.

Masson, D.G., Huggett, Q.J. and Brunsden, D. (1993) The surface texture of the Saharan debris flow deposit and some speculations on submarine debris flow processes. *Sedimentology* (forthcoming).

Masson, D.G., Kidd, R.B., Gardner, J.V., Huggett, Q.V. and Weaver, P.P.E. (1992) Saharan continental rise: facies distribution and sediment slides, in *Geologic Evolution of Atlantic Continental Rises* (eds C.W. Poag and P.C. de Graciansky), van Nostrand & Rheinhold, New York, 327–343.

Mathews, W.H. and MacKay, J.R. (1960) Deformation of soils by glacier ice and the influence of pore-pressure and permafrost. *Royal Society of Canada Transactions*, **54**, 27–36.

Mayne, P.W. (1988) Determining OCR in clays from laboratory strength. *Journal of Geotechnical Engineering*, **114**, 76–92.

Mayne, P.W. and Stewart, H.E. (1988) Pore pressure behavior of K_0 consolidated clay. *Journal of Geotechnical Engineering*, **114**, 1340–1346.

McClelland, B. (1967) Progress of consolidation in delta front and prodelta clays of the Mississippi River, in *Marine Geotechnique* (ed. A. Richards), University of Illinois Press, Urbana, 22–40.

McCrossan, R.G. (1958) Sedimentary 'boundinage' structure in the Upper Devonian Ireton Formation of Alberta. *Journal of Sedimentary Petrology*, **28**, 316–320.

McEwen, A.S. (1989) Mobility of large rock avalanches: evidence from Valles Marineris, Mars. *Geology*, **17**, 1111–1114.

McGarr, A. (1980) Some constraints on levels of shear stress in the crust from observations and theory. *Journal of Geophysical Research*, **85**, 6231–6238.

McGarr, A. (1988) On the state of lithospheric stress in the absence of applied tectonic forces. *Journal of Geophysical Research*, **96**, 13609–13617.

McGown, A., Marsland, A., Radwan, A.M. and Gabr, A.W.A. (1980) Recording and interpreting soil macrofabric data. *Geotechnique*, **30**, 417–47.

McKean, J.A., Dietrich, W.E., Finkel, R.C., Southon, J.R. and Caffee, M.W. (1993) Quantification of soil production and downslope creep rates from cosmogenic ^{10}Be accumulations on a hillslope profile. *Geology*, **21**, 343–346.

McKee, E.D. (ed.) (1979) A study of global sand seas. *United States Geological Survey Professional Paper*, **1052**, 429 pp.

McKee, E.D. and Bigarella, J.J. (1972) Deformational structures in Brazilian coastal dunes. *Journal of Sedimentary Petrology*, **42**, 670–681.

Means, W.D. (1987) A newly recognised type of slickenside striation. *Journal of Structural Geology*, **9**, 585–590.

Mehta, A.J. (1991) Understanding fluid mud in a dynamic enviroment. *Geo-Marine Letters*, **11**, 113–118.

Meier, M.F. (1989) Relation between water input, basal water pressure, and sliding of Columbia Glacier, Alaska, U.S.A. (abstr.). *Annals of Glaciology*, **12**, 214–215.

Meier, M.F. and Post, A. (1969) What are glacier surges? *Canadian Journal of Earth Sciences*, **6**, 807–817.

Mellor, M. (1978) Dynamics of snow avalanches, in *Rockslides and Avalanches I: Natural Phenomena* (ed. B. Voigt), *Developments in Geotechnical Engineering*, **14A**, Elsevier, Amsterdam, 753–792.

Menzies, J. (1981) Temperatures within subglacial debris – a gap in our knowledge. *Geology*, **9**, 271–273.

Menzies, J. (1989a) Drumlins – products of controlled or uncontrolled dynamic response? *Quaternary Science Reviews*, **8**, 151–158.

Menzies, J. (1989b) Subglacial hydraulic conditions and their possible impact on subglacial bed deformation. *Sedimentary Geology*, **62**, 125–150.

Menzies, J. and Maltman, A.J. (1992) Microstructures in diamictons – evidence of subglacial bed conditions. *Geomorphology*, **6**, 27–40.

Mesri, G. and Mayat, T.M. (1993) The coefficient of earth pressure at rest. *Canadian Geotechnical Journal*, **30**, 647–666.

Middleton, G.V. (1967) Experiments on density and turbidity currents. III. Deposition of sediment. *Canadian Journal of Earth Sciences*, **4**, 475–505.

Middleton, G.V. (1970) Experimental studies related to problems of flysch sedimentation, in *Flysch Sedimentology in North America* (ed. J. Lajoie), *Geological Association of Canada Special Paper*, **7**, 253–272.

Middleton, G.V. and Hampton, M.A. (1973) Sediment gravity flows: mechanics of flow and deposition, in *Turbidites and Deep-water Sedimentation, Society of Economic Paleontologists and Mineralogists Pacific Section, Short Course Lecture Notes*, Los Angeles, CA.

Middleton, G.V. and Hampton, M.A. (1976) Subaqueous sediment transport and deposition by sediment gravity flows, in *Marine Sediment Transport and Environmental Management* (eds D.J. Stanley and D.J.P. Swift), John Wiley & Sons, New York, 197–218.

Middleton, G.V. and Southard, J.B. (1978) Mechanics of sediment movement. *Society of Economic Paleontologists and Mineralogists Short Course*, **3**, Tulsa, OK.

Milici, R.C., Gathright, T.M., Gwin, M.R. and Miller, B.W. (1982) Subtle bedding plane faults: a major factor of coal mine roof falls in southwestern Virginia. *Geological Society of America, Northeastern and Southeastern Combined Section Meetings, Abstracts with Programs*, **14**, 65.

Miller, W.J. (1908) Highly folded strata between non-folded strata at Trenton Falls, N.Y. *Journal of Geology*, **16**, 428–433.

Miller, W.J. (1909) Geology of the Remsen Quadrangle, including Trenton Falls and vicinity, in Oneida and Herkimer Counties. *New York State Museum Bulletin*, **126**, 5–51.

Miller, W.J. (1915) Notes on the intraformational contorted strata at Trenton Falls. *New York State Museum Bulletin*, **177**, 135–143.

Miller, W.J. (1922) Intraformational corrugated rocks. *Journal of Geology*, **30**, 587–610.

Milligan, V. (1976) Geotechnical aspects of glacial tills, in *Glacial Till* (ed. R.F. Leggett), *Royal Society of Canada Special Publication*, **12**, 269–291.

Mills, P.C. (1983) Genesis and diagnostic value of soft-sediment deformation structures – review. *Sedimentary Geology*, **35**, 83–104.

Milodowski, A.E. and Zalasiewicz, J.A. (1991) The origin and sedimentary, diagenetic and metamorphic evolution of chlorite–mica stacks in Llandovery sediments of central Wales, U.K. *Geological Magazine*, **128**, 263–278.

Minell, H. (1973) An investigation of drumlins in the Narvik area of Norway. *Bulletin of the Geological Institutions of the University of Uppsala*, **5**, 133–138.

Minshull, T.A. and White, R. (1989) Sediment compaction and fluid migration in the Makran accretionary prism. *Journal of Geophysical Research*, **94**, 7387–7402.

Mitchell, J.K. (1976) *Fundamentals of Soil Behavior*. John Wiley and Sons, Inc., New York, 422 pp.

Moeyersons, J. (1989) A possible causal relationship between creep and sliding on Rwaza Hills, Southern Rwanda. *Earth Surface Processes and Landforms*, **14**, 597–614.

Mohl, K.L. and Bakken, W.E. (1968) Geologic road log. *Twentieth Annual Field Conference, Wyoming Geological Association Guidebook*, 215–225.

Montenat, C., Barrier, P. and Ott d'Estevou, P. (1991) Some aspects of the Recent tectonics in the Strait of Messina, Italy. *Tectonophysics*, **194**, 203–215.

Moon, C.F. and Hurst, C.F. (1984) Fabrics of muds and shales: an overview, in *Fine Grained Sediments: Deep-water Processes and Facies* (eds D.A.V. Stow and D.J.W. Piper), *Geological Society Special Publication*, **15**, 579–593.

Mooney, W.D. and Ginzburg, A. (1986) Seismic measurements of the internal properties of fault zones. *Pure and Applied Geophysics*, **124**, 141–157.

Moore, C.A. and Krepfl, M. (1991) Using fractals to model soil fabric. *Geotechnique*, **41**, 123–134.

Moore, D.G. (1978) Submarine slides, in *Rockslides and Avalanches I: Natural Phenomena*, (ed. B. Voight), *Developments in Geotechnical Engineering*, **14A**, Elsevier, Amsterdam, pp. 563–604.

Moore, G.F. and Karig, D.E. (1980) Structural geology of Nias Island, Indonesia: implications for subduction zone tectonics. *American Journal of Science*, **280**, 193–223.

Moore, G.F., Shipley, T.H., Stoffa, P.L., *et al.* (1990) Structure of the Nankai Trough accretionary zone from multichannel seismic reflection data. *Journal of Geophysical Research*, **95**, 8753–8765.

Moore, G.W. (1980) Slickensides in deep sea cores near the Japan Trench, Deep Sea Drilling Project Leg 57. *Initial Reports of the Deep Sea Drilling Project*, vol. 56–57, part 2, United States Government Printing Office, Washington, DC, pp. 1107–1116.

Moore, J.C. and Byrne, T. (1987) Thickening of fault zones: a mechanism of mélange formation in accreting sediments. *Geology*, **15**, 1040–1043.

Moore, J.C. and Geigle, J.E. (1974) Slaty cleavage: incipient occurrences in the deep sea. *Science*, **183**, 509–510.

Moore, J.C. and Karig, D.E. (1976) Sedimentology, structural geology, and tectonics of the Shikoku subduction zone, southwestern Japan. *Geological Society of America Bulletin*, **87**, 1259–1268.

Moore, J.C. and Lundberg, N. (1986) Tectonic overview of Deep Sea Drilling Project transects of forearcs. *Geological Society of America Special Paper*, **166**, 1–12.

Moore, J.C. and Vrolijk, P. (1992) Fluids in accretionary prisms. *Reviews in Geophysics*, **30**, 113–135.

Moore, J.C. and Wheeler, R. (1978) Structural fabric of a mélange, Kodiak Island, Alaska. *American Journal of Science*, **278**, 739–765.

Moore, J.C., Roeske, S., Lundberg, N., *et al.* (1986) Scaly fabrics from Deep Sea Drilling Project cores from forearcs. *Geological Society of America Memoir*, **166**, 55–73.

Moore, J.C., Mascle, A., *et al.* (1988) Tectonics and hydrogeology of the northern Barbados Ridge: results from Ocean Drilling Program Leg 110. *Geological Society of America Bulletin*, **100**, 1578–1593.

Moore, J.C., Orange, D.L. and Kulm, L.D. (1991) Interrelationship of fluid venting and structural evolution, Oregon margin. *Journal of Geophysical Research*, **95**, 8795–8808.

Moore, J.G., Clague, D.A., Holcomb, R.T., *et al.* (1989) Prodigious submarine landslides on the Hawaiian Ridge. *Journal of Geophysical Research*, **94**, 17465–17484.

Moore, J.C., Brown, K.M., Horath, F., *et al.* (1991) Plumbing accretionary prisms: effects of permeability variations. *Philosophical Transactions of the Royal Society of London*, **335**, 275–288.

Moore, R. (1991) The chemical and mineralogical controls upon the residual strength of pure and natural clays. *Geotechnique*, **41**, 35–47.

Moore, T.C., van Andel, T.H., Blow, W.H. and Heath, G.R. (1970) Large submarine slide off northeastern continental margin of Brazil. *Bulletin of the American Association of Petroleum Geologists*, **54**, 125–128.

Morgan, J.K. and Karig, D.E. (1993) Ductile strains in clay-rich sediments from Hole 808C: preliminary results using x-ray pole figure goniometry. *Proceedings of the Ocean Drilling Program, Scientific Results*, **131**, Ocean Drilling Program, College Station, TX, 141–155.

Morgan, J.K. and Karig, D.E., in press. Diffuse deformation within the toe of the eastern Nankai accretionary prism: balanced and restored sections based on numerical strain calculations. *Journal of Structural Geology*.

Morgan, J.K., Karig, D.E. and Maniatty, A., in press. The estimation of seismically diffuse strains in the toe of the Nankai accretionary prism: a numerical solution. *Journal of Geophysical Research*.

Morgan, J.P., Coleman, J.M. and Gagliano, S.M. (1968) Mudlumps: diapiric structures in Mississippi delta sediments, in *Diapirism and Diapirs* (eds J. Braunstein and G.D. O'Brien), *Memoir of the American Association of Petroleum Geologists*, **8**, 145–162.

Morgenstern, N.R. and Tchalenko, J.S. (1967) Microscopic structures in kaolin subjected to direct shear. *Geotechnique*, **17**, 309–328.

Morin, R.H. (1992) *In situ* measurements of fluid flow in DSDP Holes 395A and 534A; results from the Dianaut Program. *Geophysical Research Letters*, **19**, 509–512.

Morin, R. and Silva, A.J. (1984) The effects of high pressure and high temperature on some physical properties of ocean sediments. *Journal of Geophysical Research*, **89**, 511–526.

Moriya, I. (1980) 'Bandian Eruption' and landforms associated with it. *Collection of Articles in Memory of Retirement of Prof. K. Nishimura from Tohoku University, Japan*, 214–219.

Morrow, C.A. and Byerlee, J.D. (1989) Experimental studies of compaction and dilatancy during frictional sliding on faults containing gouge. *Journal of Structural Geology*, **7**, 815–825.

Morrow, C.A., Shi, L.Q. and Byerlee, J.D. (1981) Permeability and strength of fault gouge under high pressure. *Geophysical Research Letters*, **8**, 325–328.

Morrow, C.A., Shi, L.Q. and Byerlee, J.D. (1982) Strain hardening and strength of clay-rich fault gouges. *Journal of Geophysical Research*, **87**, 6771.

Morrow, C.A., Shi, L.Q. and Byerlee, J.D. (1984) Permeability of fault gouge under confining pressure and shear stress. *Journal of Geophysical Research*, **89**, 3193–3200.

Morton, R.A. (1993) Attributes and origins of ancient submarine slides and filled embayments: examples from the Gulf Coast basin. *Bulletin of the American Association of Petroleum Geologists*, **77**, 1064–1081.

Mozley, P.S. and Burns, S.J. (1993) Oxygen and carbon isotope compositions of marine carbonate concretions: an overview. *Journal of Sedimentary Petrology*, **63**, 73–83.

Mudford, B.S. (1990) A one-dimensional two phase model of overpressure generation in the Venture gas field, offshore Nova Scotia. *Bulletin of Canadian Petroleum Geology*, **38**, 246–250.

Mulhaus, H-B. and Vardoulakis, I. (1987) The thickness of shear bands in granular materials. *Geotechnique*, **37**, 271–283.

Müller, L. (1964) The rock slide in the Vaiont valley. *Rock Mechanics and Engineering Geology*, **2**, 148–212.

Müller, L. (1968) New considerations on the Vaiont slide. *Felsmechanischen Ingenieurgeologie*, **6**, 1–91.

Mullins, H.T. and Cook, H.E. (1986) Carbonate apron models: alternative to the submarine fan model for paleoenvironmental analysis and hydrocarbon exploration. *Sedimentary Geology*, **48**, 37–79.

Murchison, R.I. (1829) On the coal-field of Brora in Sutherlandshire, and some other stratified deposits in the north of Scotland. *Transactions of the Geological Society, Second Series*, **2**, 293–326.

Murdoch, L.C. (1992) Hydraulic fracturing of soil during laboratory experiments. Part 1. Methods and observations. Part 2. Propagation. Part 3. Theoretical analysis. *Geotechnique*, **43**, 255–287.

Murray, T. (1990) Deformable glacier beds: measurement and modelling. Unpublished PhD thesis, University of Wales, Aberystwyth, 321 pp.

Murray, T. and Dowdeswell, J.A. (1992) Water throughflow and the effects of deformation on sedimentary glacier beds. *Journal of Geophysical Research*, **97**, 8993–9002.

Mutti, E. and Ricci-Lucchi, F. (1972) Le torbiditi dell'Appennino settentrionale: introduzione all'analisi di facies. *Geological Society of Italy Memoir*, **11**, 161–199.

Nagaraj, T.S., Vatsala, A. and Srinivasa Murthy, B.R. (1990) Change in pore size distribution due to consolidation of clays. *Geotechnique*, **40**, 303–309.

Nardin, T.R., Hein, F.J., Gorsline, D.S. and Edwards, B.D. (1979) A review of mass movement processes, sediment and acoustic characteristics, and contrasts in slope and base-of-slope systems versus canyon-fan–basin-floor systems, in *Geology of Continental Slopes* (eds L.H. Doyle and O.H. Pilkey) *Society of Economic Paleontologists and Mineralogists Special Publication*, **27**, 61–73.

Naylor, M.A. (1982) The Casanova complex of the Northern Apennines: a mélange formed on a distal passive continental margin. *Journal of Structural Geology*, **4**, 1–18.

Needham, R.S. (1978) Giant-scale hydroplastic deformation structures formed by the loading of basalt onto water-saturated sand, middle Proterozoic, Northern Territory, Australia. *Sedimentology*, **25**, 285–294.

Nelson, R.B. and Lindsley-Griffin, N. (1987) Biopressured carbonate turbidite sediments: a mechanism for submarine slumping. *Geology*, **15**, 817–820.

Nemčok, A. (1972) Gravitational slope deformation in high mountains. *Proceedings 24th International Geology Conference, Section 13, Montreal*, 132–141.

Nemec, W. (1990) Aspects of sediment movement on steep delta slopes, in *Coarse-grained Deltas* (eds A. Colella and D.B. Prior), *Special Publications of the International Association of Sedimentologists*, **10**, 29–73.

Nemec, W. and Steel, R.J. (eds) (1988) *Fan Deltas – sedimentology and Tectonic Setting*. Blackie, Glasgow, 444 pp.

Nemec, W. and Steel, R.J., Porębski, S.J. and Spinnangr, Å. (1984) Dorrba Conglomerate, Devorian, Norway: process and lateral variability in a mass flow-dominated, lacustrine far-delta, in *Sedimentology of Gravels and Conglomerates* (eds E.H. Koster and R.J. Steel), Canadian Society of Petroleum Geologists, Memoir, **10**, 295–320.

Neumann, M.P. (1976) Recent sand volcanoes in the sand of a dyke under construction. *Sedimentology*, **23**, 421–425.

Newton, S.K. and Flanagan, K.P. in press, The Alba Field: evolution of the depositional model, in *Petroleum geology of NW Europe – Proceedings of the 4th Conference*. Geological Society, London.

Nickling, W.G. and Bennett, L. (1984) The shear strength characteristics of frozen coarse granular debris. *Journal of Glaciology*, **83**, 348–357.

Nielsen, D.M. and Johnson, A.I. (eds.) (1990) Ground water and vadose zone monitoring. *American Society for Testing and Materials, Special Technical Publication*, D-18.21, 313 pp.

Nocita, B.W. (1988) Soft-sediment deformation (fluid escape) features in a coarse grained pyroclastic-surge deposit, north-central New Mexico. *Sedimentology*, **35**, 275–285.

Nur, A. and Byerlee, J.D. (1971). An exact effective stress law for elastic deformation of rocks with fluid. *Journal of Geophysical Research*, **76**, 6414–6419.

Nye, J.F. (1957) *Physical Properties of Crystals*. Oxford University Press, Oxford, 322 pp.

Nye, J.F. (1976) Water flow in glaciers: jokulhlaups, tunnels and veins. *Journal of Glaciology*, **76**, 181–207.

O'Brian, G. (1968) Survey of diapirs and diapirism. *Memoir of the American Association of Petroleum Geologists*, **8**, 1–9.

O'Connor, B. (1980) An axial-plane cleavage in some slump-like folds. *Geological Society of America, Abstracts with Programs*, **12** (5), 253.

O'Neil, J.R. (1985) Water–rock interactions in fault gouge. *Pure and Applied Geophysics*, **122**, 440–446.

Ogawa, Y. and Miyata, Y. (1985) Vein structure and its deformational history in the sedimentary rocks of the Middle-America Trench slope off Guatemala. *Initial Reports of the Deep Sea Drilling Project*, **84**, United States Government Printing Office, Washington, DC, 811–829.

Ogawa, Y., Ashi, J. and Fujioka, K. (1992) Vein structures and their tectonic implications for the development of the Izu–Bonin forearc, Leg 126. *Proceedings of the Ocean Drilling Program, Scientific Results*, **126**, Ocean Drilling Program, College Station, TX, 195–207.

Ohlmacher, G.C. and Baskerville, C.A. (1991) Landslides on fluidlike zones in the deposits of glacial Lake Hitchcock, Windsor county, Vermont. *Bulletin of the Association of Engineering Geologists*, **28**, 31–43.

Okada, H. and Whittaker, J.H. McD. (1979) Sand volcanoes of the Paleogene Kumage Group, Tanegashima, southwest Japan. *Journal of the Geological Society of Japan*, **85**, 187–196.

Okusa, S. (1985) Measurements of wave-induced pore pressure in submarine sediments under various marine conditions. *Marine Geotechnology*, **6**, 119–144.

Orange, D.L. and Breen, N.A. (1992) The effects of fluid escape on accretionary wedges, 2, seepage force, slope failure, headless submarine canyons and vents. *Journal of Geophysical Research*, **97**, 9277–9295.

Orange, D., Geddes, D. and Moore, J.C. (1993) Structural and fluid evolution of a young accretionary complex: the Hoh rock assemblage of the western Olympic Peninsula, Washington. *Geological Society of America Bulletin*, **105**, 1053–75.

Osipov, V.I., Nikolaeva, S.K. and Sokolov, V.N. (1984) Microstructural changes associated with thixotropic phenomena in clay soils. *Geotechnique*, **34**, 293–303.

Ostry, R.C. and Deane, R.E. (1963) Microfabric analyses of till. *Geological Society of America Bulletin*, **74**, 165–168.

Owen, G. (1987) Deformation processes in unconsolidated sands, in *Deformation of Sediments and Sedimentary Rocks* (eds M.E. Jones and R.M.F. Preston), *Geological Society of London Special Publication*, **29**, 11–24.

Owen, L.A. (1988a) Terraces, uplift and climate, Karakoram Mountains, northern Pakistan. Unpublished PhD thesis, University of Leicester, 399 pp.

Owen, L.A. (1988b) Wet-sediment deformation of Quaternary and recent sediments in the Sardu Basin, Karakoram Mountains, in *Glaciotectonics: Proceedings of the Glaciotectonics Working Group* (ed. D.G. Croot), A.A. Balkema, Rotterdam, 123–142.

Owen, L.A. (1989) Neotectonics and glacial deformation in the Karakoram Mountains and Nanga Parbat Himalaya. *Tectonophysics*, **163**, 227–265.

Owen, L.A. and Derbyshire, E. (1988) Glacially deformed diamictons in the Karakoram Mountains, northern Pakistan, in *Glaciotectonics: Forms and Processes* (ed. D.G. Croot), A.A. Balkema, Rotterdam, 149–176.

Owens, W.H. (1993) Magnetic fabric studies of samples from Hole 808C, Nankai Trough. *Proceedings of the Ocean Drilling Program, Scientific Results*, **131**, Ocean Drilling Program, College Station, TX, 301–310.

Palmer, B.A. and Neall, V.E. (1991) Contrasting lithofacies architecture in ring-plain deposits related to edifice construction and destruction, the Quaternary Stratford and Opunake Formations, Egmont Volcano, New Zealand. *Sedimentary Geology*, **74**, 71–88.

Palmer, S.N. and Barton, M.E. (1986) Avoiding microfabric disruption during impregnation of friable, uncemented sands with dyed epoxy. *Journal of Sedimentary Petrology*, **56**, 556–557.

Palmer, S.N. and Barton, M.E. (1987) Porosity reduction, microfabric, and resultant lithification in UK uncemented sands, in *Diagenesis of Sedimentary Sequences* (ed. J.D. Marshall). *Geological Society of London Special Publication*, **36**, 29–40.

Pane, V. and Schiffman, R.L. (1985) A note on sedimentation and consolidation. *Geotechnique*, **35** 69–72.

Pantin, H.M. (1979) Interaction between velocity and effective density in turbidity flow: phase plane analysis, with criteria for autosuspension. *Marine Geology*, **31**, 59–99.

Parker, J.C. (1989) Multiphase flow and transport in porous media. *Reviews of Geophysics*, **27**, 311–328.

Parker, W.R. (1991) Quality control in mud coring. *Geo-Marine Letters*, **11**, 132–137.

Parry, R. (1960) Undrained shear strength in clays. *Proceedings of the First Australian – New Zealand Conference on Geomechanics*, 11–15.

Peterson, W.S.B. (1976) Vertical strain-rate measurements in an Arctic ice cap and deductions from them. *Journal of Glaciology*, **75**, 3–12.

Paul, M.A. and Eyles, N. (1990) Constraints on the preservation of diamict facies (melt-out tills) at the margins of stagnant glaciers. *Quaternary Science Reviews*, **9**, 51–69.

Paull, C.K. and Neumann, A.C. (1987) Continental margin brine seeps: their geological consequences. *Geology*, **15**, 545–548.

Pavlides, S.B. (1989) Looking for a definition of neotectonics. *Terra Nova*, **1**, 233–235.

Peacock, S.M. (1989) Numerical constraints on rates of metamorphism, fluid production, and fluid flux, during regional metamorphism. *Geological Society of America Bulletin*, **101**, 476–485.

Pedersen, S.A.S. (1987) Comparative studies of gravity tectonics in Quaternary sediments and sedimentary rocks related to fold belts, in *Deformation of Sediments and Sedimentary Rocks* (eds M.E. Jones and R.M.F. Preston), *Geological Society of London Special Publication*, **29**, 165–180.

Perla, R.I. (1978) Failure and snow slopes, in *Rockslides and Avalanches I: Natural Phenomena* (ed. B. Voight), *Developments in Geotechnical Engineering*; **14A**, Elsevier, Amsterdam, 731–52.

Petit, J.P. and Laville, E. (1987) Morphology and microstructures of hydroplastic slickensides in sandstone, in *Deformation of Sediments and Sedimentary Rocks* (eds M.E. Jones and R.M.F. Preston), *Geological Society of London Special Publication*, **29**, 107–121.

Petley, D.N., Jones, M.E., Leddra, M.J. and Kageson-Loe, N. in press, On changes in fabric and pore geometry due to compaction and shear deformation of weak North Sea sedimentary rocks, in *North Sea Oil and Gas Reservoirs III*, Graham & Trotman, London.

Phillips, C.J. and Davies, T.R.H. (1991) Determining rheological parameters of debris flow material. *Geomorphology*, **4**, 101–110.

Pickering, D.J. (1970) Anisotropic elastic parameters for soil. *Geotechnique* **20**, 271–6.

Pickering, K.T. (1983) Small scale syn-sedimentary faults in the Upper Jurassic 'Boulder Beds'. *Scottish Journal of Geology*, **19**, 169–181.

Pickering, K.T., Agar, S.M. and Prior, D.J. (1990) Vein structure and the role of pore fluids in early wet-sediment deformation; Late Miocene volcaniclastic rocks, Miura Group, SE Japan, in *Deformation Mechanisms, Rheology and Tectonics* (eds R.J. Knipe and E. Rutter), *Geological Society of London Special Publication*, **54**, 417–430.

Pickering, K.T. and Taira, A. (1994) Tectonosedimentation: with examples from the Tertiary-Recent of southeast Japan, in Continental Deformation (ed. P.L. Hancock), Pergamon Press, Oxford, 320–354.

Pickering, K.T., Stow, D.A.V., Watson, M. and Hiscott, R.N. (1986) Deep-water facies, processes and models: a review and classification scheme for modern and ancient sediments. *Earth Science Reviews*, **23**, 75–174.

Pierson, T.C. and Costa, J.E. (1987) A rheologic classification of subaerial sediment–water flows, in *Debris Flows/Avalanches; Process, Recognition, and Mitigation* (eds J.E. Costa and G.F. Wieczorek), *Geological Society of America, Reviews in Engineering Geology*, **VIII**, 1–12.

Pierson, T.C. and Scott, K.M. (1985) Downstream dilution of a lahar; transition from debris flow to hyperconcentrated streamflow. *Water Resources Research*, **21**, 1511–1524.

Piper, D.J.W. (1978) Turbidite muds on deep-sea fans and abyssal plains, in *Sedimentation in Submarine Canyons, Fans and Trenches* (eds D.J. Stanley and G. Kelling), Dowden, Hutchinson and Ross, Stroudsburg, PA, 163–175.

Pittman, E.D. (1981) Effect of fault-related granulation on porosity and permeability of quartz sandstones, Simpson Group (Ordovician), Oklahoma. *Bulletin of the American Association of Petroleum Geologists*, **65**, 2381–2387.

Pittman, E.D. and Larese, R.E. (1991) Compaction of lithic sands: experimental results and applications. *American Association of Petroleum Geologists Bulletin*, **75**, 1279–99.

Pittman, E.D. and Reynolds, R.C. (1989) The thermal transformation of smectite to illite, in *Thermal History of Sedimentary Basins: Methods and Case Histories* (eds N.D. Naeser and T.H. McCulloh), Springer-Verlag, New York, 33–40.

Plafker, G. and Ericksen, G.E. (1978) Nevado Huascarán avalanches, Peru, in *Rockslides and Avalanches, I: Natural Phenomena* (ed. B. Voight), *Developments in Geotechnical Engineering*, **14A**, Elsevier, Amsterdam, 277–314.

Plumb, R.A., Evans, K.F. and Engelder, T. (1991) Geophysical log responses and their correlation with bed-to-bed stress contrasts in Paleozoic rocks, Appalachian Plateau, New York. *Journal of Geophysical Research*, **96**, 14509–14528.

Plumley, W.J. (1980) Abnormally high fluid pressure – survey of some basic principles. *Bulletin of the American Association of Petroleum Geologists*, **51**, 1240–1254.

Pollard, D.D. and Aydin, A. (1988) Progress in understanding jointing over the past century. *Geological Society of America Bulletin*, **100**, 1181–1204.

Postma, G. (1984) Mass-flow conglomerates in a submarine canyon: Abrioja fan-delta, Pliocene, southeast Spain, in *Sedimentology of Gravels and Conglomerates* (eds E.H. Koster and R.J. Steel), *Canadian Society of Petroleum Geologists Memoir*, **10**, 237–258.

Postma, G., Nemec, W. and Kleinspehn, K. (1988) Large floating clasts in turbidites: a mechanism for their emplacement. *Sedimentary Geology*, **58**, 47–61.

Potts, D.M., Jones, M.E. and Berget, O.P. (1988) Subsidence above the Ekofisk Oil Reservoirs, in *Proceedings of the International Conference on Behaviour of Offshore Structures*, **1**: Geotechnics, Tapir, Trondheim, Norway, pp. 113–127.

Poulos, H.G. (1988) *Marine Geotechnics*. Unwin Hyman, London, 269 pp.

Powley, D.E. (1990) Pressures and hydrogeology in petroleum basins. *Earth Science Reviews*, **29**, 215–226.

Press, F. and Siever, R. (1978) *Earth*, 2nd edn, W.H. Freeman and Company, San Francisco, 656 pp.

Price, N.J. (1974) The development of stress systems and fracture patterns in undeformed sediments. *Advances in Rock Mechanics, Proceedings 3rd International Conference, Society of Rock Mechanics, Denver, CO*, **1A**, 487–498.

Price, N.J. and Cosgrove, J.W. (1990) *Analysis of Geological Structures*, Cambridge University Press, Cambridge, 502 pp.

Prior, D.B. and Bornhold, B.D. (1990) The underwater development of Holocene fan deltas. In: *Coarse-grained Deltas* (eds A. Colella and D.B. Prior), *Special Publications of the International Association of Sedimentologists*, **10**, Blackwell Scientific Publications, Oxford, 79–90.

Prior, D.B. and Coleman, J.M. (1978a) Submarine landslides on the Mississippi River delta-front slope. *Geoscience and Man*, **18**, 41–53.

Prior, D.B. and Coleman, J.M. (1978b) Disintegrating retrogressive landslides on very low-angle subaqueous slopes, Mississippi delta. *Marine Geotechnology*, **3**, 37–60.

Prior, D.B. and Coleman, J.M. (1982) Active slides and flows in underconsolidated marine sediments on the slopes of the Mississippi delta, in *Marine Slides and Other Mass Movements* (eds S. Saxov and J.K. Nieuwenhuis), NATO Workshop, Plenum Press, New York, 21–49.

Prior, D.B., Bornhold, B.D. and Johns, M.W. (1984) Depositional characteristics of a submarine debris flow. *Journal of Geology*, **92**, 707–727.

Prior, D.B., Coleman, J.M., Suhayada, J.N. and Garrison, L.E. (1981) Subaqueous landslides as they affect bottom structures, in *Offshore Geologic Hazards* (eds A. Bouma, D.A. Sangrey, J.M. Coleman *et al.*), *American Association of Petroleum Geologists Education Course Note Series*, **18** (Appendix 5-4), 5/112–5/134.

Prior, D.J. and Behrmann, J.H. (1990a) Backscatter imagery of fine-grained sediments from Hole 671B, ODP Leg 110; Preliminary results. *Proceedings of the Ocean Drilling Program, Scientific Results* **110**, Ocean Drilling Program, College Station, TX, 257–275.

Prior, D.J. and Behrmann, J.H. (1990b) Thrust-related mudstone fabrics from the Barbados Forearc: a backscattered scanning electron microscope study. *Journal of Geophysical Research*, **95**, 9055–9068.

Pulham, A. (1989) Controls on internal structure and architecture of sandstone bodies within Upper Carboniferous fluvial-dominated deltas, County Clare, western Ireland, in *Deltas – sites and traps of fossil fuels* (eds M.H. Whateley and K.T. Pickering), *Geological Society of London Special Publication*, **41**, 179–203.

Pytte, A.M. and Reynolds, R.C. (1989) The thermal transformation of smectite to illite, in *Thermal History of Sedimentary Basins, Methods and Case Histories* (eds N.D. Naesar and T.H. McCulloh) US Geological Survey, Denver, CO, 133–40.

Radbruch-Hall, D.H. (1978) Gravitational creep of rock masses on slopes, in *Rockslides and Avalanches I: Natural Phenomena* (ed. B. Voight), *Developments in Geotechnical Engineering*, **14A**, Elsevier, Amsterdam, 607–657.

Rafalovich, A. and Chaney, R.C. (1991) Correlation between OCR and depth for deep-sea sediments. *Journal of Geotechnical Engineering*, **117**, 1744–1749.

Railsback, L.B. (1993) Lithologic controls on morphology of pressure-dissolution surfaces (stylolites and dissolution seams) in Palaeozoic carbonate rocks from the mideastern United States. *Journal of Sedimentary Petrology*, **63**, 513–522.

Ramberg, H. (1981) *Gravity, deformation and the Earth's crust: in theory, experiments and geological application.* 2nd edition. Academic Press, London, 425 pp.

Ramsay, J.G. (1967) *Folding and Fracturing of Rocks.* McGraw-Hill, New York, 568 pp.

Rapp, A. (1960) Recent developments of mountain slopes in Karkevagge and surroundings. *Geografiska Annaler*, **42**, 1–158.

Rascoe, B. (1975) Tectonic origin of preconsolidation deformation in upper Pennsylvanian rocks near Bartlesville, Oklahoma. *Bulletin of the American Association of Petroleum Geologists*, **59**, 1626–1638.

Raymond, C.F. (1987) How do glaciers surge? A review. *Journal of Geophysical Research*, **92**, 9121–9134.

Raymond, L. (ed.) (1984) Mélanges: their nature, origin, and significance. *Geological Society of America Special Paper*, **198**, 170 pp.

Read, W.F. (1988) Deformation in soft Ordovician sediment produced by large masses of sandstone from a nearby, but as yet undiscovered, impact crater. *51st Annual Meeting of the Meteoritical Society, Lunar and Planetary Institute, Houston, TX*, 665, p. D-3.

Reid, M.E. and Baum, R.L. (1992) Mechanism of rainfall-induced pore-pressure increase in a clayey landslide (abstr.) *Geological Society of America Abstracts with Programs*, **24**, 203.

Reid, M.E., Nielsen, H.P. and Dreiss, S.J. (1988) Hydrologic factors triggering a shallow hillslope failure. *Bulletin of the Association of Engineering Geologists*, **25**, 349–361.

Reimnitz, E.R. and Marshall, N.F. (1965) The effects of the Alaska earthquake and tsunami on Recent deltaic sediments. *Journal of Geophysical Research*, **70**, 2363–2376.

Reynolds, O. (1885) On the dilatancy of media composed of rigid particles in contact. *The London, Edinburgh and Dublin Philosophical Magazine and Journal of Science*, **20**, 468–481.

Reynolds, S. and Gorsline, D.S. (1992) Clay microfabric of deep sea detrital mud(stones), California continental borderland. *Journal of Sedimentary Petrology*, **62**, 41–53.

Rezak, R. and Lavoie, D.L. (1990) Consolidation-related fabric changes of periplatform sediments. *Geo-Marine Letters*, **10**, 101–109.

Rhett, D.W. (1990) Long term effects of water injection on strain in North Sea Chalks. *Proceedings of the Third North Sea Chalk Symposium, Copenhagen.*

Rice, J.R. (1975) On the stability of dilatant hardening for saturated rock masses. *Journal of Geophysical Research*, **80**, 1531–1537.

Rice, J.R. (1992) Fault stress states, pore pressure distributions and weakness of the San Andreas fault, in *Fault Mechanics and Transport Properties of Rocks: A Festshrift in Honor of N.R. Brace* (eds Brian Evans and Teng-Fong Wong), Academic Press, New York, 475–503.

Rice, J.R. and Clearly, M.P. (1976) Some basic stress diffusion solutions for fluid-saturated elastic porous media with compressible constituents. *Reviews in Geophysics*, **14**, 227–241.

Rice, J.R. and Rudnicki, J.W. (1979) Earthquake precursory effects due to pore fluid stabilisation of a weakening fault zone. *Journal of Geophysical Research*, **84**, 2177–2193.

Richards, R., Elms, D.G. and Budhu, M. (1990) Dynamic fluidization of soils. *Journal of Geotechnical Engineering*, **116**, 740–759.

Ridd, M.F. (1970) Mud volcanoes in New Zealand. *Bulletin of the American Association of Petroleum Geologists*, **54**, 601–616.

Rider, M.H. (1974) The Namurian of West County Clare. *Proceedings of the Royal Irish Academy*, **74**B, 125–142.

Rider, M.H. (1978) Growth faults in Carboniferous of western Ireland. *Bulletin of the American Association of Petroleum Geologists*, **62**, 2191–2132.

Rieke, H.H. and Chilangarian, G.V. (1974) *Compaction of Argillaceous Sediments. Developments in Sedimentology*, **16**, Elsevier, Amsterdam, 424 pp.

Ringrose, P. (1988) Palaeoseismic (?) liquefaction event in late Quaternary lake sediment at Glen Roy, Scotland. *Terra Nova*, **1**, 57–62.

Ritger, S., Carson, B. and Suess, E. (1987) Methane-derived authigenic carbonates formed by subduction-induced pore-water expulsion along the Oregon Margin. *Geological Society of America Bulletin*, **98**, 147–156.

Robb, J.M. (1990) Groundwater processes in the submarine environment, in *Groundwater Geomorphology: The Role of Subsurface Water in Earth-surface Processes and Landforms* (eds C.G. Higgins and D.R. Coates), *Geological Society of America Special Paper*, **252**, 267–281.

Roberts, A. (1989) Fold-and-thrust structures in the Kintradwell 'Boulder Beds', Moray Firth. *Scottish Journal of Geology*, **25**, 173–186.

Roberts, H.H., Cratsley, D.W. and Whelan, T. III (1976) Stability of Mississippi delta sediments as evaluated by analysis of structural features in sediment borings. *Proceedings of the Offshore Technology Conference, Dallas, TX*, 9–28.

Roberts, H.H., Suhayda, J.N. and Coleman, J.M. (1980) Sediment deformation and transport on low-angle slopes: Mississippi river delta, in *Thresholds in Geomorphology* (eds D.R. Croates and J.D. Vitek), George Allen & Unwin, London, 131–167.

Roberts, J.L. (1972) The mechanics of overthrust faulting: a critical review. Tectonics-technique, Section 3. *International Geological Congress Proceedings*, Programme **24**, 593–8.

Robin, G. de Q. (1986) A soft bed is not the whole answer. *Nature*, **323**, 490–491.

Robin, G. de Q. (1989) Hard till glaciology: fast sliding/grinding. *Abstract, IGS British Branch Meeting, Cambridge.*

Robinson A.G. in press, Diagenesis and basin development. *Memoir of the American Association of Petroleum Geologists.*

Rodine, J.D. and Johnson, A.M. (1976) The ability of debris, heavily freighted with coarse clastic material, to flow on gentle slopes. *Sedimentology*, **23**, 213–234.

Roeloffs, E.A. (1988) Hydrogeologic precursors to earthquakes: a review. *Pure and Applied Geophysics*, **126**, 175–209.

Roeloffs, E.A. and Rudnicki, J.W. (1985) Coupled-deformation diffusion effects on water level changes due to propagating creep events. *Pure and Applied Geophysics*, **122**, 560–582.

Roeloffs, E.A., Burford, S.S., Riley, F.S. and Records, A.W. (1989) Hydrologic effects on water level changes associated with episodic fault creep near Parkfield, California. *Journal of Geophysical Research*, **94**, 12387–12402.

Rogers, J.L. (1972) The mechanics of overthrust faulting: a critical review. *Reports, Section 3, 24th International Geological Congress, Ottawa*, 593–598.

Ronnert, L. and Mickelson, D.M. (1992) High porosity of basal till at Burrow's Glacier, southeastern Alaska. *Geology*, **20**, 849–852.

Ronnlund, P. (1989) Viscosity ratio estimates from natural Raleigh–Taylor instabilities. *Terra Nova*, **1** 344–348.

Roscoe, K.H. (1970) The influence of strains in soil mechanics. *Geotechnique*, **20**, 129–170.

Roscoe, K.H., Schofield, A.N. and Wroth, C.P. (1958) On the yielding of soils. *Geotechnique*, **8**, 22–53.

Ross, D.A. (1971) Mass physical properties and slope stability of sediments of the northern Middle America Trench. *Journal of Geophysical Research*, **76**, 704–712.

Rothenburg, L. and Bathurst, R.J. (1989) Analytical study of induced anisotropy in idealized granular materials. *Geotechnique*, **39** 601–614.

Röthlisberger, H. (1972) Water pressure in intra and subglacial channels. *Journal of Glaciology*, **62**, 177–203.

Röthlisberger, H. and Iken, A. (1981) Plucking as an effect of water-pressure variations at the glacier bed. *Annals of Glaciology*, **2**, 57–62.

Rouse, W.C. (1984) Flowslides, in *Slope Instability* (eds D. Brunsden and D.B. Prior), John Wiley & Sons, Chichester, 491–522.

Rowe, P.W. (1962) The stress–dilatancy relation for static equilibrium of an assembly of particles in contact. *Proceedings of the Royal Society of London, Series A*, **269**, 500–527.

Rudnicki, J.W. (1986) Slip on an impermeable fault in a fluid-saturated rock mass, in *Earth Source Mechanics* (eds S. Das, J. Boatwright and C. Scholtz), *American Geophysical Union Geophysical Monograph, Maurice Ewing Series*, **6**, 86–89.

Rudnicki, J.W. and Hsu, T.C. (1988) Pore pressure changes induced by slip on permeable and impermeable faults. *Journal of Geophysical Research*, **93**, 3275–3285.

Rudnicki, J.W. and Rice, J.R. (1975) Conditions for the localization of deformation in pressure–sensitive dilatant materials. *Journal of the Mechanics and Physics of Solids*, **23**, 371–394.

Rumsey, I.A.P. (1971) Relationship of fractures in unconsolidated superficial deposits to those in the underlying bedrock. *Modern Geology*, **3**, 25–41.

Rupke, N.A. (1978) Deep clastic seas, in *Sedimentary Environments and Facies*, 1st edn (ed. H.G. Reading), Blackwell Scientific Publications, Oxford, 372–415.

Russ, D.P. (1979) Late Holocene faulting and earthquake recurrence in the Reelfoot area, northwestern Tennessee. *Geological Society of America Bulletin*, **90**, 1013–1018.

Ruszczynska-Szenajch, H. (1988) Glaciotectonics and its relationship to other glaciogenic processes, in *Glaciotectonics: Forms and Processes* (ed. D.G. Croot), A.A. Balkema, Rotterdam, 191–193.

Rutter, E.H. (1976) The kinetics of rock deformation by pressure solution. *Philosophical Transactions of the Royal Society of London, Series A*, **283**, 203–219.

Rutter, E.H. (1983) Pressure solution in nature, theory and experiment. *Journal of Geological Society of London*, **140**, 725–740.

Rutter, E.H. and Hadizadeh, J. (1991) On the influence of porosity on the low-temperature brittle–ductile transition in siliciclastic rocks. *Journal of Structural Geology*, **13**, 609–614.

Salisbury, R.D. and Atwood, W.W. (1897) Direct phenomena in the vicinity of Devil's Lake and Baraboo, Wisconsin, *Journal of Geology*, **5**, 131–147.

Salter, J.W. (1866) On faults in the drift-gravel at Hitchin, Herts. *Quarterly Journal of the Geological Society of London*, **22**, 565–566.

Sample, J.C. (1990) The effect of carbonate cementation of underthrust sediments on deformation styles during underplating. *Journal of Geophysical Research*, **95**, 9111–9121.

Sangrey, D.A. (1978) Marine geotechnology – state of the art. *Marine Geotechnology*, **2**, 45–80.

Sangrey, D.A. (1982) Geotechnical engineering link between offshore processes and hazards, in *Offshore Geologic Hazards: a Short Course Presented at Rice University, May 1981.* (eds A.H. Bouma, W.R. Parker and D.J.J. Kinsman), *American Association of Petroleum Geologists Continuing Education Course Note Series*, **18**, 3.1–3-3.75.

Sarkar, S., Chanda, S.K. and Bhattacharya, A. (1982) Soft-sediment deformation fabric in the Precambrian Bhander oolite, central India. *Journal of Sedimentary Petrology*, **52**, 95–107.

Satake, M. and Jenkins, J.T. (eds) (1988) *Micromechanics of Granular Materials*. Elsevier, Amsterdam, 366 pp.

Sauer, E.K. and Christiansen, E.A. (1988) Preconsolidation pressures in intertill glaciolacustrine clay near Blaine Lake, Saskatchewan. *Canadian Geotechnical Journal*, **25**, 831–839.

Savina, M. (1983) Engineering soil mechanics in hillslope geomorphology, in *Revolution in the Earth Sciences; Advances in the Past Half-century* (ed. S.T. Boardman), Kendall-Hunt, Dubuque, IA, 188–196.

Saxov, S. and Nieuwenhuis, J.K. (eds) (1982) *Marine Slides and other Mass Movements*. NATO Workshop, Plenum Press, New York, 353 pp.

Scarpelli, G. and Wood, D.M. (1982) Experimental observations of shear band patterns in direct shear tests. *Proceedings IUTAM Conference on Deformation and Failure of Granular Materials*, A.A. Balkema, Rotterdam, 473–484.

Schaeffer, D.G. (1990) Instability and ill-posedness in the deformation of granular materials. *International Journal for Numerical Models in Geomechanics*, **14**, 253–278.

Scheidegger, A.E. and Ai, N.S. (1986) Tectonic processes and geomorphological design. *Tectonophysics*, **126**, 285–300.

Schiffman, R.L. (1982) The consolidation of soft marine sediments. *Geo-Marine Letters*, **2**, 199–203.

Schlager, W. and Ginsburg, R.N. (1981) Bahama carbonate platforms – the deep and the past. *Marine Geology*, **44**, 1–24.

Schmertmann, J.H. (1991) The mechanical ageing of soils. *Journal of Geotechnical Engineering*, **117**, 1288–1330.

Schmoker, J.W. and Gautier, D.L. (1988) Sandstone porosity as a function of thermal maturity. *Geology*, **16**, 1007–1010.

Schmoker, J.W. and Gautier, D.L. (1989) Compaction of basin sediments: modelling based on time–temperature history. *Journal of Geophysical Research*, **94**, 7379–7386.

Schofield, A. and Wroth, P. (1968) *Critical State Soil Mechanics*. Mcgraw-Hill, London, 310 pp.

Scholtz, C.H. and Kranz, R.(1974) Notes on dilatancy recovery. *Journal of Geophysical Research*, **79**, 2132–2135.

Schokking, F. (1990) A sub-glacial sediment deformation model from geotechnical and structural properties of an overconsolidated lacustro-glacial clay. *Geologie en Mijnbouw*, **69**, 291–304.

Scholz C.H. (1987) Wear and gouge formation in brittle faulting. *Geology*, **15**, 493–495.

Schowalter, T.T. (1979) Mechanics of secondary hydrocarbon migration and entrapment. *Bulletin of the American Association of Petroleum Geologists*, **63**, 723–760.

Schultz, A.P. (1986) Ancient, giant rockslides, Sinking Creek Mountain, southern Appalachians, Virginia. *Geology*, **14**, 11–14.

Schumm, S.A. and Mosley, M.P. (eds) (1973) *Slope Morphology. Benchmark Papers in Geology*, Dowden, Hutchinson & Ross, Stroudsburg, PA, 454 pp.

Schutjens, P.M.T.M. (1991) Experimental compaction of quartz sand at low effective stress and temperature conditions. *Journal of the Geological Society*, **148**, 527–539.

Schwab, W.C., Lee, H.J., Kayen, R.E., Quinterno, P.J. and Tate, G.B. (1988) Erosion and slope instability on Horizon Guyot, Mid-Pacific Mountains. *Geo-Marine Letters*, **8**, 1–10.

Schwan, J., Van Loon, A.J., Steenbeck, R. and Van der Gaauw, P. (1980) Intraformational clay diapirism and extrusion in Weichselian sediments at Ormehoj (Funen, Denmark). *Geologie de Mijnbouw*, **59**, 241–250.

Scott, B. and Price, S. (1988) Earthquake-induced structures in young sediments. *Tectonophysics*, **147**, 165–170.

Scott, J.S. (1976) Geology of Canadian tills, in *Glacial Till* (ed. R.F. Leggett), *Royal Society of Canada Special Publication*, **12**, 11–49.

Screaton, E.J., Wuthrich, D.R. and Dreiss, S.J. (1990) Permeabilities, fluid pressures and flow rates in the Barbados Ridge complex. *Journal of Geophysical Research*, **95**, 8997–9008.

Secor, D.T. (1965) Role of fluid pressure in jointing. *American Journal of Science*, **263**, 633–646.

Seed, H.B. (1968) Landslides during earthquakes due to soil liquefaction. *Journal of Soil Mechanics Foundation, Proceedings of the American Society of Civil Engineers*, **94**, 1055–1122,

Seed, H.B. (1985) A note on earthquake-induced liquefaction (discussion of a paper by Morris, D.V., in Geotechnique, 33). *Geotechnique*, **35**, 451–454.

Seed, H.B. and Idriss, I.M. (1982) *Ground Motions and Soil Liquefaction during Earthquakes*. Earthquake Engineering Research Institute, Berkeley, CA, 134 pp.

Seed, H.B. and Rahman, M.S. (1978) Wave-induced pore pressure in relation to ocean floor stability of cohesionless soils. *Marine Geotechnology*, **3**, 123–150.

Selby, M.J. (1977) On the origin of sheeting and laminae in granitic rocks: evidence from Antarctica and the Namib Desert and central Sahara. *Madoqua*, **10**, 171–179.

Semple, R. (1988) State of the art report on engineering properties of carbonate soils. *Proceedings of the International Conference on Calcareous Sediments, Perth, Western Australia*, **2**, 807–836.

Shahinpoor, M. (ed.) (1983) *Advances in the Mechanics and the Flow of Granular Materials*, Gulf Publishing Company, Houston, TX, 2 vols.

Shaller, P.J. (1991) Analysis of a large landslide, Lost River Range, Idaho, U.S.A. *Canadian Geotechnical Journal*, **28**, 584–600.

Shanmugan, G., Moiola, R.J. and Sales, J.K. (1988) Duplex-like structures in submarine fan channels, Ouachita Mountains, Arkansas. *Geology*, **16**, 229–232.

Sharp, M.J. (1984) Annual moraine ridges at Skalafellsjokull, south-east Iceland. *Journal of Glaciology*, **104**, 82–93.

Sharp, M.J. (1985) Crevasse-fill ridges: a landform type characteristic of surging glaciers? *Geografiska Annaler*, **67**A, 213–220.

Sharp, M.J. (1988) Surging glaciers: geomorphic effects. *Progress in Physical Geography*, **12**, 533–559.

Sharp, M.J., Lawson, W. and Anderson, R.S. (1988) Tectonic processes in a surge-type glacer. *Journal of Structural Geology*, **10**, 499–516.

Shaw, E.W. (1914) The mud lumps at the mouth of the Mississippi. *United States Geological Survey Professional Paper*, **85**, 11–28.

Shearman, D.J. (1980) Sebkha facies evaporites, in *Evaporite deposits: Illustration and Interpretation of some Environmental Sequences*, Editions Technip, Paris, 19.

Shepard, F.P. (1932) Landslide modifications of submarine valleys. *Transactions of the American Geophysical Union*, **13**, 226–230.

Shephard, L.E. and Bryant, W.R. (1983) Geotechnical properties of lower trench inner-slope sediments. *Tectonophysics*, **99**, 279–312.

Shi, Y. and Wang, C.Y. (1985) High pore pressure generation in sediments in front of the Barbados Ridge complex. *Geophysical Research Letters*, **12**, 773–776.

Shi, Y. and Wang, C.Y. (1988) Generation of high pore pressures in accretionary prisms – inferences from the Barbados subduction complex. *Journal of Geophysical Research*, **93**, 8893–8910.

Shibuya, S. and Hight, D.W. (1987) On the stress path in simple shear. *Geotechnique*, **37**, 511–515.

Shimizu, M. (1982) Effect of overconsolidation on dilatancy of a cohesive soil. *Soils and Foundations*, **22**, 121–135.

Shoemaker, E.M. (1986) Sub-glacial hydrology for an ice sheet resting on a deformable aquifer. *Journal of Glaciology*, **110**, 20–30.

Shook, C.A. and Daniel, S.M. (1965) Flow of suspensions of solids in pipelines: Part I. Flow with a stable stationary deposit. *Canadian Journal of Chemical Engineering*, **43**, 56–61.

Shreve, R.L. (1968) The Blackhawk landslide. *Geological Society of America Special Paper*, **108**, 47 pp.

Sibson, R.H. (1981a) Controls on low-stress hydrofracture dilatancy in thrust, wrench and normal fault terrains. *Nature*, **289**, 665–667.

Sibson, R.H. (1981b) Fluid flow accompanying faulting: field evidence and models, in *Earthquake Prediction: an International Review* (eds R.W. Sibson and P.G. Richards), American Geophysical Union Geophysical Monograph, Maurice Ewing Series, **4**, 217–247.

Sibson, R.H. (1985) Stopping of earthquake ruptures at dilational fault jogs. *Nature*, **316**, 248–251.

Sibson, R.H. (1988) High-angle reverse faults, fluid-pressure cycling, and mesothermal gold-quartz deposits. *Geology*, **16**, 551–555.

Siebert, L. (1984) Large volcanic debris avalanches: characteristics of source areas, deposits and associated eruptions. *Journal of Volcanology and Geothermal Research*, **22**, 163–197.

Silliman, B., Wilcox, C. and Baldwin, T. (1829) Miscellaneous notices of mountain scenery and of slides and avalanches in the White and Green Mountains. *American Journal of Science*, **15**, 217–232.

Sills, G.C. and Been, K. (1984) Escape of pore fluid from consolidating sediment, in *Transfer Processes in Cohesive Sediment Systems* (Eds A.H. Bouma, D.A. Sangrey and J. Coleman *et al.*), Plenum Press, New York, 109–125.

Sills, G.C., Wheeler, S.J., Thomas, S.D. and Gardner, T.N. (1991) Behavior of offshore soils containing gas bubbles. *Geotechnique*, **41**, 227–241.

Silva, A.J. and Booth, J.S. (1985) Creep behavior of submarine sediments. *Geo-Marine Letters*, **4**, 215–219.

Silver, E. and Beutner, E. (1980) Mélanges. *Geology*, **8**, 32–34.

Sims, J.D. (1975) Determining earthquake recurrence intervals from deformational structures in young lacustrine sediments. *Tectonophysics*, **29**, 141–152.

Sims, J.D. (1978) Annotated bibliography of penecontemporaneous deformational structures in sediments, 1819–April 1978. *United States Geological Survey Open File Report*, **78–510**, 79 pp.

Sitler, R.F. and Chapman, C.A. (1955) Microfabrics of till from Ohio and Pennsylvania. *Journal of Sedimentary Petrology*, **25**, 262–269.

Skempton, A.W. (1964) Long-term stability of clay slopes. *Geotechnique*, **14**, 77–101.

Skempton, A.W. (1966) Some observations of tectonic shear zones. *Proceedings of the 1st International Conference on Rock Mechanics, Lisbon*, **6**, 329–335.

Skempton, A.W. (1970) The consolidation of clays by gravitational compaction. *Journal of the Geological Society*, **125**, 373–441.

Skempton, A.W. (1985) Residual strength of clays in landslides, folded strata, and the laboratory. *Geotechnique*, **35**, 3–18.

Skempton, A.W. and Petley, D. (1967) Strength along structural discontinuities in stiff, fissured clay. *Proceedings of the Oslo Geotechnical Conference*, **2**, 3–30.

Sladen, J.A. and Oswell, J.M. (1989) The behavior of very loose sand in the triaxial compression test. *Canadian Geotechnical Journal*, **26**, 103–113.

Slemmons, D.B. (1991) Introduction, in *Neotectonics of North America* (eds D.B. Slemmons, E. R. Engdahl, M.D. Zoback and D.D. Blackwell), *Geological Society of America Decade Map*, **1**, Boulder, CO, 1–20.

Smalley, I.J. and Piotrowski, J.A. (1987) Critical strength/stress ratios at the ice-bed interface in the drumlin forming process; from 'dilatancy' to 'cross-over', in *Drumlin Symposium* (eds J. Menzies and J. Rose), A.A. Balkema, Rotterdam, 81–86.

Smalley, I.J. and Unwin, D.J. (1968) The formation and shape of drumlins and their distribution and orientation in drumlin fields. *Journal of Glaciology*, **51**, 377–390.

Smart, P. (1985) Classification by texture and structure, in *Proceedings of the International Conference on Construction in Glacial Hills and Boulder Clays* (ed. M.C. Forde) Engineering Technical Press, Edinburgh, 227–234.

Smart, P. and Tovey, K. (1982) *Electron Microscopy of Soils and Sediments: Techniques.* Oxford University Press, Oxford, 264 pp.

Smart, P., Tovey, N.K., Leng, X., Hounslow, M.K. and McConnochie, I. (1991) Automatic analysis of microstructure of cohesive sediments, in *Microstructure of the Fine-grained Sediments* (eds R.H. Bennett, W.R. Bryant and M.H. Hulbert), Springer-Verlag, New York, 359–366 pp.

Smith, L. and Chapman, D.S. (1983) On the thermal effects of ground water flow, 1. Regional scale systems. *Journal of Geophysical Research*, **88**, 593–608.

Smith, R.C.M. (1991) Post-eruption sedimentation on the margin of a caldera lake, Taupo volcanic centre, New Zealand. *Sedimentary Geology*, **74**, 89–138.

Snavely, P.D. and Pearl, J.E. (1979) Clastic dykes: a key to Tertiary regional stress fields in the Northwest Olympic Peninsula. *United States Geological Survey Professional Paper*, **1150**, 88 pp.

Snow, D.T. (1968) Rock fracture spacings, openings, and porosities. *Journal of Soil Mechanics and Foundations Division, *Proceedings of the American Society of Civil Engineers*, **94**, 73–91.

Sorby, H.C. (1859) On the contorted stratification of the drift of the coast of Yorkshire, *Proceedings of the Geological and Polytechnic Society, West Riding, Yorkshire, 1849–1859*, 220–224.

Sousa, J. and Voight, B. (1992) Computational flow modelling for long-runout landslide hazard assessment, with an example from Clapiète, France. *Bulletin of the Association of Engineering Geologists*, **29**, 131–150.

Spencer, C.W. (1987) Hydrocarbon generation as a mechanism for overpressuring in the Rocky Mountain region. *Bulletin of the American Association of Petroleum Geologists*, **71**, 368–388.

Sprunt, E.S. and Nur, A. (1976) Reduction of porosity by pressure solution: experimental verification. *Geology*, **4**, 463–467.

Sridharan, A. and Rao, G.V. (1979) Shear strength behaviour of saturated clays and the role of the effective stress concept. *Geotechnique*, **29**, 177–193.

Sridharan, A. and Rao, G.V. (1982) Mechanisms controlling the secondary compression of clays. *Geotechnique*, **32**, 249–260.

Srinivasa Murthy, B.R., Vatsala, A. and Nagaraj, T.S. (1991) Revised Cam-clay model. *Journal of Geotechnical Engineering*, **117**, 851–871.

Stanford, S.D. and Mickelson, D.M. (1985) Till fabric and deformational structures in drumlins near Waukesha, Wisconsin, U.S.A. *Journal of Glaciology*, **109**, 220–228.

Stauffer, M.R., Gendzwill, D.J. and Sauer, E.K. (1990) Ice-thrust features and the Maymont landslide in the North Saskatchewan River valley. *Canadian Journal of Earth Sciences*, **27**, 229–242.

Stauffer, P.H. (1967) Grain flow deposits and their implications, Santa Ynez Mountains, California. *Journal of Sedimentary Petrology*, **37**, 487–498.

Steiger, R.P. and Leung, P.K. (1991) Critical state shale mechanics, in *Rock Mechanics as a Multidisciplinary Science* (ed. J-C. Roegiers), A.A. Balkema, Rotterdam, 293–302.

Stell, J.H. (1976) Clay diapirism in the lower Emsian La Vid shales near Colle, Cantabrian Mountains, NW Spain. *Geologie en Mijnbouw*, **55**, 110–116.

Stephenson, L.P., Plumley, W.J. and Palciauskas, V.V. (1992) A model for sandstone compaction by grain interpenetration. *Journal of Sedimentary Petrology*, **62**, 11–22.

Stevenson, D.J. and Scott, D.R. (1991) Mechanics of fluid–rock systems. *Annual Reviews of Fluid Mechanics*, **23**, 305–40.

Stewart, I.S. and Hancock, P.L. (1994) Neotectonics, in *Continental Deformation* (ed. P.L. Hancock), Pergamon Press, Oxford, 370–409.

Stoffa, P.L., Wood, W.T., Shipley, T.H., et al. (1992) Deep water high-resolution expanding spread and split-spread seismic profiles in the Nankai trough. *Journal of Geophysic Research*, **97**, 1687–1713.

Stoneley, R. (1983) Fibrous calcite veins overpressures and primary oil migration. *American Association of Petroleum Geologists Bulletin*, **67**, 1427–8.

Stow, D.A.V. (1979) Distinguishing between fine-grained turbidites and contourites on the Nova Scotian deep-water margin. *Sedimentology*, **26**, 371–387.

Stow, D.A.V. (1986) Deep clastic seas, in *Sedimentary Environments and Facies*, 2nd edn (ed. H.G. Reading), Blackwell Scientific Publications, Oxford, 399–444.

Stow, D.A.V. and Bowen, A.J. (1980) A physical model for the transport and sorting of fine-grained sediments by turbidity currents. *Sedimentology*, **27**, 31–46.

Stow, D.A.V. and Shanmugam, G. (1980) Sequence of structures in fine-grained turbidites: comparison of recent deep-sea and ancient flysch sediments. *Sedimentary Geology*, **25**, 23–42.

Strangways, W.T.H.F. (1821) Description of strata in the Brook Pulcova, near the village of Great Pulcova, in the neighbourhood of St Petersburg. *Transactions of the Geological Society, First Series*, **5**, 382–458.

Strickland, H.E. (1840) On some remarkable dikes of Calcareous Grit, at Ethie in Ross-shire. *Transactions of the Geological Society, Second Series*, **5**, 599–600.

Suess, E., Carson, B., Ritger, S., et al. (1985) Biological communities at vent sites along the subduction zones off Oregon, in *The Hydrothermal Vents of the Eastern Pacific: An Overview* (ed. M.L. Jones), *Bulletin of the Biological Society of Washington*, **6**, 475–484.

Suess, E. von Huene, R. and the shipboard scientific party (1988) *Proceedings of the Ocean Drilling Program, Scientific Reports*, **112**, Ocean Drilling Program, College Station, TX, 1015 pp.

Suhayda, J.N., Whelan III, T., Coleman, J.M., Booth, J.S. and Garrison, L.E. (1976) Marine sediment instability: interaction of hydrodynamic forces and bottom sediments. *Offshore Technology Conference Proceedings*, **1**, 29–40.

Surlyk, F. (1987) Slope and deep shelf gully sandstones, Upper Jurassic, East Greenland. *Bulletin of the American Association of Petroleum Geologists*, **71**, 464–475.

Swanson, F.A. (1964) Groundwater phenomena associated with the Hebgen Lake earthquake. *United States Geological Survey Professional Paper*, **435**, 159–165.

Swarbrick, R.E. and Naylor, M.A. (1980) The Kathikas mélange, S.W. Cyprus: late Cretaceous submarine debris flows. *Sedimentology*, **27**, 63–78.

Swartz, J.F. and Lindsley-Griffin, N. (1990) An improved impregnation technique for studying structure of unlithified cohesive sediments. *Proceedings of the Ocean Drilling Program, Scientific Results*, **112**, Ocean Drilling Program, College Station, TX, 87–91.

Swolfs, H.S. (1984) The triangle stress diagram – a graphical representation of crustal stress measurements. *United States Geological Survey Professional Paper*, **1291**, 19 pp.

Tada, R. and Siever, R. (1989) Pressure solution during diagenesis. *Annual Reviews Earth and Planetary Sciences*, **17**, 89–118.

Taira, A., Byrne, T. and Ashi, J. (1992) *Photographic Atlas of an Accretionary Prism. Geologic Structures of the Shimanto Belt, Japan*. University of Tokyo Press, Tokyo, 124 pp.

Taira, A., Hill, I.I. and the shipboard scientific party (1991) *Proceedings of the Ocean Drilling Program, Initial Reports*, **131**, Ocean Drilling Program, College Station, TX, 306 pp.

Taira, A., Katto, J., Tashiro, M., Okamura, M. and Kodama, K. (1988) The Shimanto Belt in Shikoku, Japan – evolution of Cretaceous to Miocene accretionary prism. *Modern Geology*, **12**, 5–46.

Talbot, C.J. and von Brunn, V. (1987) Intrusive and extrusive (micro) mélange couplets as distal effects of tidal pumping by a marine ice sheet. *Geological Magazine*, **124**, 513–525.

Tan, T.S., Yong, K.Y., Leong, A.C. and Lee, S.L. (1990) Sedimentation of clayey slurry. *Journal of Geotechnical Engineering*, **116**, 885–898.

Ter-Stepanian, G. (1969) Types of depth creep of slopes in rock masses. *Proceedings of the 1st Congress of the International Society of Rock Mechanics (1966), Lisbon*, 157–160.

Terwilliger, V.J. (1990) Effects of vegetation on soil slippage by pore pressure modification. *Earth Surface Processes and Landforms*, **15**, 553–570.

Terzaghi, K. (1936) The shearing resistance of saturated soils and the angle between the planes of shear. *Proceedings of the First International Conference on Soil Mechanics and Foundation Engineering, Cambridge, MA*, **1**, 54–56.

Terzaghi, K. (1943) *Theoretical Soil Mechanics*. John Wiley and Sons, New York, 510 pp.

Terzaghi, K. (1947) Shear characteristics of quicksand and soft clay. *7th Texas Conference on Soil Mechanics and Foundation Engineering, Houston, TX*, 1–10.

Terzaghi, K. (1955) Influence of geological factors on the engineering properties of sediments. *Economic Geology*, **50**, 557–618.

Terzaghi, K. (1962) Stability of steep slopes on hard unweathered rock. *Geotechnique*, **12**, 251–270.

Teufel, L.W. and Warpinski, N.R. (1984) Determination of in situ stress from anelastic strain recovery measurement of oriented core, in *Comparison of Hydraulic Fracture Stress Measurements in the Rollins Sandstone* (eds C. Dowling and S. Singh). Proceedings of the 25th U.S. Symposium on Rock Mechanics, Northwestern University, Evanston, IL.

Thomas, G.S.P. and Connell, R.J. (1985) Iceberg drop, dump, and grounding structures from Pleistocene glacio-lacustrine sediments, Scotland. *Journal of Sedimentary Petrology*, **55**, 243–249.

Thomas, I.W., Collinson, J.D. and Jones, C.M. (1992) Depositional modelling of UK North Sea Alba field reservoir. *Energy Exploration and Exploitation*, **10**, 300–320.

Thomas, W.A. and Baars, D.L. (convenors) (1988) Synsedimentary tectonics, Penrose Conference report. *Geology*, **16**, 190–191.

Tobin, H.J., Moore, J.C., MacKay, M.E., Orange, D.L. and Kulm, L.D. (1993) Fluid flow along a strike-slip fault at the toe of the Oregon accretionary prism: implications for the geometry of frontal accretion. *Geological Society of America Bulletin*, **105**, 569–582.

Tobisch, O.T. (1984) Development of foliation and fold interference patterns produced by sedimentary processes. *Geology*, **12**, 441–444.

Toolan, F.E. (1988) Preface to 'The Engineering Application of Direct and Simple Shear Testing' Symposium. *Geotechnique*, **37**, 1–2.

Toorman, E.A. and Berlamont, J.E. (1991) A hindered settling model for the prediction of settling and consolidation of cohesive sediment. *Geo-Marine Letters*, **11**, 179–183.

Torrance, J.K. (1983) Towards a general model of quick clay development. *Sedimentology*, **30**, 547–555.

Tremlett, C.R. (1982) The structure and structural history of the Lower Palaeozoic rocks of north Dyfed, Wales. Unpublished PhD thesis. University College of Wales, Aberystwyth.

Trewin, N.H. (1992) 'Subaqueous shrinkage cracks' in the Devonian of Scotland reinterpreted – discussion. *Journal of Sedimentary Petrology*, **62**, 921–922.

Tribble, J.S. (1990) Clay diagenesis in the Barbados accretionary complex: potential impact on hydrology and subduction dynamics. *Proceedings of the Ocean Drilling Program, Scientific Results*, **110**, Ocean Drilling Program, College Station, TX, 97–110.

Trincardi, F. and Argnani, A. (1991) Gela submarine slide: a major basin-wide event in the Plio-Quaternary foredeep of Sicily. *Geo-Marine Letters*, **10**, 13–21.

Truswell, J.F. (1972) Sandstone sheets and related intrusions from Coffee Bay, Transkei, South Africa. *Journal of Sedimentary Petrology*, **42**, 578–583.

Tsuchida, T., Kobayashi, M. and Mizukami, J.I. (1991) Effect of ageing of marine clay and its duplication by high temperature consolidation. *Soils and Foundations, Japanese Society of Soil Mechanics and Foundation Engineering*, **31**, 133–147.

Tsui, P.C., Cruden, D.M. and Thomson, M. (1989) Fabric studies of ice-thrust shear zones as applied to problems in geotechnical engineering, in *Applied Quaternary Research* (eds F. J. de Mulder and B.P. Hageman), A.A. Balkema, Rotterdam, 147–164.

Tuttle, M. and Seeber, L. (1991) Historic and prehistoric earthquake-induced liquefaction in Newbury, Massachusetts. *Geology*, **19**, 594–597.

U.S. Corps of Engineers, (1973) *Interim Report on Foundation Treatment, Laurel Dam*. United States Army Engineer District, Corps of Engineers, Nashville, TN, 1–18.

Ui, T. (1983) Volcanic dry avalanche deposits – identification and comparison with non-volcanic debris-stream deposits, in *Arc Volcanism* (eds S. Aramaki, and I. Kushiro), *Journal of Volcanology and Geothermal Research*, **18**, 135–150.

Uriel, S. and Serrano, A.A. (1973) Geotechnical properties of two collapsible volcanic soils of low bulk density at the site of two dams in the Canary Islands. *Proceedings of the Eighth International Conference on Soil Mechanics and Foundation Engineering, Moscow*, 251–264.

Vaid, Y.P., Chung, E.K.F. and Keurbis, R.H. (1990) Stress path and steady state. *Canadian Geotechnical Journal*, **27**, 1–7.

Vaid, Y.P., Fisher, J.M., Kuerbis, R.H. and Negussey, D. (1990) Particle gradation and liquefaction. *Journal of Geotechnical Engineering*, **116**, 698–709.

Valent, P.J. and Lee, H.J. (1976) Feasibility of sub-seafloor emplacement of nuclear waste. *Marine Geotechnology*, **1**, 267–293.

Van den Bark, E. and Thomas, O.D. (1980) Ekofisk: first of the giant oil fields of Western Europe, in *Giant Oil and Gas Fields of the Decade 1968–1978* (ed. M. Halbouty) *Memoir of the American Association of Petroleum Geologists*, **30**, 195–224.

Van den Berg, L. (1987) Experimental re-deformation of naturally deformed scaly clays. *Geologie en Mijnbouw*, **65**, 309–315.

Van der Meer, J.J.M. (1985) Sedimentology and genesis of glacial deposits in the Goudsberg, Central Netherlands. *Mededelingen Rijks Geologische Dienst*, **39**, 1–29.

Van der Meer, J.J.M. (ed.), (1987) *Tills and Glaciotectonics. Proceedings of an INQUA Meeting on the Genesis and Lithology of Glacial Deposits, Amsterdam*. A.A. Balkema, Rotterdam, 280 pp.

Van Genuchten, P.M.B. and de Rijke, H. (1989) On pore water pressure variations causing slide velocities and accelerations observed in a seasonally active landslide. *Earth Surface Processes and Landforms*, **14**, 577–586.

Van Loon, A. J. (1992) The recognition of soft-sediment deformation as early-diagenetic features – a literature review, in *Diagenesis, III* (eds K.H. Wolf and G.V. Chilingarian), *Developments in Sedimentology*, **47**, Elsevier, Amsterdam, 135–189.

Van Loon, A. J. and Wiggers, A. J. (1976) Metasedimentary 'graben' and associated structures in the lagoonal Almere member (Groningen Formation, The Netherlands). *Sedimentary Geology*, **16**, 237–254.

Van Loon, A.J., Brodzikowski, K. and Gotowala, R. (1984) Structural analysis of kink bands in unconsolidated sands. *Tectonophysics*, **104**, 351–374.

Van Loon, A.J., Brodzikowski, K. and Gotowala, R. (1985) Kink structures in unconsolidated fine-grained sediments. *Sedimentary Geology*, **41**, 283–300.

Van Steijn, H. and Coutard, J.P. (1989) Laboratory experiments with small debris flows: physical properties related to sedimentary characteristics. *Earth Surface Processes and Landforms*, **14**, 587–596.

Vanuxem, L. (1842) *Geology of New York. Part III, Survey of the 3rd District*, State Geological Survey of New York, Albany, NY, 306 pp.

Varnes, D.J. (1978) Slope movements and types of processes, in *Landslides: Analysis and Control, National Academy of Sciences, Transportation Research Board, Special Report* **176**, Washington DC, 11–33.

Vaughan, P.R. (1985) Mechanical and hydraulic properties of *in situ* residual soils. *First International Conference in Geomechanics of Tropical and Saprolitic Soils, Brasilia*, **3**, 231–263.

Vaughan, P.R., Maccarini, M. and Mokhtar, S.M. (1988) Indexing the engineering properties of residual soils. *Quarterly Journal of Engineering Geology*, **21**, 69–86.

Vermeer, P.A. (1982) A simple shear-band analysis using compliances. *Proceedings of the International Union of Theoretical and Applied Mechanics Conference on Deformation and Failure of Granular Materials, Delft*. A.A. Balkema, Rotterdam, 493–499.

Vermeer, P.A. (1990) The orientation of shear bands in biaxial tests. *Geotechnique*, **40**, 223–236.

Vermeer, P.A. and Luger, H.J. (eds.) (1982) *Deformation and Failure of Granular Materials*, A.A. Balkema, Rotterdam, 661 pp.

Vesajoki, H. (1982) Deformation of soft sandy sediments during deglaciation and subsequent emergence of land areas: examples from northern Karelia, Finland. *Boreas*, **11**, 11–28.

Vickers, B. (1983) *Laboratory Work in Soil Mechanics*, 2nd edn. Granada, London, 170 pp.

Virkkala, K. (1969) On the lithology and provenance of the till of a gabbro area in Finland. *INQUA (International Union for Quaternary Research) VIII International Geological Congress General Session*, 711–714.

Visser, J.N.J. and Joubert, A. (1990) Possible earthquake induced sediment liquefaction in thermal spring deposits of Florisbad, Orange Free State. *South African Journal of Geology*, **93**, 525–30.

Visser, J.N.J., Colliston, W.P. and Tereblanche, J.C. (1984) The origin of soft-sediment structures in Permo-Carboniferous glacial and proglacial beds, South Africa. *Journal of Sedimentary Petrology*, **54**, 1183–1196.

Voight, B. (1976) Editor's comments on Paper 38. *Mechanics of Thrust Faults and Décollement* (ed. B. Voight), *Benchmark Papers in Geology*, **32**, Dowden, Hutchinson, and Ross, Stroudsburg, PA, 349–350.

Voight, B. (ed.) (1978a) *Rockslides and Avalanches I: Natural Phenomena. Developments in Geotechnical Engineering*, **14A**. Elsevier, Amsterdam, 825 pp.

Voight, B. (ed.) (1978b) *Rockslides and Avalanches II: Engineering Sites. Developments in Geotechnical Engineering*, vol. 14B. Elsevier, Amsterdam, 844 pp.

Voight, B., Glicken, H., Janda, R.J. and Douglass, P.M. (1981) Catastrophic rockslide avalanche of May 18, in *The 1980 Eruptions of Mount St. Helens, Washington* (eds P.W. Lipman and D.R. Mullineaux), *United States Geological Survey Professional Paper*, **1250**, 347–378.

Voight, B., Janda, R.J., Glicken, H. and Douglass, P.M. (1983) Nature and mechanics of the Mount St. Helens rockslide avalanche of 18 May, 1980. *Geotechnique*, **33**, 243–273.

Von Brunn, V. and Talbot, C.J. (1986) Intrusive clastic sheets and their subglacial deformation in the Dwyka Formation on Northern Natal, South Africa. *Journal of Sedimentary Petrology*, **56**, 35–44.

Von Huene, R. and Kulm, L.D. (1973) Tectonic summary of Leg 18. *Initial Reports of the Deep Sea Drilling Project*, **18**, United States Government Printing Office, Washington, DC, 961–976.

Vrolijk, P.J. (1987) Tectonically driven fluid flow in the Kodiak accretionary complex, Alaska. *Geology*, **15**, 466–469.

Vrolijk, P.J., Meyers, G. and Moore, J. (1988) Warm fluid migration along tectonic mélanges in the Kodiak accretionary complex, Alaska. *Journal of Geophysical Research*, **93**, 10313–10324.

Vrolijk, P.J., Chambers, S., Gieskes, J. and O'Neil, J. (1990) Stable isotope ratios of interstitial fluids from the nothern Barbados accretionary prism, ODP Leg 110. *Proceedings of the Ocean Drilling Program, Scientific Results*, **110**, Ocean Drilling Program, College Station, TX, 189–205.

Vyalov, S.S. (1986) *Rheological Fundamentals of Soil Mechanics. Developments in Geotechnical Engineering*. Elsevier, Amsterdam, 564 pp.

Walder, J. (1986) Hydraulics of subglacial cavities. *Journal of Glaciology*, **112**, 439–445.

Walder, J. and Fowler, A. (1989) Channelized subglacial drainage over a deformable bed. *Eos, (Transactions of the American Geophysical Union)*, **70**, 1084.

Wahrhaftig, C. and Cox, A. (1959) Rock glaciers in the Alaska Range. *Geological Society of America Bulletin*, **70**, 383–436.

Walker, R.G. (1965) The origin and significance of the internal sedimentary structures of turbidites. *Proceedings of the Yorkshire Geological Society*, **35**, 1–32.

Walker, R.G. (1975) Generalized facies models for resedimented conglomerates of turbidite association. *Geological Society of America Bulletin*, **86**, 737–748.

Walker, R.G. (1977) Deposition of upper Mesozoic resedimented conglomerates and associated turbidites in southwestern Oregon. *Geological Society of America Bulletin*, **88**, 273–285.

Walker, R.G. (1978) Deep-water sandstone facies and ancient submarine fans: models for exploration for stratigraphic traps. *Bulletin of the American Association of Petroleum Geologists*, **62**, 932–966.

Wallis, G.B. (1969) *One-dimensional Two-phase Flow*. McGraw-Hill, New York, 408 pp.

Walsh, J.B. (1981) Effect of pore pressure and confining pressure on fracture permeability. *International Journal of Rock Mechanics, Mining Science, and Geomechanics*, **18**, 429–435.

Walther, J.V. and Orville, P.M. (1982) Volatile production and transport in regional metamorphism. *Contributions to Mineralogy and Petrology*, **79**, 252–257.

Wang, C.Y., Shi, Y., Hwang, W. and Chen, H. (1990) Hydrogeologic processes in the Oregon–Washington accretionary complex. *Journal of Geophysical Research*, **95**, 9009–9024.

Ward, W.H. (1945) The stability of natural slopes. *Geological Journal*, **105**, 170–197.

Wardlaw, N.C. (1972) Sedimentary folds and associated structures in Cretaceous salt deposits of Sergipe, Brazil. *Journal of Sedimentary Petrology*, **42**, 175–188.

Warpinski, N.R. (1986) Elastic and viscoelastic calculations of stresses in sedimentary basins. *Proceedings, Unconventional Gas Technology Symposium, Louisville, KY*, 409–417.

Warpinski, N.R. (1989) Determining the minimum *in situ* stress from hydraulic fracturing through perforations. *International Journal of Rock Mechanics and Mining Science*, **26**, 523–531.

Warpinski, N.R. and Teufel, L.W. (1987) *In situ* stresses in low-permeability, non-marine rocks. *Society of Petroleum Engineers/Department of Energy Paper*, **16402**.

Warpinski, N.R., Branagan, P. and Wilmer, R. (1985) *In situ* stress measurements at U.S. DOE's Multiwell Experiment Site, Mesaverde Group, Rifle, Colorado. *Journal of Petroleum Technology*, **37**, 527–536.

Warren, P.T., Price, D., Nutt, M.J.C. and Smith, E.G. (1984) *Geology of the Country around Rhyl and Denbigh*. Memoirs of the Geological Survey of Great Britain, HMSO, London, 271 pp.

Watts, N.L. (1983) Microfractures in chalk of the Albuskjell Field, Norwegian Sector, North Sea: possible origin and distribution. *Bulletin of the American Association of Petroleum Geologists*, **67**, 201–234.

Webb, B.C. and Cooper, A.H. (1988) Slump folds and gravity slide structures in a lower Palaeozoic marginal basin sequence (the Skiddaw Group) NW England. *Journal of Structural Geology*, **10**, 463–472.

Weber, K.J. and Daukoru, E. (1975) Petroleum geology of the Niger delta. *Proceedings of the 9th World Petroleum Conference*, 209–221.

Weertman, J. (1969) Water lubrication mechanism of glacier surges. *Canadian Journal of Earth Sciences*, **6**, 929–942.

Weertman, J. (1972) General theory of water flow at the base of a glacier or ice sheet. *Reviews of Geophsics*, **10**, 287–333.

Wesley, L.D. (1990) Influence of structure and composition on residual soils. *Journal of Geotechnical Engineering*, **116**, 589–603.

Westbrook, G.K. and Smith, M.J. (1983) Long décollements and mud volcanoes: evidence from the Barbados Ridge complex for the role of high pore-fluid pressures in the development of an accretionary complex. *Geology*, **11**, 279–283.

Wetzel, A. (1990) Interrelationships between porosity and other geotechnical properties of slowly deposited, fine-grained marine surface sediments. *Marine Geology*, **92**, 105–113.

Whalley, W.B. (1984) Rockfalls, in *Slope Instability* (eds D. Brunsden and D.B. Prior), John Wiley & Sons, Chichester, 217–256.

Wheeler, S.J. (1989) A conceptual model for soils containing large gas bubbles. *Geotechnique*, **38**, 389–397.

Whelan, T., III, Coleman, J.M., Roberts, H.H. and Suhayda, J.N. (1976) The occurrence of methane in recent deltaic sediments and its effect on soil stability. *International Association of Engineering Geologists Bulletin*, **14**, 55–64.

Whitaker, F.F. and Smart, P.L. (1990) Active circulation of saline ground waters in carbonate platforms: evidence from the Great Bahama Bank. *Geology*, **18**, 200–203.

White, J.D.L. and Busby-Spera, C. (1987) Deep marine arc apron deposits and syndepositional magmatism in the Alisistos group at Punta Cono, Baja California, Mexico. *Sedimentology*, **34**, 911–927.

White, R.S. (1977) Recent fold development in the Gulf of Oman. *Earth and Planetary Science Letters*, **36**, 85–91.

White, R.S. and Klitgord, K.D. (1976) Sediment deformation and plate tectonics in the Gulf of Oman. *Earth and Planetary Science Letters*, **32**, 199–209.

White, S.E. (1976) Is frost action really only hydration shattering? A review. *Arctic and Alpine Research*, **8**, 1–6.

White, T.G. (1895) The faunas of the Upper Ordovician strata at Trenton Falls, Oneida County, New York. *Transactions of the New York Academy of Sciences*, **15**, 71–96.

Whiticar, M.J. and Werner, F. (1981) Pockmarks: submarine vents of natural gas or freshwater seeps? *Geo-Marine Letters*, **1**, 193–199.

Whittecar, G.R. and Mickelson, D.M. (1979) Composition, internal structures, and an hypothesis of formation for drumlins, Waukesha County, Wisconsin, U.S.A. *Journal of Glaciology*, **87**, 357–370.

Whittow, R. (1990) *Basic Soil Mechanics*, 2nd edn. Longman, London, 528 pp.

Wilbur, C. and Amadei, B. (1990) Flow pump measurements of fracture transmissivity as a function of normal stress, in *Rock Mechanics and Challenges* (eds W.A. Hustrulid and G.A. Johnson), A.A. Balkema, Rotterdam, 621–627.

Wilhelm, B. and Somerton, W.H. (1967) Simultaneous measurements of pore and elastic properties of rocks under trivial stress conditions. *Society of Petroleum Engineers Journal* **7**, 283–94.

Wilkinson, M. (1993) Concretions of the Valtos Sandstone Formation of Skye: geochemical indicators of palaeo-hydrology. *Journal of the Geological Society*, **150**, 57–66.

Wilkinson, W.B. and Shipley (1972) Vertical and horizontal laboratory permeability measurements in clay soils. *Developments in Soil Science, 2: Fundamentals of Transport in Porous Media.* Elsevier, London, 392 pp.

Will, T.M. and Wilson, C.J.L. (1989) Experimentally produced slickenside lineations in pyrophyllitic clay. *Journal of Structural Geology*, **11**, 657–667.

Williams, G.D. and Chapman, P. (1983) Strains developed in the hangingwalls of thrusts due to their slip/propagation rate: a dislocation model. *Journal of Structural Geology*, **5**, 563–572.

Williams, M.Y. (1927) Sandstone dykes in southeastern Alberta. *Transactions of the Royal Society of Canada, Section 4, Geological Sciences, Third Series*, **21**, 153–174.

Williams, P.F. (1983) Timing of deformation and the mechanism of cleavage development in a Newfoundland mélange. *Maritime Sediments*, **19**, 31–48.

Williams, P.F. (1985) Multiply deformed terrains – problems of correlation. *Journal of Structural Geology*, **7**, 269–280.

Williams, P.F. (1986) Critical review of criteria for distinguishing structures developed before and after lithification (abstract). *Bulletin of the American Association of Petroleum Geologists*, **70**, 663.

Williams, P.F., Collins, A.R. and Wiltshire, R.G. (1969) Cleavage and penecontemporaneous deformation structures in sedimentary rocks. *Journal of Geology*, **77**, 415–425.

Williams, P.F. Goodwin, L.B. and Raiser, S. (1994) Ductile deformation process, in *Continental Deformation* (ed. P.L. Hancock), Pergamon Press, Oxford, 1–27.

Williams, S.J. (1987) Faulting in abyssal-plain sediments, Great Meteor East, Madeira abyssal plain, in *Geology and Geochemistry of Abyssal Plains* (eds P.E. Weaver, and J. Thomson), *Geological Society of London Special Publication*, **31**, 87–104.

Wilson, C.J.L. and Will, T.M. (1990) Slickenside lineations due to ductile processes, in *Deformation Mechanisms, Rheology and Tectonics* (eds R.J. Knipe and E.H. Rutter) *Geological Society of London Special Publication*, **54**, 455–460.

Wilson, J.T. (1938) Drumlins of south-west Nova Scotia. *Transactions of the Royal Society of Canada*, **32**, 41–47.

Winker, C.D. and Edwards, M.B. (1983) Unstable progradational clastic shelf margins, in *The Shelfbreak: Critical Interface on Continental Margins* (eds D.J. Stanley and G. T. Moore), *Society of Economic Paleontologists and Mineralogists Special Publication*, **33**, 139–157.

Winn, R.D. Jr. and Dott, R.H. Jr. (1978) Submarine-fan turbidites and resedimented conglomerates in a Mesozoic arc-rear marginal basin in southern South America, in *Sedimentation in Submarine Canyons, Fans and Trenches* (eds D.J. Stanley and G. Kelling), Dowden, Hutchinson and Ross, Stroudsburg, PA, 362–373.

Winslow, M.A. (1983) Clastic dike swarms and the structural evolution of the foreland fold and thrust belt of the southern Andes. *Geological Society of America Bulletin*, **94**, 1073–1080.

Wnuk, C. and Maberry, J.O. (1990) Enigmatic eight-meter trace fossils in the Lower Pennsylvanian Lee Sandstone, Central Appalachian Basin, Tennessee. *Journal of Palaeontology*, **64**, 440–450.

Wohletz, K.H. (1986) Explosive magma–water interactions: thermodynamics, explosion mechanisms, and field studies. *Bulletin of Volcanology*, **48**, 245–264.

Wong, T.F. (1990) Mechanical compaction and the brittle–ductile transition in porous sandstones, in *Deformation Mechanisms, Rheology and Tectonics* (eds R.J. Knipe and E.H. Rutter). *Geological Society of London Special Publication*, **54**, 111–122.

Wong, T.-F., Hiram, S. and Zhang, J., in press, Effect of loading path and porosity on the failure mode of porous rocks. *Applied Mechanics Review*.

Wong, T.-F., Szeto, H. and Zhang, J. (1992) Effect of loading path and porosity on the failure mode of porous rocks. *Applied Mechanics Review*, **45**, 281–293.

Wood, D. Muir (1990) *Soil Behaviour and Critical State Soil Mechanics*. Cambridge University Press, Cambridge, 462 pp.

Wood, D. Muir (1991) Strength ratio, pore pressure parameter and effective stress change. *Soils and Foundations, Japanese Society of Soil Mechanics and Foundation Engineering*, **31**, 194–199.

Woodcock, N.H. (1976) Structural style in slump sheets: Ludlow Series, Powys, Wales. *Journal of the Geological Society*, **132**, 399–415.

Woodcock, N.H. (1979a) Sizes of submarine slides and their significance. *Journal of Structural Geology*, **1**, 137–142.

Woodcock, N.H. (1979b) The use of slump structures as palaeoslope orientation estimators *Sedimentology*, **26**, 83–99.

Xiao, H. and Suppe, J. (1992) Origin of rollover. *American Association of Petroleum Geologists Bulletin*, **76**, 509–529.

Yagishita, K. and Morris, R.C. (1979) Microfabrics of a recumbent fold in cross-bedded sandstones. *Geological Magazine*, **116**, 105–116.

Yagishita, K., Westgate, J.A. and Pearce, G.W. (1981) Remanent magnetization in penecontemporaneous structures of the Pleistocene Scarborough Formation, Ontario, Canada. *Journal of the Geological Society*, **138**, 549–557.

Yassir, N. (1989a) Mud volcanoes and the behaviour of overpressured clays and silts. Unpublished PhD thesis, University of London, 249 pp.

Yassir, N. (1989b) Undrained shear behaviour of clay at high total stresses, in *Rock at Great Depth* (eds V. Maury and D. Fourmaintraux), A.A. Balkema, Rotterdam, 907–913.

Yassir, N.A. (1990) The behaviour of overpressured clays and mudrocks, in *Deformation Mechanisms, Rheology and Tectonics* (eds R.J. Knipe and E.H. Rutter). *Geological Society of London Special Publication*, **54**, 399–404.

Yong, R.N. and Townsend, F.C. (eds) (1984) *Sedimentation Consolidation Models*. American Society of Civil Engineers, New York, 609 pp.

Yong, R.N. and Townsend, F.C. (eds) (1986) Consolidation of soils: testing and evaluation. *American Society for Testing and Materials Special Technical Publication*, **892**, 750 pp.

Yoshida, Y., Kuwano, J. and Kuwano, R. (1991) Effects of saturation on shear strength of soils. *Soils and Foundations, Japanese Society of Soil Mechanics and Foundation Engineering*, **31**, 181–186.

Young, A. and Low, P. (1965) Osmosis in argillaceous rocks. *Bulletin of the American Association of Petroleum Geologists*, **49**, 1005–1007.

Yu, Z., Lerche, I. and Lowrie, A. (1991) Modelling fracturing and faulting within sediments around rapidly moving lateral salt sheets. *Bulletin of the American Association of Petroleum Geologists*, **75**, 1772–1778.

Zaruba, Q. (1987) Landslides and other mass movements, in *Ground Engineer's Handbook* (ed. F.G. Bell), Butterworths, London, 10/1–10/14.

Zay, K. (1807) *Goldau und seine Gegend, wie sie war und was sie geworden*. Orell, Fuseli, Zurich.

Zhang, J., Wong, T.F. and Davis, D.M. (1990) Micromechanics of pressure-induced grain crushing in porous rocks. *Journal of Geophysical Research*, **95**, 341–352.

Znidaric, D. and Schiffman, R.L. (1982) On Terzaghi's concept of consolidation. *Geotechnique*, **32**, 387–389.

Zoback, M.D. and Byerlee, J.D. (1976) Effect of high pressure deformation on permeability of Ottawa Sand. *Bulletin of the American Association of Petroleum Geologists*, **60**, 1531–1542.

Zoback, M.L. and Zoback, M.D. (1989) Tectonic stress field of the continental United States, in *Geophysical Framework of the Continental United States. Geological Society of America Memoir*, **172**, 523–539.

Index

Aberfan flow, Wales 66
Abnormal fluid pressure 13
  see also Overpressure
Accretionary prism
  Barbados 16, 189–90, 192, 236
  Kodiak, Alaska 16, 245–56
  Nankai, Japan 189–203, 240–5
  Olympic Peninsula, Washington 256–9
  Oyo complex, Nias 70
Accretionary prisms (general)
  Brittle deformation in toes 192
  deformation structures in 31
  diffuse strain in toes 190–5
  ductile strain in toes 192–3
  experimental deformation 195–200
  fluid flow in 211
  protothrust zones 190
  scaly clay in 275
  stress paths 187–204
  structural style of 31, 187
  taper angles 189, 211
Accretionary wedges, see Accretionary prisms (general)
Active deformable bed, below glacier 75
Aeolian sediments 22, 101
Ageing of sediments 10
Agricultural soils, sampling techniques 261
Alaska
  examples of subglacial deformation 74–91
  Kodiak accretionary prism 245–56
  Trapridge glacier 74, 76, 78, 80, 83
Alleghany Plateau, New York
  in situ stress 184
  joints 186
Anelastic strain recovery 182
Angle of dilatancy 271
Angle of internal friction 8, 9, 15, 25, 271
Angle of repose 25
Anglesey, North Wales 34
Anisotropy
  mechanical 171
  permeability 224, 226
  Poisson's ratio 175
  porosity 79
  in subglacial sediments 81, 84, 87, 89

thermal 175
Antarctica
  examples of subglacial deformation 73–88
  Ice Stream B 73, 74, 76, 77, 80, 81, 82, 88
Appalachian orogen, lithologies and effective stresses 13
Aquathermal pressuring 15
Atterberg limits 7, 26
Autosuspension 133
Avalanches
  debris 135
  snow and ice 138
  tracks of 139
Axial-planar cleavage, see Axial-planar foliations
Axial-planar foliations 284
  as deformation criteria 305
  in slump folds 305

Ball-and-pillow structure 108, 288
Barbados accretionary prism 16, 189–90, 192, 236
Bed thickness, below glaciers 77
Bed types, below glaciers 74
Bedding-parallel fabrics 38, 67, 242, 268
Bedding-parallel faults 279
Beef structure 118
Bingham fluid 11
Biogenic gas 15, 25
Bilogical communities, as seepage sites 205
Biopressuring 15
Bioturbation, effect on sediment fabric 268
Black River, New York 18
Black Ven landslip, Dorset 59
Boswell, P.G.H. 295–307
Boudinage 15
  diagenesis and 285
  from nature 285
  Kodiak, Alaska 245
  in slumps 145
Bouma sequence, in turbidites 134
Breidamerkurjökull, Iceland 76, 88, 91
Brittle deformation, in prism toes 191–2
  paths 60, 168–87

Burial
  effects on sediment fabric 268
  pressure 6
  stress 40, 48, 50
    paths 60, 168–87

Cam-clay theory 10
Carbonate minerals
  concretions 117, 119
  fault crusts 215
  as fracture-fill 306
  septarian nodules 111
Carbonate sediments
  Cam-clay theory 10
  compaction 68
  critical state 55
  friction 8
  pressure solution 171
  stress paths 56
  see also Chalk
Carbonate veins 250, 259
Carbowax 262
CAT scanning 262
Cataclastic shear zones 295, 301
Cell pressure 6
Cementation
  compaction resistance 67
  concretions 118, 120
  destruction effects 47
  effect on deformation 60
  fluid flow effects 215
  Kodiak, Alaska 16
  meaning of, 1
  Olympic Peninsula, Washington 259
  overconsolidation similarity 66
  peak strength re-establishment 57
  pore-filling effects 48
  porosity preservation 59
  in shear zones 273
  stiffness increase 56
  see also Diagenesis
Central graben, North Sea 38
Chalk 38, 60
  deformation behaviour of 56–9
  see also Carbonate sediments
Chaotic unit, in mass-movement slide 157
Chemical precipitation 1, 98, 118, 121
Chemical sources of fluid 218–20
Chicken-wire texture 121

China, Urumqui glacier 76
Chlorite-mica stacks, as deformation criteria 302
Civil engineering, sampling techniques 261
Clastic sediments, microfabrics of 265
Clasts, fabric in rockslides etc. 277
Clathrate decomposition, as overpressure source 15
Clays 98
 angle of internal friction 9
 Cam-clay theory 10
 cohesion 96, 98
 compaction 68
 concretions 120
 consolidation 50, 172
  fabrics 38
 contractive behaviour 10
 compaction curves 46
 critical state 55
 deformation microstructures 271
 dehydration 218
 deposition 96
 diagenesis 16
 diapirs 104
 dilative behaviour 10
 dish-and-pillar structure 108
 effect on permeability 221
 fabrics 67
 fault gouge 217
 fluid pressure dissipation 50
 glaciolacustrine consolidation 17
 hydroplasticity 98
 in situ stress 184
 influence on sediment strength 8
 K_0 values 61, 173, 231
 mass-movement slides in 26
 microfabrics of 265
 mud volcanoes 51
 overconsolidated 65
 peak strength 9
 permeability changes in faults 225
 permeability of 221
 plasticity 7
 Poisson's ratio 175
 pressure solution 171
 quick 8, 15, 66
 reloading of overconsolidated 61
 residual strength of 9
 rheological creep 162
 rheology 12
 salinity effects on strength 9
 scaly, *see* Scaly clay
 silty (Nankai) 195–200
 slides (mass movements) 156
 subglacial 80
 swelling, elasticity effect 46
 synaeresis 116
 temperature effect on permeability 209
 turbidites 133
 veins 253
 see also London clay; Gault clay
Cleavage, axial-planar, *see* Axial-planar foliations

Climate, influence on rock falls 131
Coarse-tail grading 135
Coefficient of consolidation 49
Coefficient of earth pressure at rest 7, 61,
 see also K_0
Cohesion 8, 9, 13, 22, 25, 46, 80, 83, 84, 96, 129, 145
Cohesionless debris flows 131, 144
Compaction
 around concretions 119
 definition of 6
 deformation related to 117
 differential 67
 effects of lithology 117
 initial 40
 meaning of 37
Compaction curves 46
Complete coupling, below glaciers 76
Complete decoupling, below glaciers 76
Compressibility 40, 47, 176
Compression, as synonym for consolidation 37
Compressive stress, below glaciers 78
Computer-assisted tomography 262
Concretions 118
 as deformation criteria 303
 see also Nodules
Cone-in-cone structure 120
Confining pressure 6
Consistency of sediments 7
Consolidation 6–8, 12
 below glaciers 86
 clays 50, 172
 coefficient of 49
 definition of 6
 diagenesis effects 173–5
 differential 22
 due to igneous intrusion 33
 duration of 51
 effect on permeability anisotropy 222
 experimental 172–5
 fabrics due to 268
 fluid behaviour 210
 historical usage 5
 ice loading, due to 17
 isotropic 51
 meaning of 6, 37
 natural 50
 normal 7, 37, 55
 over- 7
 primary 6, 48–51
 sand 172
 secondary 6, 50, 171, 177
 self-weight 50
 silts 50
 stress paths 60
 stress, meaning of 173
 subglacial structures due to 87
 tectonic 31
 under- 7
Consolidometer, *see* Oedometer

Constrained modulus 174
Contractive behaviour 10, 12
Convolute bedding 101
Convolute lamination 101
Coring disturbance 261
Coulomb behaviour
 in accretionary prisms 206
 in debris flow 140
County Clare, Ireland 104, 114
Coupling, below glacier 76
Cracks
 desiccation 116
 synaeresis 116
Creep 10
 depth creep and folding 281
 downslope 10, 27
 gelifluction 163
 low strain 10
 rates 162
 rock 163
  rates 163
  Goldau Switzerland example 165
 secondary consolidation 11, 50, 48–51, 171, 177
 solifluction 162
 subaerial
  physical 162
  rheological 162
 subaqueous 163
 Vaiont, Italy, example 165
 varied usage of term 10
Criteria for recognizing sediment deformation 249, 301
Critical state 10
 lithology, role of 55
 meaning of 51, 195
 models 10
 Nankai prism experiments 195–9
Critical state line 53
Crosby, W.O. 19
Cross-bedding, deformation of 105
Cross-bedding, overturned 105
CT scanning 262
Cubic law, of fluid flow 227
Cyclic loading 21, 22, 25, 91, 147
Cyclic mobility 21
Cylindrical structures 295

Dana, James 28
Darcy's law 296
Darcy, Henry 27
Darwin, Charles 27
Debris avalanches 135
Debris falls 131
Debris flows 84, 140
 cohesionless 131, 144
 subaerial 142
 subaqueous 142
décollement
 Barbados accretionary prism 16, 190
 fluid flow in prisms 190, 225
 Nankai accretionary prism 190, 242
Decoupling, below glaciers 76

Deformable bed, below glaciers 75
Deformation
 chalk, behaviour of 56–9
 contraction during 12
 dilation during 12
 due to coring 261
 effects on permeability 235–6
 elastic 9, 41
 glaciotectonic 87–91
 meaning of 1
 organic traces 117
 tectonic 3
 trace fossils 117
 undrained 54
Deformation bands, see Shear zones
Deformation paths, see Stress paths
Density inversion 99
 structures due to 99–105
Density modified grain flows 144
Deposition
 debris falls 132
 density instabilities 21
 general 96
 hindered settling 8
 subglacial 82, 89, 91, 92
 synchronous deformation 95
 turbidity currents 134
Depth creep 281
Desiccation cracks 116
Detrital remnant magnetization, as deformation criterion 302
Deviatoric stress 7
 meaning of 51
dewatering 13
 igneous intrusion, due to 33
 in nature 29
 in subglacial sediments 85–6
 structures due to 29, 108–13, 295–307
 subglacial structures due to 87
 tectonic 13, 31
Diagenesis 1, 15, 37, 66, 118
 as fluid source 218
 Barbados accretionary prism, effects in 16
 boudinage and 284
 carbonate veins 250
 carbonate, at Kodiak, Alaska 16
 clay 16
 consolidation interaction 173–5
 deformation interaction 15
 effects on sediment fabric 268
 force of crystallization 16
 in shear zones 271
 K_0 effects 178
 meaning of 1
 overconsolidation, role in 8
 slope failure reduction 25
 timing and deformation 240
 see also Cementation
Diapirs 28, 34, 35, 69, 103
 Kodiak, Alaska 250
 mud 69, 206, 250
 serpentinite 206

shale 69
Diatremes 206
Differential consolidation 22
Differential stress 7
 meaning of 51
Diffusion mass transfer 1
 see also Pressure solution
Diffusivity 213
Dilatancy, see Dilation
Dilatancy angle, see Angle of dilatancy
Dilation 10, 12
 below glaciers 86
 subglacial structures due to 88
Dilative behaviour 10
 see also Dilation
Dish-and-pillar structures 108, 143
Distribution grading 135
Disturbance due to coring 261
Diverticulation, in mass-movement slides 156
Downslope creep 162–5
Drainage 13, 45, 49
Drained test 9
Drumlins 89
Drying techniques for samples 262
Ductile strain, in prism toes 193
Dykes, Neptunean 28
Dykes, sedimentary, see Sedimentary dykes
Dynamic loading 21

Earth pressure at rest, coefficient of 61
 see also K_0
Effective stress 12, 13, 44–8, 209, 210, 222
Ekofisk, North Sea 38, 67
Elastic deformation 9, 40, 41, 46, 169
Elastic rheology 11
Elastic swell line 46
Elastic volumetric strain 211
Elasticity 40
 see also Elastic deformation
Electron microscopy
 of clay fabrics 265
 preparation techniques for 262
Elevation head 13, 207
Engineering geology 5
 see also Soil mechanics
Enterolithic structure 121
Epoxy-resin, for sample impregnation 261
Equilibrium line, in subglacial transport 81
Equivalent permeability 224
Eulerian description of deformation 188
Evaporites 23, 98, 121
Excess fluid pressure 13
 see also Overpressure
Excess pore pressure 13
 see also Overpressure
Experimental deformation 6, 16, 50, 56–9, 168
 accretionary prism sediments 195–200
 consolidation 172–5
Extensional fracture 230–5
Extruded sand sheets 113
 see also Sedimentary intrusions

Fabrics 295–307
 ageing 10
 axial planar to folds 305
 bedding-parallel 268
 chalk 38, 56
 consolidation 10, 38
 glacially-deformed sediments 17
 influence of lithology 67
 Joe's River, Barbados 65
 Kodiak, Alaska 247
 Nankai accretionary prism 193, 242
 permeability effects 226
 permeability influence 222
 quantification in electron microscopy 264
 residual state 10
 sediment deformation effects 10
 shear 44, 60
 till 17
 tortuosity effects, 226
Factor of safety, in slope stability analysis 24
Falls (mass movements) 130–3
Fault gouge 217
Fault valving 206
Faults
 bedding-parallel 279
 differential consolidation, due to 22
 examples from nature 295–307
 extensional 233, 230–5
 extensional (Atlantic) 31
 fluid sources and sinks 212–18
 imbricate 218
 Kodiak, Alaska 247
 load parallel 233
 microfractures 233
 neotectonic 30
 normal, stress indicators 186
 Olympic Peninsula, Washington 258
 in slumps 145
 strike-slip 30
 syn-sedimentary 113, 186
 thrust 233
 tortuosity effects 235
 valving mechanism 206
 see also Shear zones
Fisherstreet slump, W. Ireland 148
Fissility 117
Flame structures 99
Flow laws
 for sediment deformation 11, 12
 for subglacial sediments 83
Flows (mass movements) 133–44
 avalanches 138
 debris avalanches 135
 debris flows 140
 grain flows 143

Flows (mass movements) contd.
    lahars 138
    liquefied 142
    plastic behaviour 27
    volcanic 135–8
Flowslides 142
Fluid flow, basic concepts 209
Fluid movements 27–9
    channelized 29
    historical aspects 27
    structures due to 108–13
Fluid potential gradient 13
Fluid pressure 12, 45, 47, 44–8, 176, 180–1
    abnormal 13
    excess 13
    in subglacial sediments 85–6
    negative 13, 179
    normal 13
    sealed compartments 179
    subnormal 13
    supernormal 13
Fluid pressure ratio (λ) 13
Fluid sinks 209
    transient 212
Fluid sources 209
    chemical 218–20
    transient 212
Fluidal flows (mass movements) 133–40
Fluidization 22, 97
    adjacent to igneous intrusions 32
Fluidized flows 140
Fluids
    consolidation effects 210
    faults as sources and sinks 218
    role in sediment deformation 12–16
    two-phase 15
Flutings, in subglacial sediments 89
Folds
    due to depth creep 281
    from nature 285–8
        scales of 281
        styles of 281
    kink-bands 282
    Kodiak, Alaska 245
    neotectonic 30
    Olympic Peninsula, Washington 258
    slump 105, 145
        axial-planar foliations in 305
        from nature 284
        palaeoslope interpretation 284
Force 39–42
    seepage 15
Forms, in subglacial sediments 89–91
Fractures, see Faults, Joints
Friction angle, see Angle of internal friction
Frozen bed, below glacier 76, 92
Frozen sediments 92

Gas, biogenic 15
Gas hydrates 205

    in liquefaction 22
Gaspe, Quebec 18
Gault Clay 51
Gela slide, Sicily 157
Gelifluction 163
Geopressure 13
    see also Overpressure
Geotechnics, see Soil mechanics
Ghost Rocks Formation, Alaska 245, 260
Ghost veins 293
Glacial deformation 16–17
    structures resulting from 17
    see also Glaciotectonics
Glaciers, rock 163
Glaciotectonics
    deformation 87–91
    structures 87–91
        preservation of 91
    usage of term 3
Goldau, Switzerland, rock creep example 165
Gouge, glacial 17
Gouge, fault 217, 228
    permeability of 229
Grabau, Amadeus 20
Grading, in turbidites 135
    coarse-tail 135
    distribution 135
Grain flows 143
    density modified 144
    modified 143
Grain size
    effect on shear zone thickness 270
    in subglacial sediments 79
Gravitational instability 17
Gravitational loading, structures due to 99–105
Gravitational mass-movements, see Mass movements
Gravity, as body force 3
Gravity spreading 156
Greenly, Edward 34
Griffith/Navier–Coulomb criterion 230
Guaymas basin, Gulf of California 33
Gulf Coast sedimentary basin 114, 156, 211, 218, 232
    in situ stress 182

Hahn, Fritz 20
Hall, James 4
Hard bed, below glaciers 75
Head 207
    elevation 207
    gradient 207
    hydraulic 207
    pressure 207
Head, of mass-movement slide 157
Heim, Albert 20
Helmsdale, NE Scotland 34
Hindered settling 8
Hooke's Law 11
Horizontal burial stress 7, 169

Hutton, James 4
Hvorslev failure criterion 82
    surface 55
Hydraulic conductivity 13, 49, 208, 213
    units of 208
Hydraulic head 207
Hydraulic processes, in subglacial sediments 85–6
Hydraulics, see Hydrogeology
Hydrocarbons
    as fluid sources 219
    overpressure 219
Hydrofracture 230, 233, 235
Hydrofracturing, stress measurement in borehole 182
Hydrogeology, basic concepts 207–9
Hydroplastic 3, 98
Hydropressure 13
    see also Overpressure
Hydrostatic gradient 209
Hydrostatic pressure 7, 13, 45

Ice
    as cause of sediment deformation 16–17
    in sediments 92
Iceland, examples of subglacial deformation 73–91
Igneous activity, effect on sediments 247
Igneous intrusion
    loading effect 21
    effect on sediments 32–33
    structures in sediments due to 32
Ignimbrites 137
Illite
    angle of internal friction 9,
    temperature effects on permeability 209
Imbricate faults 279
Impregnation of samples 261
In place deformation, see In place disturbance
In place disturbance 21–3
    structures due to 99–103
Independent particulate flow 2
Inhomogeneity, in subglacial sediments 84
Inter-grain friction 8
Intergranular friction 8
Intraformational structures, as deformation criteria 303
Intrusions, sedimentary, see Sedimentary dykes
Ionic diffusion, in pore fluids 15
Isolated load balls 100
Isotropic consolidation 51
Isotropic stress 51

Joe's River Formation, Barbados 65, 70
Joints 30, 131, 224, 235
    glacial sediments 17
    stress indicators 186
Jones, O.T. 295–307

K value 7
Kaolinite, angle of internal friction 9
Kelut, Java, lahar example 138
Kidnappers Slide, New Zealand 159
Kink-bands 30
  from nature 282
  in association with shear zones 271
  Nankai prism 191
Kleszczow Graben, Poland 17
K_0 7, 60, 64, 231
  for clays 231
  condition 60
  diagenesis 178
  in nature 67
  stress path 64, 169, 172
  Kodiak accretionary prism, Alaska 16, 245–56

Lagrangian description of deformation 189
Lahars 138
Landforms, glacial 17
Landslides
  meaning of term 23
  scaly clay in 275
  see also Rockslides
Lateral burial stress 7
Lateral ramps, in mass-movement slide 157
Lesser Antilles accretionary prism, see Barbados accretionary prism
Lineations
  morphology of 274
  due to shear 274
  on slickensides 274
  in slump folds 284
Liquefaction 15, 22, 26, 66, 97, 103, 112
  structures due to 295–307
Liquefaction structures
  as deformation criteria 304
Liquefied flows 142
Liquid limit 7
Lithification 1–3, 37
  effect on igneous intrusion 32
  fluid flow effects 214–18
  influence on consolidation 8
  meaning of 1, 2
  partial 2, 37, 245, 250
  usage of term 2
Lithology
  critical state, effect on 55
  depositional aspects 96
  fabric, influence on 67
  liquefaction, effect on 22
  microfabric variations 265
  permeability, effect on 220–2
  plasticity, influence on 7
  subglacial sediments, effect in 79–81
  rock falls, influence on 130
  strength, influence on 8
  see also Specific lithological names
Lithostatic pressure 6
Load balls, isolated 100
Load casts 99

Loading
  cyclic 21
  dynamic 21
  gravitational 99
  monotonic 21
  shock 21
  static 21
  see also Gravitational loading
Logan, William 18
London Clay 50, 51, 226
Lowe divisions, of turbidities 135
Lubrication 14

Magma
  effect of sediments 32–3
  explosive mixing with sediment 32
Magnetic susceptibility 262
Mammal footsteps, loading effect 21
Mariana Trench 30
Mass movements 23–7, 53, 67, 105
  avalanches 138
  classification 127
  clast orientations 277
  creep, 162–5
  debris avalanches 135
  debris flows 140
  falls 130–3
  flows 133–44
  flowslides 142
  fluidal flows 133–40
  fluidized flows 140
  grain flows 143
  historical aspects 23
  interpretation of physical conditions 281
  lahars 138
  liquefied flows 142
  mechanics of 24, 127
  melanges 156
  mudflows 140, 151
  mudslide, Dorset 25
  mudslides, 92
    Point of Relief, Ireland, example 158
  periglacial 163
  plastic flows 140–4
  pyroclastic flows 137
    Mont Pelee example 138
  rheology 128
  slides 152–62
  slumps 144–52
  turbidity currents 133–5
  see also Creep; Slides; Slumps
Mean effective stress, meaing of 45, 51
Melanges
  as mass movements 156
  historical aspects 34
  Kodiak, Alaska, example 249–56
  shale-matrix 35, 239
Metamorphic dehydration, as fluid source 219
Meteorite impact, loading effect 21
Microfabric, see Fabric
Microfabrics

  in clastic sediments 265
  in clays 265
Microfaults, see Shear zones
Microscaliness 275
Microstructures 268–71
  in deformed clays 271
  in subglacial sediments 88
Middle America Trench, slope stability at 25
Miller, W.J. 19
Mineral dehydration
  as fluid source 218
  as overpressure source 15
Mineralization, as deformation criterion 304–5
Mineralized fractures, as deformation criteria 304–5
Mississippi Delta 103
  diapirism in 70
  mudlumps 17
  overpressures and slopes 26
  permeabilities and strengths 13
  slumps 151
Mobility, cyclic 21
Modified grain flows 143
Mohr circles 10, 43, 51, 52, 53, 197
Mohr–Coulomb equation 8, 12, 24, 25, 96
Mohr–Coulomb failure criterion 82
Mohr–Coulomb failure surface 55, 57, 63
Monotonic loading 21
Mont Pelée, Martinique, lahar example 138
Montmorillonite, angle of internal friction 9
Mt Etna, Sicily 30
Mud
  permeability 221
  permeability in faults 225
Mud diapirs, 206
  see also Clays
  see also Diapirs; Mud volcanoes
Mud volcanoes 69, 206
  clays in 51
Mud-filled veins 291
Mudflows 138, 140, 151
Mudlumps 17, 28, 103
Muds
  diapirs 28, 103, 113
  flame structures 99
  igneous intrusion into 32
  Mississippi Delta 103
  sands comparison 13
  shrinkage 114
  slides (mass movements) 156
  turbidity currents 134
  volcanoes 29
Mudslides 25, 92, 127, 153
  Point of Relief, Ireland, example 158

Nankai accretionary prism, Japan 189–203, 240–5
  experimental deformation 195–200

Negative fluid pressure 13
Neotectonic deformation 17
Neotectonics 30
Neptunian dykes 28, 30, 59, 110
Newtonian fluid 11
  mass movements 128
Niger Delta 104, 114, 156
Nodules, 118–21
  septarian, 117, 120
  see also Concretions
Nomogram, for slope failure 279
Normal consolidation 7, 37, 55
Normal fluid pressure 12, 13
Normal stress 42
North Sea 38, 59, 66, 67, 70, 205
Nuées ardentes 137

Ocean Drilling Program 31
  Leg 110 (Barbados) 31
  Leg 131 (Nankai) 31, 192–7, 242
  Leg 146 (Cascadia) 31
Oedometer 40
Olistoliths 156
Olistostrome 34
Olympic Peninsula, Washington,
  accretionary sequence 256–9
Organic traces, deformation of 117
Organic creep 162
Overburden pressure 6
Overconsolidation 7, 55, 64–6, 214
Overpressure 13, 15, 21, 25, 50, 53, 65,
  152, 207, 209, 211, 232
  duration of 70
  effect on burial stress 180–1
  effect on sediment strength 13
  effects of 96
  origins of 15, 70–1
  see also Fluid pressure
Overturned cross-bedding 105
Oyo complex, Nias 70

p–q diagrams 10
Pacific Rim melange, Vancouver 35
Palaeoseismicity, interpretation of 289
Palaeoslope 284
  see also folds, slump
Partial decoupling, below glaciers 76
Partial lithification, see Lithification,
  partial
Partially frozen bed, below glaciers 76
Peak strength 9, 53
  in gravitational slides 26
Penecontemporaneous structures 95
Periglacial mass movements 163
Permeability 12, 15, 29, 48, 209
  active deformation effects 235–6
  anisotropy of 15, 224, 226
  anisotropy from consolidation 222
  definition of 208
  dilation, effect of 15
  engineering hydrology 15
  equivalent 224
  examples from Mississippi Delta 13
  inhomogeneities, effects of 15

lithology effects 220–2
measurement of 16
muddy faults 225
sandy faults 226
subglacial sediments 85–6
temperature effects in illite 209
tortuosity effects 235
units of 208
Physical creep (mass movement) 162
Piceance Basin, Colorado, *in situ* stress
  183
Pipes (drainage conduits) 29
Plastic flows (mass movements) 140–4
Plastic limit 7
Plastic range 7, 27
Plastic rheology 11
Plasticity index 7, 27
Pockmarks 29
Point of Relief slide, Ireland 158
Poisson's ratio 11, 169, 174, 175, 232,
  235
Pore pressure 12
  see also Fluid pressure
Pore pressure, excess 13
  see also Overpressure
Porosity 44, 45
  Barbados prism, 193
  in consolidation 37, 168–87
  definition of 7
  Nankai prism 192
  on initial deposition 96
Potential gradient fluid 13, 15, 207
Pouisseille flow 227
Pouisseille's law 222
Power law, for viscous flow 11
Pressure
  burial 6
  cell 6
  confining 6
  fluid 12, 44–8
  fluid, subglacial 85–6
  hydrostatic 7, 13
  ionic diffusion in pore fluid 15
  lithostatic 6
  normal fluid 12
  overburden 6
  pore 12
Pressure head 13, 207
Pressure ridges, in slumps 148
Pressure solution 51, 171
  meaning of 1
Primary consolidation 6, 48–51
Principal stress 42–4
Proglacial deformation 92
Propagation zone, in mass-movement
  slide 157
Protothrust zone, in accretionary
  prisms 190, 194
Pseudonodules 100
Pseudoplastics 128
Pyroclastic flows 137

Quartz, as fracture-fill 306–7
Quaternary studies, sampling

  techniques 261
Quick clay 8, 15, 66

Raleigh–Taylor instabilities 22
Ramsey Island, SW Wales 32
Ramps, lateral, in mass-movement
  slides 157
Ratchetting 79
Recognition of sediment deformation
  structures 295–307
Recrystallization, meaning of 1
Remoulded sediments 8
Repose, angle of 25
Residual strength 9, 26, 53
  in gravitational slides 26
Resin, for sample impregnation 261
Retrogressive failure, of slope 16, 26
Rheological creep 162
Rheology, 98
  mass movements 128
  sediments 11, 26
  subglacial sediments 83
Rio Santa, Peru, flowslide example 142
Rock, usage of term 1, 5
Rock falls 130
Rock glaciers 163
Rockslides, clast fabrics in 277
Rocky Mountain arsenal, Denver,
  fluid-induced earthquakes 206
Roscoe surface 55, 60, 63
Rotational slides 153

Sacramento River, California 28
Safety factor, in slope stability analysis
  24
Salinity, effect on
  clay microfabric 265
  clay strength 9
Sampling of sediments 261
Sand
  sheets, extruded 112
  turbidites 133
  volcanoes 112
Sand dunes
  deformation in 22
  loading effect 21
Sand volcanoes 112
  see also Sedimentary dykes
Sand-blows 30
Sands 229
  aeolian 101
  Cam-clay theory 10
  channels 118
  cohesion 80
  compaction 68
  consolidation 8, 50, 172, 179
  contractive behaviour 10
  critical state 55
  cross-bedded 101
  cyclic loading 22
  deltaic 103
  desiccation structures 110, 116
  dilative behaviour 10
  disaggregated 250

dish-and-pillar structure 108
dyke injection 28
extrusion structures 112
flows 144
in situ stress 183
injection structures 110
intrusive, *see* Sedimentary dykes
K_0 values 61, 173
load-casts 99
loading structures 101
mud comparison 13
oil well invasion 59
packing density 221
peak strength 9
permeability in faults 226
permeability of 220
Poisson's ratio 175
pressure solution 171
sheet dewatering structures 109
stress paths 56
volcanoes 29, 112
wetting structures 117
see also Sedimentary dykes
Sandstone dykes, *see* Sedimentary dykes
Sapping 26
Sarawak 69
Scaly clay 65, 239, 295–307
 in accretionary prisms 275
 fabric of 275
 in landslides 275
 Nankai décollement 242
Scaly fabric, *see* Scaly clay
Scaly foliation, *see* Scaly clay
Scotian shelf, *in situ* stress 183
Secondary consolidation 6, 50, 177
Sediment deformation
 criteria for recognition of 301
 diagnostic property of 1
 influence of early thinking 4–5
 principles 9–12
Sediment deformation structures, recognition of 295–307
Sediment transport, below glaciers 81–2
Sedimentary basins, stress paths in 61
Sedimentary dykes 110–12
 as deformation criteria 301
 from nature 288
 historical aspects 27
 uses of 28
Sediment injection structures 110–12
Sedimentary intrusions 28, 110–12
Sediments
 examination 261
 techniques of 261
 meaning of 1, 37, 41
 sampling of 261
 strength of 8–9
Seepage force 15
Seeps, ocean floor 205
Seismic waves, loading effect 21
Self-weight consolidation 50
Semi-lithification, *see* lithification,
 partial
Sensitivity
 in marine sediments 26
 meaning of 8
Septarian nodules 117, 120
Serpentinite diapirs 206
Shale diapirs, *see* Diapirs
Shale-matrix melanges 35, 239
Shear
 structures due to 105–8
 subglacial structures due to 88–9
Shear bands, *see* Shear zones
Shear modulus 175
Shear strength 8, 53, 96
 examples from Mississippi Deta 13
Shear stress 42–4, 51
 below glacier 79
Shear surfaces, *see* Shear zones *and* Slickensides
Shear thickening 11
Shear thinning 11
Shear zones 268–71
 cataclastic 295, 301
 diagenetic effects 271
 fluid flow in 226
 glacial sediments 17
 kink band association 271
 Kodiak, Alaska 250
 morphology of 270
 Nankai prism 191–92
 in nature 271
 orientation of 269
 origin of 269
 in sub-glacial sediments 88
 thickness of 270
Sheet dewatering structures 109
Shimanto Belt, Japan 30
Shock loading 21
Shrinkage, structures due to 114–17
Shrinkage cracks 116
Silts
 cohesion 80
 consolidation 50
 deposition 98
 K_0 values 61
 igneous intrusion into 32
 Nankai silty clays 195–200
 plasticity 7
 residual strength behaviour 9
Slickensides 295–307
Slides (mass movements) 105, 147, 152–62
 chaotic unit 157
 component parts 154, 157
 diverticulation 156
 Gela, Sicily, example 157
 head region 154
 Kidnappers, New Zealand, example 159
 lateral ramps 157
 Point of Relief, Ireland, example 158
 propagational zone 157
 rotational 153
 toe region 154
 translational 153
 translational zone 157
Slope failure
 causes of 25
 nomogram for 279
 see also Mass movement
Slope stability
 analysis of 24
 see also Mass movement
Slump deformation, historical controversy 301
Slump folds 105, 145
 see also Folds, slump
Slump sheet, loading effect 21
Slumping, recognition of 302
Slumps (mass movements) 144–52
 faults, 145
 Fisherstreet, Ireland example 148
 folds 105, 145
 mechanics 145
 Mississippi Delta example 151
 origin 147
 pressure ridges 148
 size 146
 Storegga, Norway example 150
 see also Folds, slump
Snout, of debris flows 141
Soft-sediment deformation, discussion of term 3
Soil mechanics 6, 10, 12, 29, 37, 41
 application to submarine installations 6
 application to submarine situation 15
 standard tests 16
 see also Civil engineering; Geotechnics
Solifluction 162
Solution transfer, *see* Pressure solution
Southern Uplands, Scotland 30
Specific storage 213
State boundary surface 53
Static loading 21
Storegga slumps, Norway 150
Storm waves
 overpressuring source 15
 slope instability 25
Strain 39, 42–4
 diffuse in prism toes 195
 horizontal elastic 180
 lateral 177
 meaning of 39
 rate, in subglacial sediments 83
 thermal 175
 uniaxial, *see* K_0
 volumetric 3, 6, 51
 volumetric, during deformation 10
Strait of Messina, Italy (extensional structures) 30
Strangways, W.T.H.F. 27
Strength
 of sediments 8–9
 peak 9, 53
 residual 9, 53

Strength *contd.*
  shear 8, 53
Stress 39, 42–4
  amplification of 234
  below glacier 78
  burial 40, 48, 50
  compressive, below glacier 78
  deviatoric 7
    meaning of 51
  differential 7
    meaning of 51
  effective 12, 44–8
  horizontal burial 7
  in situ measurement 181–5
  isotropic 51
  lateral burial 7
  mean effective 45
  meaning of 39
  normal 42
  principal 42–4
  shear 42–4, 51
    below glacier 79
  tectonic 29–32, 68
  tectonic, meaning of 3
  tensile 114
    in nature 282
  tensional 230–5
Stress paths 10, 56, 210
  accretionary prisms 187–204
  unloading 180
Stress path diagrams 10, 52, 53
  Nankai prism examples 198, 199, 241
Stress, differential, *see* Deviatoric stress
Striations, on slickensides 274
Structures
  Chemical precipitation 118–21
  Density inversion 99–105
  dewatering 108–13, 291
  fluid movement 113
  glaciotectonic 87–91
  gravitational loading 99–105
  in-place disturbance 103
  Kodiak, Alaska 245–56
  liquefaction, due to 288
  Nankai accretionary prism 242–4
  shear 105–8
  shrinkage 114–17
  wetting 117
Subaerial creep 162
Subaerial debris flow 142
Subnormal fluid pressure 13
Sulphides, as fracture-fill 306
Sunda Trench 30
Sunnmøre, W. Norway, avalanche example 139
Supernormal fluid pressure 13
  see also Overpressure

Surge-type glaciers 91
Suspension sedimentation stage, of deposition 135
Swell line, elastic 46
Swelling 210, 211
Synsedimentary faults 113
Synaeresis cracks 116

Taiwan 68, 70
Taper angle, of accretionary prisms 189
Techniques, of examining sediments 261
Tectonic
  deformation 29–32
  stress 29–32
    meaning of term 3
    usage of term 3–4
Tectonic consolidation 31
Tectonic deformation
  as preserved in rocks 30
  meaning 3
Tectonic dewatering 31
Tectonic stress 68
Tectonic structures in sediments 30
Temperature
  below glaciers 75
  effect on burial 175–6
  effect on consolidation 178
  effect on deformation 176
  smectite–illite transition 218
Tensile strength 114
Tensile stress 114
  in nature 281
Terzaghi, Karl 14, 45, 48, 49
Thermal compressibility 175
Thermal regime, below glacier 75
Thermal strain 175
Thick bed, below glacier 77
Thickness, of sediments below glacier 77
Thin-sections, preparation techniques 262
Thixotropy 22, 269
Till, glacial 17
Time, role in experimental deformation 177
Toe, in mass-movement slide 157
Tortuosity
  definition of 222
  of flow paths 235
Trace fossils, deformation of 117
Tractional stage, of deposition 134
Translational slides 153
Translational zone, in mass-movement slides 157
Trenton, New York 18–21
Triaxial tests 9, 54
Turbidites 134

Turbidity currents 133–5
Two-phase fluids 15

Underconsolidation 7, 64–6
Underpressure 13, 179
Undrained deformation 54
Undrained test 9
Unfrozen bed, below glacier 76
Uniaxial strain, *see* K_0
Unlithified bed, below glacier 75
Unloading stress paths 180

Vaiont slide, Italy 165
Vanuxem, Lardner 18
Vein structure (mud-filled) 291
Veins
  as deformation criteria 304
  ghost 293
  as lithification criteria 250–6, 259
  mud-filled 291
  Olympic Peninsula, Washington 259
Viscosity, coefficient of 11
Viscous rheology 11
Void ratio 46
  in consolidation 168–87
  definition of 7
Volcanic flows 135–8
Volcanoes
  sand, 112
  sedimentary, 29, 112
Volume change, *see* Volumetric strain
Volumetric strain 3, 6, 43, 51
  during burial 169–81
  during deformation 10, 43
  elastic 211
  Nankai prism 191–203

Water content, definition of 7
Waves
  ocean 103
  loading effect 15, 21
Weak rock
  typical behaviour of 56
  usage of term 2
Web structure 293
Welded tuffs 137
Wet water content, definition of 7
Wetting of sediments, structures due to 117
White, Theodore 19

X-radiography 262

Yapp Trench 30
Yield 46
Yield surface 55
Young's modulus 11, 169, 174, 175, 185